Spring Boot 编程思想（核心篇）

小马哥（mercyblitz）/ 著

電子工業出版社
Publishing House of Electronics Industry
北京•BEIJING

内 容 简 介

本书是《Spring Boot编程思想》的核心篇，开篇总览 Spring Boot 核心特性，接着讨论自动装配（Auto-Configuration）与 SpringApplication。全书的讨论以 Spring Boot 为中心，议题发散至 Spring 技术栈、JSR 及 Java。希望透过全局的视角，帮助读者了解 Spring Boot 变迁的历程；经过多方的比较，帮助读者理解 Spring Boot 特性的原理；整合标准的规范，帮助读者掌握 Spring Boot 设计的哲学。

本书适合对 Spring Boot 感兴趣的读者阅读。

图书在版编目（CIP）数据

Spring Boot 编程思想：核心篇 / 小马哥著. —北京：电子工业出版社，2019.3
ISBN 978-7-121-36039-8

Ⅰ. ①S… Ⅱ. ①小… Ⅲ. ①JAVA 语言—程序设计 Ⅳ. ①TP312.8

中国版本图书馆 CIP 数据核字（2019）第 025869 号

责任编辑：陈晓猛
印　　刷：三河市华成印务有限公司
装　　订：三河市华成印务有限公司
出版发行：电子工业出版社
　　　　　北京市海淀区万寿路 173 信箱　　邮编：100036
开　　本：787×980　1/16　印张：39.25　字数：879.2 千字
版　　次：2019 年 3 月第 1 版
印　　次：2022 年 1 月第 11 次印刷
定　　价：118.00 元

凡所购买电子工业出版社图书有缺损问题，请向购买书店调换。若书店售缺，请与本社发行部联系，联系及邮购电话：（010）88254888，88258888。

质量投诉请发邮件至 zlts@phei.com.cn，盗版侵权举报请发邮件至 dbqq@phei.com.cn。

本书咨询联系方式：010-51260888-819，faq@phei.com.cn。

自序

非常感谢您阅读本书，在成长道路上，我们从此不再孤单。

大约在三年前，我有幸参与全集团微服务架构的演进及基础设施的构建，在此期间痛苦和受益并存。二〇一六年十二月，经朋友引荐，作为“SFDC 2016 杭州开发者大会”的嘉宾，进行了一场名为“微服务实践之路”的演讲，从此正式开始了我的微服务布道师之路。次年三月，segmentfault“讲堂”栏目上线，我再次受邀，作为 Java 讲师。同年六月二日，“Java 微服务实践”系列讲座正式直播，我主讲 Spring Boot 和 Spring Cloud。无独有偶，当月正好我工作满十周年，也萌生了著书的意向，计划写一本关于 Spring Boot 微服务开发实践的书籍，希望借此机会与诸君分享我的微服务实践经验。然而，随后的变故将此念头变为了现实。当月九日上午，正值当差，父亲传来一通电话，告知外婆于八点左右过世，听此噩耗，悲从中来，不可断绝。即刻带着身怀六甲的妻子，启程回湘。

外婆一直陪伴着我的成长，直到我远赴杭州求职，才分隔两地。现如今祖孙二人天各一方，生死茫茫，无处话凄凉，子欲养而亲不待的痛楚莫过于此。我曾向上天祷告，愿她能安享西方极乐。若非外婆的离世，我绝对不会有坚定意志和足够勇气来完成此书，书籍的内容也不会有颠覆性的变化，讨论的议题从过去的“Spring Boot 微服务开发实践”逐渐转变为“Spring Boot 编程思想”。希望竭尽所能，将技术积累、学习方法、实战经验，以及所思所想和盘托出。每当自己午夜梦回，脑海中浮现外婆的容貌时，总会潸然泪下，所有的思想动摇和行为慵懒立即烟消云散。外婆是虔诚的佛教徒，平日乐善好施。从小耳濡目染的我也尽一点绵薄之力，将书籍五成以上的稿费作为公益基金，支持贫困地区的青少年教育，并且不定期地公开账目信息，供广大读者监督。这或许杯水车薪，但仍希望他们能够感到一丝温暖。

祸兮福所倚，福兮祸所伏，生死轮回，自然之理。外婆去世后的两个月，我的儿子降临人间。作为一名新晋的父亲，自然会以更高的标准来要求自己，对书籍的质量同样趋于严苛，将早期已完成的部分“付之一炬”，推倒重来，内容篇幅剧增。作为我儿的表率，著书只是“立言”的开始，捐赠作为“立德”的发端，而“立行”则需身体力行，持之以恒。或许“著作等身”是一种不错的选择，然而现代科技的进步，尤其是文字载体的革新，要做到这一点，难度实在不小。不过，“为者常成，行者常至”，实现从“小马哥”到“马三立”先生的华丽转身并非遥不可及。

已故南京大学历史系教授高华先生曾引述凯斯·詹京斯的观点，“历史乃论述过去，但绝不等于过去”。既然是论述，那么或多或少会存在偏差，不但受限于论述者的知识、能力及记忆等主观因素，而且取决于当时的时空环境。为了遵照原著，在功能特性的介绍上，本书将引述官方文档的英文原文，并做出适当的解释。由于文档的编写者或许不是代码的实现者，即使是实现者本人，也难免会站在自己的立场和高度，抑或章节安排及文字组织等诸多因素影响阅读和理解。因此，针对官方文档语焉不详的部分，本书将补充说明；对其错误的结论，将加以修正。由于本人能力和水平的局限，不敢妄言理解“格物致知”的奥义，难免有主观臆断和谬论之处，且仅一家之言，供诸君参考，切莫将此奉为圭臬，书云亦云，不假思索。老子有言，“上士闻道，勤而行之”，希望读者能学以致用，若能在实践中激发出创新的灵感，善莫大焉。

最后，借此机会，由衷地感谢我的太太，没有她背后默默地付出，我不会有如此多的精力投入，更无法专注写作。同时，向陈晓猛编辑致敬，他是一位谦谦君子，极富耐心，在书籍编写的过程中，给予我不少的帮助和鼓励。再次向各位朋友送上我诚挚的歉意，由于个人的原因，使得书籍出版时间一再跳票。

小马哥
公元二〇一九年一月于杭州

前言

本书全名为《Spring Boot 编程思想（核心篇）》，以 Spring Boot 2.0 为讨论的主线，讨论的范围将涵盖 Spring Boot 1.x 的所有版本，以及所关联的 Spring Framework 版本，致力于：

- 场景分析——掌握技术选型；
- 系统学习——拒绝浅尝辄止；
- 重视规范——了解发展趋势；
- 源码解读——理解设计思想；
- 实战演练——巩固学习成果。

内容总览

由于本书的内容跨度广，所以分“核心篇”“运维篇”和“Web 篇”三册分别讨论 Spring Boot 的功能特性。“核心篇”开篇总览 Spring Boot 核心特性，逐一讨论 Spring Boot 官网所罗列之六大特性，然而其中两点并非 Spring Boot 专属，故点到为止，而将讨论聚焦在其五大特性，分别为自动装配（Auto-Configuration）、SpringApplication、外部化配置、Spring Boot Actuator 和嵌入式 Web 容器。其中，前两者是“核心篇”讨论的议题，后两者则是 Spring Boot 官方定义的 Production-Ready 特性，均偏向 Spring Boot 应用运维，因此纳入“运维篇”的讨论范畴。至于嵌入式 Web 容器，将结合传统 Java EE Servlet、Spring Web MVC 和 Spring 5 WebFlux 的有关内容放至“Web 篇”探讨，具体章节安排如下。

- 核心篇
  - 总览 Spring Boot
  - 走向自动装配
  - 理解 SpringApplication

- 运维篇
  - 超越外部化配置
  - 简化 Spring 应用运维体系
- Web 篇
  - “渐行渐远”的 Servlet
  - 从 Servlet 到 Web MVC
  - 从 Reactive 到 WebFlux
  - 嵌入式 Web 容器

目前，“核心篇”和“运维篇”已编写完毕，“Web 篇”正在同步更新，其目录安排可能发生变更，请读者以最终发行版本为准。

在内容结构上，本书采用“总—分—总”的方式，首先总体介绍讨论范围，随后深入展开细节的讨论，最后予以总结。同时，为了避免先入为主的影响，本书将会针对官方文档的描述内容提出疑问或假设，大胆地猜测其可能实现的方式，再结合实现源码加以验证，随后通过示例代码巩固理解。在写作手法上，本书效仿中国历史书籍的传统编著手法，将纪传体和编年体予以综合。从功能特性来看，它属于纪传体，如自动装配、SpringApplication 和外部化配置等。如此表述方式更容易系统地介绍 Spring Boot 和 Spring Framework 的核心特性。从特性的发展历程来观察，它则属于编年体，如 Spring Framework 注解驱动编程模型从 1.x 到 5.0 的发展与 Spring Boot 自动装配之间的关联，以及 Spring Boot 1.0 到 1.4 的外部化配置源是怎样利用 Spring Environment 抽象逐步完善的。更重要的是，在论述方式上，增加了论点、论证和论据，从而做到知其然也知其所以然。在对特性的讨论中，会穿插一些补充说明。在特性讨论的结尾处，将总结所论议题。

所谓“兼听则明，偏听则暗”，本书的讨论范围并不会局限在 Spring Boot 或 Spring Framework，会将 Spring Cloud 甚至 Spring Cloud Data Flow 纳入参考，探讨 Spring Boot 在两者中的运用。站在更宏观的角度，在整个 Java EE 的生态中，Spring 技术栈并非独此一家，也不完全是“开山之作”，不少相关的特性可在 JSR 规范和其他 Java EE 实现中找到原型。换言之，Spring 技术栈可被认为是一种非常成功的“重复发明轮子”，不仅适配了 JSR 实现，而且“借鉴”了他山之石，逐步实现了自身的生态系统。

总而言之，全书的讨论将以 Spring Boot 为中心，议题发散至 Spring 技术栈、JSR 及 Java。希望读者透过全局的视角，了解变迁的历程；通过多方的比较，理解特性的原理；整合标准的规范，掌握设计的哲学。当您纵览全书之后，或许会明白为什么说“Spring Boot 易学难精”。因为它的核心是 Spring Framework，而对后者的理解程度又取决于对 JSR 规范及 Java 的熟悉度。

版本范围

为了系统性地讨论 Spring Boot 的发展脉络，本书会将 Spring Boot 2.0 与 1.x 的版本加以对比，探索从 1.0 到 2.0 版本的重要变化，便于读者后续架构、整合及迁移等工作。由于编写时间恰逢 Spring Boot 2.0.2.RELEASE 版本发布，为统一源码分析，将讨论的 Spring Boot 2.0 版本固定为 2.0.2.RELEASE，相同时间点的 Spring Boot 1.5 的版本则是 1.5.10.RELEASE。同时由于更低的版本已停止维护，所以可选择最后发行的小版本作为参考，故所有涉及的 Spring Boot 版本如下所示。

- Spring Boot 2.0：2.0.2.RELEASE；
- Spring Boot 1.5：1.5.10.RELEASE；
- Spring Boot 1.4：1.4.7.RELEASE；
- Spring Boot 1.3：1.3.8.RELEASE；
- Spring Boot 1.2：1.2.8.RELEASE；
- Spring Boot 1.1：1.1.9.RELEASE；
- Spring Boot 1.0：1.0.2.RELEASE。

由于 Spring Boot 2.0 最低依赖的 Java 版本为 8，而 Spring Boot 1.x 最低兼容 Java 1.6，因此 Java 1.8 是 Spring Boot 各个版本兼容的交集，也是实例工程 `thinking-in-spring-boot-samples` 的运行时环境。除此之外，讨论将更多地关注 Spring Boot 与其核心依赖 Spring Framework 之间的版本映射关系，如下表所示。

Spring Boot 版本	Spring Framework 版本	JDK 版本
2.0.2.RELEASE	5.0.6.RELEASE	1.8+
1.5.10.RELEASE	4.3.14.RELEASE	1.6+
1.4.7.RELEASE	4.3.9.RELEASE	1.6+
1.3.8.RELEASE	4.2.8.RELEASE	1.6+
1.2.8.RELEASE	4.1.9.RELEASE	1.6+
1.1.9.RELEASE	4.0.8.RELEASE	1.6+
1.0.2.RELEASE	4.0.3.RELEASE	1.6+

不难看出，Spring Boot 2.0 对应的 Spring Framework 版本是 5.0，而 Spring Boot 1.x 则依赖 Spring Framework 4.x。之所以要具体到 Spring Framework 的某个版本号，除了避免版本差异所导致源码分析失准的情况，更多的是由于 Spring 技术栈特殊的版本号管理。按照传统 Java 版本的约定，第一位数字表示主版本号，控制大版本更新，第二位代表次版本号，可小范围地引入新的特性和相关 API，而第三位则用于问题修正或安全补丁等。Spring 技术栈不时地利用第三位版本号引入新的

API，比如 Spring Framework 3.0.1 引入的 API BeanDefinitionRegistryPostProcessor，Spring Boot 1.3.2 引入的 API ExitCodeEvent。同时，在框架 API 的兼容性方面，从 Spring Framework 到 Spring Cloud 逐渐降低，Spring Boot 处于比上不足比下有余的状况。因此，本书在深入讨论的过程中反复地强调 API 兼容的重要性，也希望读者在自研的过程中多加关注。

为了理解 Spring Boot 特性发展的过程，将 Spring Framework 版本讨论的范围设定为从 1.x 到 5.0。换言之，本书将涵盖几乎所有的 Spring Framework 和 Spring Boot 版本，包括两者涉及的 JSR（Java Specification Requests），如 Servlet、Bean Validation 和 JAX-RS 等规范。

> 更多的 JSR 资讯，请参考官方网页 https://jcp.org/en/jsr/overview，或者访问小马哥 JSR 收藏页面 https://github.com/mercyblitz/jsr，下载归档的 JSR PDF 文件。

相关约定

本书在议题的讨论中，将在文档引用、示例代码、日志输出、源码版本、源码省略等方面做出约定。

文档引用约定

在文档引用方面，Spring Boot 官方文档的默认版本为 2.0.2.RELEASE，地址为：https://docs.spring.io/spring-boot/docs/2.0.2.RELEASE/reference/htmlsingle/。

Spring Framework 官方文档则选择 5.0.6.RELEASE，地址为：https//docs.spring.io/spring/docs/5.0.6.RELEASE/spring-framework-reference/。

为了遵照原文，讨论的过程中将援引原文，如 Spring Boot 官方文档在“11.5 Creating an Executable Jar”章节中介绍此等构建方式，即构建可执行 JAR，又称之为“fat jars”。

> We finish our example by creating a completely self-contained executable jar file that we could run in production. Executable jars (sometimes called “fat jars”) are archives containing your compiled classes along with all of the jar dependencies that your code needs to run.

此处的 Spring Boot 官方文档版本为 2.0.2.RELEASE，否则在讨论的内容中会特别说明其他版本信息。

示例代码约定

为了减少代码冗余和内容篇幅，通常会省略 Java 示例代码的 package 和 import 部分，如下所示。

```
@SpringBootApplication
public class FirstAppByGuiApplication {

    public static void main(String[] args) {
        SpringApplication.run(FirstAppByGuiApplication.class, args);
    }

    @Bean
    public RouterFunction<ServerResponse> helloWorld() {
        return route(GET("/hello-world"),
                request -> ok().body(Mono.just("Hello,World"), String.class)
        );
    }
}
```

如果该示例被重构或调整多次，则其不变的代码将被省略，仅关注变更或核心代码：

```
@SpringBootApplication
public class FirstAppByGuiApplication {
    ...

    /**
     * {@link ApplicationRunner#run(ApplicationArguments)}方法在
     * Spring Boot 应用启动后回调
     *
     * @param context WebServerApplicationContext
     * @return ApplicationRunner Bean
     */
    @Bean
    public ApplicationRunner runner(WebServerApplicationContext context) {
        return args -> {
            System.out.println("当前 WebServer 实现类为："
                    + context.getWebServer().getClass().getName());
```

```
        };
    }
}
```

不难发现，为了方便理解，在示例代码中会添加必要的注释加以说明。同时，如果在调整的过程中出现颠覆性变化，那么通常将注释不需要的功能，方便后续回顾：

```
//@Configuration
//@ComponentScan
@EnableAutoConfiguration
//@SpringBootApplication(scanBasePackages = "thinking.in.spring.boot.config")
public class FirstAppByGuiApplication {

    public static void main(String[] args) {
        SpringApplication.run(FirstAppByGuiApplication.class, args);
    }

//    /**
//     * {@link ApplicationRunner#run(ApplicationArguments)} 方法在
//     * Spring Boot 应用启动后回调
//     *
//     * @param context WebServerApplicationContext
//     * @return ApplicationRunner Bean
//     */
//    @Bean
//    public ApplicationRunner runner(WebServerApplicationContext context) {
//        return args -> {
//            System.out.println("当前 WebServer 实现类为："
//                    + context.getWebServer().getClass().getName());
//        };
//    }
}
```

同样的原则适用于 XML 文件或其他配置文件：

```
<?xml version="1.0" encoding="UTF-8"?>
<project xmlns="http://maven.apache.org/POM/4.0.0"
xmlns:xsi="http://www.w3.org/2001/XMLSchema-instance"
      xsi:schemaLocation="http://maven.apache.org/POM/4.0.0
http://maven.apache.org/xsd/maven-4.0.0.xsd">
```

```
<modelVersion>4.0.0</modelVersion>

<groupId>thinking-in-spring-boot</groupId>
<artifactId>first-app-by-gui</artifactId>
<version>0.0.1-SNAPSHOT</version>
<packaging>war</packaging>
...
    <!--&lt;!&ndash; Use Jetty instead &ndash;&gt;-->
    <!--<dependency>-->
        <!--<groupId>org.springframework.boot</groupId>-->
        <!--<artifactId>spring-boot-starter-jetty</artifactId>-->
    <!--</dependency>-->

    <!--&lt;!&ndash; Use Undertow instead &ndash;&gt;-->
    <!--<dependency>-->
        <!--<groupId>org.springframework.boot</groupId>-->
        <!--<artifactId>spring-boot-starter-undertow</artifactId>-->
    <!--</dependency>-->

    <dependency>
        <groupId>org.springframework.boot</groupId>
        <artifactId>spring-boot-starter-tomcat</artifactId>
    </dependency>
...
</project>
```

日志输出约定

为了精简示例运行时的日志输出，书中的内容并不一定完全与实际情况相同，主要的差异在于移除了重复及相关时间等非重要信息，例如：

```
$ mvn spring-boot:run
(...部分内容被省略...)
[           main] o.s.w.r.f.s.s.RouterFunctionMapping       : Mapped (GET &&
/hello-world) ->
thinkinginspringboot.firstappbygui.FirstAppByGuiApplication$$Lambda$276/708609190@7
af17431
(...部分内容被省略...)
```

```
当前 WebServer 实现类为：org.springframework.boot.web.embedded.tomcat.TomcatWebServer
[           main] t.f.FirstAppByGuiApplication             : Started
FirstAppByGuiApplication in 2.119 seconds (JVM running for 5.071)
```

当内容中出现“(...部分内容被省略...)”时，说明其中省略了数行的日志内容，并且几乎所有的日志输出到标准输出（System.out）。

源码版本约定

在源码分析过程中，为了理清不同版本中的实现细节和变化，通常在目标代码下方带有 Spring Boot 版本信息，以及 Maven 依赖的 GAV 坐标信息（GAV = groupId、artifactId 和 version），如下所示。

```
@Target(ElementType.TYPE)
@Retention(RetentionPolicy.RUNTIME)
@Documented
@Inherited
@Configuration
@EnableAutoConfiguration
@ComponentScan
public @interface SpringBootApplication {

    /**
     * Exclude specific auto-configuration classes such that they will never be applied.
     * @return the classes to exclude
     */
    Class<?>[] exclude() default {};

}
```

> 以上实现源码源于 Spring Boot 1.2.8.RELEASE。
> Maven GAV 坐标为：org.springframework.boot:spring-boot-autoconfigure:1.2.8.RELEASE。

如果以上版本信息没有出现，则 Spring Boot 默认采用 2.0.2.RELEASE，Spring Framework 选择 5.0.6.RELEASE，JDK 源码的版本是 1.8.0_172。

源码省略约定

在源码分析的过程中，考虑到实现代码可能相对繁杂，为观其大意，便于记忆，会注释或移除部分无关痛痒的内容，例如：

```
public class AutoConfigurationImportSelector
        implements DeferredImportSelector, BeanClassLoaderAware, ResourceLoaderAware,
        BeanFactoryAware, EnvironmentAware, Ordered {
    ...
    protected List<String> getCandidateConfigurations(AnnotationMetadata metadata,
            AnnotationAttributes attributes) {
        List<String> configurations = SpringFactoriesLoader.loadFactoryNames(
                getSpringFactoriesLoaderFactoryClass(), getBeanClassLoader());
        ...
        return configurations;
    }
    ...
    protected Class<?> getSpringFactoriesLoaderFactoryClass() {
        return EnableAutoConfiguration.class;
    }
    ...
}
```

当然，其省略的部分并非一无是处，而是根据小马哥的个人经验来筛选的，必然受个人主观想法的影响，建议读者结合对应的版本源码，整体把握，逐步形成选读的意识。

表达约定

本书的讨论内容可能对相同事物出现不同的表述方式。

- 注解：Annotation；
- 配置 Class：`@Configuration` 类、`@Configuration` Class、Configuration Class；
- 包：`package`；
- 类路径：Class Path、class-path、类路径。
- 事件/监听：事件/监听、事件/监听器、事件监听器。

示例工程

本书所有的示例代码均存放在 https://github.com/mercyblitz/thinking-in-spring-boot-samples，该工程为标准的 Maven 多模块工程，运行时要求为 Java 1.8+和 Maven 3.2.5+。其协议为 Apache License Version 2.0，不必担心商业用途所带来的风险。由于本系列图书尚未完全定稿，工程结构未来可能存在微调。因此，当前内容无法确保百分之百匹配，请读者定期关注 README.md 文件，确保内容的更新。

工程结构

示例工程 `thinking-in-spring-boot-samples` 的结构如下图所示。

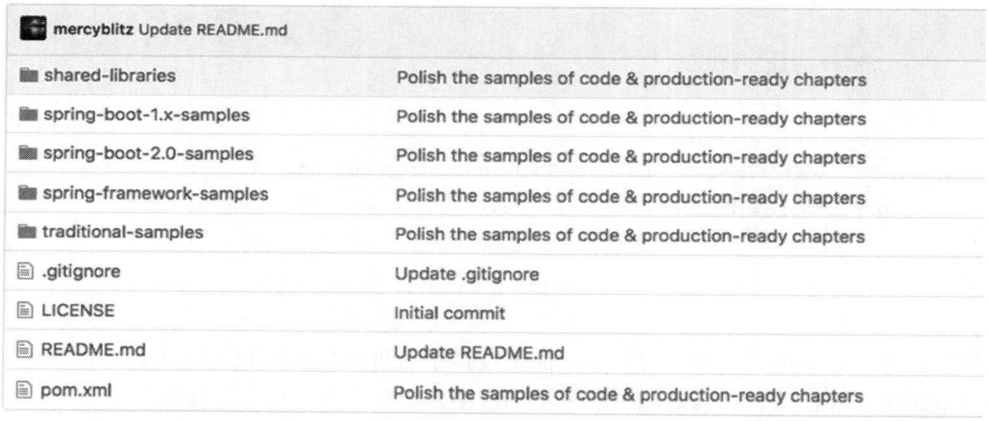

mercyblitz Update README.md

shared-libraries	Polish the samples of code & production-ready chapters
spring-boot-1.x-samples	Polish the samples of code & production-ready chapters
spring-boot-2.0-samples	Polish the samples of code & production-ready chapters
spring-framework-samples	Polish the samples of code & production-ready chapters
traditional-samples	Polish the samples of code & production-ready chapters
.gitignore	Update .gitignore
LICENSE	Initial commit
README.md	Update README.md
pom.xml	Polish the samples of code & production-ready chapters

该工程包含五个子模块和四个文件，分别如下。

子模块及说明如下表所示。

子 模 块	说 明
shared-libraries	共享类库，为其他工程提供基础 API 或依赖
spring-boot-1.x-samples	Spring Boot 1.x 示例工程，包含六个子模块，主要用于参考和对比 Spring Boot 1.x 各版本的实现差异，并且提供章节示例代码实现
spring-boot-2.0-samples	Spring Boot 2.0 示例工程，也是主示例工程，以 2.0.2.RELEASE 作为基础版本
spring-framework-samples	Spring Framework 示例工程，作为 Spring Boot 底层实现框架，版本范围从 2.0 到 5.0
traditional-samples	传统 Java EE 示例工程，用于理解 Java EE 与 Spring Boot 的关联和差异

文件及其说明如下表所示。

文 件	说 明
.gitignore	Git 版本控制文件
LICENSE	工程许可文件
README.md	工程说明文件
pom.xml	示例工程 Maven pom.xml 文件

其中，又以 spring-boot-2.0-samples、spring-boot-1.x-samples 和 spring-framework-samples 为本示例工程最核心的子模块，对此将详细说明。

子模块 spring-boot-2.0-samples

spring-boot-2.0-samples 作为《Spring Boot 编程思想（核心篇）》的主示例工程，基于 Spring Boot 2.0.2.RELEASE 实现，由若干子模块组成，这些模块与章节所讨论的议题紧密关联：

```
├── spring-boot-2.0-samples
│   ├── auto-configuration-sample
│   ├── externalized-configuration-sample
│   ├── first-app-by-gui
│   ├── first-spring-boot-application
│   ├── formatter-spring-boot-starter
│   ├── pom.xml
│   ├── production-ready-sample
│   ├── spring-application-sample
│   ├── spring-boot-2.0-samples.iml
│   └── traditional-web-sample
```

按照本书的安排，模块与章节所对应的关系如下表所示。

子 模 块	说 明	篇 章
first-spring-boot-application	基于 Maven 插件构建的第一个 Spring Boot 应用	核心篇——总览 Spring Boot
first-app-by-gui	基于图形化界面 https://start.spring.io/构建的第一个 Spring Boot 应用	核心篇——总览 Spring Boot
auto-configuration-sample	Spring Boot 自动装配示例	核心篇——走向自动装配
formatter-spring-boot-starter	Spring Boot 自动装配 Starter 示例	核心篇——走向自动装配
spring-application-sample	Spring Boot SpringApplication 示例	核心篇——理解 SpringApplication

续表

子 模 块	说　明	篇　章
externalized-configuration-sample	Spring Boot 外部化配置示例	运维篇——超越外部化配置
production-ready-sample	Spring Boot Production-Ready	运维篇——简化 Spring 应用运维体系
traditional-web-sample	Spring Boot 应用部署到传统 Servlet 容器示例	Web 篇——“渐行渐远”的 Servlet

除此之外，相关示例代码部分也可能放置在其他子模块，如“走向自动装配”章节中，大量的实例代码在子模块 spring-framework-samples 之中。

子模块 spring-boot-1.x-samples

前文提到，该子模块包含六个子模块，它们对应了所有的 Spring Boot 1.x 实现版本：

```
├── spring-boot-1.x-samples
│   ├── pom.xml
│   ├── spring-boot-1.0.x-project
│   ├── spring-boot-1.1.x-project
│   ├── spring-boot-1.2.x-project
│   ├── spring-boot-1.3.x-project
│   ├── spring-boot-1.4.x-project
│   ├── spring-boot-1.5.x-project
```

截至当前编写时间，恰逢 Spring Boot 1.5 的发行版本为 `1.5.10.RELEASE`，而 1.5 之前的版本已停止维护，可选择其最后发行的版本作为参考，故子模块、Spring Boot 1.x 版本及对应 Spring Framework 的关系如下表所示。

子 模 块	Spring Boot 1.x 版本	Spring Framework 版本
spring-boot-1.0.x-project	1.0.2.RELEASE	4.0.3.RELEASE
spring-boot-1.1.x-project	1.1.9.RELEASE	4.0.8.RELEASE
spring-boot-1.2.x-project	1.2.8.RELEASE	4.1.9.RELEASE
spring-boot-1.3.x-project	1.3.8.RELEASE	4.2.8.RELEASE
spring-boot-1.4.x-project	1.4.7.RELEASE	4.3.9.RELEASE
spring-boot-1.5.x-project	1.5.10.RELEASE	4.3.14.RELEASE

值得注意的是，子模块 spring-boot-1.x-samples 并非主示例工程，各章节讨论的特性示例并非面面俱到。相反，读者应重点关注子模块 spring-boot-2.0-samples。

子模块 spring-framework-samples

spring-framework-samples 作为示例实现的辅助分析子模块，涵盖从 2.0 到 5.0 的版本，由于 spring-boot-1.x-samples 间接引入了 Spring Framework 4.0～4.3 的依赖，因此当前模块并未将 4.x 版本细分：

```
├── spring-framework-samples
│   ├── pom.xml
│   ├── spring-framework-2.0.x-sample
│   ├── spring-framework-2.5.6-sample
│   ├── spring-framework-3.0.x-sample
│   ├── spring-framework-3.1.x-sample
│   ├── spring-framework-3.2.x-sample
│   ├── spring-framework-4.3.x-sample
│   ├── spring-framework-5.0.x-sample
│   └── spring-webmvc-3.2.x-sample
```

原则上，以上模块所选择的 Spring Framework 版本与 spring-boot-1.x-samples 类似，故子模块与 Spring Framework 的依赖关系如下表所示。

子 模 块	Spring Framework 版本
spring-framework-2.0.x-sample	2.0.8
spring-framework-2.5.6-sample	2.5.6.SEC03
spring-framework-3.0.x-sample	3.0.0.RELEASE
spring-framework-3.1.x-sample	3.1.4.RELEASE
spring-framework-3.2.x-sample	3.2.18.RELEASE
spring-framework-4.3.x-sample	4.3.17.RELEASE
spring-framework-5.0.x-sample	5.0.6.RELEASE
spring-webmvc-3.2.x-sample	3.2.18.RELEASE

当读者发现子模块工程仅包含 `pom.xml` 文件时，说明它引入的目的在于 Spring Framework 源码分析，用于比对版本间 Spring 特性的变迁和实现的差异。

示例代码说明

由于本书几乎覆盖所有的 Spring Framework 和 Spring Boot 版本，通常在配合章节说明时，绝大多数示例代码的结尾部分带有“源码位置”的信息，相对于 https://github.com/mercyblitz/thinking-in-spring-boot-samples 工程路径，例如：

```
public class GenericEventListenerBootstrap {

    public static void main(String[] args) {
        // 创建注解驱动 Spring 应用上下文
        AnnotationConfigApplicationContext context = new AnnotationConfigApplicationContext();
        // 注册 UserEventListener，即实现 ApplicationListener ，也包含 @EventListener 方法
        context.register(UserEventListener.class);
        // 初始化上下文
        context.refresh();
        // 构造泛型事件
        GenericEvent<User> event = new GenericEvent(new User("小马哥"));
        // 发送泛型事件
        context.publishEvent(event);
        // 发送 User 对象作为事件源
        context.publishEvent(new User("mercyblitz"));
        // 关闭上下文
        context.close();
    }
    ...
}
```

> 源码位置：以上示例代码可通过查找 spring-framework-samples/spring-framework-5.0.x-sample 工程获取。

如果以上信息尚未提供，那么在默认情况下，书中的 Spring Framework 示例存放在 **spring-framework-samples/spring-framework-5.0.x-sample** 工程中，而 Spring Boot 示例存放在 **spring-boot-2.0-samples/**所对应的章节工程中。

UML 源文件说明

本书引入的相关 UML 图的源文件位于子项目工程的 src/main/resources/uml 目录中，文件扩展名为.ucls。比如子模块 spring-application-sample：

```
src/main/resources/uml/
├── ApplicationContext.ucls
├── ApplicationContextEvent.ucls
├── ApplicationEventPublisher.ucls
```

```
├── ConfigurableApplicationContext_ApplicationEventPublisher.ucls
└── SimpleApplicationEventMulticaster.ucls
```

请读者使用 Eclipse 插件“ObjectAid UML Explorer”将其打开，插件下载地址为：http://www.objectaid.com/update/current，比如文件 `ApplicationContext.ucls`，如下图所示。

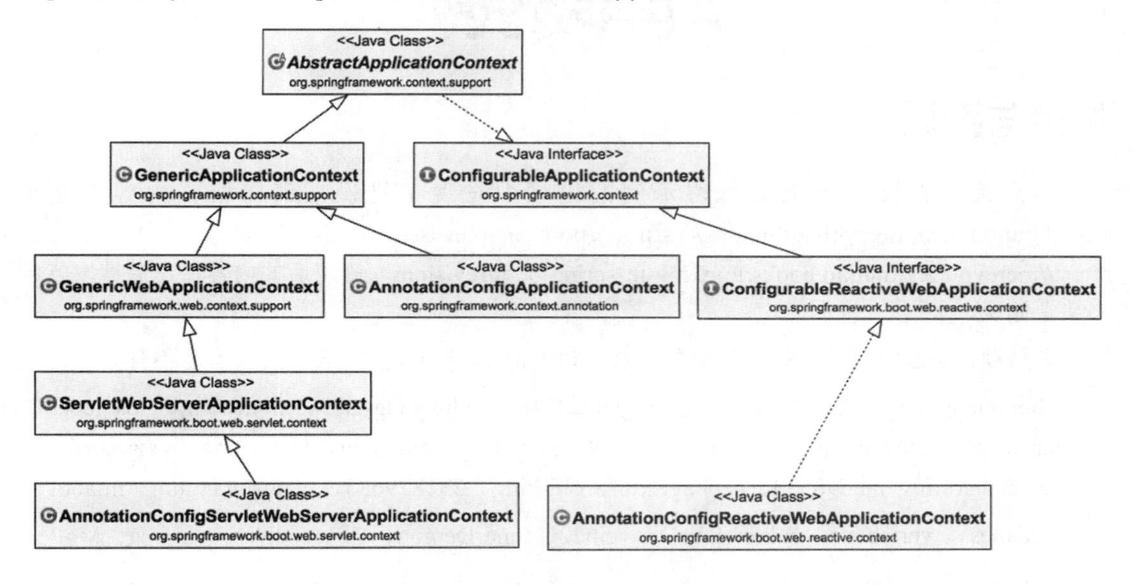

配套视频

尽管本书细致地讨论了 Spring Boot 和 Spring Framework 相关特性，不过它并非快速上手教程，如果读者具备三年以上的开发经验，或者是资深的 Spring 用户，那么阅读起来会相对顺畅。反之，如果读者出现阅读困难的情况，则不妨先参考官方文档，掌握基本使用技能，或者观看小马哥在慕课网的免费视频：

- Spring Boot 2.0 深度实践——初遇 Spring Boot（https://www.imooc.com/learn/933）；
- Spring Boot 2.0 深度实践之系列总览（https://www.imooc.com/learn/1058）。

其中视频“初遇 Spring Boot”先介绍 Spring Boot 2.0 的基本特性，随后创建第一个 Spring Boot 应用，再将其改造成多 Maven 模块应用，这有助于对本书示例工程结构的理解。而“系列总览”则是“核心篇——总览 Spring Boot”章节对应的视频。换言之，慕课网的“Spring Boot 2.0 深度实践”系列属于本书的配套视频。读者可访问 https://www.imooc.com/t/5387391 或扫描下面的二维码参与报名。

纠错与勘误

如果读者在阅读本书或练习示例的过程中发现了错误，请将错误内容提交至 https://github.com/mercyblitz/thinking-in-spring-boot-samples/issues，小马哥将勘误内容汇总到 https://mercyblitz.github.io/books/thinking-in-spring-boot/revision/，修正后的内容将在后续发行的书籍中体现。

参与勘误贡献者名单包括未收录的人员，以 GitHub 为 ID 的方式记录（排名不分先后）：

Zhengjiangming、RockFishChina、wqyblue3316、Yuhuiyang-Dev、alonecong、nosqlcoco、stackfing、liaozan、porscheYong、THyyy、xkcoding、hulanhui、bilaisheng、yxzjerryfish、feichangxinfu、landy8530、cabbageXun、old-king、331829683、jiangshuangjun、miaoo92、Redaness、verichenn、caixingjava、myejb22、InnerDemon、alonecong、codingma、MoPei、Runyan、loupXing、zhongqinzhen、punisherj、morningcat2018、bfchengnuo、JamesDrago、Zhuangjinjin、a974389757、huitu1991、LJaer、tuablove、jobcxq、lwx19960428、Dovelol、Darling012、innerpeacez、mzanthem、liangxiong3403、liqi19950722、CmdSmith、9526xu、zhangzl0119、zyandu、AK47Sonic、nzmc663、codewaltz1994、Isaac-Zhang、useaname、Fvames、illman2016、outpaces、heqiang、robert95998、wanglforever、McJayFeng、dwlsxj。

Fvames 公益资金流向

本书五成以上的稿费将捐赠给贫困地区小朋友作为教育公益基金，读者访问 https://mercyblitz.github.io/books/thinking-in-spring-boot/donate/来关注和监督资金流向。

关于我

“我是谁？”是一个不错的哲学问题。

在“江湖”上，大家亲切地称我为“小马哥”，我做公益，也做生意；在社区中，我又以 mercyblitz 的身份出没在众多开源项目中，“mercy”符合我的性格，“blitz”说明我的风格。

承蒙错爱，不少朋友对我过去的分享称赞有加，然而“千夫诺诺，不如一士谔谔”，时常又让自己陷入一种迷思，到底是平台的帮衬，还是个人的确禁得起考验？于是我选择隐匿真名，希望能够听到更真实的声音。尽管在互联网时代，个人信息几乎无处遁形，无可讳言，我所属的公司及职业头衔必然会形成“舞台效应”，如此一来，不但违背了写书的初衷，而且模糊了讨论的焦点。所以，本书既不会出现这些信息，又不会搞“个人崇拜”。它的价值应该体现在知识的传播上，至于它的优劣则由诸君来评判。

个人简介

Java 劝退师，Apache 和 Spring Cloud 等知名开源架构成员。目前主要负责开源项目、微服务架构演进、基础设施构建等。通过 SUN Java（SCJP、SCWCD、SCBCD）和 Oracle OCA 等认证。

交流社区

- 微信公众号：次灵均阁。

- 知识星球：

- GitHub：http://github.com/mercyblitz。

更多个人信息，请使用 Google 搜索“mercyblitz”。

主要开源项目

- Spring Cloud Alibaba：https://github.com/spring-cloud-incubator/spring-cloud-alibaba。
- Apache Dubbo：https://github.com/apache/incubator-dubbo。
- Nacos：https://github.com/alibaba/nacos。

谨以此书纪念已故外婆

解厚群

目录

第 1 部分　总览 Spring Boot

第 2 部分 走向自动装配

第 3 部分 理解 SpringApplication

第 1 部分　总览 Spring Boot

> “靡不有初，鲜克有终”——《诗经 · 大雅 · 荡》

在编写本书的过程中，这句古训一直在脑海中萦绕。回想起 2015 年到 2018 年这三年，小马哥一直参与集团内部微服务架构的设计和基础设施建设，就个人而言，阶段性任务已经完成，尽管有不少应用已经走向微服务架构，然而从全局来看，这不过只是一个开始。起初本书想讨论微服务架构及运用，然而发现在微服务的粒度划分上很难达成共识，为了避免争议，后来将重心放在 Spring Boot 上。

Spring Boot 作为目前非常流行的微服务框架，深受互联网企业的青睐，几乎成为微服务中间件事实上的标准。有趣的是，Spring Boot 项目的原意并非为微服务架构而生。从时间轴上分析，Spring Boot 早在 2013 年就已问世，而微服务架构则由 Martin Fowler 等人于 2014 年才提出。更有意思的是，微服务架构和 Spring Boot 并非新兴事物，不过在业界大肆宣传后，两者变得异常受人瞩目，使得坐而论道者甚繁，作而行之者盖寡，从而模糊了讨论焦点。微服务架构作为一种细粒度的 SOA，无论用何种方式表述，不过是名词之争，都无法解决现实中业务场景的复杂性。尽管 Spring Boot 是一种不错的选择，然而并非唯一的选择，即使运用传统的 Java EE 技术，或者使用 Vert.x 这类相对小众的框架照样也能实现。换言之，架构设计的好坏不在于理论和技术，而在于实施者对业务的理解和专业水平。如果不能做到以其昭昭，如何使人昭昭呢？

在过去的三年里，小马哥有幸直接或间接地参与数以万计的 Spring Boot 应用迁移，发现很多人基本停留在使用层面，并未深入其中，当遇到棘手的问题时仍旧束手无策，一方面由于 Spring Boot 材料不足，且讨论不够深入，即使如同官方文档，同样语焉不详。另外一方面则是缺乏深入探讨的机会和勇气。因此，小马哥著此书，希望通过各章节的讨论，再结合实际的经验，向读者全面且系统地介绍 Spring Boot 的特性，理解其中的设计，最终掌握实现背后的哲学思想。本书的脉络大致从功能介绍到特性关联；从提出疑问到结论推导；从源码分析到示例验证。

第 1 部分将站在一定高度总览 Spring Boot 的特性，后续的章节将在此基础上展开探讨，实现以“技”会友，共同进步。

chapter 1

第 1 章 初览 Spring Boot

1.1 Spring Framework 时代

大约在 2007 年，小马哥当时还是一名在校大学生，在学校图书馆看到了一本名为 *Expert One-on-One J2EE Development without EJB* 的书籍，作者是 Rod Johnson，后来得知他也是 Spring Framework 的作者之一。

大概因为该书的缘故，使得不少像我一样的年轻人，对 Spring Framework 产生了先入为主的好感。同时，也人云亦云般地认为 EJB 如何笨拙。

Spring Framework 的成功一定程度上颠覆了 J2EE 的开发模式，原来除了 J2EE 标准，还有另外一条“康庄大道”可以走。而颠覆的方式也比较有意思，基本上依赖于新的概念，即你我熟知的两种技术：IoC（Inversion of Control，控制反转）和 DI（Dependency Inject，依赖注入）。

大学毕业后，我幸运地进入一家外企，全球有几处分公司，有相当多的外国开发人员。在与他们技术交流中，他们时不时地会谈到 JSR（JSR - Java Specification Requests），也就是常说的 Java 标准规范。随着技术探讨的深入，虽说仍是“半缘修道半缘君”，可不再“取次花丛懒回顾”。回眸 Java EE 标准，发现 Spring Framework 也是一种重复发明的轮子。在 Java EE 技术体系中，IoC 的实现方式为 JNDI（Java Naming and Directory Interface，Java 命名和目录接口），而 DI 是 EJB 容器注入。

那时的 Web 表示层有 Struts，模板有 JSP、Velocity 等，Hibernate、iBatis（后来改名为 MyBatis）更是持久层的“明星”。所以，Spring Framework 被“尴尬地”作为集成层的代表，国外的开发人员戏称它为“胶水”框架，意在黏合各家技术。

读者不难发现，前面对 Spring Framework 的描述基本上是负面的。那么，Spring Framework 有哪些优点呢？

小马哥经常与同仁们开玩笑说：“开发人员在技术领域要摒除内心的感性认知，不要‘技术感性化’”。也就是说，不要偏颇地认为某种技术是无出其右的，甚至说它是“万能”的。Spring Framework 自然也是如此。

在小马哥看来，Spring Framework 的优点非常多，如 Spring Framework 对事务（Transaction）的抽象，统一了数据库事务和分布式事务，“润物细无声”地利用 AOP（面向切面编程）技术使得事务编程极简化。另外一个实例是 Spring Web MVC，使整体框架的发展达到了顶峰。框架作者对 Java EE 技术体系的理解达到了“炉火纯青”的地步，框架具备简化抽象的能力。

无论 Spring Framework 使接口如何简化，设计如何优美，始终无法摆脱被动的境况——由于它自身并非容器，基本上不得不随 Java EE 容器启动而装载。然而 Spring Boot 的出现，改变了 Spring Framework 甚至整体 Spring 技术体系的现状。

1.2 Spring Boot 简介

子曰：“名不正则言不顺”，因此，项目的发起总有其意义和目的，Spring Boot 项目自然也不例外。在 Spring 官网的首页，Spring Boot 被官方定义为“BUILD ANYTHING”。

官方又进一步做出了解释：

> Spring Boot is designed to get you up and running as quickly as possible, with minimal upfront configuration of Spring. Spring Boot takes an opinionated view of building production ready applications.

文中的前半段相对好理解，Spring Boot 为快速启动且最小化配置的 Spring 应用而设计，而后半段令人费解，其中“**opinionated**”有固执己见之意，是一个贬义词。可是放在以上语境中，好像也不太合适。小马哥试图将其解释为“固化”。为了更好地确认其意义，在 **stackoverflow** 上找到一处相关问答（https://stackoverflow.com/questions/802050/what-is-opinionated-software），小马哥觉得解释得比较清楚。

问题：

> I often see people saying that certain software is "very opinionated" or that Microsoft tends to write "un-opinionated" frameworks. What does this actually mean?

回答（节选）：

> If a framework is opinionated, it lock or guides you into their way of doing things.
>
> ...
>
> Un-opinionated software design is more like PERL/PHP. It allows the developer and trusts the developer to make the right decisions and puts more control in their hands.

结合以上的内容，小马哥更加确信了“**opinionated view**”可翻译为“固化的视图”。那么，“**production-ready**”又是什么意思呢？字面意思是“为生产准备的”，似乎有点拗口，小马哥偏好将其解释成“预备生产的”。

因此，“**Spring Boot takes an opinionated view of building production ready applications.**”可解释为“Spring Boot 具有一套固化的视图，该视图用于构建生产级别的应用”。这些应用具备开发（Dev）和运维（Ops）双重特性，即业界常提及的 DevOps。近几年 DevOps 成为业界热议的话题之一，不少企业纷纷声称已响应 DevOps 的潮流。不过据小马哥的臆断，其中不少企业仍停留在意识流阶段。原因很简单，首先，响应不代表实施或落实。其次，基本上没有看到颠覆性的产品出现。Spring Boot 的出现，着实让人眼前一亮。“Production Ready”的含义将在后面的章节中展开讨论。

单击首页顶部菜单“PROJECTS”，进入 Spring 主要项目页面。在所有项目中，Spring Boot 项目排名第一，Spring 官方对其重视程度可见一斑。

单击“SPRING BOOT”区块进入其项目页面：

在 Spring Boot 项目页面中，官方对 Spring Boot 的定义做出了细微调整：

> Spring Boot makes it easy to create stand-alone, production-grade Spring based Applications that you can "just run".
>
> We take an opinionated view of the Spring platform and third-party libraries so you can get started with minimum fuss. Most Spring Boot applications need very little Spring configuration.

此处强调的是 Spring Boot 的价值，它使构建独立的 Spring 生产级别应用变得简单。随后的文字中，官方再次使用“opinionated view”来固化 Spring Boot 所依赖的 Spring 平台和第三方类库，从而减少管理它们的烦恼。换言之，开发人员只需关注 Spring Boot 的版本，不需要关心下游依赖的版本信息，如本书中讨论的 Spring Boot 版本为 `2.0.2.RELEASE`，而开发人员不需要显式地声明它所依赖的 Spring Framework 版本为 `5.0.6.RELEASE`，因此减少了依赖管理的打扰。同时，官方最后补充道：“大多数 Spring Boot 应用只需少量的 Spring 配置”。此句的描述是比较含糊的，所谓的 Spring 配置既包括属性配置，也包括 Java 代码配置。在官方网页的“Features”部分，总结了 Spring Boot 的六大特性，接下来将逐一讨论。

1.3 Spring Boot的特性

将官网“Features”部分的六大特性复制如下：

- Create stand-alone Spring applications
- Embed Tomcat, Jetty or Undertow directly (no need to deploy WAR files)
- Provide opinionated 'starter' dependencies to simplify your build configuration
- Automatically configure Spring and 3rd party libraries whenever possible
- Provide production-ready features such as metrics, health checks and externalized configuration
- Absolutely no code generation and no requirement for XML configuration

将其逐一解释为：

- 创建独立的 Spring 应用；
- 直接嵌入 Tomcat、Jetty 或 Undertow 等 Web 容器（不需要部署 WAR 文件）；
- 提供固化的“starter”依赖，简化构建配置；
- 当条件满足时自动地装配 Spring 或第三方类库；
- 提供运维（Production-Ready）特性，如指标信息（Metrics）、健康检查及外部化配置；
- 绝无代码生成，并且不需要 XML 配置。

同时，为了深入讨论的便利性，接下来需要做一些预备工作。

1.4 准备运行环境

Spring Boot 2.0 基于 Spring Framework 5.0 开发，运行 Spring Framework 5.0 的最低版本要求为 Java 8。同时，Spring Boot 采用模块化设计，其模块类库管理依赖于 Apache Maven，官方的说明是 Spring Boot 兼容 Apache Maven 3.2 或更高版本。

如果已经熟悉或已安装相关的运行环境，则可忽略这部分内容。

1.4.1 装配 JDK 8

前面已提及 Spring Boot 2.0 依赖 Java 8 ，读者可自行确认本地机器安装的 JDK 版本。比如，小马哥当前使用的机器为 MacBook Pro，操作系统是 macOS Sierra 10.13.6，在控制台执行 `java -version` 命令确认 Java 的版本：

```
$ java -version
java version "1.8.0_172"
Java(TM) SE Runtime Environment (build 1.8.0_172-b11)
Java HotSpot(TM) 64-Bit Server VM (build 25.172-b11, mixed mode)
```

如果读者的机器尚未安装 JDK 8，那么可以前往 Oracle 官方网站下载最新的 JDK：http://www.oracle.com/technetwork/java/javase/downloads/index.html。

关于下载和安装 JDK 的部分，此处不再说明。

1.4.2 装配 Maven

相对于装配 JDK 8，装配 Maven 稍显复杂，小马哥稍微花一点篇幅说明一下。

第一步：下载 Maven

前往 Maven 官方下载页面（http://maven.apache.org/download.cgi），下载 Maven 二进制归档 zip 文件（**Binary zip archive**），当前最新的版本为 `3.5.0`。

第二步：安装 Maven

将 Maven 二进制归档 zip 文件（`apache-maven-3.5.0.zip`）解压到安装目录，以小马哥的安装目录为例，文件路径：`/Users/mercyblitz/software/java/open-source/apache-maven-3.5.0`。

解压后的文件列表如下：

```
$ ls -ls
total 64
48 -rw-r--r--  1   mercyblitz    staff    20934    4    3    2017     LICENSE
 8 -rw-r--r--  1   mercyblitz    staff    182      4    3    2017     NOTICE
 8 -rw-r--r--  1   mercyblitz    staff    2544     4    3    2017     README.txt
 0 drwxr-xr-x  8   mercyblitz    staff    256      4    3    2017     bin
 0 drwxr-xr-x  3   mercyblitz    staff    96       4    3    2017     boot
 0 drwxr-xr-x  6   mercyblitz    staff    192      7    23   20:57    conf
 0 drwxr-xr-x  83  mercyblitz    staff    2656     4    3    2017     lib
```

第三步：设置 M2_HOME 环境变量

设置 `M2_HOME` 环境变量采用的是 Maven 官方推荐的方式，利用该变量替换安装目录，类似于 `JAVA_HOME` 环境变量。

切换到控制台，编辑当前用户的`.bash_profile` 文件：

```
$ vi ~/.bash_profile
```

追加 M2_HOME 环境变量内容到该文件，即：

```
export M2_HOME=/Users/mercyblitz/software/java/open-source/apache-maven-3.5.0
```

该设置立即生效：

```
$ source ~/.bash_profile
```

第四步：关联 mvn 命令

mvn 命令即 Maven 执行文件，该文件存放在$M2_HOME/bin 目录下，关联该文件到用户 bin 目录：

```
$ sudo ln -s $M2_HOME/bin/mvn /usr/local/bin/mvn
```

完成关联后，执行命令，观察其是否生效：

```
$ mvn -version
Apache Maven 3.5.0 (ff8f5e7444045639af65f6095c62210b5713f426; 2017-04-
04T03:39:06+08:00)
Maven home: /Users/mercyblitz/software/java/open-source/apache-maven-3.5.0
Java version: 1.8.0_172, vendor: Oracle Corporation
Java home: /Library/Java/JavaVirtualMachines/jdk1.8.0_172.jdk/Contents/Home/jre
Default locale: zh_CN, platform encoding: UTF-8
OS name: "mac os x", version: "10.13.6", arch: "x86_64", family: "mac"
```

如果控制台输出以上信息，则说明关联成功。

第五步：配置 Maven 仓库

配置 Maven 仓库主要关注两个方面：Maven 下载的类库必然存在来源，即远程 Maven 仓库。下载后的类库需要地方存储，即本地 Maven 仓库。两者的设置需要配置 Maven settings.xml 文件（该文件存放在$M2_HOME/conf 目录下）。

由于 Spring Boot 相关 Maven 依赖库基本存放在海外镜像中，小马哥推荐使用阿里云的 Maven 镜像，以提升下载速度和稳定性。

本地 Maven 仓库路径和 Maven 镜像的配置如下所示。

```
<settings>
  <!-- 本地 Maven 仓库 -->
  <localRepository>/Users/mercyblitz/.m2/repository</localRepository>


  <!-- Maven 镜像 -->
```

```
    <mirrors>
      <mirror>
        <id>alimaven</id>
        <name>aliyun maven</name>
        <url>http://maven.aliyun.com/nexus/content/groups/public/</url>
        <mirrorOf>central</mirrorOf>
      </mirror>
    </mirrors>
  </settings>
```

更多 Maven 配置可参考官方文档：http://maven.apache.org/configure.html。

1.4.3 装配 IDE（集成开发环境）

目前业界常用的 Java IDE 工具为 Eclipse 和 IntelliJ IDEA。小马哥推荐使用 IntelliJ IDEA（下面简称 IDEA）。相比于 Eclipse 而言，IDEA 的开发效率更高。

IDEA 官方网站提供两种下载版本，分别为“Ultimate”（旗舰版）和“Community”（社区版），下载地址：https://www.jetbrains.com/idea/download/。

其中，IDEA 旗舰版属于收费产品，功能非常强大，尽管小马哥拥有其 License（使用许可），不过仅限于研发开源项目。

> 欢迎读者一同参与开源项目。
>
> Apache Dubbo：https://dubbo.apache.org。
>
> 成为项目 Committer 后，可向 JetBrains 公司免费申请 IDEA 旗舰版 Open Source License。

因此，本书的内容将使用免费的 IDEA 社区版，这并不影响讨论的质量。同时，也希望读者保护知识产权，支持正版软件。

当以上工作准备妥当后，接下来针对官方罗列的特性逐一讨论，以理解其中深层次的含义。

第 2 章 理解独立的 Spring 应用

创建独立的 Spring 应用作为 Spring Boot 的首要特性，其中有两层含义值得思考，首先，为什么要独立的应用？其次，为什么是 Spring 应用，而非 Spring Boot 应用？

在大多数 Spring Boot 应用场景中，程序直接或间接地使用 `SpringApplication` API 引导应用。其中又结合嵌入式 Web 容器，对外提供 HTTP 服务。如果从应用类型上划分，那么 Spring Boot 应用包括 Web 应用和非 Web 应用。其中非 Web 应用主要用于服务提供、调度任务、消息处理等场景，而 Web 应用类型在 Spring Boot 1.x 中有且仅有 Servlet 容器实现，包括传统的 Servlet 和 Spring Web MVC。而从 2.0 版本开始新增了 Reactive Web 容器实现，即 Spring 5.0 WebFlux，故 Spring Boot 2.0 在 `SpringApplication` API 上新增了 `setWebApplicationType(WebApplicationType)`方法，允许程序显式地设置 Web 应用的枚举类型，其中成员 `NONE` 表示非 Web 类型，`SERVLET` 和 `REACTIVE` 分别代表 Servlet Web 和 Reactive Web。光指定应用类型还不够，还需要搭配 Spring Boot Starter 技术，直接或间接地引入相关的依赖，结合 Spring Boot 自动装配，再利用 Spring Boot 和 Spring Framework 的生命周期，创建并启动嵌入式的 Web 容器，如 Servlet Web 的 Maven 依赖为 `org.springframework.boot:spring-boot-starter-web`，而 Reactive Web 则需依赖 `org.springframework.boot:spring- boot-starter-webflux`，相关细节在“**理解 SpringApplication**”中的“**推断 Web 应用类型**”一节中有详细的描述。换言之，Spring Boot 应用无须再像传统的 Java EE 应用那样，将

应用打包成 WAR 文件或 EAR 文件，并部署到 Java EE 容器中运行。不过，Spring Boot 也支持传统的 Web 部署方式，官方文档在“87. Traditional Deployment”章节中有使用方法的说明。然而这种部署方式属于一种兼容或过渡的手段，并非独立的 Spring 应用，其核心实现原理请参考“**走向自动装配——Spring Web 自动装配**”的内容。总而言之，Spring Boot 应用采用嵌入式容器，独立于外部容器，对应用生命周期拥有完全自主的控制，从此改变了“寄人篱下”的境地。

> 不少开发人员存在一个理解上的误区，认为 Spring Boot 嵌入式 Web 容器启动时间少于传统的 Servlet 容器，实际上并没有证据证明这种情况，而应将其理解为方便快捷的启动方式，可以提升开发和部署效率。

口头上，开发人员常将 Spring Boot 框架构建的应用称为 Spring Boot 应用，然而实际上是 Spring 应用。在上述的 Web 容器中，无论 Servlet 容器，还是 Netty Web 容器，均不属于 Spring 家族的产品。在传统的 Spring 应用中，外置容器需要启动脚本将其引导，随其生命周期回调执行 Spring 上下文的初始化。比较有代表性的是 Spring Web 中的 `ContextLoaderListener` 和 Web MVC 中的 `DispatcherServlet`，前者利用 `ServletContext` 生命周期构建 Web ROOT Spring 应用上下文，后者结合 `Servlet` 生命周期创建 `DispatcherServlet` 的 Spring 应用上下文。无论何种方式，均属于被动的回调执行，这也是为什么它们并没有完整的应用主导权的原因。然而当 Spring Boot 出现嵌入式容器启动方式后，嵌入式容器则成为应用的一部分，从本质上来说，它属于 Spring 应用上下文中的组件 Beans，这些组件和其他组件均由自动装配特性组装成 Spring Bean 定义（`BeanDefinition`），随 Spring 应用上下文启动而注册并初始化。而驱动 Spring 应用上下文启动的核心组件则是 Spring Boot 核心 API `SpringApplication`，所以是 Spring 应用，也可以称为 Spring Boot 应用。其中细节的讨论，请参考“**理解 SpringApplication**”部分。

> 由于本章为 Spring Boot 总览性的内容，所以各部分的讨论直接给出结论，其结论的推导、证明和示例将安排在专属章节中深入探讨。

引用 Linus 的名言“Talk is cheap. Show me the code.”，马上创建第一个独立的 Spring 应用吧！

2.1 创建 Spring Boot 应用

建议读者先阅读 Spring Boot 官方文档的“11. Developing Your First Spring Boot Application”章节，尽管本节的内容与官方案例多少存在重复，不过对于官方文档中的缺失将进行补充。同时，需要对程序运行前后的变化仔细观察，随之进行思考。

本节将介绍两种创建 Spring Boot 应用的方法，即“**命令行创建方式**”和“**图形化创建方式**”，以及两种运行方式，即“**生产环境运行方式**”和“**开发阶段运行方式**”。

2.1.1 命令行方式创建 Spring Boot 应用

命令行（Command-Line）是最传统的人机交互方式。在官方文档的“11.1 Creating the POM”章节中并没有直接说明创建 Spring Boot 应用的命令，而是告诉读者，首先创建一个 Maven pom.xml 文件，然后通过自己偏好的文本编辑器来添加以下内容：

```
<?xml version="1.0" encoding="UTF-8"?>
<project xmlns="http://maven.apache.org/POM/4.0.0"
xmlns:xsi="http://www.w3.org/2001/XMLSchema-instance"
xsi:schemaLocation="http://maven.apache.org/POM/4.0.0
http://maven.apache.org/xsd/maven-4.0.0.xsd">
<modelVersion>4.0.0</modelVersion>

<groupId>com.example</groupId>
<artifactId>myproject</artifactId>
<version>0.0.1-SNAPSHOT</version>

<parent>
 <groupId>org.springframework.boot</groupId>
 <artifactId>spring-boot-starter-parent</artifactId>
 <version>2.0.2.RELEASE</version>
</parent>

<!-- Additional lines to be added here... -->

</project>
```

Spring Boot 参考文档并不完全是为初学者准备的，而是假设读者非常熟悉 Spring 和 Java，或者有选择性地省略细枝末节。下面通过的 Maven Archetype 插件创建第一个 Spring Boot 应用。

1. 使用 Maven Archetype 插件

假设当前操作系统为非图形化界面（GUI），并且安装了 Maven。可借助 Maven 插件在控制台执行相关命令来完成项目创建。其中，maven-archetype-plugin 是最具代表性的插件之一。

执行前，请确保执行命令的目录存在写权限。比如本例的目录为/Users/mercyblitz/thinking-in-spring-boot-samples，命令如下：

```
$ mvn archetype:generate -DgroupId=thinking-in-spring-boot
-DartifactId=first-spring-boot-application -Dversion=1.0.0-SNAPSHOT -DinteractiveMode=false
```

```
-Dpackage=thinking.in.spring.boot
```

控制台输出内容：

```
    [INFO] Scanning for projects...
    (...部分内容被省略...)
    [INFO] Parameter: basedir, Value: /Users/mercyblitz/thinking-in-spring-boot-samples
    [INFO] Parameter: package, Value: thinking.in.spring.boot
    [INFO] Parameter: groupId, Value: thinking-in-spring-boot
    [INFO] Parameter: artifactId, Value: first-spring-boot-application
    [INFO] Parameter: packageName, Value: thinking.in.spring.boot
    [INFO] Parameter: version, Value: 1.0.0-SNAPSHOT
    [INFO] project created from Old (1.x) Archetype in dir:
/Users/mercyblitz/thinking-in-spring-boot-samples/first-spring-boot-application
    [INFO] ------------------------------------------------------------------------
    [INFO] BUILD SUCCESS
    [INFO] ------------------------------------------------------------------------
    (...部分内容被省略...)
```

结合控制台输出内容，解读构建命令：

- Maven 命令——`mvn`
- 插件简称——`archetype`

  在命令行内容中，我们看不到插件 `maven-archetype-plugin` 的部分，它却出现在控制台输出中。原因是 Maven 为了降低使用成本，允许简化插件的名称，所以 `archetype` 为 `maven-archetype-plugin` 的简称。更确切地说，完整的插件名称为 `org.apache.maven.plugins:maven-archetype-plugin:3.0.1`，也包括 `groupId`（org.apache.maven.plugins）、`artifactId`（maven-archetype-plugin）和 `version`（3.0.1）。
- 插件目标（Goal）——`archetype:generate`

  我们已了解 `archetype` 为 `maven-archetype-plugin` 的简称，`archetype:generate` 是 `maven-archetype-plugin` 插件的目标（Goal）。详细介绍可参考官方文档：http://maven.apache.org/archetype/maven-archetype-plugin/generate-mojo.html。
- 插件参数（Parameters）

  在 Java 启动命令中，通过`-D` 命令行参数设置 Java 的系统属性：`System.getProperties()`。Maven 插件也通过此方式获取所需的参数。在上例中，传递了五个参数，其中包括项目的 GAV 信息：

```
[INFO] Parameter: package, Value: thinking.in.spring.boot
[INFO] Parameter: groupId, Value: thinking-in-spring-boot
[INFO] Parameter: artifactId, Value: first-spring-boot-application
[INFO] Parameter: packageName, Value: thinking.in.spring.boot
[INFO] Parameter: version, Value: 1.0.0-SNAPSHOT
```

- 交互模式参数：interactiveMode

当参数值为 false 时，表示非交互式构建，也就是常说的“静默方式”。

2. 观察应用目录结构

mvn archetype:generate 执行完毕后，创建一个名为 first-spring-boot-application 的标准 Maven 工程：

```
$ ls -ls
total 8
8 -rw-r--r--  1 mercyblitz  staff  696 10 30 13:16 pom.xml
0 drwxr-xr-x  4 mercyblitz  staff  128 10 30 13:16 src
```

采用默认的 macOS ls 命令观察其子目录的方式并不是太直观，推荐使用 Homebrew（https://brew.sh/）安装 tree：

```
$ brew install tree
```

tree 的安装结果：

```
$ brew install tree
Updating Homebrew...
==> Downloading
https://homebrew.bintray.com/bottles/tree-1.7.0.high_sierra.bottle.1.tar.gz
(...部分内容被省略...)
```

在当前目录下执行 tree 命令：

```
.
├── pom.xml
└── src
    ├── main
    │   └── java
    │       └── thinking
    │           └── in
    │
```

```
    |           └── spring
    |               └── boot
    |                   └── App.java
    └── test
        └── java
            └── thinking
                └── in
                    └── spring
                        └── boot
                            └── AppTest.java

13 directories, 3 files
```

其中，`App.java` 与 `AppTest.java` 分别为“引导类”和“单元测试”Java 源码，`pom.xml` 为主 POM 文件：

```
<project xmlns="http://maven.apache.org/POM/4.0.0"
xmlns:xsi="http://www.w3.org/2001/XMLSchema-instance"
  xsi:schemaLocation="http://maven.apache.org/POM/4.0.0
http://maven.apache.org/maven-v4_0_0.xsd">
  <modelVersion>4.0.0</modelVersion>
  <groupId>thinking-in-spring-boot</groupId>
  <artifactId>first-spring-boot-application</artifactId>
  <packaging>jar</packaging>
  <version>1.0.0-SNAPSHOT</version>
  <name>first-spring-boot-application</name>
  <url>http://maven.apache.org</url>
  <dependencies>
    <dependency>
      <groupId>junit</groupId>
      <artifactId>junit</artifactId>
      <version>3.8.1</version>
      <scope>test</scope>
    </dependency>
  </dependencies>
</project>
```

默认情况下，`maven-archetype-plugin` 为主 POM 文件仅增添了 `junit` 的依赖，因此，需要再添加 Spring Boot 依赖。

3．增加 Spring Boot 依赖

向 first-spring-boot-application 项目的 pom.xml 文件中增加 spring-boot-starter-web 依赖，并在<name>元素中为项目增加名称：

```
    <project xmlns="http://maven.apache.org/POM/4.0.0"
xmlns:xsi="http://www.w3.org/2001/XMLSchema-instance"
         xsi:schemaLocation="http://maven.apache.org/POM/4.0.0
http://maven.apache.org/maven-v4_0_0.xsd">
        <modelVersion>4.0.0</modelVersion>
        <groupId>thinking-in-spring-boot</groupId>
        <artifactId>first-spring-boot-application</artifactId>
        <packaging>jar</packaging>
        <version>1.0.0-SNAPSHOT</version>
        <name>《Spring Boot 编程思想》第一个 Spring Boot 应用</name>
        <url>http://maven.apache.org</url>
        <dependencies>
            <!-- 增加 Spring Boot Web 依赖 -->
            <dependency>
                <groupId>org.springframework.boot</groupId>
                <artifactId>spring-boot-starter-web</artifactId>
                <version>2.0.2.RELEASE</version>
            </dependency>

            <dependency>
                <groupId>junit</groupId>
                <artifactId>junit</artifactId>
                <version>3.8.1</version>
                <scope>test</scope>
            </dependency>
        </dependencies>
    </project>
```

添加 spring-boot-starter-web 依赖后，执行 mvn dependency:tree -Dincludes=org.springframework*，观察项目的依赖树发生了哪些变化（仅关注 Spring 相关的依赖）：

```
    (...部分内容被省略...)
    [[INFO] --- maven-dependency-plugin:2.8:tree (default-cli) @ first-spring-boot-application
---
```

```
[INFO] thinking-in-spring-boot:first-spring-boot-application:jar:1.0.0-SNAPSHOT
[INFO] \- org.springframework.boot:spring-boot-starter-web:jar:2.0.2.RELEASE:compile
[INFO]    +- org.springframework.boot:spring-boot-starter:jar:2.0.2.RELEASE:compile
[INFO]    |  +- org.springframework.boot:spring-boot:jar:2.0.2.RELEASE:compile
[INFO]    |  +-
org.springframework.boot:spring-boot-autoconfigure:jar:2.0.2.RELEASE:compile
[INFO]    |  +-
org.springframework.boot:spring-boot-starter-logging:jar:2.0.2.RELEASE:compile
[INFO]    |  \- org.springframework:spring-core:jar:5.0.6.RELEASE:compile
[INFO]    |     \- org.springframework:spring-jcl:jar:5.0.6.RELEASE:compile
[INFO]    +- org.springframework.boot:spring-boot-starter-json:jar:2.0.2.RELEASE:compile
[INFO]    +- org.springframework.boot:spring-boot-starter-tomcat:jar:2.0.2.RELEASE:compile
[INFO]    +- org.springframework:spring-web:jar:5.0.6.RELEASE:compile
[INFO]    |  \- org.springframework:spring-beans:jar:5.0.6.RELEASE:compile
[INFO]    \- org.springframework:spring-webmvc:jar:5.0.6.RELEASE:compile
[INFO]       +- org.springframework:spring-aop:jar:5.0.6.RELEASE:compile
[INFO]       +- org.springframework:spring-context:jar:5.0.6.RELEASE:compile
[INFO]       \- org.springframework:spring-expression:jar:5.0.6.RELEASE:compile
[INFO] ------------------------------------------------------------------------
[INFO] BUILD SUCCESS
[INFO] ------------------------------------------------------------------------
(...部分内容被省略...)
```

尽管当前项目仅添加了 `org.springframework.boot:spring-boot-starter-web:2.0.2.RELEASE` 的依赖，不过整体项目的依赖树发生了变化。如果读者熟悉 Spring Web MVC 开发，那么从控制台输出的内容分析，不难发现 `spring-webmvc` 和 Tomcat 等依赖被添加了进来。同时，以上结果更直观地说明了 Spring Boot 的第三点特性，即“Provide opinionated 'starter' dependencies to simplify your build configuration”，而 `spring-boot-starter-web` 就是文中所述的“starter”。

当依赖添加完毕，下一步就是为该 Spring Boot 补充执行代码。

4. 增加执行代码

前面增加的依赖为程序执行提供了基础，接下来为 `first-spring-boot-application` 项目增加可执行代码。已知在当前项目创建后，引导类源文件 `App.java` 被创建，存放于项目的 `src/main/java/thinking/in/spring/boot` 目录下，其类全名为 `thinking.in.spring.boot.App`，源码如下：

```
package thinking.in.spring.boot;

/**
 * Hello world!
 *
 */
public class App
{
    public static void main( String[] args )
    {
        System.out.println( "Hello World!" );
    }
}
```

调整该引导类如下：

```
package thinking.in.spring.boot;

import org.springframework.boot.SpringApplication;
import org.springframework.boot.autoconfigure.SpringBootApplication;
import org.springframework.web.bind.annotation.RequestMapping;
import org.springframework.web.bind.annotation.RestController;

/**
 * Hello world!
 */
@RestController
@SpringBootApplication
public class App {

    @RequestMapping("/")
    public String index() {
        return "Welcome , My Buddy!";
    }

    public static void main(String[] args) {
        SpringApplication.run(App.class, args);
    }
}
```

如果读者熟悉 Spring Web MVC，那么对以上代码肯定不会觉得陌生。从 Spring Web MVC 2.5 版本开始，@RequestMapping 注解在代码中的出镜率极高，用作请求映射。从上述代码来看，当 HTTP 请求路径为“/”时，index()方法会被调用执行。如果读者对@RestController 注解的语义不清楚，则可认为@RestController = @Controller + @ResponseBody。因此 index()方法不需要标注@ResponseBody，@RestController 能够直接输出文本内容，而非像传统@Controller 那样导向某个页面，故 HTTP 请求“/”的响应内容为“**Welcome , My Buddy!**”，不过前提是如何引导当前 Spring Boot 应用。由于当前环境为非图形化界面，因此需要将以上工程编译并运行，方可提供 HTTP 服务。

Spring Boot 官方文档在“11.4 Running the Example”一节中向开发人员介绍了 Spring Boot Maven 插件的运行样例工程，该插件同样适用于 first-spring-boot-application 项目，故在项目根目录执行 mvn spring-boot:run 命令：

```
    [INFO] Scanning for projects...
    Downloading:
http://maven.aliyun.com/nexus/content/groups/public/org/apache/maven/plugins/maven-metadata.x
ml
    (...部分内容被省略...)
    [INFO] BUILD FAILURE
    (...部分内容被省略...)
    [ERROR] No plugin found for prefix 'spring-boot' in the current project and in the plugin groups
[org.apache.maven.plugins, org.codehaus.mojo] available from the repositories [local
(/Users/mercyblitz/.m2/repository), alimaven
(http://maven.aliyun.com/nexus/content/groups/public/)] -> [Help 1]
    (...部分内容被省略...)
```

运行结果显示插件“spring-boot”无法在本地仓库中找到，因此 Spring Boot 官方文档也间接地提示该插件声明在 spring-boot-starter-parent POM 文件中：

> Since you used the spring-boot-starter-parent POM, you have a useful run goal that you can use to start the application. Type mvn spring-boot:run from the root project directory to start the application.

顾名思义，该文件用作 Maven Parent POM，而 first-spring-boot-application 项目并未将其声明在 pom.xml 文件中，故需要将其添加到 pom.xml 文件中。

5. 使用 Spring Boot Maven 插件引导 Spring Boot 应用

添加 spring-boot-starter-parent GAV 信息到 first-spring-boot-application 项目的

pom.xml 文件中：

```xml
    <project xmlns="http://maven.apache.org/POM/4.0.0"
xmlns:xsi="http://www.w3.org/2001/XMLSchema-instance"
          xsi:schemaLocation="http://maven.apache.org/POM/4.0.0
http://maven.apache.org/maven-v4_0_0.xsd">
        <modelVersion>4.0.0</modelVersion>
        <!-- 添加 Spring Boot Parent POM -->
        <parent>
            <groupId>org.springframework.boot</groupId>
            <artifactId>spring-boot-starter-parent</artifactId>
            <version>2.0.2.RELEASE</version>
        </parent>

        <groupId>thinking-in-spring-boot</groupId>
        <artifactId>first-spring-boot-application</artifactId>
        <packaging>jar</packaging>
        <version>1.0.0-SNAPSHOT</version>
        <name>《Spring Boot 编程思想》第一个 Spring Boot 应用</name>
        <url>http://maven.apache.org</url>
        <dependencies>
            <!-- 增加 Spring Boot Web 依赖 -->
            <dependency>
                <groupId>org.springframework.boot</groupId>
                <artifactId>spring-boot-starter-web</artifactId>
                <version>2.0.2.RELEASE</version>
            </dependency>

            <dependency>
                <groupId>junit</groupId>
                <artifactId>junit</artifactId>
                <version>3.8.1</version>
                <scope>test</scope>
            </dependency>
        </dependencies>
    </project>
```

再次在项目根目录执行 mvn spring-boot:run 命令：

```
[INFO] Scanning for projects...
[INFO]
[INFO] ------------------------------------------------------------------------
[INFO] Building 《Spring Boot 编程思想》第一个 Spring Boot 应用 1.0.0-SNAPSHOT
[INFO] ------------------------------------------------------------------------
(...部分内容被省略...)
  .   ____          _            __ _ _
 /\\ / ___'_ __ _ _(_)_ __  __ _ \ \ \ \
( ( )\___ | '_ | '_| | '_ \/ _` | \ \ \ \
 \\/  ___)| |_)| | | | | || (_| |  ) ) ) )
  '  |____| .__|_| |_|_| |_\__, | / / / /
 =========|_|==============|___/=/_/_/_/
 :: Spring Boot ::        (v2.0.2.RELEASE)
(...部分内容被省略...)
INFO 38812 --- [           main] thinking.in.spring.boot.App              : Starting App on bogon with PID 38812
(...部分内容被省略...)
INFO 38812 --- [           main] s.w.s.m.m.a.RequestMappingHandlerMapping : Mapped "{[/]}" onto public java.lang.String thinking.in.spring.boot.App.index()
(...部分内容被省略...)
INFO 38812 --- [           main] o.s.b.w.embedded.tomcat.TomcatWebServer  : Tomcat started on port(s): 8080 (http) with context path ''
INFO 38812 --- [           main] thinking.in.spring.boot.App              : Started App in 2.437 seconds (JVM running for 6.27)
```

以上控制台日志提供了不少信息，部分信息被有意识地省略，剩下的内容至少透漏出三方面的信息：

- 进程 ID

  日志 `PID 38812` 说明了当前应用的进程 ID 为 `38812`。

- 请求映射

  日志“Mapped "{[/]}" onto public java.lang.String thinking.in.spring.boot.App.index()”显示 `App.index()`方法映射的 HTTP 请求路径为“`/`”。

- Web 服务端口

  日志“Tomcat started on port(s): 8080 (http)”提示当前 Spring Boot 应用使用 Tomcat 作为嵌入式 Web 容器，并且暴露了 `8080` 端口作为 HTTP 服务。

当 `first-spring-boot-application` 应用启动后，下一步就是检验 `http://127.0.0.1:`

8080/ HTTP 服务响应内容是否为 “**Welcome , My Buddy!**”。

6. 检验 Spring Boot 应用 HTTP 服务

由于当前非图形化环境无法通过浏览器访问资源 http://127.0.0.1:8080/，因此可执行 curl 命令进行检验，如下所示。

```
$ curl http://127.0.0.1:8080/
Welcome , My Buddy!
```

该 HTTP 请求响应的结果正好是 “Welcome, My Buddy!”，说明 App.index()方法的执行结果返回了客户端。以上步骤完整地演示了在非图形化环境下创建并运行 Spring Boot 应用的过程。接下来将继续讨论如何通过图形化界面创建 Spring Boot 应用。

2.1.2 图形化界面创建 Spring Boot 应用

相对于命令行创建方式，Spring 官方提供了一个更高效和快捷的在线创建 Spring Boot 应用的图形化工具：https://start.spring.io/。

在页面的表单中，修改 **Group**（thinking-in-spring-boot）、**Artifact**（first-app-by-gui），选择 Spring Boot 版本（2.0.2），以及在 **Search for dependencies** 中搜索关键字 “Web”，如下图所示。

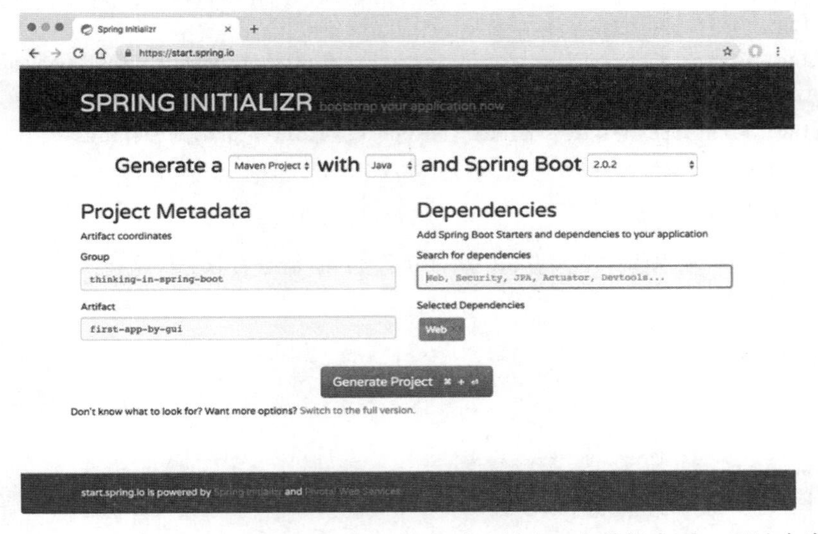

设置完毕后，单击 “Generate Project” 按钮完成项目创建。表单提交后，网站会生成一份名为 first-app-by-gui.zip 的压缩文档，选择合适的保存路径，比如示例工程根路径/Users/mercyblitz/thinking-in-spring-boot-samples。在该目录下执行 unzip 命令解压文件：

```
$ unzip first-app-by-gui.zip
Archive:  first-app-by-gui.zip
   creating: first-app-by-gui/
  inflating: first-app-by-gui/mvnw
   creating: first-app-by-gui/.mvn/
   creating: first-app-by-gui/.mvn/wrapper/
   creating: first-app-by-gui/src/
   creating: first-app-by-gui/src/main/
   creating: first-app-by-gui/src/main/java/
   creating: first-app-by-gui/src/main/java/thinkinginspringboot/
   creating: first-app-by-gui/src/main/java/thinkinginspringboot/firstappbygui/
   creating: first-app-by-gui/src/main/resources/
   creating: first-app-by-gui/src/test/
   creating: first-app-by-gui/src/test/java/
   creating: first-app-by-gui/src/test/java/thinkinginspringboot/
   creating: first-app-by-gui/src/test/java/thinkinginspringboot/firstappbygui/
  inflating: first-app-by-gui/.gitignore
  inflating: first-app-by-gui/.mvn/wrapper/maven-wrapper.jar
  inflating: first-app-by-gui/.mvn/wrapper/maven-wrapper.properties
  inflating: first-app-by-gui/mvnw.cmd
  inflating: first-app-by-gui/pom.xml
  inflating: first-app-by-gui/src/main/java/thinkinginspringboot/firstappbygui/
FirstAppByGuiApplication.java
  inflating: first-app-by-gui/src/main/resources/application.properties
  inflating: first-app-by-gui/src/test/java/thinkinginspringboot/firstappbygui/
FirstAppByGuiApplicationTests.java
```

或许解压后控制台输出同样不太直观，再次执行 tree 命令查看文件路径：

```
$ tree -a first-app-by-gui
first-app-by-gui
├── .gitignore
├── .mvn
│   └── wrapper
│       ├── maven-wrapper.jar
│       └── maven-wrapper.properties
├── mvnw
├── mvnw.cmd
├── pom.xml
```

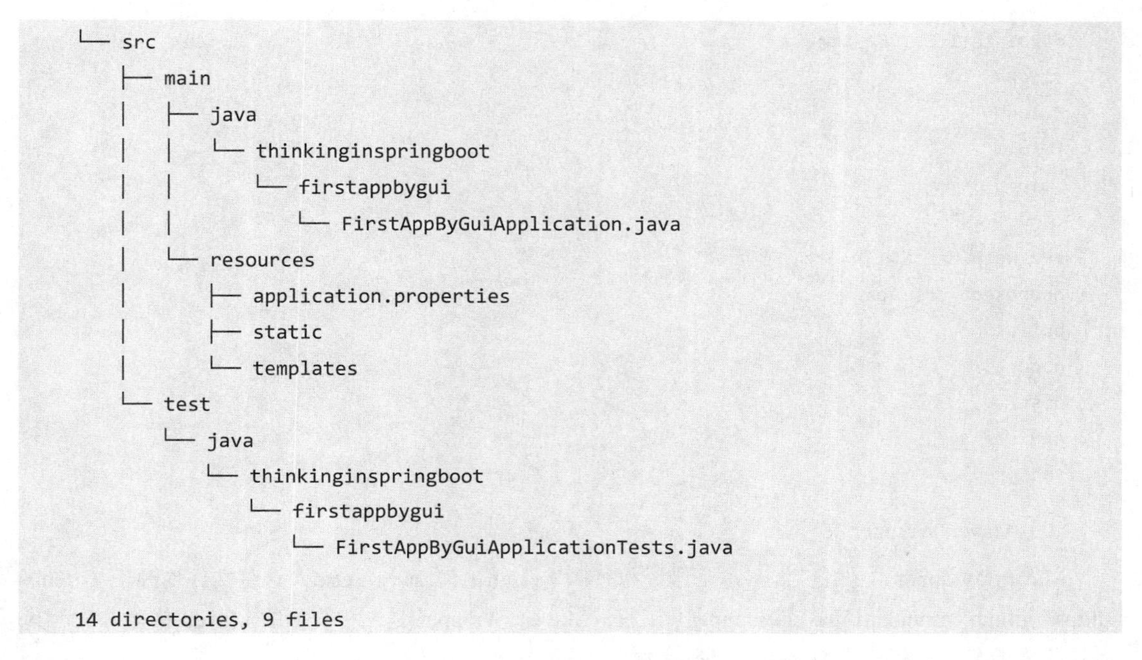

```
└── src
    ├── main
    │   ├── java
    │   │   └── thinkinginspringboot
    │   │       └── firstappbygui
    │   │           └── FirstAppByGuiApplication.java
    │   └── resources
    │       ├── application.properties
    │       ├── static
    │       └── templates
    └── test
        └── java
            └── thinkinginspringboot
                └── firstappbygui
                    └── FirstAppByGuiApplicationTests.java

14 directories, 9 files
```

相比于 `first-spring-boot-application` 项目，当前工程多了一些文件和目录，下面从上至下依次分类说明。

1. Spring Boot 模板应用目录结构

（1）`.gitignore` 文件。

引入`.gitignore` 的目的是，目前大多数的现代应用均使用 Git 作为代码版本控制系统，它定义了最常见的文件或目录的版本控制忽略名单，包括基于 Eclipse 的 STS（Spring Tool Suite）、IDEA 和 NetBeans 等项目元信息资源：

```
target/
!.mvn/wrapper/maven-wrapper.jar

### STS ###
.apt_generated
.classpath
.factorypath
.project
.settings
.springBeans
```

```
### IntelliJ IDEA ###
.idea
*.iws
*.iml
*.ipr

### NetBeans ###
nbproject/private/
build/
nbbuild/
dist/
nbdist/
.nb-gradle/
```

（2）Maven Wrapper 文件。

Maven Wrapper 文件包括`.mvn` 目录、执行脚本 `mvnw` 和 `mvnw.cmd`，这些文件均源于 GitHub（https://github.com/takari/maven-wrapper）工程。Maven Wrapper 是一种简单的 Maven 构建方式，其运行环境不需要提前安装 Maven 二进制文件：

> The Maven Wrapper is an easy way to ensure a user of your Maven build has everything necessary to run your Maven build.

- `.mvn/wrapper/maven-wrapper.jar`

  该文件由脚本引导，用于从 Maven 官方下载 Maven 二进制文件：

  > By default, the Maven Wrapper JAR archive is added to the using project as small binary file `.mvn/wrapper/maven-wrapper.jar`. It is used to bootstrap the download and invocation of Maven from the wrapper shell scripts.

- `.mvn/wrapper/maven-wrapper.properties`

  此 Properties 文件定义了文件下载的 URL，当其 URL 不可用时，由 `wrapperUrl` 属性配置：

  > If the JAR is not found to be available by the scripts they will attempt to download the file from the URL specified in `.mvn/wrapper/maven-wrapper.properties` under `wrapperUrl` and put it in place.

- `mvnw` 和 `mvnw.cmd`

两个文件具有相同的职责，引导`.mvn/wrapper/maven-wrapper.jar`下载 Maven 二进制文件，前者用于*nix 平台，后者工作于 Windows 操作系统。

同时，文档对比了传统 Maven 命令和 Maven Wrapper 脚本的使用方法：

> Normally you instruct users to run the `mvn` command like the following:
>
> `$ mvn clean install`
>
> But now, with a Maven Wrapper setup, you can instruct users to run wrapper scripts:
>
> `$ ./mvnw clean install`
>
> or
>
> `$ ./mvnw.cmd clean install`

简言之，`mvnw` 或 `mvnw.cmd` 脚本相当于 `mvn` 命令。因此，可在当前 macOS 环境中执行 `mvnw` 脚本，并配合 `spring-boot:run` 插件使用。

（3）Spring Boot 应用属性配置文件——`application.properties`。

`application.properties` 是 Spring Boot 默认的应用外部配置文件，其配置属性可以控制 Spring Boot 应用的行为，如调整 Web 服务端口等，在后续的讨论中，将通过 `server.port` 属性调整 Web 服务端口。关于 `application.properties` 文件的所有细节将在“**超越外部化配置**”中深入讨论。

（4）Spring Boot 应用 JUnit 测试文件。

`FirstAppByGuiApplicationTests.java` 为 Spring Boot 应用的模板 JUnit 测试文件，与其引导的 Java 文件对应，故其名称并不确定。不过“麻雀虽小，五脏俱全”，它提供了 Spring Boot 集成测试的基本模式：

```
package thinkinginspringboot.firstappbygui;

import org.junit.Test;
import org.junit.runner.RunWith;
import org.springframework.boot.test.context.SpringBootTest;
import org.springframework.test.context.junit4.SpringRunner;

@RunWith(SpringRunner.class)
@SpringBootTest
public class FirstAppByGuiApplicationTests {
```

```
    @Test
    public void contextLoads() {
    }

}
```

了解了模板 Spring Boot 应用目录结构后，下一步将结合 mvnw 脚本执行 spring-boot:run 插件。

2．使用 mvnw 脚本执行 Spring Boot Maven 插件

首先执行 cd first-app-by-gui 命令，将当前目录切换至目标路径，随后执行 sh mvnw spring-boot:run，启动 Spring Boot 应用。由于控制台日志输出相对复杂，下面逐步说明：

```
$ sh mvnw spring-boot:run
/Users/mercyblitz/thinking-in-spring-boot-samples/first-app-by-gui
Downloading
https://repo1.maven.org/maven2/org/apache/maven/apache-maven/3.5.0/apache-maven-3.5.0-bin.zip
Unzipping
/Users/mercyblitz/.m2/wrapper/dists/apache-maven-3.5.0-bin/jl1qqhdeineh9qg83dtj7i91c/apache-maven-3.5.0-bin.zip to
/Users/mercyblitz/.m2/wrapper/dists/apache-maven-3.5.0-bin/jl1qqhdeineh9qg83dtj7i91c
Set executable permissions for:
/Users/mercyblitz/.m2/wrapper/dists/apache-maven-3.5.0-bin/jl1qqhdeineh9qg83dtj7i91c/apache-maven-3.5.0/bin/mvn
```

诚如 Maven Wrapper 官方文档描述，当执行 mvnw 命令后，首先从 Maven 的中央仓库下载 Maven 的二进制 zip 文件，随后将其解压到/Users/mercyblitz/.m2/wrapper/目录，即~/.m2/wrapper 目录，再为 bin 目录下的 mvn 脚本赋予执行权限。而日志输出的后半段则与 first-spring-boot-application 应用类似：

```
[INFO] Scanning for projects...
(...部分内容被省略...)

  .   ____          _            __ _ _
 /\\ / ___'_ __ _ _(_)_ __  __ _ \ \ \ \
( ( )\___ | '_ | '_| | '_ \/ _` | \ \ \ \
 \\/  ___)| |_)| | | | | || (_| |  ) ) ) )
  '  |____| .__|_| |_|_| |_\__, | / / / /
 =========|_|==============|___/=/_/_/_/
```

```
  :: Spring Boot ::        (v2.0.2.RELEASE)

    (...部分内容被省略...)
    INFO 44201 --- [ost-startStop-1] o.s.b.w.servlet.ServletRegistrationBean  : Servlet
dispatcherServlet mapped to [/]
    (...部分内容被省略...)
    Caused by: java.net.BindException: Address already in use
    (...部分内容被省略...)
    ***************************
    APPLICATION FAILED TO START
    ***************************

    Description:

    The Tomcat connector configured to listen on port 8080 failed to start. The port may already
be in use or the connector may be misconfigured.

    Action:

    Verify the connector's configuration, identify and stop any process that's listening on port
8080, or configure this application to listen on another port.
    (...部分内容被省略...)
```

发现当前应用 `first-app-by-gui` 启动失败，这是 Spring Boot 常见的错误之一，是由于 Web 端口冲突导致的，其 **8080** 端口被 `first-spring-boot-application` 进程占用。不过相较于传统的 Spring 应用，Spring Boot 的错误提示更友好，便于开发人员排查问题。值得注意的是，该特性并非 Spring Boot 与生俱来，而是从 1.4 版本才予以支持的，称为错误分析报告器，用 API `FailureAnalysisReporter` 表示。相关细节请参考"**理解 SpringApplication**"中的"**错误分析报告器——FailureAnalysisReporter**"部分的讨论。

因此，将 `first-spring-boot-application` 进程退出，再次执行 `sh mvnw spring- boot:run`：

```
    $ sh mvnw spring-boot:run
    /Users/mercyblitz/thinking-in-spring-boot-samples/first-app-by-gui
    (...部分内容被省略...)

      .   ____          _            __ _ _
     /\\ / ___'_ __ _ _(_)_ __  __ _ \ \ \ \
    ( ( )\___ | '_ | '_| | '_ \/ _` | \ \ \ \
```

```
  \\/  ___)| |_)| | | | | || (_| |  ) ) ) )
   '  |____| .__|_| |_|_| |_\__, | / / / /
  =========|_|==============|___/=/_/_/_/
  :: Spring Boot ::        (v2.0.2.RELEASE)

  (...部分内容被省略...)
  INFO 44638 --- [           main] s.w.s.m.m.a.RequestMappingHandlerMapping : Mapped
"{[/error]}" onto public org.springframework.http.ResponseEntity<java.util.Map<java.lang.String,
java.lang.Object>>
org.springframework.boot.autoconfigure.web.servlet.error.BasicErrorController.error(javax.ser
vlet.http.HttpServletRequest)
  (...部分内容被省略...)
  INFO 44638 --- [           main] o.s.b.w.embedded.tomcat.TomcatWebServer  : Tomcat started on
port(s): 8080 (http) with context path ''
  INFO 44638 --- [           main] t.f.FirstAppByGuiApplication             : Started
FirstAppByGuiApplication in 3.296 seconds (JVM running for 6.613)
```

下一步执行 `curl` 命令，访问当前应用首页 `http://127.0.0.1:8080/`：

```
  $ curl http://127.0.0.1:8080/
  {"timestamp":"...","status":404,"error":"Not Found","message":"No message
available","path":"/"}
```

不难发现，`first-app-by-gui` 应用并没有默认的首页 HTTP 服务，故返回 `404` 的 JSON 响应结果。同时，该内容的输出是由 Spring Boot 框架内建 HTTP Endpoint 实现的，即 `BasicErrorController#error(HttpServletRequest)`方法，该方法的签名曾出现在 Spring Boot 应用启动日志中：

```
  INFO 44638 --- [           main] s.w.s.m.m.a.RequestMappingHandlerMapping : Mapped
"{[/error]}" onto public org.springframework.http.ResponseEntity<java.util.Map<java.lang.String,
java.lang.Object>>
org.springframework.boot.autoconfigure.web.servlet.error.BasicErrorController.error(javax.ser
vlet.http.HttpServletRequest)
```

为后续方便查看 JSON HTTP 响应内容，可通过 `brew` 安装 `json_pp` 插件：

```
  $ brew install json_pp
```

重新执行 `curl` 并配合 `json_pp` 插件——`curl http://127.0.0.1:8080/ | json_pp`：

```
  {
```

```
    "message" : "No message available",
    "path" : "/",
    "timestamp" : "...",
    "error" : "Not Found",
    "status" : 404
}
```

至此，`first-spring-boot-application` 和 `first-app-by-gui` 应用均能直接或间接地执行 `mvn spring-boot:run` 命令来引导应用，不过线上环境基本不会安装 Maven 或临时地下载 Maven 来执行该插件，并且这种运行方式需要编译执行源码，这两个前置条件同时成立的可能性几乎为零。如果按照传统 Web 应用的运行模式，则需要将代码工程打包成 WAR 包，并部署到 Tomcat 容器中。已知 Spring Boot 改变此等现状，它支持独立 JAR 或 WAR 文件直接执行。Spring Boot 官方文档在“11.5 Creating an Executable Jar”章节中介绍此等构建方式，即构建可执行 JAR，又称为“fat jars”：

> We finish our example by creating a completely self-contained executable jar file that we could run in production. Executable jars (sometimes called “fat jars”) are archives containing your compiled classes along with all of the jar dependencies that your code needs to run.

2.1.3 创建 Spring Boot 应用可执行 JAR

官方文档也告知了开发人员构建可执行 JAR 的前提，即需要添加 `spring-boot-maven-plugin` 到 `pom.xml` 文件中，而该插件默认地被追加到由 `https://start.spring.io` 构建的 Spring Boot 应用中，因此 `first-app-by-gui` 项目天然地存在该插件，相反，`first-spring-boot-application` 则没有。故打开并观察 `first-app-by-gui` 项目的 `pom.xml` 文件：

```
<?xml version="1.0" encoding="UTF-8"?>
<project xmlns="http://maven.apache.org/POM/4.0.0"
xmlns:xsi="http://www.w3.org/2001/XMLSchema-instance"
    xsi:schemaLocation="http://maven.apache.org/POM/4.0.0
http://maven.apache.org/xsd/maven-4.0.0.xsd">
    <modelVersion>4.0.0</modelVersion>
        ...
    <parent>
        <groupId>org.springframework.boot</groupId>
        <artifactId>spring-boot-starter-parent</artifactId>
        <version>2.0.2.RELEASE</version>
        <relativePath/> <!-- lookup parent from repository -->
```

```
</parent>
    ...
<dependencies>
    <dependency>
        <groupId>org.springframework.boot</groupId>
        <artifactId>spring-boot-starter-web</artifactId>
    </dependency>

    <dependency>
        <groupId>org.springframework.boot</groupId>
        <artifactId>spring-boot-starter-test</artifactId>
        <scope>test</scope>
    </dependency>
</dependencies>

<build>
    <plugins>
        <plugin>
            <groupId>org.springframework.boot</groupId>
            <artifactId>spring-boot-maven-plugin</artifactId>
        </plugin>
    </plugins>
</build>


</project>
```

对于 `first-spring-boot-application` 项目的 `pom.xml` 文件，除 `spring-boot-maven-plugin` 插件不存在外，`first-app-by-gui` 项目的 `pom.xml` 中的 Maven 依赖和插件均不需要指定版本。换言之，其版本等信息继承自 Parent `spring-boot-starter-parent`，同样从工程层面解释了官方对 Spring Boot 的描述：

> "We take an opinionated view of the Spring platform and third-party libraries so you can get started with minimum fuss."

既然 `spring-boot-maven-plugin` 插件存在于 `first-app-by-gui` 项目中，那么可放心地按照 Spring Boot 官方文档的指示来打包项目工程：

> Save your pom.xml and run mvn package from the command line

退出 sh mvnw spring-boot:run 的运行，执行 mvn package：

```
$ mvn package
[INFO] Scanning for projects...
[INFO]
[INFO] ------------------------------------------------------------------------
[INFO] Building first-app-by-gui 0.0.1-SNAPSHOT
[INFO] ------------------------------------------------------------------------
(...部分内容被省略...)
[INFO] -------------------------------------------------------
[INFO]  T E S T S
[INFO] -------------------------------------------------------
(...部分内容被省略...)
[INFO] Tests run: 1, Failures: 0, Errors: 0, Skipped: 0, Time elapsed: 2.519 s - in
thinkinginspringboot.firstappbygui.FirstAppByGuiApplicationTests
(...部分内容被省略...)
[INFO] --- maven-jar-plugin:3.0.2:jar (default-jar) @ first-app-by-gui ---
[INFO] Building jar: /Users/mercyblitz/thinking-in-spring-boot-samples/first-app-by-gui/
target/first-app-by-gui-0.0.1-SNAPSHOT.jar
[INFO]
[INFO] --- spring-boot-maven-plugin:2.0.2.RELEASE:repackage (default) @ first-app-by-gui ---
[INFO] ------------------------------------------------------------------------
[INFO] BUILD SUCCESS
[INFO] ------------------------------------------------------------------------
(...部分内容被省略...)
```

在打包的过程中，执行了项目仅存的单个用例类，即 FirstAppByGuiApplicationTests。当测试通过后，spring-boot-maven-plugin 插件执行了 repackage goal，最终生成 first-app-by-gui-0.0.1-SNAPSHOT.jar，位于 first-app-by-gui 工程的 target 目录下。接下来，执行该 FAT JAR。

2.2 运行 Spring Boot 应用

Spring Boot 官方文档在"11.5 Creating an Executable Jar"章节中告知开发人员怎样执行 Spring Boot 应用可执行 JAR：

> To run that application, use the java -jar command

2.2.1 执行 Spring Boot 应用可执行 JAR

按照 Spring Boot 官方文档指示，执行如下：

```
$ java -jar target/first-app-by-gui-0.0.1-SNAPSHOT.jar

  .   ____          _            __ _ _
 /\\ / ___'_ __ _ _(_)_ __  __ _ \ \ \ \
( ( )\___ | '_ | '_| | '_ \/ _` | \ \ \ \
 \\/  ___)| |_)| | | | | || (_| |  ) ) ) )
  '  |____| .__|_| |_|_| |_\__, | / / / /
 =========|_|==============|___/=/_/_/_/
 :: Spring Boot ::        (v2.0.2.RELEASE)
(...部分内容被省略...)
INFO 55162 --- [           main] t.f.FirstAppByGuiApplication             : Started
FirstAppByGuiApplication in 2.462 seconds (JVM running for 2.969)
```

观察控制台日志，`java -jar` 命令与 `mvn spring-boot:run` 命令基本无异，并且其运行速度也不受影响。两者的区别在于，前者为“**生产环境运行方式**”，而后者为“**开发阶段运行方式**”。

2.2.2 Spring Boot 应用可执行 JAR 资源结构

将目录切换至 `target` 目录，并执行 `tree -h` 命令：

```
$ tree -h
.
├── [ 128]  classes
│   ├── [   0]  application.properties
│   └── [  96]  thinkinginspringboot
│       └── [  96]  firstappbygui
│           └── [ 796]  FirstAppByGuiApplication.class
├── [ 15M]  first-app-by-gui-0.0.1-SNAPSHOT.jar
├── [3.0K]  first-app-by-gui-0.0.1-SNAPSHOT.jar.original
├── [  96]  generated-sources
│   └── [  64]  annotations
├── [  96]  generated-test-sources
│   └── [  64]  test-annotations
├── [  96]  maven-archiver
```

```
│   └── [ 114]  pom.properties
├── [  96]  maven-status
│   └── [ 128]  maven-compiler-plugin
│       ├── [  96]  compile
│       │   └── [ 128]  default-compile
│       │       ├── [  66]  createdFiles.lst
│       │       └── [ 146]  inputFiles.lst
│       └── [  96]  testCompile
│           └── [ 128]  default-testCompile
│               ├── [  71]  createdFiles.lst
│               └── [ 151]  inputFiles.lst
├── [ 128]  surefire-reports
│   ├── [ 18K]  TEST-thinkinginspringboot.firstappbygui.FirstAppByGuiApplicationTests.xml
│   └── [ 377]  thinkinginspringboot.firstappbygui.FirstAppByGuiApplicationTests.txt
└── [  96]  test-classes
    └── [  96]  thinkinginspringboot
        └── [  96]  firstappbygui
            └── [ 685]  FirstAppByGuiApplicationTests.class

18 directories, 12 files
```

以上命令输出了各个文件的路径和大小，first-app-by-gui-0.0.1-SNAPSHOT.jar 文件占用磁盘空间为 15MB。同时注意一个几乎同名的文件 first-app-by-gui-0.0.1-SNAPSHOT.jar.original，该文件仅包含应用本地资源（如编译后的 `classes` 目录下的资源文件），未引入第三方依赖资源，所以空间占用仅为 3.0KB。

通过文件名称分析，first-app-by-gui-0.0.1-SNAPSHOT.jar.original 属于原始 Maven 打包 JAR 文件，而 first-app-by-gui-0.0.1-SNAPSHOT.jar 则是"加工"后的文件。因此，可以大胆地猜测，执行 `spring-boot-maven-plugin` Maven 插件后，first-app-by-gui-0.0.1-SNAPSHOT.jar.original 被"`repackage`"成了 first-app-by-gui-0.0.1-SNAPSHOT.jar，并且引入相关的第三方依赖资源。

为了检验此猜测正确性，解压 first-app-by-gui-0.0.1-SNAPSHOT.jar 文件，探究其中的资源结构。

> JAR 文件有时也称为"Fat JAR"，采用 zip 压缩格式存储，因此凡是能解压 zip 压缩文件的软件，均可将 JAR 包解压。

回到控制台，执行解压命令，将 first-app-by-gui-0.0.1-SNAPSHOT.jar 文件解压到临时目录 `temp` 下：

```
$ unzip first-app-by-gui-0.0.1-SNAPSHOT.jar -d temp
Archive:  first-app-by-gui-0.0.1-SNAPSHOT.jar
   creating: temp/META-INF/
  inflating: temp/META-INF/MANIFEST.MF
   creating: temp/org/
(...部分内容被省略...)
 extracting: temp/BOOT-INF/lib/slf4j-api-1.7.25.jar
 extracting: temp/BOOT-INF/lib/spring-core-5.0.6.RELEASE.jar
 extracting: temp/BOOT-INF/lib/spring-jcl-5.0.6.RELEASE.jar
```

再执行 tree 命令，以树形结构展示 temp 目录：

```
$ tree temp/
temp/
├── BOOT-INF
│   ├── classes
│   │   ├── application.properties
│   │   └── thinkinginspringboot
│   │       └── firstappbygui
│   │           └── FirstAppByGuiApplication.class
│   └── lib
│       ├── ...
│       ├── spring-beans-5.0.6.RELEASE.jar
│       ├── spring-boot-2.0.2.RELEASE.jar
│       ├── spring-boot-autoconfigure-2.0.2.RELEASE.jar
│       ├── ...
├── META-INF
│   ├── MANIFEST.MF
│   └── maven
│       └── thinking-in-spring-boot
│           └── first-app-by-gui
│               ├── pom.properties
│               └── pom.xml
└── org
    └── springframework
        └── boot
            └── loader
                ├── ExecutableArchiveLauncher.class
                ├── JarLauncher.class
```

```
            ├── ...

17 directories, 91 files
```

不难看出，first-app-by-gui-0.0.1-SNAPSHOT.jar 解压后的目录比传统 JAR 文件更复杂，不过也有规律可循，其中：

- BOOT-INF/classes 目录存放应用编译后的 class 文件；
- BOOT-INF/lib 目录存放应用依赖的 JAR 包；
- META-INF/目录存放应用相关的元信息，如 MANIFEST.MF 文件；
- org/目录存放 Spring Boot 相关的 class 文件。

对此，Spring Boot 官方文档在“E.1.1 The Executable Jar File Structure”章节中也做出了相关的说明。

同时，将 first-app-by-gui-0.0.1-SNAPSHOT.jar.original 文件解压，比较两者的目录结构差异：

```
$ unzip first-app-by-gui-0.0.1-SNAPSHOT.jar.original -d original
Archive:  first-app-by-gui-0.0.1-SNAPSHOT.jar.original
  inflating: original/META-INF/MANIFEST.MF
   creating: original/thinkinginspringboot/
   creating: original/thinkinginspringboot/firstappbygui/
   creating: original/META-INF/maven/
   creating: original/META-INF/maven/thinking-in-spring-boot/
   creating: original/META-INF/maven/thinking-in-spring-boot/first-app-by-gui/
  inflating: original/thinkinginspringboot/firstappbygui/FirstAppByGuiApplication.class
  inflating: original/application.properties
  inflating: original/META-INF/maven/thinking-in-spring-boot/first-app-by-gui/pom.xml
  inflating:
original/META-INF/maven/thinking-in-spring-boot/first-app-by-gui/pom.properties
```

以树形展示目录：

```
$ tree original/
original/
├── META-INF
│   ├── MANIFEST.MF
│   └── maven
│       └── thinking-in-spring-boot
│           └── first-app-by-gui
```

```
|               ├── pom.properties
|               └── pom.xml
├── application.properties
└── thinkinginspringboot
    └── firstappbygui
        └── FirstAppByGuiApplication.class

6 directories, 5 files
```

无论文件数量，还是结构复杂度，first-app-by-gui-0.0.1-SNAPSHOT.jar.original 都不是和前者同一级别的。除了 `META-INF` 目录，其他资源均发生了变化，“`repackage`”后的 JAR，将应用 class 文件和 application.properties 文件从根目录放置到了 `BOOT-INF/classes`，所依赖的 JAR 均存放到了 `BOOT-INF/lib` 目录，由此可证明我们前面的猜想。

回到 `temp` 解压目录，其中目录 `BOOT-INF/classes` 和 `BOOT-INF/lib` 看起来非常眼熟。在标准的 Java EE Web 应用中，class 文件不正是存放在 `WEB-INF/classes` 目录，而依赖的 JAR 包不也是存放在 `WEB-INF/lib` 目录下的吗？原来 Spring Boot 打包后的 JAR 文件也在模仿 Java EE 的 WAR 文件。不过令人不解的是，为何 `java -jar` 命令能够执行 FAT JAR 文件呢？其中原委有待进一步分析。

2.2.3 FAT JAR 和 WAR 执行模块——spring-boot-loader

当 Spring Boot 应用可执行 JAR 文件被 `java -jar` 命令执行时，其命令本身对 JAR 文件是否来自 Spring Boot 插件打包并不感知。换言之，该命令引导的是标准可执行 JAR 文件，而按照 Java 官方文档的规定，`java -jar` 命令引导的具体启动类必须配置在 `MANIFEST.MF` 资源的 `Main-Class` 属性中：

> If the `-jar` option is specified, its argument is the name of the JAR file containing class and resource files for the application. The startup class must be indicated by the `Main-Class` manifest header in its source code.
> 参考文档地址：https://docs.oracle.com/javase/8/docs/technotes/tools/windows/java.html。

同时，根据“JAR 文件规范”，`MANIFEST.MF` 资源必须存放在`/META-INF/`目录下：

> The META-INF directory
> The following files/directories in the META-INF directory are recognized and interpreted by the Java 2 Platform to configure applications, extensions, class loaders and services:
> - `MANIFEST.MF`

因此，切换到 temp 解压目录，并将/META-INF/MANIFEST.MF 资源输出到控制台：

```
$ cat META-INF/MANIFEST.MF
Manifest-Version: 1.0
Implementation-Title: first-app-by-gui
Implementation-Version: 0.0.1-SNAPSHOT
Built-By: mercyblitz
Implementation-Vendor-Id: thinking-in-spring-boot
Spring-Boot-Version: 2.0.2.RELEASE
Main-Class: org.springframework.boot.loader.JarLauncher
Start-Class: thinkinginspringboot.firstappbygui.FirstAppByGuiApplicati
 on
Spring-Boot-Classes: BOOT-INF/classes/
Spring-Boot-Lib: BOOT-INF/lib/
Created-By: Apache Maven 3.5.0
Build-Jdk: 1.8.0_172
Implementation-URL: https://projects.spring.io/spring-boot/#/spring-bo
 ot-starter-parent/first-app-by-gui
```

发现 Main-Class 属性指向的 Class 为 org.springframework.boot.loader.JarLauncher，而该类存放在 org/springframework/boot/loader/ 目录下，并且项目的引导类定义在 Start-Class 属性中，该属性并非 Java 平台标准 META-INF/MANIFEST.MF 属性。这些信息在 Spring Boot 官方文档的“E.3.1 Launcher Manifest”章节中也有描述：

> You need to specify an appropriate Launcher as the Main-Class attribute of META-INF/MANIFEST.MF. The actual class that you want to launch (that is, the class that contains a main method) should be specified in the Start-Class attribute.
> The following example shows a typical MANIFEST.MF for an executable jar file:
>
> ```
> Main-Class: org.springframework.boot.loader.JarLauncher
> Start-Class: com.mycompany.project.MyApplication
> ```
>
> For a war file, it would be as follows:
>
> ```
> Main-Class: org.springframework.boot.loader.WarLauncher
> Start-Class: com.mycompany.project.MyApplication
> ```

文档提到了 org.springframework.boot.loader.JarLauncher 对应的 WAR 文件的实现，即 org.springframework.boot.loader.WarLauncher。前者是可执行 JAR 的启动器，后者是可执行 WAR 的启动器。

可执行 JAR 文件启动器——`JarLauncher`

对比 first-app-by-gui-0.0.1-SNAPSHOT.jar.original 文件，启动类 `org.springframework.boot.loader.JarLauncher` 并非项目中的文件，似乎是由 `spring-boot-maven-plugin` 插件 `repackage` 追加进去的。同时，Spring Boot 官方文档并没有直接说明 `org.springframework.boot.loader.JarLauncher` 类的所属，如 GAV 等信息。因此，可通过 Class 名称搜索 Maven 中心仓库，查找其 GAV 信息，在浏览器中输入 https://search.maven.org/，并且单击“**Classic Search**”链接，跳转到新页面。

单击“**Advanced Search**”链接，浏览器展示新的表单页面。

在“**Classname:**”输入框中输入 `org.springframework.boot.loader.JarLauncher`，如下图所示。

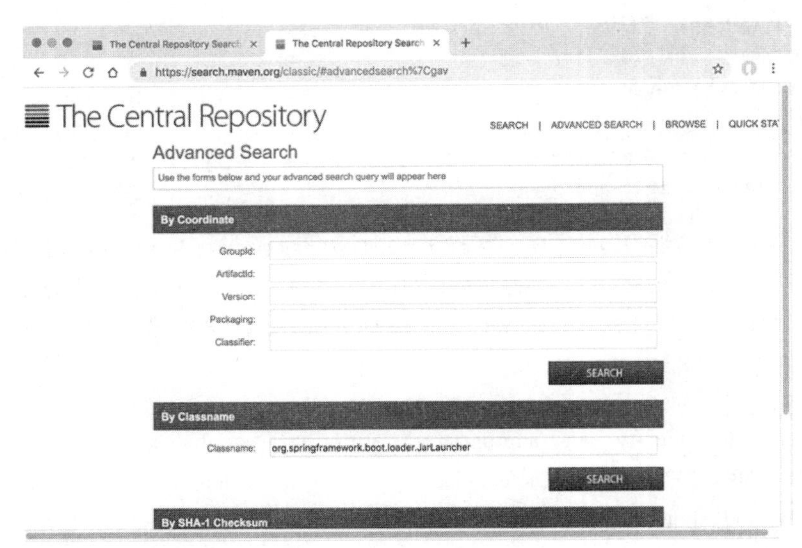

单击“SEARCH”按钮，执行搜索，直接定位到 Maven GA “org.springframework.boot:spring-boot-loader”，如下图所示。

org.springframework.boot	spring-boot-loader	2.0.2.RELEASE	09-May-2018	pom jar javadoc.jar sources.jar
org.springframework.boot	spring-boot-loader	1.5.13.RELEASE	09-May-2018	pom jar javadoc.jar sources.jar
org.springframework.boot	spring-boot-loader	1.5.12.RELEASE	10-Apr-2018	pom jar javadoc.jar sources.jar
org.springframework.boot	spring-boot-loader	2.0.1.RELEASE	05-Apr-2018	pom jar javadoc.jar sources.jar
org.springframework.boot	spring-boot-loader	1.5.11.RELEASE	05-Apr-2018	pom jar javadoc.jar sources.jar
org.springframework.boot	spring-boot-loader	2.0.0.RELEASE	01-Mar-2018	pom jar javadoc.jar sources.jar
org.springframework.boot	spring-boot-loader	1.5.10.RELEASE	31-Jan-2018	pom jar javadoc.jar sources.jar

> 由于 `org.springframework.boot.loader.JarLauncher` 来自 Spring 官方，所以那些重制的三方包予以排除，应直接锁定在 GroupId 为 `org.springframework.boot` 的资源上，最终定位到 Maven GA `org.springframework.boot:spring-boot-loader`。其中 Maven GA 即 Maven GroupId 和 ArtifactId。

综上所述，`org.springframework.boot.loader.JarLauncher` 所在的 JAR 文件的 Maven GAV 信息为 `org.springframework.boot:spring-boot-loader:2.0.2.RELEASE`。通常情况下，这个依赖没有必要引入 Spring Boot 项目的 `pom.xml` 文件。既然 `org.springframework.boot.loader.JarLauncher` 类文件存在于 FAT JAR 中，那么也可以通过 `java` 命令在解压根目录中引导该类文件：

```
$ java org.springframework.boot.loader.JarLauncher

  .   ____          _            __ _ _
 /\\ / ___'_ __ _ _(_)_ __  __ _ \ \ \ \
( ( )\___ | '_ | '_| | '_ \/ _` | \ \ \ \
 \\/  ___)| |_)| | | | | || (_| |  ) ) ) )
  '  |____| .__|_| |_|_| |_\__, | / / / /
 =========|_|==============|___/=/_/_/_/
 :: Spring Boot ::        (v2.0.2.RELEASE)
(...部分内容被省略...)
INFO 65401 --- [           main] t.f.FirstAppByGuiApplication             : Started
FirstAppByGuiApplication in 2.91 seconds (JVM running for 3.546)
```

实际运行的结果证实了刚才的推测，同时从控制台输出分析，项目引导类 `thinkinginspringboot.firstappbygui.FirstAppByGuiApplication` 被 `JarLauncher` 装载并执行。换言之，`META-INF/MANIFEST.MF` 资源中的 `Start-Class` 属性被 `JarLauncher` 关联项目引导，这个推论暂缓证明。反之，如果直接使用 `java` 命令引导 `thinkinginspringboot.firstappbygui.FirstAppByGuiApplication` 类又会怎么样呢？根据前面的讨论，已知该类存放在 `BOOT-INF/classes` 目录下，故先切换路径，再执行 `java` 命令：

```
$ cd BOOT-INF/classes
$ java thinkinginspringboot.firstappbygui.FirstAppByGuiApplication
Exception in thread "main" java.lang.NoClassDefFoundError:
org/springframework/boot/SpringApplication
    at thinkinginspringboot.firstappbygui.FirstAppByGuiApplication.main
(FirstAppByGuiApplication.java:10)
```

```
Caused by: java.lang.ClassNotFoundException: org.springframework.boot.SpringApplication
 at java.net.URLClassLoader.findClass(URLClassLoader.java:381)
 at java.lang.ClassLoader.loadClass(ClassLoader.java:424)
 at sun.misc.Launcher$AppClassLoader.loadClass(Launcher.java:349)
 at java.lang.ClassLoader.loadClass(ClassLoader.java:357)
 ... 1 more
```

命令执行后，JVM 运行失败，提示类 `org.springframework.boot.SpringApplication` 无法找到。不难理解，这是由于 `java` 命令未指定 Class Path。而当前 Spring Boot 依赖的 JAR 文件均存放在 `BOOT-INF/lib` 目录下。`org.springframework.boot.loader.JarLauncher` 会将这些 JAR 文件作为 `FirstAppByGuiApplication` 的类库依赖，所以 `JarLauncher` 能够引导，反之 `FirstAppByGuiApplication` 则不行。除此之外，还存在一个疑惑，`FirstAppByGuiApplication` 的执行进程到底属于 `JarLauncher` 的子进程，还是与 `JarLauncher` 处于同一进程？就此疑惑和前面未证实的结论，一同探讨 `JarLauncher` 的实现原理。

2.2.4 JarLauncher 的实现原理

在正式展开讨论前，需将 `first-app-by-gui` 工程导入 IDEA。在 IDEA 顶部菜单选择“File”→“Open...”，具体操作如下图所示。

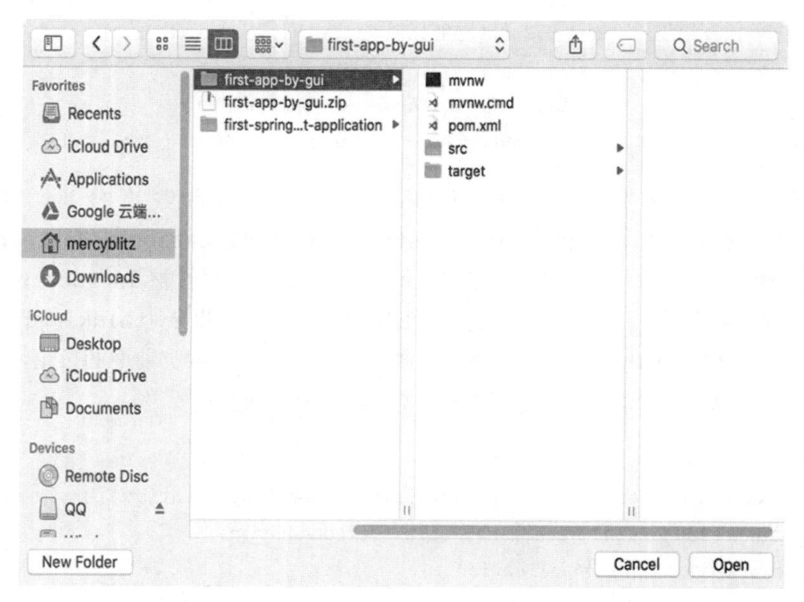

然后打开 `pom.xml`，如下图所示。

由于 org.springframework.boot.loader.JarLauncher 类存在于 org.springframework.boot:spring-boot-loader 中，为了便于源码分析，需要将该依赖添加至当前 pom.xml 文件中，而且因为 spring-boot-starter-parent 固化依赖的关系，所以不需要指定版本，配置如下所示。

```
    <?xml version="1.0" encoding="UTF-8"?>
    <project xmlns="http://maven.apache.org/POM/4.0.0"
xmlns:xsi="http://www.w3.org/2001/XMLSchema-instance"
            xsi:schemaLocation="http://maven.apache.org/POM/4.0.0
http://maven.apache.org/xsd/maven-4.0.0.xsd">
        <modelVersion>4.0.0</modelVersion>

        <groupId>thinking-in-spring-boot</groupId>
        <artifactId>first-app-by-gui</artifactId>
        <version>0.0.1-SNAPSHOT</version>
        <packaging>jar</packaging>
        ...
        <dependencies>
            ...
            <!-- spring-boot-loader 用于源码分析 -->
```

```
        <dependency>
            <groupId>org.springframework.boot</groupId>
            <artifactId>spring-boot-loader</artifactId>
            <scope>provided</scope>
        </dependency>
        ...
    </dependencies>
    ...
</project>
```

由于运行时 `spring-boot-loader` 存在于 FAT JAR 中，所以 `spring-boot-loader` 的依赖 `scope` 为 `provided`。当 IDEA 下载 `spring-boot-loader` 完毕后，单击右侧“Maven Projects”按钮，查看是否完成，如下图所示。

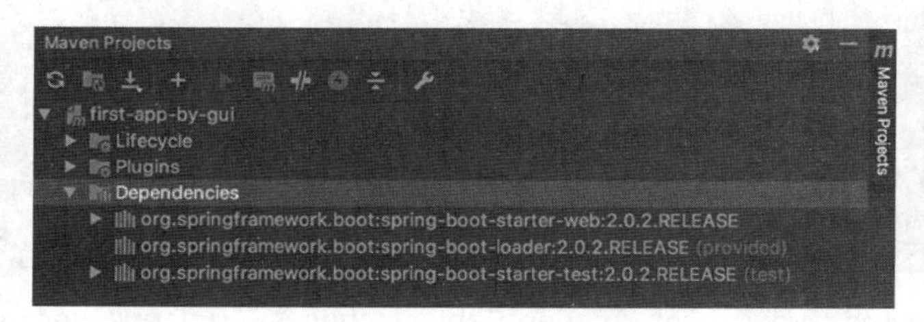

图中的第二个依赖就是 `spring-boot-loader`。下一步，在 IDEA 中同时按下 `command+N` 键，查找 `org.springframework.boot.loader.JarLauncher` 类，如下图所示。

按下回车键确认选择，打开实现源码：

```
public class JarLauncher extends ExecutableArchiveLauncher {

static final String BOOT_INF_CLASSES = "BOOT-INF/classes/";
```

```
    static final String BOOT_INF_LIB = "BOOT-INF/lib/";

    public JarLauncher() {
    }

    protected JarLauncher(Archive archive) {
        super(archive);
    }

    @Override
    protected boolean isNestedArchive(Archive.Entry entry) {
        if (entry.isDirectory()) {
            return entry.getName().equals(BOOT_INF_CLASSES);
        }
        return entry.getName().startsWith(BOOT_INF_LIB);
    }

    public static void main(String[] args) throws Exception {
        new JarLauncher().launch(args);
    }

}
```

明显地发现，BOOT-INF/classes/和BOOT-INF/lib/路径分别用常量BOOT_INF_ CLASSES和BOOT_INF_LIB表示，并且用于isNestedArchive(Archive.Entry)方法判断，从该方法的实现分析，方法参数Archive.Entry对象看似为JAR文件中的资源，比如application.properties，不过该对象与Java标准的java.util.jar.JarEntry对象类似，其name属性（getName()方法）为JAR资源的相对路径。当application.properties资源位于FAT JAR时，实际的Archive.Entry#getName()为/BOOT-INF/classes/application.properties，故符合xxxx.startWith(BOOT_INF_CLASSES)的判断，即isNestedArchive(Archive.Entry)方法返回true。反之，该方法返回false时，说明FAT JAR被解压至文件目录，因此从侧面说明了Spring Boot应用能直接通过java org.springframework.boot.loader.JarLauncher命令启动的原因。换言之，Archive.Entry与java.util.jar.JarEntry也存在差异，它也可表示文件或目录。实际上，Archive.Entry存在两种实现，其中一种为JarFileArchive.JarFileEntry，基于java.util.jar.JarEntry实现，表示FAT JAR嵌入资源：

```
public class JarFileArchive implements Archive {
```

```
    ...
    /**
   * {@link Archive.Entry} implementation backed by a {@link JarEntry}.
   */
  private static class JarFileEntry implements Entry {

      private final JarEntry jarEntry;

      JarFileEntry(JarEntry jarEntry) {
          this.jarEntry = jarEntry;
      }

      public JarEntry getJarEntry() {
          return this.jarEntry;
      }

      @Override
      public boolean isDirectory() {
          return this.jarEntry.isDirectory();
      }

      @Override
      public String getName() {
          return this.jarEntry.getName();
      }

  }

}
```

另外一种实现则是 `ExplodedArchive.FileEntry`，基于文件系统实现：

```
public class ExplodedArchive implements Archive {
    ...
    /**
   * {@link Entry} backed by a File.
   */
  private static class FileEntry implements Entry {
```

```
        private final String name;

        private final File file;

        FileEntry(String name, File file) {
            this.name = name;
            this.file = file;
        }

        public File getFile() {
            return this.file;
        }

        @Override
        public boolean isDirectory() {
            return this.file.isDirectory();
        }

        @Override
        public String getName() {
            return this.name;
        }

    }

}
```

所以，这也从实现层面证明了 JarLauncher 支持 JAR 和文件系统两种启动方式。

同时，JarLauncher 同样作为引导类，当执行 java -jar 命令时，/META-INF/资源的 Main-Class 属性将调用其 main(String[])方法，实际上调用的是 JarLauncher#launch(args)方法，而该方法继承于基类 org.springframework.boot.loader.Launcher，它们之间的继承关系如下图所示。

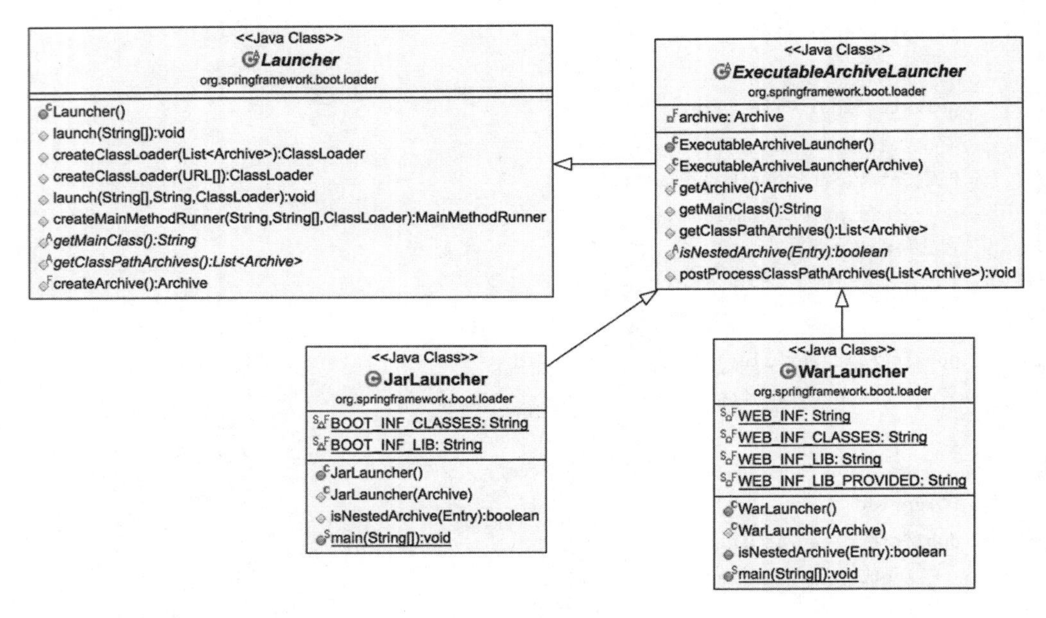

简化其层次：

- `org.springframework.boot.loader.Launcher`
  - `org.springframework.boot.loader.ExecutableArchiveLauncher`
    - `org.springframework.boot.loader.JarLauncher`
    - `org.springframework.boot.loader.WarLauncher`

前面也提到，`WarLauncher` 是可执行 WAR 文件的启动器，两者的具体区别后面会讨论。

下面分析 `Launcher#launch(args)`方法实现：

```
public abstract class Launcher {
    ...
 protected void launch(String[] args) throws Exception {
     JarFile.registerUrlProtocolHandler();
     ClassLoader classLoader = createClassLoader(getClassPathArchives());
     launch(args, getMainClass(), classLoader);
 }
    ...
}
```

该方法首先执行 `JarFile.registerUrlProtocolHandler()`方法：

```
public class JarFile extends java.util.jar.JarFile {
```

```
    ...
private static final String PROTOCOL_HANDLER = "java.protocol.handler.pkgs";

private static final String HANDLERS_PACKAGE = "org.springframework.boot.loader";
    ...
public static void registerUrlProtocolHandler() {
    String handlers = System.getProperty(PROTOCOL_HANDLER, "");
    System.setProperty(PROTOCOL_HANDLER, ("".equals(handlers) ? HANDLERS_PACKAGE
            : handlers + "|" + HANDLERS_PACKAGE));
    resetCachedUrlHandlers();
}

private static void resetCachedUrlHandlers() {
    try {
        URL.setURLStreamHandlerFactory(null);
    }
    catch (Error ex) {
        // Ignore
    }
}
    ...
}
```

JarFile.registerUrlProtocolHandler()方法利用了 java.net.URLStreamHandler 扩展机制，其实现由 URL#getURLStreamHandler(String)提供：

```
public final class URL implements java.io.Serializable {
    ...
    private static final String protocolPathProp = "java.protocol.handler.pkgs";
    ...
    static URLStreamHandler getURLStreamHandler(String protocol) {

        URLStreamHandler handler = handlers.get(protocol);
        if (handler == null) {

            boolean checkedWithFactory = false;

            // Use the factory (if any)
            if (factory != null) {
```

```
            handler = factory.createURLStreamHandler(protocol);
            checkedWithFactory = true;
        }

        // Try java protocol handler
        if (handler == null) {
            String packagePrefixList = null;

            packagePrefixList
                = java.security.AccessController.doPrivileged(
                new sun.security.action.GetPropertyAction(
                    protocolPathProp,""));
            if (packagePrefixList != "") {
                packagePrefixList += "|";
            }

            // REMIND: decide whether to allow the "null" class prefix
            // or not.
            packagePrefixList += "sun.net.www.protocol";

            StringTokenizer packagePrefixIter =
                new StringTokenizer(packagePrefixList, "|");

            while (handler == null &&
                   packagePrefixIter.hasMoreTokens()) {

                String packagePrefix =
                  packagePrefixIter.nextToken().trim();
                try {
                    String clsName = packagePrefix + "." + protocol +
                      ".Handler";
                    Class<?> cls = null;
                    try {
                        cls = Class.forName(clsName);
                    } catch (ClassNotFoundException e) {
                        ClassLoader cl = ClassLoader.getSystemClassLoader();
                        if (cl != null) {
                            cls = cl.loadClass(clsName);
```

```
                    }
                }
                if (cls != null) {
                    handler  =
                      (URLStreamHandler)cls.newInstance();
                }
            } catch (Exception e) {
                // any number of exceptions can get thrown here
            }
        }
    }
    ...
  }
  return handler;

}
}
```

以上方法实现较为复杂，简言之，URL 的关联协议（Protocol）对应一种 URLStreamHandler 实现类，JDK 默认支持文件（file）、HTTP、JAR 等协议，故 JDK 内建了对应协议的实现。这些实现类均存放在 sun.net.www.protocol 包下，并且类名必须为 Handler，其类全名模式为 sun.net.www.protocol.${protocol}.Handler，其中${protocol}表示协议名，常见的协议实现如下。

- FILE：sun.net.www.protocol.file.Handler
- JAR：sun.net.www.protocol.jar.Handler
- HTTP：sun.net.www.protocol.http.Handler
- HTTPS：sun.net.www.protocol.https.Handler
- FTP：sun.net.www.protocol.ftp.Handler

以上类均为 java.net.URLStreamHandler 实现类，换言之，如果需要扩展，则继承 URLStreamHandler 类为必选项，通常配置 Java 系统属性（System#getProperties()）java.protocol.handler.pkgs，追加 URLStreamHandler 实现类的 package，多个 package 以“|”分割。因此，JarFile.registerUrlProtocolHandler() 方法将 package org.springframework.boot.loader 追加到 Java 系统属性 java.protocol.handler.pkgs 中。也就是说，org.springframework.boot.loader 包下存在协议对应的 Handler 类，即 org.springframework.boot.loader.jar.Handler，按照类名模式，其实现协议为 JAR。令人疑惑的是，JAR 协议不是内建实现了吗？它是如何覆盖的呢？请注意 URL#getURLStreamHandler

(String)方法的实现，其处理器包名先读取 Java 系统属性 java.protocol.handler.pkgs，无论是否存在，再追加 sun.net.www.protocol 包，所以 JDK 内建实现是默认（或兜底）实现的，故 org.springframework.boot.loader.jar.Handler 可覆盖 sun.net.www.protocol.jar.Handler。问题的关键在于为什么 Spring Boot 要选择覆盖？不难发现，Spring Boot FAT JAR 除包含传统 Java JAR 中的资源外，还包含依赖的 JAR 文件。换言之，它是一个独立的应用归档文件。当 Spring Boot FAT JAR 被 java -jar 命令引导时，其内部的 JAR 文件无法被内建实现 sun.net.www.protocol.jar.Handler 当作 Class Path，故需要替换实现。

> 由于 Spring Boot 扩展 JAR 协议实现相当复杂，在此不做过多的说明，请读者自行参考 org.springframework.boot.loader.jar.JarURLConnection#get(URL,JarFile) 方法和 org.springframework.boot.loader.jar.JarFile 类。

讨论完 JarFile.registerUrlProtocolHandler() 扩展 URL 协议的目的，下一步执行 createClassLoader(List) 方法，创建 ClassLoader，其中 getClassPathArchives() 方法返回值作为方法参数：

```
public abstract class Launcher {
    ...
protected abstract List<Archive> getClassPathArchives() throws Exception;
    ...
}
```

该方法为抽象方法，具体实现由子类 ExecutableArchiveLauncher 提供：

```
public abstract class ExecutableArchiveLauncher extends Launcher {
    ...
@Override
protected List<Archive> getClassPathArchives() throws Exception {
    List<Archive> archives = new ArrayList<>(
            this.archive.getNestedArchives(this::isNestedArchive));
    postProcessClassPathArchives(archives);
    return archives;
}
    ...
    protected abstract boolean isNestedArchive(Archive.Entry entry);
    ...
}
```

同样，isNestedArchive(Archive.Entry)方法需要由子类 JarLauncher 或 WarLauncher 实现，以 JarLauncher 为例：

```
public class JarLauncher extends ExecutableArchiveLauncher {
    ...
    @Override
    protected boolean isNestedArchive(Archive.Entry entry) {
        if (entry.isDirectory()) {
            return entry.getName().equals(BOOT_INF_CLASSES);
        }
        return entry.getName().startsWith(BOOT_INF_LIB);
    }
    ...
}
```

该方法又回到前文的讨论，即过滤 Archive.Entry 实例是否匹配 BOOT-INF/classes/的名称或 BOOT-INF/lib/的名称前缀。换句话说，无论 Archive.Entry 的实现类是 JarFileArchive.JarFileEntry 还是 ExplodedArchive.FileEntry，只要它们的名称符合以上路径即可，故 getClassPathArchives()返回值还取决于 archive 属性对象的内容：

```
public abstract class ExecutableArchiveLauncher extends Launcher {

    private final Archive archive;

    public ExecutableArchiveLauncher() {
        try {
            this.archive = createArchive();
        }
        ...
    }
    ...
}
```

该 archive 属性来源于父类 Launcher 的 createArchive()方法：

```
public abstract class Launcher {
    ...
    protected final Archive createArchive() throws Exception {
        ProtectionDomain protectionDomain = getClass().getProtectionDomain();
```

```
        CodeSource codeSource = protectionDomain.getCodeSource();
        URI location = (codeSource != null ? codeSource.getLocation().toURI() : null);
        String path = (location != null ? location.getSchemeSpecificPart() : null);
        if (path == null) {
            throw new IllegalStateException("Unable to determine code source archive");
        }
        File root = new File(path);
        if (!root.exists()) {
            throw new IllegalStateException(
                    "Unable to determine code source archive from " + root);
        }
        return (root.isDirectory() ? new ExplodedArchive(root)
                : new JarFileArchive(root));
    }

}
```

此处主要通过当前 `Launcher` 所在的介质，判断是否为 JAR 归档文件实现（`JarFileArchive`）或解压目录实现（`ExplodedArchive`），不过该方法为 `final` 实现，因此其子类 `JarLauncher` 或 `WarLauncher` 均继承该实现。获取当前 Spring Boot FAT JAR 或解压其内容后，下一步调用实际的引导类 `launch(String[],String,ClassLoader)`：

```
public abstract class Launcher {
    ...
    protected void launch(String[] args, String mainClass, ClassLoader classLoader)
            throws Exception {
        Thread.currentThread().setContextClassLoader(classLoader);
        createMainMethodRunner(mainClass, args, classLoader).run();
    }
    ...
    protected MainMethodRunner createMainMethodRunner(String mainClass, String[] args,
            ClassLoader classLoader) {
        return new MainMethodRunner(mainClass, args);
    }
    ...
}
```

该方法的实际执行者为 `MainMethodRunner#run()`方法：

```
public class MainMethodRunner {
    ...
    public MainMethodRunner(String mainClass, String[] args) {
        this.mainClassName = mainClass;
        this.args = (args != null ? args.clone() : null);
    }

    public void run() throws Exception {
        Class<?> mainClass = Thread.currentThread().getContextClassLoader()
                .loadClass(this.mainClassName);
        Method mainMethod = mainClass.getDeclaredMethod("main", String[].class);
        mainMethod.invoke(null, new Object[] { this.args });
    }

}
```

MainMethodRunner 对象需要关联 mainClass 及 main 方法参数 args，而 mainClass 来自 ExecutableArchiveLauncher#getMainClass()方法：

```
public abstract class ExecutableArchiveLauncher extends Launcher {
    ...
    @Override
    protected String getMainClass() throws Exception {
        Manifest manifest = this.archive.getManifest();
        String mainClass = null;
        if (manifest != null) {
            mainClass = manifest.getMainAttributes().getValue("Start-Class");
        }
        if (mainClass == null) {
            throw new IllegalStateException(
                    "No 'Start-Class' manifest entry specified in " + this);
        }
        return mainClass;
    }
    ...
}
```

> 以上实现源码源于 Spring Boot 2.0.2.RELEASE。
> Maven GAV 坐标为：org.springframework.boot:spring-boot-loader:2.0.2.RELEASE。

类名称来自/META-INF/MANIFEST.MF 资源中的 Start-Class 属性，其子类并未覆盖该方法实现，故无论 JAR 还是 WAR，读取 Spring Boot 启动类均来自此属性。获取 mainClass 之后，MainMethodRunner#run()方法将读取 mainClass 类中的 main(String[])方法，并且作为静态方法调用。因此，JarLauncher 实际上是同进程内调用 Start-Class 类的 main(String[])方法，并且在启动前准备好 Class Path。

同理，WarLauncher 大部分实现逻辑与 JarLauncher 类似，接下来将讨论差异性部分。

可执行 WAR 文件启动器——WarLauncher

WarLauncher 与 JarLauncher 的差异甚小，两者均继承于 ExecutableArchiveLauncher，并使用 JarFileArchive 和 ExplodedArchive 分别管理归档文件和解压目录两种资源，其主要区别在于项目类文件和 JAR Class Path 路径的不同：

```
public class WarLauncher extends ExecutableArchiveLauncher {

    private static final String WEB_INF = "WEB-INF/";

    private static final String WEB_INF_CLASSES = WEB_INF + "classes/";

    private static final String WEB_INF_LIB = WEB_INF + "lib/";

    private static final String WEB_INF_LIB_PROVIDED = WEB_INF + "lib-provided/";

    public WarLauncher() {
    }

    protected WarLauncher(Archive archive) {
        super(archive);
    }

    @Override
    public boolean isNestedArchive(Archive.Entry entry) {
        if (entry.isDirectory()) {
            return entry.getName().equals(WEB_INF_CLASSES);
        }
```

```
        else {
            return entry.getName().startsWith(WEB_INF_LIB)
                    || entry.getName().startsWith(WEB_INF_LIB_PROVIDED);
        }
    }

    public static void main(String[] args) throws Exception {
        new WarLauncher().launch(args);
    }

}
```

根据前面的讨论，可以确定的是，WEB-INF/classes/（用常量 WEB_INF_CLASSES 表示）、WEB-INF/lib（用常量 WEB_INF_LIB 表示）和 WEB-INF/lib-provided/（用常量 WEB_INF_LIB_PROVIDED 表示）均为 WarLauncherClassLoader 的 Class Path。其中 WEB-INF/classes/和 WEB-INF/lib 是传统的 Servlet 应用的 Class Path 路径，而 WEB-INF/lib-provided/属于 Spring Boot WarLauncher 定制实现，那么，是否意味着其路径存放 Maven 依赖为<scope>provided</scope>的 JAR 呢？回顾 first-app-by-gui 项目，其中 scope 是 provided 的直接依赖为 org.springframework.boot:spring-boot-loader，因此，可将当前 pom.xml 文件的<packaging>元素从 jar 调整为 war，改变当前 Spring Boot 项目的打包归档类型：

```
    <?xml version="1.0" encoding="UTF-8"?>
    <project xmlns="http://maven.apache.org/POM/4.0.0"
xmlns:xsi="http://www.w3.org/2001/XMLSchema-instance"
          xsi:schemaLocation="http://maven.apache.org/POM/4.0.0
http://maven.apache.org/xsd/maven-4.0.0.xsd">
        <modelVersion>4.0.0</modelVersion>

        <groupId>thinking-in-spring-boot</groupId>
        <artifactId>first-app-by-gui</artifactId>
        <version>0.0.1-SNAPSHOT</version>
        <packaging>war</packaging>
        ...
    </project>
```

重新使用 Maven 打包当前工程：

```
    $ mvn clean package
    [INFO] Scanning for projects...
```

```
(...部分内容被省略...)
[INFO] Building war:
/Users/mercyblitz/thinking-in-spring-boot-samples/first-app-by-gui/target/first-app-by-gui-0.0.1-SNAPSHOT.war
[INFO]
[INFO] --- spring-boot-maven-plugin:2.0.2.RELEASE:repackage (default) @ first-app-by-gui ---
[INFO] ------------------------------------------------------------------------
[INFO] BUILD SUCCESS
[INFO] ------------------------------------------------------------------------
(...部分内容被省略...)
```

执行日志显示 first-app-by-gui-0.0.1-SNAPSHOT.war 被创建，故切换到 target 目录，将其内容同样解压到 temp 目录：

```
$ cd target/
$ unzip first-app-by-gui-0.0.1-SNAPSHOT.war -d temp
Archive:  first-app-by-gui-0.0.1-SNAPSHOT.war
(...部分内容被省略...)
 extracting: temp/WEB-INF/lib-provided/spring-boot-loader-2.0.2.RELEASE.jar
```

再执行 tree temp，观察其目录结构：

```
$ cd temp/
$ tree temp/
temp/
├── META-INF
│   ├── MANIFEST.MF
│   └── maven
│       └── thinking-in-spring-boot
│           └── first-app-by-gui
│               ├── pom.properties
│               └── pom.xml
├── WEB-INF
│   ├── classes
│   │   ├── application.properties
│   │   └── thinkinginspringboot
│   │       └── firstappbygui
│   │           └── FirstAppByGuiApplication.class
│   ├── lib
```

```
│   │   (...部分内容被省略...)
│   │   ├── spring-boot-starter-web-2.0.2.RELEASE.jar
│   │   (...部分内容被省略...)
│   │   ├── spring-boot-starter-tomcat-2.0.2.RELEASE.jar
│   │   (...部分内容被省略...)
│   └── lib-provided
│       └── spring-boot-loader-2.0.2.RELEASE.jar
└── org
    └── springframework
        └── boot
            └── loader
                ├── ExecutableArchiveLauncher.class
                (...部分内容被省略...)

18 directories, 92 files
```

相比 FAT JAR 的解压目录，WAR 增加了 `WEB-INF/lib-provided`，并且该目录仅有一个 JAR 文件，即 `spring-boot-loader-2.0.2.RELEASE.jar`，由此证明 `WEB-INF/lib-provided` 目录存放的是`<scope>provided</scope>`的 JAR 文件。不过 `spring-boot-loader` 如此设计以上目录结构的意图是什么呢？前面提到，传统的 Servlet 应用的 Class Path 路径仅关注 `WEB-INF/classes/`和 `WEB-INF/lib` 目录，因此，`WEB-INF/lib-provided/`中的 JAR 将被 Servlet 容器忽略，如 Servlet API，该 API 由 Servlet 容器提供。因此，这样设计的好处在于，打包后的 WAR 文件能够在 Servlet 容器中兼容运行，当然 Spring Boot WebFlux 应用除外。总而言之，**打包 WAR 文件是一种兼容措施，既能被 `WarLauncher` 启动，又能兼容 Servlet 容器环境。换言之，`WarLauncher` 与 `JarLauncher` 并无本质差别，所以建议 Spring Boot 应用使用非传统 Web 部署时，尽可能地使用 JAR 归档方式**。因此，以上 JAR 和 WAR 归档应用均为 Spring Boot 独立应用。接下来继续探讨 Spring Boot 固化的 Maven 依赖。

第 3 章 理解固化的 Maven 依赖

3.1 spring-boot-starter-parent 与 spring-boot-dependencies 简介

在前面的讨论中，曾引入相关的 Starter 的依赖，比如 `org.springframework.boot:spring-boot-starter-web`，`org.springframework.boot:spring-boot-loader` 同样如此。又已知两者的版本信息均继承于 `org.springframework.boot:spring-boot-starter-parent`，而这些特性属于 Maven 依赖管理的范畴，这样就降低了 Spring Boot 应用管理依赖的成本，因此，Spring Boot 官方的描述为“We take an opinionated view of the Spring platform and third-party libraries so you can get started with minimum fuss.”。不过通过配置 `org.springframework.boot:spring-boot- starter-parent` 的方式存在一定的限制，因为采用单继承的方式，所以限制了其固化 Maven 依赖（仅限于 Spring Boot 相关），并且很有可能应用 `pom.xml` 拥有自定义 Parent。如果需要固化其他类型的依赖则较为繁琐，如 Spring Cloud 依赖。Spring Boot 官方文档在“13.2.2 Using Spring Boot without the Parent POM”章节中有相关说明：

> Not everyone likes inheriting from the `spring-boot-starter-parent` POM. You may have your own corporate standard parent that you need to use or you may prefer to explicitly declare all your Maven configuration.

同时，官方介绍了单独导入 pom 的方式：

> If you do not want to use the `spring-boot-starter-parent`, you can still keep the benefit of the dependency management (but not the plugin management) by using a `scope=import` dependency, as follows:
>
> ```
> <dependencyManagement>
>         <dependencies>
>         <dependency>
>             <!-- Import dependency management from Spring Boot -->
>             <groupId>org.springframework.boot</groupId>
>             <artifactId>spring-boot-dependencies</artifactId>
>             <version>2.0.2.RELEASE</version>
>             <type>pom</type>
>             <scope>import</scope>
>         </dependency>
>     </dependencies>
> </dependencyManagement>
> ```

因此，可对 `first-app-by-gui` 项目的 `pom.xml` 做出调整：

```
    <?xml version="1.0" encoding="UTF-8"?>
    <project xmlns="http://maven.apache.org/POM/4.0.0"
xmlns:xsi="http://www.w3.org/2001/XMLSchema-instance"
             xsi:schemaLocation="http://maven.apache.org/POM/4.0.0
http://maven.apache.org/xsd/maven-4.0.0.xsd">
        <modelVersion>4.0.0</modelVersion>

        <groupId>thinking-in-spring-boot</groupId>
        <artifactId>first-app-by-gui</artifactId>
        <version>0.0.1-SNAPSHOT</version>
        <packaging>war</packaging>

        <name>first-app-by-gui</name>
```

```
    <description>Demo project for Spring Boot</description>

    <properties>
        <project.build.sourceEncoding>UTF-8</project.build.sourceEncoding>
        <project.reporting.outputEncoding>UTF-8</project.reporting.outputEncoding>
        <java.version>1.8</java.version>
    </properties>

    <dependencyManagement>
        <dependencies>
            <dependency>
                <!-- Import dependency management from Spring Boot -->
                <groupId>org.springframework.boot</groupId>
                <artifactId>spring-boot-dependencies</artifactId>
                <version>2.0.2.RELEASE</version>
                <type>pom</type>
                <scope>import</scope>
            </dependency>
        </dependencies>
    </dependencyManagement>
    ...
</project>
```

将当前项目重新打包：

```
$ mvn clean package
[INFO] Scanning for projects...
(...部分内容被省略...)
[ERROR] Failed to execute goal org.apache.maven.plugins:maven-war-plugin:2.2:war
(default-war) on project first-app-by-gui: Error assembling WAR: webxml attribute is required (or
pre-existing WEB-INF/web.xml if executing in update mode)
(...部分内容被省略...)
```

Maven 构建日志提示 `maven-war-plugin` 插件执行失败，这是因为 Web 应用部署描述文件 `WEB-INF/web.xml`（Deployment Descriptor）在项目中不存在的缘故。当项目 `pom.xml` 设置 `<packaging>war</packaging>`后，它将执行 `maven-war-plugin`，而 `WEB-INF/web.xml` 默认为必选项，可以通过插件配置属性将其忽略。然而为何调整之前的打包会成功呢？或许问题的答案在于 `spring-boot-starter-parent` 与 `spring-boot-dependencies` 的差别。

3.2 理解 spring-boot-starter-parent 与 spring-boot-dependencies

先从 spring-boot-starter-parent 的定义入手，观察其主要内容：

```
<?xml version="1.0" encoding="UTF-8"?>
<project xsi:schemaLocation="http://maven.apache.org/POM/4.0.0
http://maven.apache.org/xsd/maven-4.0.0.xsd" xmlns="http://maven.apache.org/POM/4.0.0"
    xmlns:xsi="http://www.w3.org/2001/XMLSchema-instance">
  <modelVersion>4.0.0</modelVersion>
  <parent>
    <groupId>org.springframework.boot</groupId>
    <artifactId>spring-boot-dependencies</artifactId>
    <version>2.0.2.RELEASE</version>
    <relativePath>../spring-boot-dependencies</relativePath>
  </parent>
  <groupId>org.springframework.boot</groupId>
  <artifactId>spring-boot-parent</artifactId>
  <version>2.0.2.RELEASE</version>
  <packaging>pom</packaging>
  ...
  <prerequisites>
    <maven>3.5</maven>
  </prerequisites>
</project>
```

不难发现，spring-boot-dependencies 是 spring-boot-starter-parent 的 parent。换言之，spring-boot-dependencies 也可以直接作为 parent：

```
<?xml version="1.0" encoding="utf-8"?><project xmlns="http://maven.apache.org/POM/4.0.0"
xmlns:xsi="http://www.w3.org/2001/XMLSchema-instance"
xsi:schemaLocation="http://maven.apache.org/POM/4.0.0
http://maven.apache.org/xsd/maven-4.0.0.xsd">
    <modelVersion>4.0.0</modelVersion>
    <groupId>org.springframework.boot</groupId>
    <artifactId>spring-boot-dependencies</artifactId>
    <version>2.0.2.RELEASE</version>
```

```
    <packaging>pom</packaging>
    ...
    <properties>
    ...
        <maven-war-plugin.version>3.1.0</maven-war-plugin.version>
    ...
    </properties>
    ...
    <dependencyManagement>
        <dependencies>
            ...
            <dependency>
                <groupId>org.springframework.boot</groupId>
                <artifactId>spring-boot-starter-web</artifactId>
                <version>2.0.2.RELEASE</version>
            </dependency>
            ...
            <dependency>
                <groupId>org.springframework.boot</groupId>
                <artifactId>spring-boot-test</artifactId>
                <version>2.0.2.RELEASE</version>
            </dependency>
            ...
        </dependencies>
    </dependencyManagement>
    ...
    <build>
        <pluginManagement>
            <plugins>
                <plugin>
                    <artifactId>maven-war-plugin</artifactId>
                    <version>${maven-war-plugin.version}</version>
                </plugin>
            </plugins>
        </pluginManagement>
    </build>
    ...
</project>
```

无论 Spring Boot 依赖 org.springframework.boot:spring-boot-starter-web 和 org.springframework.boot:spring-boot-test，还是插件 maven-war-plugin，均定义在 org.springframework.boot:spring-boot-dependencies pom 中。当项目 pom.xml 文件中的 <parent>引用 spring-boot-starter-parent 时，mvn package 将使用 maven-war-plugin:3.1.0 和 spring-boot-maven-plugin:2.0.2.RELEASE。相反，<dependencyManagement> 导入 spring-boot-dependencies 的方式尽管与前方法同源，然而本方式仅关注 <dependencyManagement>，所以 maven-war-plugin 采用的版本为 2.2。因此，将 maven-war-plugin:3.1.0 添加到 first-app-by-gui 项目的 pom.xml 文件中：

```
    <?xml version="1.0" encoding="UTF-8"?>
    <project xmlns="http://maven.apache.org/POM/4.0.0"
xmlns:xsi="http://www.w3.org/2001/XMLSchema-instance"
         xsi:schemaLocation="http://maven.apache.org/POM/4.0.0
http://maven.apache.org/xsd/maven-4.0.0.xsd">
    ...
    <build>
        <plugins>
            <!-- 保持与 spring-boot-dependencies 版本一致 -->
            <plugin>
                <groupId>org.apache.maven.plugins</groupId>
                <artifactId>maven-war-plugin</artifactId>
                <version>3.1.0</version>
            </plugin>

            <plugin>
                <groupId>org.springframework.boot</groupId>
                <artifactId>spring-boot-maven-plugin</artifactId>
            </plugin>
        </plugins>
    </build>
</project>
```

打包当前项目：

```
    $ mvn clean package
    [INFO] Scanning for projects...
    (...部分内容被省略...)
    [INFO] --- maven-war-plugin:3.1.0:war (default-war) @ first-app-by-gui ---
```

```
    [INFO] Packaging webapp
    [INFO] Assembling webapp [first-app-by-gui] in
[/Users/mercyblitz/thinking-in-spring-boot-samples/first-app-by-gui/target/first-app-by-gui-0
.0.1-SNAPSHOT]
    [INFO] Processing war project
    [INFO] Webapp assembled in [127 msecs]
    [INFO] Building war:
/Users/mercyblitz/thinking-in-spring-boot-samples/first-app-by-gui/target/first-app-by-gui-0.
0.1-SNAPSHOT.war
    [INFO] ------------------------------------------------------------------------
    [INFO] BUILD SUCCESS
    [INFO] ------------------------------------------------------------------------
    (...部分内容被省略...)
```

启动当前应用：

```
    $ java -jar target/first-app-by-gui-0.0.1-SNAPSHOT.war
    target/first-app-by-gui-0.0.1-SNAPSHOT.war 中没有主清单属性
```

控制台提示在 WAR 文件中没有找到主类，无法启动。若对之前的构建日志进行分析，则不难看出，maven-war-plugin 插件的确执行了，然而 spring-boot-maven-plugin 插件却没有执行。原因在于该插件未指定版本，故进行版本号 2.0.2.RELEASE 的配置操作：

```
    <?xml version="1.0" encoding="UTF-8"?>
    <project xmlns="http://maven.apache.org/POM/4.0.0"
xmlns:xsi="http://www.w3.org/2001/XMLSchema-instance"
        xsi:schemaLocation="http://maven.apache.org/POM/4.0.0
http://maven.apache.org/xsd/maven-4.0.0.xsd">
        ...
        <build>
            <plugins>
                <!-- 保持与 spring-boot-dependencies 版本一致 -->
                <plugin>
                    <groupId>org.apache.maven.plugins</groupId>
                    <artifactId>maven-war-plugin</artifactId>
                    <version>3.1.0</version>
                </plugin>

                <plugin>
```

```
            <groupId>org.springframework.boot</groupId>
            <artifactId>spring-boot-maven-plugin</artifactId>
            <version>2.0.2.RELEASE</version>
        </plugin>
    </plugins>
  </build>
</project>
```

再次执行 mvn clean package：

```
(...部分内容被省略...)
[INFO] --- maven-war-plugin:3.1.0:war (default-war) @ first-app-by-gui ---
[INFO] Packaging webapp
[INFO] Assembling webapp [first-app-by-gui] in
[/Users/mercyblitz/thinking-in-spring-boot-samples/first-app-by-gui/target/first-app-by-gui-0
.0.1-SNAPSHOT]
[INFO] Processing war project
[INFO] Webapp assembled in [111 msecs]
[INFO] Building war:
/Users/mercyblitz/thinking-in-spring-boot-samples/first-app-by-gui/target/first-app-by-gui-0.
0.1-SNAPSHOT.war
[INFO] ------------------------------------------------------------------------
[INFO] BUILD SUCCESS
[INFO] ------------------------------------------------------------------------
(...部分内容被省略...)
```

实际上，spring-boot-maven-plugin 插件仍旧没有执行，为此官方文档在“68.1 Including the Plugin”章节进行了说明：

> To use the Spring Boot Maven Plugin, include the appropriate XML in the plugins section of your pom.xml, as shown in the following example:
>
> ```
> <?xml version="1.0" encoding="UTF-8"?>
> <project xmlns="http://maven.apache.org/POM/4.0.0"
> xmlns:xsi="http://www.w3.org/2001/XMLSchema-instance"
>    xsi:schemaLocation="http://maven.apache.org/POM/4.0.0
> http://maven.apache.org/xsd/maven-4.0.0.xsd">
>     <modelVersion>4.0.0</modelVersion>
>     <!-- ... -->
> ```

```
    <build>
        <plugins>
            <plugin>
                <groupId>org.springframework.boot</groupId>
                <artifactId>spring-boot-maven-plugin</artifactId>
                <version>2.0.2.RELEASE</version>
                <executions>
                    <execution>
                        <goals>
                            <goal>repackage</goal>
                        </goals>
                    </execution>
                </executions>
            </plugin>
        </plugins>
    </build>
</project>
```

需添加<goal>为 repackage，故将其配置到 first-app-by-gui 项目的 pom.xml 文件中：

```
<?xml version="1.0" encoding="UTF-8"?>
<project xmlns="http://maven.apache.org/POM/4.0.0"
xmlns:xsi="http://www.w3.org/2001/XMLSchema-instance"
         xsi:schemaLocation="http://maven.apache.org/POM/4.0.0
http://maven.apache.org/xsd/maven-4.0.0.xsd">
    ...
    <build>
        <plugins>
            <!-- 保持与 spring-boot-dependencies 版本一致 -->
            <plugin>
                <groupId>org.apache.maven.plugins</groupId>
                <artifactId>maven-war-plugin</artifactId>
                <version>3.1.0</version>
            </plugin>

            <plugin>
                <groupId>org.springframework.boot</groupId>
                <artifactId>spring-boot-maven-plugin</artifactId>
```

```
                <version>2.0.2.RELEASE</version>
                <executions>
                    <execution>
                        <goals>
                            <goal>repackage</goal>
                        </goals>
                    </execution>
                </executions>
            </plugin>
        </plugins>
    </build>
</project>
```

重新打包当前项目：

```
$ mvn clean package
[INFO] Scanning for projects...
(...部分内容被省略...)
[INFO] --- maven-war-plugin:3.1.0:war (default-war) @ first-app-by-gui ---
[INFO] Packaging webapp
[INFO] Assembling webapp [first-app-by-gui] in
[/Users/mercyblitz/thinking-in-spring-boot-samples/first-app-by-gui/target/first-app-by-gui-0
.0.1-SNAPSHOT]
[INFO] Processing war project
[INFO] Webapp assembled in [90 msecs]
[INFO] Building war:
/Users/mercyblitz/thinking-in-spring-boot-samples/first-app-by-gui/target/first-app-by-gui-0.
0.1-SNAPSHOT.war
[INFO]
[INFO] --- spring-boot-maven-plugin:2.0.2.RELEASE:repackage (default) @ first-app-by-gui ---
[INFO] ------------------------------------------------------------------------
[INFO] BUILD SUCCESS
[INFO] ------------------------------------------------------------------------
(...部分内容被省略...)
```

控制台日志中清晰地显示了 `spring-boot-maven-plugin` 插件的执行记录。启动 `first-app-by-gui` 应用：

```
$ java -jar target/first-app-by-gui-0.0.1-SNAPSHOT.war
```

```
  .   ____          _            __ _ _
 /\\ / ___'_ __ _ _(_)_ __  __ _ \ \ \ \
( ( )\___ | '_ | '_| | '_ \/ _` | \ \ \ \
 \\/  ___)| |_)| | | | | || (_| |  ) ) ) )
  '  |____| .__|_| |_|_| |_\__, | / / / /
 =========|_|==============|___/=/_/_/_/
 :: Spring Boot ::        (v2.0.2.RELEASE)
(...部分内容被省略...)
[           main] t.f.FirstAppByGuiApplication             : Started
FirstAppByGuiApplication in 5.631 seconds (JVM running for 6.107)
```

尽管应用正常启动，不过以上依赖和插件的配置方式需要开发人员清晰地认识到其中的差异，其一是 `maven-war-plugin` 插件在新老版本上的差异。在老版本 `maven-war-plugin:2.2` 中，默认的打包规则是必须存在 Web 应用部署描述文件 `WEB-INF/web.xml`，而 `maven-war-plugin:3.1.0` 调整了该默认行为。其二是单独引入 `spring-boot-maven-plugin` 插件时，需要配置 `repackage` `<goal>`元素，否则不会添加 Spring Boot 引导依赖，进而无法引导当前应用。其三是根据使用习惯，通常不会将 `spring-boot-dependencies` 作为 Maven 项目的`<parent>`，尽管 `spring-boot-starter-parent` 只是简单地继承了 `spring-boot-dependencies`。

总而言之，Spring Boot 利用 Maven 的依赖管理特性，进而固化其 Maven 依赖。所以，前面提到的该特性并非 Spring Boot 专属，然而 Spring 技术栈却将其利用得相当充分。为此，Spring 特别为所有 Spring 技术栈提供了统一的 Maven 依赖管理项目，即 Spring IO Platform，不过该项目将在 2019 年 4 月 9 日后不再维护，官方将使用 `spring-boot-starter-parent` 或 `spring-boot-dependencies` 替代：

> The Platform will reach the end of its supported life on 9 April 2019. Maintenence releases of both the Brussels and Cairo lines will continue to be published up until that time. Users of the Platform are encourage to start using Spring Boot's dependency management directory, either by using `spring-boot-starter-parent` as their Maven project's parent, or by importing the `spring-boot-dependencies` bom.
>
> Spring IO Platorm 原本在 Spring 项目中排名第一，目前已调整为最后。由于未来它不再维护，建议读者从遗留项目中移除其依赖管理。

无论 `first-app-by-gui` 应用以何种方式启动，其类型都属于 Spring Boot Servlet Web 应用，同时，控制台日志提示其 Web 容器是 Tomcat：

```
    [           main] o.s.b.w.embedded.tomcat.TomcatWebServer  : Tomcat started on port(s): 8080
(http) with context path ''
```

然而其项目 `pom.xml` 文件没有直接依赖 Tomcat JAR，不过在展示 WAR 解压目录结构时，曾出现一个名为 `spring-boot-starter-tomcat-2.0.2.RELEASE.jar` 的文件，位于 `WEB-INF/lib/` 目录下。由此说明该 JAR 文件由 `org.springframework.boot:spring-boot-starter-web` 间接引入，而按照 Maven JAR 文件的命名规则，`spring-boot-starter-tomcat` 为 artifactId，而 version 则是 `2.0.2.RELEASE`，packaging 则是 `jar`，故利用 Maven `dependency` 插件分析其依赖树结构：

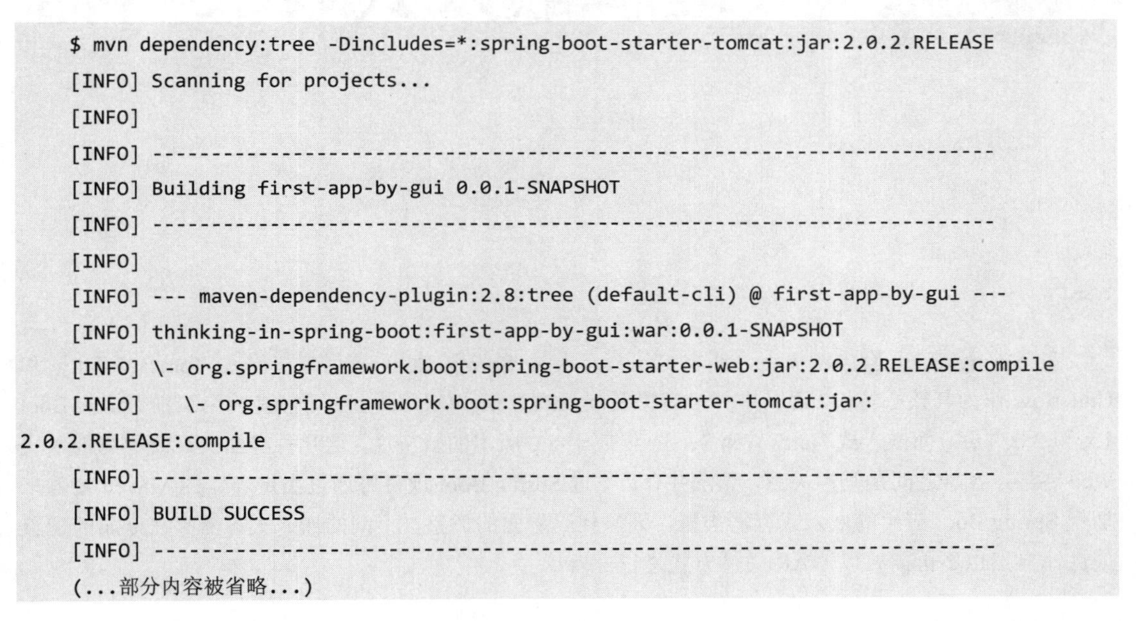

```
    $ mvn dependency:tree -Dincludes=*:spring-boot-starter-tomcat:jar:2.0.2.RELEASE
    [INFO] Scanning for projects...
    [INFO]
    [INFO] ------------------------------------------------------------------------
    [INFO] Building first-app-by-gui 0.0.1-SNAPSHOT
    [INFO] ------------------------------------------------------------------------
    [INFO]
    [INFO] --- maven-dependency-plugin:2.8:tree (default-cli) @ first-app-by-gui ---
    [INFO] thinking-in-spring-boot:first-app-by-gui:war:0.0.1-SNAPSHOT
    [INFO] \- org.springframework.boot:spring-boot-starter-web:jar:2.0.2.RELEASE:compile
    [INFO]    \- org.springframework.boot:spring-boot-starter-tomcat:jar:
2.0.2.RELEASE:compile
    [INFO] ------------------------------------------------------------------------
    [INFO] BUILD SUCCESS
    [INFO] ------------------------------------------------------------------------
    (...部分内容被省略...)
```

由此依赖树分析报告证实了 `spring-boot-starter-tomcat` 是由 `spring-boot-starter-web` 间接依赖的结论。然而新的疑惑又产生了，为什么当前 Spring Boot 应用仅依赖 `spring-boot-starter-tomcat` 就能引导 Tomcat 容器，并且该容器嵌入当前应用，不需要预备安装？这个问题的答案将在下一章揭晓。

第4章 理解嵌入式Web容器

在 Spring Boot 官方网页上，明文提示开发人员 Spring Boot 应用直接嵌入 Tomcat、Jetty 和 Undertow 作为其核心特性，而其中并没有提及嵌入式 Netty Web Server，可能是因为当前 Spring Boot 1.x 和 2.0 并存，而嵌入式 Netty Web Server 仅属于 2.0 版本的新特性。这几种容器实现统称为嵌入式 Web 容器，容器之间是互斥关系，无法并存。不过 Spring Boot 项目可通过指定容器的 Maven 依赖来切换 Spring Boot 应用的嵌入式容器类型，无须代码层面的调整，不同的嵌入式容器存在专属的配置属性，自然也不再需要以 WAR 文件方式进行部署：

> Embed Tomcat, Jetty or Undertow directly (no need to deploy WAR files)

在互联网场景中，与终端用户交互的应用大多数是 Web 应用，其通信协议基本为 HTTP。由于多年 Java Web 技术栈的发展，使得 Java 开发人员仅有两种选择，即 Servlet 与其他，而前者几乎垄断了 Java Web 开发。Tomcat 和 Jetty 为 Servlet 容器的经典实现，后来者的 Undertow 作为 JBoss 社区推出的新一代兼容 Servlet 3.1+规范的容器，成为嵌入式 Servlet 容器的新选择。

最新的版本支持 HTTP/2 和 Servlet 4.0 规范，且核心 JAR 体积仅占 2.2 MB 左右。早在 Tomcat 5.x 和 Jetty 5.x 就已支持嵌入式容器。Servlet 规范实现与这三种 Servlet 容器的版本对应关系如下表所示。

Servlet 规范	Tomcat	Jetty	Undertow
4.0	9.x	9.x	2.x
3.1	8.x	8.x	1.x
3.0	7.x	7.x	N/A
2.5	6.x	6.x	N/A

至于嵌入式 Reactive Web 容器，默认实现为 Netty Web Server，与业界其他基于 Netty 实现的 Web Server 类似，如 Eclipse vert.x。不过由于 Spring WebFlux 基于 Reactor 框架实现，因此在 Spring Boot 中，Netty Web Server 属于 Reactor 与 Netty 的整合实现，由 `spring-boot-starter-reactor-netty` 引入，通常被 `spring-boot-starter-webflux` 间接引入。不过，以上三种嵌入式 Servlet 容器也能作为 Reactive Web Server，并允许替换默认实现 Netty Web Server，因为 Servlet 3.1+容器同样满足 Reactive 异步非阻塞的特性。换言之，热门的 Web 容器实现均支持嵌入式容器方式，这并非 Spring Boot 的独创，只不过是在整合上的技术创新。

因此，下面将分别讨论嵌入式 Servlet Web 容器和嵌入式 Reactive Web 容器。

4.1　嵌入式 Servlet Web 容器

根据 Spring Boot 2.0 官方文档在“9.1 Servlet Containers”章节的介绍，Spring Boot 支持三种嵌入式 Servlet 3.1+ 容器。

> Spring Boot supports the following embedded servlet containers:

Name	Servlet Version
Tomcat 8.5	3.1
Jetty 9.4	3.1
Undertow 1.4	3.1

> You can also deploy Spring Boot applications to any Servlet 3.1+ compatible container.

类似的说明在 Spring Boot 1.5 官方文档的“9.1 Servlet Containers”章节也有体现：

> The following embedded servlet containers are supported out of the box:

Name	Servlet Version	Java Version
Tomcat 8	3.1	Java 7+
Tomcat 7	3.0	Java 6+
Jetty 9.3	3.1	Java 8+
Jetty 9.2	3.1	Java 7+
Jetty 8	3.0	Java 6+
Undertow 1.3	3.1	Java 7+

> You can also deploy Spring Boot applications to any Servlet 3.0+ compatible container.

Spring Boot 2.0 延续了 Spring Boot 1.5 支持的三种嵌入式 Servlet 容器类型的特性。不过 Spring Boot 1.x 兼容的 Servlet 和 Java 版本更低一点，即 Servlet 3.0+和 Java 6+。

> 嵌入式 Undertow 容器是从 Spring Boot 1.2 才予以支持的，而 Spring Boot 1.5 默认需要 Java 7，同样兼容 Java 6，然而需要一些额外的操作，具体详情请参考官方文档的“9. System Requirements”章节：
> By default, Spring Boot 1.5.10.RELEASE requires Java 7 and Spring Framework 4.3.14. RELEASE or above. You can use Spring Boot with Java 6 with some additional configuration. See Section 84.11, “How to use Java 6” for more details. Explicit build support is provided for Maven (3.2+), and Gradle 2 (2.9 or later) and 3.

接下来将逐一展开对 Tomcat、Jetty 和 Undertow 作为嵌入式 Servlet Web 容器的讨论。

4.1.1 Tomcat 作为嵌入式 Servlet Web 容器

目前 Apache Tomcat 官网仍旧提供 7.0.x 版本的嵌入式版本下载（https://tomcat.apache.org/download-70.cgi），如下图所示。

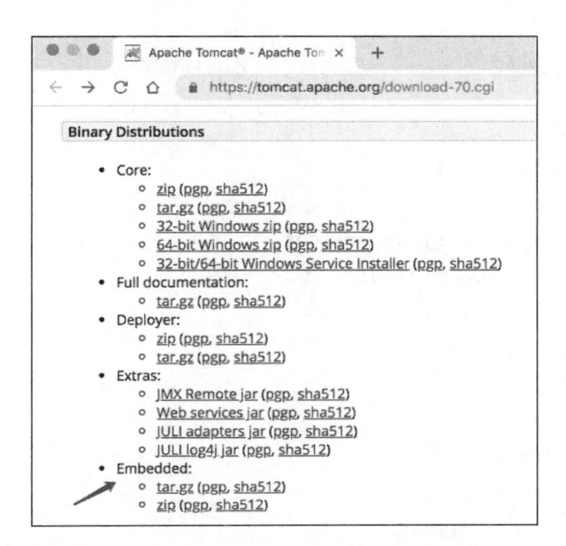

图中箭头指向的内容就是嵌入式 Tomcat 容器二进制文件。以此类推，Tomcat 8.x 和 9.x 均延续了嵌入式特性。通常嵌入式 Tomcat 作为 Web 应用的一部分，结合其 API 实现 Servlet 容器的引导。除此之外，在 Apache Maven 插件的官方网页中，Apache Tomcat 同样提供了 Maven 插件。

该插件用于快速开发 Servlet Web 应用，与嵌入式 Tomcat 的差异在于，它不需要编码，也不需要外置 Tomcat 容器，将当前应用直接打包为可运行的 JAR 或 WAR 文件，通过 `java -jar` 命令启动，修正 Spring Boot FAT JAR 或 FAT WAR。根据该插件发布的时间分析，不得不让人怀疑 Spring Boot 在独立应用和嵌入式容器的设计上或多或少地"借鉴"了其中实现。

不过两者的差异也是比较明显的，Tomcat Maven 插件在打包时，它将完整的 Tomcat 运行时资源添加至当前 JAR 或 WAR 文件中，当该 JAR 或 WAR 文件被 `java -jar` 引导后：

```
$ java -jar servlet-sample-1.0.0-SNAPSHOT-war-exec.jar
(...部分内容被省略...)
org.apache.coyote.http11.Http11NioProtocol start
信息: Starting ProtocolHandler ["http-nio-8080"]
```

先将其归档内容解压到`.extract` 目录：

```
$ tree .extract/
.extract/
├── conf
│   └── web.xml
├── logs
│   (...部分内容被省略...)
├── temp
```

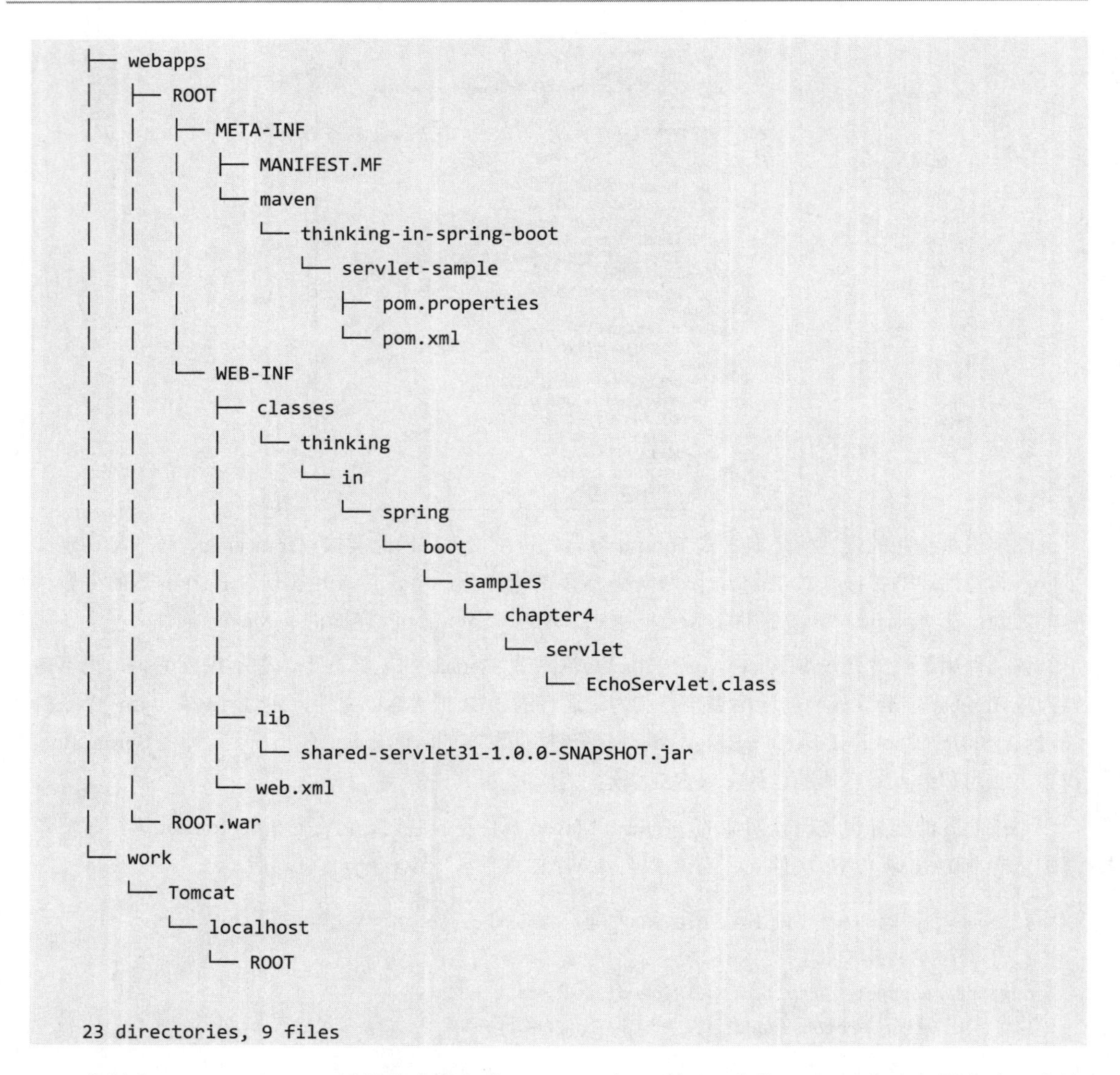

```
├── webapps
│   ├── ROOT
│   │   ├── META-INF
│   │   │   ├── MANIFEST.MF
│   │   │   └── maven
│   │   │       └── thinking-in-spring-boot
│   │   │           └── servlet-sample
│   │   │               ├── pom.properties
│   │   │               └── pom.xml
│   │   └── WEB-INF
│   │       ├── classes
│   │       │   └── thinking
│   │       │       └── in
│   │       │           └── spring
│   │       │               └── boot
│   │       │                   └── samples
│   │       │                       └── chapter4
│   │       │                           └── servlet
│   │       │                               └── EchoServlet.class
│   │       ├── lib
│   │       │   └── shared-servlet31-1.0.0-SNAPSHOT.jar
│   │       └── web.xml
│   └── ROOT.war
└── work
    └── Tomcat
        └── localhost
            └── ROOT

23 directories, 9 files
```

实际上，Tomcat Maven 插件并非嵌入式 Tomcat，仍旧利用了传统 Tomcat 容器部署方式，先将 Web 应用打包为 `ROOT.war` 文件，然后在 Tomcat 应用启动的过程，将 `ROOT.war` 文件解压至 `webapps` 目录。当然，该插件支持指定 `ServletContext` 路径。反观 Spring Boot 2.0 的实现，它利用嵌入式 Tomcat API 构建为 `TomcatWebServer` Bean，由 Spring 应用上下文将其引导，其嵌入式 Tomcat 组件的运行（如 `Context`、`Connector` 等），以及 `ClassLoader` 的装载均由 Sprig Boot 框架代码实现。

> Spring Boot 1.x 嵌入式 Servlet 容器 Bean 为 `EmbeddedServletContainer` 实例，其中 Tomcat 的实现类为 `TomcatEmbeddedServletContainer`。

除此之外，Tomcat Maven 插件打包后的 JAR 或 WAR 文件属于非 FAT 模式，因为归档文件存在压缩的情况，然而 Spring Boot Maven 插件 `spring-boot-maven-plugin` 采用零压缩模式，将应用目录归档到 JAR 或 WAR 文件，相当于在 `jar` 命令归档的过程中添加`-0` 参数：

```
$ jar
用法: jar {ctxui}[vfmn0PMe] [jar-file] [manifest-file] [entry-point] [-C dir] files ...
选项:
(...部分内容被省略...)
    -0  仅存储; 不使用任何 zip 压缩
(...部分内容被省略...)
示例 1: 将两个类文件归档到一个名为 classes.jar 的档案中
       jar cvf classes.jar Foo.class Bar.class
示例 2: 使用现有的清单文件 mymanifest 并将 foo/目录中的所有文件归档到 classes.jar 中
       jar cvfm classes.jar mymanifest -C foo/ .
```

不过`-0` 参数在大多数场景中不会出现。

传统 Servlet 容器将压缩的 WAR 文件解压到对应目录，再加载该目录中的资源。而 Spring Boot 可执行 WAR 文件则需要在不解压当前 WAR 文件的前提下读取其中的资源，这就是为什么 `spring-boot-loader` 需要覆盖内建 `JAR` 协议的 `URLStreamHandler` 实现的原因。

不过遗憾的是，最新 Tomcat 官方维护的插件最高支持 Tomcat 7。而 Tomcat 8 Maven 插件则由外部社区维护，如以下 Maven GAV 所示。

```
<plugin>
    <groupId>org.apache.tomcat.maven</groupId>
    <artifactId>tomcat8-maven-plugin</artifactId>
    <version>3.0-r1655215</version>
    <executions>
        <execution>
            <id>tomcat-run</id>
            <goals>
                <!-- 最终打包成可执行的 JAR 包 -->
                <goal>exec-war-only</goal>
            </goals>
            <phase>package</phase>
            <configuration>
                <!-- ServletContext 路径 -->
                <path>/</path>
```

```
            </configuration>
        </execution>
    </executions>
</plugin>
```

该版本插件的导入需要增加外部 Maven 仓库：

```
<pluginRepositories>
    <pluginRepository>
        <!-- tomcat8-maven-plugin 所在仓库 -->
        <id>Alfresco</id>
        <name>Alfresco Repository</name>
        <url>https://artifacts.alfresco.com/nexus/content/repositories/public/</url>
        <snapshots>
            <enabled>false</enabled>
        </snapshots>
    </pluginRepository>
</pluginRepositories>
```

> 部分地区访问 Alfresco Maven 仓库不是非常稳定，可能需要借助 VPN 等工具来提高访问速度。

总而言之，Tomcat Maven 插件的版本支持 6～8 版本，对应的 Servlet 规范支持 2.5～3.1 版本。同时，从 Servlet 3.0 开始，Servlet 组件均能通过 `ServletContext` API 在运行时装配，如 `Servlet`、`Filter` 和 `Listener`，再结合 `ServletContainerInitializer` 生命周期回调，可实现 Servlet 组件的自动装配，Spring Framework 3.1 同样运用了这些特性，抽象出 `WebApplicationInitializer` 接口，降低 `ServletContainerInitializer` 接口的理解成本。在"**走向自动装配**"部分的"**Spring Web 自动装配**"章节中将有详细的讨论。

换言之，Tomcat 7+ Maven 插件能构建可执行 JAR 或 WAR 文件，实现独立的 Web 应用程序，也支持 Servlet 组件的自动装配，所以 Spring Boot 独立应用并非是嵌入式 Servlet 容器的"开山之祖"，而是站在巨人的肩膀上，加以简化和创新。

> 在本书的后续讨论中，读者将看到不少的事例证明 Spring Boot 与 Spring Framework 并非一流的技术，至少在设计哲学上，基本上参考了 JSR 规范设计和其他开源项目的实现。

除了 Tomcat，Jetty 也存在 Maven 插件，并且其 API 天然的可插拔性对于嵌入式 Servlet 容器来说更是"如虎添翼"，因此，Google 的 GAE（Google App Engine）将其 Servlet 容器从 Apache Tomcat 转向 Jetty。Webtide 团队在接受 InfoQ 采访时，曾提到（http://www.infoq.com/cn/news/2009/08/

google-chose-jetty）：

> **InfoQ**：Google App Engine 为何选择 Jetty 而放弃了 Tomcat，或是其他的服务器？
> 答：Google 之所以选择 Jetty，主要是看中了其大小和灵活性。在云中，服务器大小是非常重要的，因为你可能会在 10 秒内运行几千个 Jetty 实例（就像 Google 那样），如果每个服务器能省下 1MB，那么 10 秒内就会节省 GB 量级的内存（省下的内存可供应用使用），这个规模相当可观。
> Jetty 还具有可插拔和可扩展的特性，这样 Google 就能最大限度地对其进行定制了。他们已经插入了自己的 HTTP 连接器、Google 认证及 Session 集群了。这些特性说明 Jetty 不仅适合于云环境，对嵌入式设备如电话和机顶盒等同样适用。

想必 GAE 在迁移的过程中一定遇到了不小的麻烦，如果当时业界出现了 Spring Boot，或许难度会降低不少。

4.1.2 Jetty 作为嵌入式 Servlet Web 容器

将默认的嵌入式容器 Tomcat 切换至 Jetty 的步骤相当简单，Spring Boot 2.0 官方文档在“75.1 Use Another Web Server”章节介绍了切换方法：

```
<dependency>
<groupId>org.springframework.boot</groupId>
<artifactId>spring-boot-starter-web</artifactId>
<exclusions>
    <!-- Exclude the Tomcat dependency -->
    <exclusion>
        <groupId>org.springframework.boot</groupId>
        <artifactId>spring-boot-starter-tomcat</artifactId>
    </exclusion>
</exclusions>
</dependency>
<!-- Use Jetty instead -->
<dependency>
<groupId>org.springframework.boot</groupId>
<artifactId>spring-boot-starter-jetty</artifactId>
</dependency>
```

前面曾分析，`spring-boot-starter-tomcat` 为 `spring-boot-starter-web` 间接引入，故切

换 Jetty 时，需要排除该依赖，再添加新的依赖 `spring-boot-starter-jetty`。据此调整 `first-app-by-gui` 的嵌入式 Servlet 容器：

```
    <?xml version="1.0" encoding="UTF-8"?>
    <project xmlns="http://maven.apache.org/POM/4.0.0"
xmlns:xsi="http://www.w3.org/2001/XMLSchema-instance"
            xsi:schemaLocation="http://maven.apache.org/POM/4.0.0
http://maven.apache.org/xsd/maven-4.0.0.xsd">
        <modelVersion>4.0.0</modelVersion>

        <groupId>thinking-in-spring-boot</groupId>
        <artifactId>first-app-by-gui</artifactId>
        <version>0.0.1-SNAPSHOT</version>
        <packaging>war</packaging>

        <name>first-app-by-gui</name>
        ...
            <dependency>
                <groupId>org.springframework.boot</groupId>
                <artifactId>spring-boot-starter-web</artifactId>
                <exclusions>
                    <!-- Exclude the Tomcat dependency -->
                    <exclusion>
                        <groupId>org.springframework.boot</groupId>
                        <artifactId>spring-boot-starter-tomcat</artifactId>
                    </exclusion>
                </exclusions>
            </dependency>

            <!-- Use Jetty instead -->
            <dependency>
                <groupId>org.springframework.boot</groupId>
                <artifactId>spring-boot-starter-jetty</artifactId>
            </dependency>

            <!-- spring-boot-loader 用于源码分析 -->
            <dependency>
                <groupId>org.springframework.boot</groupId>
```

```
            <artifactId>spring-boot-loader</artifactId>
            <scope>provided</scope>
        </dependency>
    ...
</project>
```

再次使用 Maven 插件 `spring-boot` 引导当前应用：

```
$ mvn spring-boot:run
[INFO] Scanning for projects...
[INFO]
[INFO] ------------------------------------------------------------------------
[INFO] Building first-app-by-gui 0.0.1-SNAPSHOT
[INFO] ------------------------------------------------------------------------
(...部分内容被省略...)
  .   ____          _            __ _ _
 /\\ / ___'_ __ _ _(_)_ __  __ _ \ \ \ \
( ( )\___ | '_ | '_| | '_ \/ _` | \ \ \ \
 \\/  ___)| |_)| | | | | || (_| |  ) ) ) )
  '  |____| .__|_| |_|_| |_\__, | / / / /
 =========|_|==============|___/=/_/_/_/
 :: Spring Boot ::        (v2.0.2.RELEASE)

(...部分内容被省略...)
[           main] o.s.b.web.embedded.jetty.JettyWebServer  : Jetty started on port(s) 8080
(http/1.1) with context path '/'
[           main] t.f.FirstAppByGuiApplication             : Started
FirstAppByGuiApplication in 2.27 seconds (JVM running for 5.114)
```

命令执行后，`first-app-by-gui` 应用运行如故，不同的是容器切换至 Jetty。其中，`JettyWebServer` 就是 Spring Boot 结合 Jetty API 实现的 `org.springframework.boot.web.server.WebServer` Bean 实现。Spring Boot 官方文档在同章节中也提示开发人员，若将 Spring Boot Servlet 容器切换至 Undertow，则只需添加或替换依赖为 `spring-boot-starter-undertow` 即可：

> `spring-boot-starter-web` includes Tomcat by including `spring-boot-starter-tomcat`, but you can use `spring-boot-starter-jetty` or `spring-boot-starter-undertow` instead.

下面将讨论如何切换到 Undertow 嵌入式 Servlet 容器。

4.1.3 Undertow 作为嵌入式 Servlet Web 容器

将 first-app-by-gui 项目 pom.xml 文件中的 spring-boot-starter-jetty 替换为 spring-boot-starter-undertow：

```
<?xml version="1.0" encoding="UTF-8"?>
<project xmlns="http://maven.apache.org/POM/4.0.0"
xmlns:xsi="http://www.w3.org/2001/XMLSchema-instance"
         xsi:schemaLocation="http://maven.apache.org/POM/4.0.0
http://maven.apache.org/xsd/maven-4.0.0.xsd">
    <modelVersion>4.0.0</modelVersion>

    <groupId>thinking-in-spring-boot</groupId>
    <artifactId>first-app-by-gui</artifactId>
    <version>0.0.1-SNAPSHOT</version>
    <packaging>war</packaging>

    <name>first-app-by-gui</name>
    ...
    <dependencies>
        <dependency>
            <groupId>org.springframework.boot</groupId>
            <artifactId>spring-boot-starter-web</artifactId>
            <exclusions>
                <!-- Exclude the Tomcat dependency -->
                <exclusion>
                    <groupId>org.springframework.boot</groupId>
                    <artifactId>spring-boot-starter-tomcat</artifactId>
                </exclusion>
            </exclusions>
        </dependency>

        <!--&lt;!&ndash; Use Jetty instead &ndash;&gt;-->
        <!--<dependency>-->
            <!--<groupId>org.springframework.boot</groupId>-->
            <!--<artifactId>spring-boot-starter-jetty</artifactId>-->
```

```
        <!--</dependency>-->

        <!-- Use Undertow instead -->
        <dependency>
            <groupId>org.springframework.boot</groupId>
            <artifactId>spring-boot-starter-undertow</artifactId>
        </dependency>
        ...
    </dependencies>
    ...
</project>
```

使用 Maven 插件 `spring-boot` 引导 Undertow Web 应用：

```
z$ mvn spring-boot:run
[INFO] Scanning for projects...
[INFO]
[INFO] ------------------------------------------------------------------------
[INFO] Building first-app-by-gui 0.0.1-SNAPSHOT
[INFO] ------------------------------------------------------------------------
(...部分内容被省略...)
  .   ____          _            __ _ _
 /\\ / ___'_ __ _ _(_)_ __  __ _ \ \ \ \
( ( )\___ | '_ | '_| | '_ \/ _` | \ \ \ \
 \\/  ___)| |_)| | | | | || (_| |  ) ) ) )
  '  |____| .__|_| |_|_| |_\__, | / / / /
 =========|_|==============|___/=/_/_/_/
 :: Spring Boot ::        (v2.0.2.RELEASE)
(...部分内容被省略...)
[           main] o.s.b.w.e.u.UndertowServletWebServer     : Undertow started on port(s) 8080
(http) with context path ''
[           main] t.f.FirstAppByGuiApplication             : Started
FirstAppByGuiApplication in 2.056 seconds (JVM running for 4.825)
```

根据日志提示，`UndertowServletWebServer` 为 Spring Boot `WebServer` 的实现类，其类名模式与 Jetty 类似，不过它为什么要强调“Servlet”，是不是还有其他实现呢？答案是肯定的，其中 `UndertowWebServer` 是 Reactive Web 的实现类。接下来将继续讨论嵌入式 Reactive Web 容器。

4.2 嵌入式 Reactive Web 容器

嵌入式 Reactive Web 容器作为 Spring Boot 2.0 的新特性，通常处于被动激活状态，如增加 `spring-boot-starter-webflux` 依赖，然而当它与 `spring-boot-starter-web` 同时存在时，`spring-boot-starter-webflux` 会被忽略，这是由 `SpringApplication` 实现中的 Web 应用类型（`WebApplicationType`）推断逻辑决定的，详细的讨论请参考“**理解 SpringApplication**”部分中的“**推断 Web 应用类型**”。

4.2.1 UndertowServletWebServer 作为嵌入式 Reactive Web 容器

在 `first-app-by-gui` 项目的 `pom.xml` 文件中添加 `spring-boot-starter-webflux` 依赖：

```
    <?xml version="1.0" encoding="UTF-8"?>
    <project xmlns="http://maven.apache.org/POM/4.0.0"
xmlns:xsi="http://www.w3.org/2001/XMLSchema-instance"
         xsi:schemaLocation="http://maven.apache.org/POM/4.0.0
http://maven.apache.org/xsd/maven-4.0.0.xsd">
        <modelVersion>4.0.0</modelVersion>

        <groupId>thinking-in-spring-boot</groupId>
        <artifactId>first-app-by-gui</artifactId>
        <version>0.0.1-SNAPSHOT</version>
        <packaging>war</packaging>

        <name>first-app-by-gui</name>
        ...
        <dependencies>
            <dependency>
                <groupId>org.springframework.boot</groupId>
                <artifactId>spring-boot-starter-web</artifactId>
                <exclusions>
                    <!-- Exclude the Tomcat dependency -->
                    <exclusion>
                        <groupId>org.springframework.boot</groupId>
                        <artifactId>spring-boot-starter-tomcat</artifactId>
                    </exclusion>
                </exclusions>
```

```
        </dependency>

        <!--&lt;!&ndash; Use Jetty instead &ndash;&gt;-->
        <!--<dependency>-->
            <!--<groupId>org.springframework.boot</groupId>-->
            <!--<artifactId>spring-boot-starter-jetty</artifactId>-->
        <!--</dependency>-->

        <!-- Use Undertow instead -->
        <dependency>
            <groupId>org.springframework.boot</groupId>
            <artifactId>spring-boot-starter-undertow</artifactId>
        </dependency>

        <!-- WebFlux 依赖 -->
        <dependency>
            <groupId>org.springframework.boot</groupId>
            <artifactId>spring-boot-starter-webflux</artifactId>
        </dependency>
        ...
    </dependencies>
    ...
</project>
```

执行 spring-boot Maven 插件，观察控制台日志输出：

```
$ mvn spring-boot:run
[INFO] Scanning for projects...
[INFO]
[INFO] ------------------------------------------------------------------------
[INFO] Building first-app-by-gui 0.0.1-SNAPSHOT
[INFO] ------------------------------------------------------------------------
(...部分内容被省略...)
  .   ____          _            __ _ _
 /\\ / ___'_ __ _ _(_)_ __  __ _ \ \ \ \
( ( )\___ | '_ | '_| | '_ \/ _` | \ \ \ \
 \\/  ___)| |_)| | | | | || (_| |  ) ) ) )
  '  |____| .__|_| |_|_| |_\__, | / / / /
 =========|_|==============|___/=/_/_/_/
```

```
  :: Spring Boot ::        (v2.0.2.RELEASE)

 (...部分内容被省略...)
  [          main] o.s.b.w.servlet.FilterRegistrationBean   : Mapping filter:
'requestContextFilter' to: [/*]
  (...部分内容被省略...)
  [          main] o.s.b.w.e.u.UndertowServletWebServer     : Undertow started on port(s) 8080
(http) with context path ''
  [          main] t.f.FirstAppByGuiApplication             : Started
FirstAppByGuiApplication in 2.275 seconds (JVM running for 5.532)
```

日志显示 UndertowServletWebServer 仍旧作为 WebServer 实现，FilterRegistrationBean 同样说明了当前项目在 spring-boot-starter-web 与 spring-boot-starter-webflux 并存的情况下，后者被忽略的现象。

4.2.2 UndertowWebServer 作为嵌入式 Reactive Web 容器

将 spring-boot-starter-web 从 first-app-by-gui 项目的 pom.xml 文件依赖中移除（注释），即：

```
<?xml version="1.0" encoding="UTF-8"?>
<project xmlns="http://maven.apache.org/POM/4.0.0"
xmlns:xsi="http://www.w3.org/2001/XMLSchema-instance"
        xsi:schemaLocation="http://maven.apache.org/POM/4.0.0
http://maven.apache.org/xsd/maven-4.0.0.xsd">
    <modelVersion>4.0.0</modelVersion>

    <groupId>thinking-in-spring-boot</groupId>
    <artifactId>first-app-by-gui</artifactId>
    <version>0.0.1-SNAPSHOT</version>
    <packaging>war</packaging>

    <name>first-app-by-gui</name>
    ...
    <dependencies>
        <!--<dependency>-->
            <!--<groupId>org.springframework.boot</groupId>-->
            <!--<artifactId>spring-boot-starter-web</artifactId>-->
```

```
        <!--<exclusions>-->
            <!--&lt;!&ndash; Exclude the Tomcat dependency &ndash;&gt;-->
            <!--<exclusion>-->
                <!--<groupId>org.springframework.boot</groupId>-->
                <!--<artifactId>spring-boot-starter-tomcat</artifactId>-->
            <!--</exclusion>-->
        <!--</exclusions>-->
    <!--</dependency>-->

    <!--&lt;!&ndash; Use Jetty instead &ndash;&gt;-->
    <!--<dependency>-->
        <!--<groupId>org.springframework.boot</groupId>-->
        <!--<artifactId>spring-boot-starter-jetty</artifactId>-->
    <!--</dependency>-->

    <!-- Use Undertow instead -->
    <dependency>
        <groupId>org.springframework.boot</groupId>
        <artifactId>spring-boot-starter-undertow</artifactId>
    </dependency>

    <!-- WebFlux 依赖 -->
    <dependency>
        <groupId>org.springframework.boot</groupId>
        <artifactId>spring-boot-starter-webflux</artifactId>
    </dependency>
    ...
  </dependencies>
  ...
</project>
```

重新执行 `spring-boot` Maven 插件，观察日志输出变化：

```
$ mvn spring-boot:run
[INFO] Scanning for projects...
[INFO]
[INFO] ------------------------------------------------------------------------
[INFO] Building first-app-by-gui 0.0.1-SNAPSHOT
[INFO] ------------------------------------------------------------------------
```

```
(...部分内容被省略...)
  .   ____          _            __ _ _
 /\\ / ___'_ __ _ _(_)_ __  __ _ \ \ \ \
( ( )\___ | '_ | '_| | '_ \/ _` | \ \ \ \
 \\/  ___)| |_)| | | | | || (_| |  ) ) ) )
  '  |____| .__|_| |_|_| |_\__, | / / / /
 =========|_|==============|___/=/_/_/_/
 :: Spring Boot ::        (v2.0.2.RELEASE)

(...部分内容被省略...)
[           main] o.s.b.w.e.u.UndertowServletWebServer     : Undertow started on port(s) 8080
(http)
[           main] t.f.FirstAppByGuiApplication             : Started
FirstAppByGuiApplication in 1.793 seconds (JVM running for 4.713)
```

不难发现，即使移除 `spring-boot-starter-web`，`UndertowServletWebServer` 仍旧出现，不过 `FilterRegistrationBean` 的日志内容不复存在，单凭这一点似乎还不足以说明当前应用嵌入 Reactive Web 容器，因此，可向 `first-app-by-gui` 应用增加一些 WebFlux 函数式 Endpoint 代码：

```java
@SpringBootApplication
public class FirstAppByGuiApplication {

    public static void main(String[] args) {
        SpringApplication.run(FirstAppByGuiApplication.class, args);
    }

    @Bean
    public RouterFunction<ServerResponse> helloWorld() {
        return route(GET("/hello-world"),
                request -> ok().body(Mono.just("Hello,World"), String.class)
        );
    }
}
```

执行 `spring-boot` Maven 插件，观察日志输出变化：

```
$ mvn spring-boot:run
[INFO] Scanning for projects...
[INFO]
```

```
[INFO] ------------------------------------------------------------------------
[INFO] Building first-app-by-gui 0.0.1-SNAPSHOT
[INFO] ------------------------------------------------------------------------
(...部分内容被省略...)
[ERROR] COMPILATION ERROR :
[INFO] -------------------------------------------------------------
[ERROR]
/Users/mercyblitz/thinking-in-spring-boot-samples/first-app-by-gui/src/main/java/thinkinginsp
ringboot/firstappbygui/FirstAppByGuiApplication.java:[24,25] -source 1.5 中不支持 Lambda 表达式
  (请使用 -source 8 或更高版本以启用 Lambda 表达式)
[INFO] 1 error
[INFO] -------------------------------------------------------------
[INFO] ------------------------------------------------------------------------
[INFO] BUILD FAILURE
[INFO] ------------------------------------------------------------------------
(...部分内容被省略...)
```

以上代码会出现编译问题，因为 Maven 默认编译器插件 maven-compiler-plugin 采用的 Java 编译级别为 1.5，需要调整到 1.8。根据 spring-boot-dependencies 声明的 maven-compiler-plugin 版本信息，同步添加至 pom.xml 文件：

```
<?xml version="1.0" encoding="UTF-8"?>
<project xmlns="http://maven.apache.org/POM/4.0.0"
xmlns:xsi="http://www.w3.org/2001/XMLSchema-instance"
        xsi:schemaLocation="http://maven.apache.org/POM/4.0.0
http://maven.apache.org/xsd/maven-4.0.0.xsd">
    <modelVersion>4.0.0</modelVersion>

    <groupId>thinking-in-spring-boot</groupId>
    <artifactId>first-app-by-gui</artifactId>
    <version>0.0.1-SNAPSHOT</version>
    <packaging>war</packaging>

    <name>first-app-by-gui</name>
    <description>Demo project for Spring Boot</description>

    <properties>
        <project.build.sourceEncoding>UTF-8</project.build.sourceEncoding>
```

```
        <project.reporting.outputEncoding>UTF-8</project.reporting.outputEncoding>
        <java.version>1.8</java.version>
    </properties>
    ...
    <build>
        <plugins>
            <!-- 保持与 spring-boot-dependencies 版本一致 -->
            <plugin>
                <groupId>org.apache.maven.plugins</groupId>
                <artifactId>maven-compiler-plugin</artifactId>
                <version>3.7.0</version>
                <configuration>
                    <source>${java.version}</source>
                    <target>${java.version}</target>
                </configuration>
            </plugin>
            ...
        </plugins>
    </build>
</project>
```

运行 `spring-boot` Maven 插件，检验运行效果：

```
$ mvn spring-boot:run
[INFO] Scanning for projects...
[INFO]
[INFO] ------------------------------------------------------------------------
[INFO] Building first-app-by-gui 0.0.1-SNAPSHOT
[INFO] ------------------------------------------------------------------------
(...部分内容被省略...)
  .   ____          _            __ _ _
 /\\ / ___'_ __ _ _(_)_ __  __ _ \ \ \ \
( ( )\___ | '_ | '_| | '_ \/ _` | \ \ \ \
 \\/  ___)| |_)| | | | | || (_| |  ) ) ) )
  '  |____| .__|_| |_|_| |_\__, | / / / /
 =========|_|==============|___/=/_/_/_/
 :: Spring Boot ::        (v2.0.2.RELEASE)
(...部分内容被省略...)
[           main] o.s.w.r.f.s.s.RouterFunctionMapping      : Mapped (GET && /hello-world) ->
```

```
thinkinginspringboot.firstappbygui.FirstAppByGuiApplication$$Lambda$271/415984148@1d23e040
    (...部分内容被省略...)
    [           main] o.s.b.w.e.u.UndertowServletWebServer     : Undertow started on port(s) 8080
(http)
    [           main] t.f.FirstAppByGuiApplication             : Started
FirstAppByGuiApplication in 1.913 seconds (JVM running for 5.314)
```

新增的/hello-world 的 URI 映射已出现在日志中，说明已支持 WebFlux 函数式 Endpoint 的特性，再检验其运行结果：

```
    $ curl http://localhost:8080/hello-world
    Hello,World
```

综上讨论，当前应用属于 Spring Boot WebFlux 应用，并且使用基于 Servlet 技术所构建的 UndertowServletWebServer 实例作为嵌入式 Reactive Web 服务器。然而前面提到的 UndertowWebServer 为什么没有被创建呢？这是由 UndertowWebServer 对应的日志名称 Bug 导致的：

```
    public class UndertowWebServer implements WebServer {

     private static final Log logger = LogFactory.getLog(UndertowServletWebServer.class);
        ...
    }
```

那么如何证明该结论的正确性呢？Spring Boot 2.0 新引入了一种 ApplicationContext 实现实现 WebServerApplicationContext，它提供获取 WebServer 的接口方法 getWebServer()：

```
    public interface WebServerApplicationContext extends ApplicationContext {

     /**
      * Returns the {@link WebServer} that was created by the context or {@code null} if
      * the server has not yet been created.
      * @return the web server
      */
     WebServer getWebServer();
        ...
    }
```

该接口返回的 WebServer 实例需要在服务器中启动后才能获取。因此，无论 Servlet 实现 ServletWebServerApplicationContext，还是 Reactive 实现 ReactiveWebServerApplication

Context，只需注入 WebServerApplicationContext 对象，并且在 Spring Boot 应用启动后再输出其关联的 WebServer 实现类即可，改造如下：

```
@SpringBootApplication
public class FirstAppByGuiApplication {
    ...

    /**
     * {@link ApplicationRunner#run(ApplicationArguments)} 方法在
     * Spring Boot 应用启动后回调
     *
     * @param context WebServerApplicationContext
     * @return ApplicationRunner Bean
     */
    @Bean
    public ApplicationRunner runner(WebServerApplicationContext context) {
        return args -> {
            System.out.println("当前 WebServer 实现类为: "
                    + context.getWebServer().getClass().getName());
        };
    }
}
```

本例利用 ApplicationRunner Bean 的特性，其 run(ApplicationArguments)方法在 Spring Boot 应用启动后回调，故再次使用 spring-boot 插件运行当前项目：

```
$ mvn spring-boot:run
[INFO] Scanning for projects...
[INFO]
[INFO] ------------------------------------------------------------------------
[INFO] Building first-app-by-gui 0.0.1-SNAPSHOT
[INFO] ------------------------------------------------------------------------
(...部分内容被省略...)
  .   ____          _            __ _ _
 /\\ / ___'_ __ _ _(_)_ __  __ _ \ \ \ \
( ( )\___ | '_ | '_| | '_ \/ _` | \ \ \ \
 \\/  ___)| |_)| | | | | || (_| |  ) ) ) )
  '  |____| .__|_| |_|_| |_\__, | / / / /
 =========|_|==============|___/=/_/_/_/
```

```
 :: Spring Boot ::        (v2.0.2.RELEASE)
(...部分内容被省略...)
[           main] t.f.FirstAppByGuiApplication            : Started
FirstAppByGuiApplication in 1.789 seconds (JVM running for 4.604)
当前 WebServer 实现类为 org.springframework.boot.web.embedded.undertow.UndertowWebServer。
```

控制台输出日志显示 UndertowWebServer 为实际运行对象，换言之，当前 Spring Boot 2.0.2.RELEASE 存在 UndertowWebServer 的日志名称 Bug。

> 以上 Bug 是一种常见的低级失误，你我均有可能重蹈覆辙，可利用 Lombok @Log 注解来避免以上错误。

尽管当前应用已正常获取 UndertowWebServer 实例，然而其实现方式的健壮性并不完善，原因在于 ApplicationRunner @Bean 方法声明时，该方法依赖注入了 WebServerApplicationContext，这看似无可厚非，同时兼顾 Servlet 和 Reactive 两种 Web 场景，不过它却没有考虑非 Web 应用类型的场景。或许读者认为这有些吹毛求疵，然而小马哥认为，逻辑的严谨性是程序员必备的素养。为此，Spring Boot 在架构设计层面提供了兼容性更强的方案，即 Web 服务器已初始化事件——WebServerInitializedEvent。

4.2.3 WebServerInitializedEvent

在 Spring Boot 官方文档中，WebServerInitializedEvent 并没有直接被提到，而在“75.7 Discover the HTTP Port at Runtime”章节中，在介绍如何获取运行时的 HTTP 端口时提及 ServletWebServerInitializedEvent：

> You can access the port the server is running on from log output or from the ServletWebServerApplicationContext through its WebServer. The best way to get that and be sure that it has been initialized is to add a @Bean of type ApplicationListener<ServletWebServerInitializedEvent> and pull the container out of the event when it is published.

前半句描述的场景和前面讨论的类似，通过 ServletWebServerApplicationContext 获取 WebServer，再获取其端口。后半句指通过监听 ServletWebServerInitializedEvent 事件获取 WebServer。明显地，当前讨论仅限于 Servlet Web 场景，不过 ServletWebServerInitializedEvent 是 WebServerInitializedEvent 的子类。换言之，监听 WebServerInitializedEvent 事件所覆盖的场景更广。故将引导类 FirstAppByGuiApplication 加以改造，注释 ApplicationRunner 实

现，添加 WebServerInitializedEvent 事件监听：

```
@SpringBootApplication
public class FirstAppByGuiApplication {
    ...
    @EventListener(WebServerInitializedEvent.class)
    public void onWebServerReady(WebServerInitializedEvent event) {
        System.out.println("当前 WebServer 实现类为：" +
event.getWebServer().getClass().getName());
    }
}
```

再次使用 spring-boot 插件运行当前项目：

```
$ mvn spring-boot:run
[INFO] Scanning for projects...
[INFO]
[INFO] ------------------------------------------------------------------------
[INFO] Building first-app-by-gui 0.0.1-SNAPSHOT
[INFO] ------------------------------------------------------------------------
(...部分内容被省略...)
  .   ____          _            __ _ _
 /\\ / ___'_ __ _ _(_)_ __  __ _ \ \ \ \
( ( )\___ | '_ | '_| | '_ \/ _` | \ \ \ \
 \\/  ___)| |_)| | | | | || (_| |  ) ) ) )
  '  |____| .__|_| |_|_| |_\__, | / / / /
 =========|_|==============|___/=/_/_/_/
 :: Spring Boot ::        (v2.0.2.RELEASE)
(...部分内容被省略...)
当前 WebServer 实现类为 org.springframework.boot.web.embedded.undertow. UndertowWebServer。
[           main] t.f.FirstAppByGuiApplication             : Started
FirstAppByGuiApplication in 9.821 seconds (JVM running for 13.604)
```

日志照样输出 UndertowWebServer 的内容，不过相比之前的实现，此实现更具健壮性，即使在非 Web 应用中运行，也不至于注入 WebServerApplicationContext 失败。在此基础上，将 Reactive 容器切换至 Jetty 也不是问题。

4.2.4 Jetty 作为嵌入式 Reactive Web 容器

在 Spring Boot 官方文档的“75.1 Use Another Web Server”中，Tomcat、Jetty 和 Undertow 均能替换默认的 Netty Web Server 作为 Reactive Web 容器实现：

> `spring-boot-starter-webflux` includes Reactor Netty by including `spring-boot-starter-reactor-netty`, but you can use `spring-boot-starter-tomcat`, `spring-boot-starter-jetty`, or `spring-boot-starter-undertow` instead.

因此，将 `first-app-by-gui` 项目的 `pom.xml` 文件恢复 `spring-boot-starter-jetty` 依赖，注释 `spring-boot-starter-undertow` 依赖：

```xml
<?xml version="1.0" encoding="UTF-8"?>
<project xmlns="http://maven.apache.org/POM/4.0.0"
xmlns:xsi="http://www.w3.org/2001/XMLSchema-instance"
         xsi:schemaLocation="http://maven.apache.org/POM/4.0.0
http://maven.apache.org/xsd/maven-4.0.0.xsd">
    <modelVersion>4.0.0</modelVersion>

    <groupId>thinking-in-spring-boot</groupId>
    <artifactId>first-app-by-gui</artifactId>
    <version>0.0.1-SNAPSHOT</version>
    <packaging>war</packaging>
    ...
        <!-- Use Jetty instead -->
        <dependency>
            <groupId>org.springframework.boot</groupId>
            <artifactId>spring-boot-starter-jetty</artifactId>
        </dependency>

        <!--&lt;!&ndash; Use Undertow instead &ndash;&gt;-->
        <!--<dependency>-->
            <!--<groupId>org.springframework.boot</groupId>-->
            <!--<artifactId>spring-boot-starter-undertow</artifactId>-->
        <!--</dependency>-->
    ...
</project>
```

重新执行 `spring-boot` 插件，运行当前应用：

```
$ mvn spring-boot:run
(...部分内容被省略...)
[           main] o.s.w.r.f.s.s.RouterFunctionMapping      : Mapped (GET && /hello-world) ->
thinkinginspringboot.firstappbygui.FirstAppByGuiApplication$$Lambda$255/1886579719@566fd0d5
(...部分内容被省略...)
当前 WebServer 实现类为 org.springframework.boot.web.embedded.jetty. JettyWebServer。
[           main] t.f.FirstAppByGuiApplication             : Started
FirstAppByGuiApplication in 2.786 seconds (JVM running for 5.905)
```

当前应用切换至 `spring-boot-starter-jetty` 后，`/hello-world` URI 也映射正常，并且控制台显示当前 WebServer 实现类为 `JettyWebServer`。若非仔细观察 Spring Boot 文档，很难发现 `spring-boot-starter-jetty` 支持 Reactive Web 的特性，`spring-boot-starter-tomcat` 也是如此。

4.2.5 Tomcat 作为嵌入式 Reactive Web 容器

将 `spring-boot-starter-tomcat` 依赖添加至 `first-app-by-gui` 项目的 `pom. xml` 文件中，并注释 `spring-boot-starter-jetty` 依赖：

```xml
<?xml version="1.0" encoding="UTF-8"?>
<project xmlns="http://maven.apache.org/POM/4.0.0"
xmlns:xsi="http://www.w3.org/2001/XMLSchema-instance"
         xsi:schemaLocation="http://maven.apache.org/POM/4.0.0
http://maven.apache.org/xsd/maven-4.0.0.xsd">
    <modelVersion>4.0.0</modelVersion>

    <groupId>thinking-in-spring-boot</groupId>
    <artifactId>first-app-by-gui</artifactId>
    <version>0.0.1-SNAPSHOT</version>
    <packaging>war</packaging>
    ...
        <!--&lt;!&ndash; Use Jetty instead &ndash;&gt;-->
        <!--<dependency>-->
            <!--<groupId>org.springframework.boot</groupId>-->
            <!--<artifactId>spring-boot-starter-jetty</artifactId>-->
        <!--</dependency>-->

        <!--&lt;!&ndash; Use Undertow instead &ndash;&gt;-->
        <!--<dependency>-->
            <!--<groupId>org.springframework.boot</groupId>-->
```

```
            <!--<artifactId>spring-boot-starter-undertow</artifactId>-->
        <!--</dependency>-->

        <dependency>
            <groupId>org.springframework.boot</groupId>
            <artifactId>spring-boot-starter-tomcat</artifactId>
        </dependency>
    ...
</project>
```

重新执行 spring-boot 插件，运行当前应用：

```
$ mvn spring-boot:run
(...部分内容被省略...)
[           main] o.s.w.r.f.s.s.RouterFunctionMapping      : Mapped (GET && /hello-world) ->
thinkinginspringboot.firstappbygui.FirstAppByGuiApplication$$Lambda$276/708609190@7af17431
(...部分内容被省略...)
当前 WebServer 实现类为 org.springframework.boot.web.embedded.tomcat. TomcatWebServer。
[           main] t.f.FirstAppByGuiApplication             : Started
FirstAppByGuiApplication in 2.119 seconds (JVM running for 5.071)
```

同样地，/hello-world URI 映射正确，并且 WebServer 实现类为 TomcatWebServer。由此，将以上三种嵌入式 Web 容器进行总结，如下表所示。

容　器	Maven 依赖	WebServer 实现类
Tomcat	spring-boot-starter-tomcat	TomcatWebServer
Jetty	spring-boot-starter-jetty	JettyWebServer
Undertow	spring-boot-starter-undertow	UndertowWebServer

以上三种嵌入式 Web 容器的 Maven artifactId 在名称上存在一定的规律，它们均使用字符 spring-boot-starter-作为前缀，官方称之为“starter”。当这些“starter”添加到应用的 Class Path 中时，其关联的特性随应用的启动而自动地装载，这种机制在 Spring Boot 中称为“Automatically configure”，有些资料将其译作“自动配置”，小马哥偏好翻译为“自动装配”，这也是 Spring Boot 最大的亮点之一。

第5章 理解自动装配

Spring Boot 官方文档在“16. Auto-configuration”章节简单地介绍了自动装配特性：

> Spring Boot auto-configuration attempts to automatically configure your Spring application based on the jar dependencies that you have added.

按照该说明，自动装配是存在前提的，它取决于开发人员在应用的 Class Path 下添加的 JAR 文件依赖，同时其自动装配的实体并非一定装载，所以文档使用了“attempts”（尝试）来进行描述。不过，当前尚未说明所载之物为何？故文档举例说明：

> For example, if `HSQLDB` is on your classpath, and you have not manually configured any database connection beans, then Spring Boot auto-configures an in-memory database.

当 `HSQLDB` 存在于应用的 Class Path 中时，开发人员不需要手动配置数据库连接的 Beans，而是由 Spring Boot 自动装配一个内存型的数据库。不过官方如此轻描淡写地描述自动装配的示例，恐怕让人产生疑惑。能够理解的是，Spring Boot 自动装配的对象是 Spring Bean，所以不需要人工干预，比如通过 XML 配置文件或 Java 编码等方式组装 Bean。显然，官方认为开发人员应该知道 `HSQLDB`。同样地，它也认为 Spring Boot 用户熟悉 Spring Framework。

> 针对官方文档言之不详或描述模糊的部分，本书将做出详细的讨论和说明，并且对官方未公布的内容加以说明。

接下来，官方文档继续介绍激活自动装配的方法：

> You need to opt-in to auto-configuration by adding the `@EnableAutoConfiguration` or `@SpringBootApplication` annotations to one of your `@Configuration` classes.

确切地说，文档仅提到激活自动化装配的注解`@EnableAutoConfiguration`和`@SpringBootApplication`，将两者选其一标注在`@Configuration`类上，然而并没有说明如何装配`@Configuration`类。如果开发人员熟悉Spring Framework，那么或许会想到常见三种的方法，如XML元素`<context:component-scan>`、注解`@Import`和`@ComponentScan`，而这三种装配手段又需要Spring应用上下文引导，前者可采用`ClassPathXmlApplicationContext`加载，后两者需要`AnnotationConfigApplicationContext`注册。假设读者不熟悉这些细节，按照常理，官方文档应该介绍Spring Boot的引导办法，即通过`SpringApplication`实现，如`first-app-by-gui`项目中的引导类`FirstAppByGuiApplication`：

```
@SpringBootApplication
public class FirstAppByGuiApplication {

    public static void main(String[] args) {
        SpringApplication.run(FirstAppByGuiApplication.class, args);
    }
    ...
}
```

根据该应用的运行结果来看，WebFlux和嵌入式Web容器均已自动装配，那么是否可认为当前引导类`FirstAppByGuiApplication`充当了`@Configuration`类的角色呢？答案是肯定的，不过需要对`@SpringBootApplication`注解的语义进一步进行分析。

5.1 理解@SpringBootApplication注解语义

关于`@SpringBootApplication`注解语义，Spring Boot官方文档在“18. Using the @SpringBootApplication Annotation”中做出了解释：

> Many Spring Boot developers like their apps to use auto-configuration, component scan and be able to define extra configuration on their "application class". A single `@SpringBootApplication` annotation can be used to enable those three features, that is:
>
> - `@EnableAutoConfiguration`: enable Spring Boot's auto-configuration mechanism
> - `@ComponentScan`: enable `@Component` scan on the package where the application is located (see the best practices)
> - `@Configuration`: allow to register extra beans in the context or import additional configuration classes

`@SpringBootApplication` 被用于激活`@EnableAutoConfiguration`、`@ComponentScan` 和 `@Configuration` 三个注解的特性。其中，`@EnableAutoConfiguration` 负责激活 Spring Boot 自动装配机制，`@ComponentScan` 激活`@Component` 的扫描，`@Configuration` 声明被标注为配置类。官方文档继续告诉开发人员 `@SpringBootApplication` 注解等同于 `@Configuration`、`@EnableAutoConfiguration` 和`@ComponentScan` 注解，且它们均使用默认属性，并举例说明两者的等价关系：

> The `@SpringBootApplication` annotation is equivalent to using `@Configuration`, `@EnableAutoConfiguration`, and `@ComponentScan` with their default attributes, as shown in the following example:
>
> ```java
> package com.example.myapplication;
>
> import org.springframework.boot.SpringApplication;
> import org.springframework.boot.autoconfigure.SpringBootApplication;
>
> @SpringBootApplication // same as @Configuration @EnableAutoConfiguration @ComponentScan
> public class Application {
>
>    public static void main(String[] args) {
>         SpringApplication.run(Application.class, args);
>    }
>
> }
> ```

据此重构 `first-app-by-gui` 项目中的引导类 `FirstAppByGuiApplication`，将 `@SpringBootApplication` 注解替换为三注解声明方式：

```
@Configuration
@ComponentScan
@EnableAutoConfiguration
public class FirstAppByGuiApplication {
    ...
}
```

运行 spring-boot 插件，观察日志变化：

```
$ mvn spring-boot:run
[INFO] Scanning for projects...
[INFO]
[INFO] ------------------------------------------------------------------------
[INFO] Building first-app-by-gui 0.0.1-SNAPSHOT
[INFO] ------------------------------------------------------------------------
(...部分内容被省略...)
  .   ____          _            __ _ _
 /\\ / ___'_ __ _ _(_)_ __  __ _ \ \ \ \
( ( )\___ | '_ | '_| | '_ \/ _` | \ \ \ \
 \\/  ___)| |_)| | | | | || (_| |  ) ) ) )
  '  |____| .__|_| |_|_| |_\__, | / / / /
 =========|_|==============|___/=/_/_/_/
 :: Spring Boot ::        (v2.0.2.RELEASE)
(...部分内容被省略...)
[           main] o.s.w.r.f.s.s.RouterFunctionMapping      : Mapped (GET && /hello-world) ->
thinkinginspringboot.firstappbygui.FirstAppByGuiApplication$$Lambda$276/1472678237@64757ea
(...部分内容被省略...)
[           main] t.f.FirstAppByGuiApplication             : Started
FirstAppByGuiApplication in 2.193 seconds (JVM running for 6.231)
```

日志提示一切运行如故，再发送 GET 请求到/hello-world，检验是否正常运作：

```
$ curl http://127.0.0.1:8080/hello-world
Hello,World
```

请求结果同样符合期望，说明改造后的 first-app-by-gui 与之前的并无差别，进而证明文档的描述无误。然而实际的情况并非如此单一，以上 Spring Boot 2.0 官方文档的描述基本上与 Spring Boot 1.3 版本中的内容无异。

> 请读者参考 Spring Boot 1.3 文档“18. Using the @SpringBootApplication annotation”章节的描述，文档地址：https://docs.spring.io/spring-boot/docs/1.3.x/reference/htmlsingle/#using-boot-using-springbootapplication-annotation。

换言之，2.0 版本的文档并没有真实地反映实现情况，以 Spring Boot `2.0.2.RELEASE` 实现为例，`@SpringBootApplication` 注解的声明如下：

```
@Target(ElementType.TYPE)
@Retention(RetentionPolicy.RUNTIME)
@Documented
@Inherited
@SpringBootConfiguration
@EnableAutoConfiguration
@ComponentScan(excludeFilters = {
        @Filter(type = FilterType.CUSTOM, classes = TypeExcludeFilter.class),
        @Filter(type = FilterType.CUSTOM, classes = AutoConfigurationExcludeFilter.class) })
public @interface SpringBootApplication {
    ...
}
```

实际上，`@SpringBootApplication` 等价于`@EnableAutoConfiguration`、`@ComponentScan` 和`@SpringBootConfiguration`，不过`@ComponentScan` 并非使用了默认值，而是添加了排除的 `TypeFilter` 实现：`TypeExcludeFilter` 和 `AutoConfigurationExcludeFilter`。前者由 Spring Boot 1.4 引入，用于查找 `BeanFactory` 中已注册的 `TypeExcludeFilter` Bean，作为代理执行对象：

```
public class TypeExcludeFilter implements TypeFilter, BeanFactoryAware {
    ...
 @Override
 public boolean match(MetadataReader metadataReader,
         MetadataReaderFactory metadataReaderFactory) throws IOException {
     if (this.beanFactory instanceof ListableBeanFactory
             && getClass() == TypeExcludeFilter.class) {
         Collection<TypeExcludeFilter> delegates = ((ListableBeanFactory)
this.beanFactory)
                 .getBeansOfType(TypeExcludeFilter.class).values();
         for (TypeExcludeFilter delegate : delegates) {
             if (delegate.match(metadataReader, metadataReaderFactory)) {
                 return true;
```

```
                }
            }
        }
        return false;
    }
    ...
}
```

而后者从 Spring Boot 1.5 开始支持，用于排除其他同时标注`@Configuration` 和 `@EnableAutoConfiguration` 的类：

```
public class AutoConfigurationExcludeFilter implements TypeFilter, BeanClassLoaderAware {
    ...
    @Override
    public boolean match(MetadataReader metadataReader,
            MetadataReaderFactory metadataReaderFactory) throws IOException {
        return isConfiguration(metadataReader) && isAutoConfiguration(metadataReader);
    }

    private boolean isConfiguration(MetadataReader metadataReader) {
        return metadataReader.getAnnotationMetadata()
                .isAnnotated(Configuration.class.getName());
    }

    private boolean isAutoConfiguration(MetadataReader metadataReader) {
        return getAutoConfigurations()
                .contains(metadataReader.getClassMetadata().getClassName());
    }

    protected List<String> getAutoConfigurations() {
        if (this.autoConfigurations == null) {
            this.autoConfigurations = SpringFactoriesLoader.loadFactoryNames(
                    EnableAutoConfiguration.class, this.beanClassLoader);
        }
        return this.autoConfigurations;
    }

}
```

对比 Spring Boot 1.3.8.RELEASE @SpringBootApplication 的声明：

```
@Target(ElementType.TYPE)
@Retention(RetentionPolicy.RUNTIME)
@Documented
@Inherited
@Configuration
@EnableAutoConfiguration
@ComponentScan
public @interface SpringBootApplication {
    ...
}
```

该实现与官方文档的描述保持一致。尽管 Spring Boot 1.4 之后的@SpringBootApplication 通常不会表现出与文档相异的行为，然而这也给我们敲响了警钟，文档仅供参考，实践更需谨慎，尤其在升级 Spring Boot 的过程中，发现版本之间的实现差异是非常困难的。不过在后续的讨论中，本书将尽可能地讨论实现行为与文档描述不一致的情况。

值得关注的是，从 Spring Boot 1.4 开始，@SpringBootApplication 注解不再标注 @Configuraion，而是@SpringBootConfiguration ，不过两者在运行上的行为无异，这种类似于对象之间的继承关系，小马哥称之为“多层次 @Component ‘派生性’”，并且这种能力也允许开发人员扩展使用。简言之，@Configuraion 注解上标注了@Component：

```
@Target(ElementType.TYPE)
@Retention(RetentionPolicy.RUNTIME)
@Documented
@Component
public @interface Configuration {
    ...
}
```

@Configuration 实际上是@Component 的“派生”注解，同理，@SpringBootConfiguration 标注了@Configuration：

```
@Target(ElementType.TYPE)
@Retention(RetentionPolicy.RUNTIME)
@Documented
@Configuration
public @interface SpringBootConfiguration {
```

```
}
```

因此，三者之间的层次关系如下：

- @Component
    - @Configuration
        - @SpringBootConfiguration

尽管@ComponentScan 仅关注于@Component，然而由于 @SpringBootConfiguration 属于多层次的@Component“派生”注解，所以能够被@CompoentScan 识别。逆向思考，@CompoentScan 属于 Spring Framework，而@SpringBootConfiguration 来自 Spring Boot，那么必然存在一种机制让@CompoentScan 能够识别 Spring Boot 的注解@SpringBootConfiguration，这种机制就是前面提到的“多层次@Component‘派生性’”。以此类推，@Repository、@Service 和@Controller 均属于@Component“派生”注解，不过它们是直接“派生”，官方称它们为“Spring 模式注解（Stereotype Annotations）”，关于它们深入的讨论将在“**走向自动化装配**”部分的“**Spring 模式注解（Stereotype Annotations）**”章节展开，包括如何理解@Component“派生性”、自定义@Component“派生”注解和多层次@Component“派生性”等内容。

> Spring 模式注解（Stereotype Annotations）的官方描述记录在 GitHub Wiki ["Spring Annotation Programming Model"]（https://github.com/spring-projects/spring-framework/wiki/Spring-Annotation-Programming-Model）。

除此之外，Spring Boot 官方文档简单地提到@SpringBootApplication 注解提供了属性别名，用于自定义@EnableAutoConfiguration 和@ComponentScan 的属性：

> @SpringBootApplication also provides aliases to customize the attributes of @EnableAutoConfiguration and @ComponentScan.

明显地，官方再次认为开发人员熟悉 Spring 注解属性别名的细节，并且文档也没有实例说明。下面将展开讨论。

5.2 @SpringBootApplication 属性别名

前面的讨论并未展示@SpringBootApplication 注解的全貌，在此补充其属性方法声明：

```
public @interface SpringBootApplication {
```

```
    ...
    @AliasFor(annotation = EnableAutoConfiguration.class)
    Class<?>[] exclude() default {};

    /**
     * ...
     * @since 1.3.0
     */
    @AliasFor(annotation = EnableAutoConfiguration.class)
    String[] excludeName() default {};

    /**
     * ...
     * @since 1.3.0
     */
    @AliasFor(annotation = ComponentScan.class, attribute = "basePackages")
    String[] scanBasePackages() default {};

    /**
     * ...
     * @since 1.3.0
     */
    @AliasFor(annotation = ComponentScan.class, attribute = "basePackageClasses")
    Class<?>[] scanBasePackageClasses() default {};

}
```

了解其属性方法声明后，想必多少可以理解官方文档的描述，以上所有属性方法均标注 `@AliasFor` 注解，顾名思义，它用于桥接其他注解的属性。需要特别注意的是，`@SpringBootApplication` 大多数属性方法从 Spring Boot 1.3 才开始出现，同时，该注解从 Spring Boot 1.2 就已经引入，主要的原因在于`@AliasFor` 注解从 Spring Framework 4.2 才予以支持，而 Spring Boot 1.2 对应的 Spring Framework 版本是 4.1。

> 更多的 Spring Boot 版本信息请参考“理解 Spring Boot 版本”一节。

明显地，`@AliasFor` 注解能够将一个或多个注解的属性“别名”在某个注解中。因此，可再次重构 `first-app-by-gui` 项目中的引导类 `FirstAppByGuiApplication`，将其包名从 `thinkinginspringboot.firstappbygui` 调整为 `thinking.in.spring.boot.firstappbygui`。

再将 RouterFunction 的@Bean 定义与@EventListener 方法抽取至 thinking.in.spring.boot.config.WebConfiguration：

```
package thinking.in.spring.boot.config;
...
@Configuration
public class WebConfiguration {

    @Bean
    public RouterFunction<ServerResponse> helloWorld() {
        return route(GET("/hello-world"),
                request -> ok().body(Mono.just("Hello,World"), String.class)
        );
    }

    @EventListener(WebServerInitializedEvent.class)
    public void onWebServerReady(WebServerInitializedEvent event) {
        System.out.println("当前 WebServer 实现类为：" + 
event.getWebServer().getClass().getName());
    }
}
```

此时，引导类 FirstAppByGuiApplication 与@Configuration 类 WebConfiguration 各自存放在 thinking.in.spring.boot.firstappbygui 和 thinking.in.spring.boot.config 包下，且在 FirstAppByGuiApplication 中再次使用@SpringBootApplication，并将 scanBasePackages 属性设置为 thinking.in.spring.boot.config：

```
package thinking.in.spring.boot.firstappbygui;
...
//@Configuration
//@ComponentScan
//@EnableAutoConfiguration
@SpringBootApplication(scanBasePackages = "thinking.in.spring.boot.config")
public class FirstAppByGuiApplication {

    public static void main(String[] args) {
        SpringApplication.run(FirstAppByGuiApplication.class, args);
    }
    ...
```

```
}
```

如果/hello-world URI 映射工作正常，并且当前 WebServer 实现类也没有发生变化，那么说明@SpringBootApplication 利用@AliasFor 注解别名了@ComponentScan 注解的 basePackages()属性。再次执行 spring-boot 插件，观察控制台日志：

```
$ mvn spring-boot:run
[INFO] Scanning for projects...
[INFO]
[INFO] ------------------------------------------------------------------------
[INFO] Building first-app-by-gui 0.0.1-SNAPSHOT
[INFO] ------------------------------------------------------------------------
(...部分内容被省略...)
  .   ____          _            __ _ _
 /\\ / ___'_ __ _ _(_)_ __  __ _ \ \ \ \
( ( )\___ | '_ | '_| | '_ \/ _` | \ \ \ \
 \\/  ___)| |_)| | | | | || (_| |  ) ) ) )
  '  |____| .__|_| |_|_| |_\__, | / / / /
 =========|_|==============|___/=/_/_/_/
 :: Spring Boot ::        (v2.0.2.RELEASE)

(...部分内容被省略...)
[           main] o.s.w.r.f.s.s.RouterFunctionMapping      : Mapped (GET && /hello-world) ->
thinking.in.spring.boot.config.WebConfiguration$$Lambda$276/ 453786037@7c9da249
(...部分内容被省略...)
当前 WebServer 实现类为 org.springframework.boot.web.embedded.tomcat. TomcatWebServer。
[           main] t.i.s.b.f.FirstAppByGuiApplication       : Started
FirstAppByGuiApplication in 2.712 seconds (JVM running for 7.413)
```

单从日志上观察，组件装配和运行正常，再次访问/hello-world URI：

```
$ curl http://127.0.0.1:8080/hello-world
Hello,World
```

综合对比来看，一切运作如故。同样地，@AliasFor 的注解编程模型也被 Spring 官方 Wiki“Spring Annotation Programming Model”记录，其中“Attribute Aliases and Overrides”一节解释了注解属性别名和覆盖的规则，为此，在“**走向自动装配**”部分，特意安排“**Spring 注解属性别名和覆盖（Attribute Aliases and Overrides）**”一节从实现的角度分析其规则的语义。

综上所述，@SpringBootApplication 是一个聚合注解，包含@ComponentScan、@Configuration 和@EnableAutoConfiguration 的核心特性，类似的注解如 @RestController 等。通常@SpringBootApplication 标注在引导类上，然而并不限制于此。

5.3 @SpringBootApplication 标注非引导类

将当前 first-app-by-gui 项目的引导类 FirstAppByGuiApplication 中的@SpringBootApplication 移动到 WebConfiguration 类，由于@SpringBootApplication 包含@Configuration 的特性，故将其注释。同时，WebConfiguration 存放于 thinking.in.spring.boot.config 包下，故@SpringBootApplication 的 scanBasePackages()属性没有配置的必要，调整如下：

```
//@Configuration
@SpringBootApplication
public class WebConfiguration {
    ...
}
```

FirstAppByGuiApplication 不再包含任何 Spring 注解，因此将 SpringApplication#run(Class,String...)方法的首个参数调整为 WebConfiguration：

```
public class FirstAppByGuiApplication {

    public static void main(String[] args) {
        SpringApplication.run(WebConfiguration.class, args);
    }
}
```

运行 spring-boot 插件，检验重构效果：

```
$ mvn spring-boot:run
[INFO] Scanning for projects...
[INFO]
[INFO] ------------------------------------------------------------------------
[INFO] Building first-app-by-gui 0.0.1-SNAPSHOT
[INFO] ------------------------------------------------------------------------
(...部分内容被省略...)
  .   ____          _            __ _ _
 /\\ / ___'_ __ _ _(_)_ __  __ _ \ \ \ \
```

```
( ( )\___ | '_ | '_| | '_ \/ _` | \ \ \ \
 \\/  ___)| |_)| | | | | || (_| |  ) ) ) )
  '  |____| .__|_| |_|_| |_\__, | / / / /
 =========|_|==============|___/=/_/_/_/
 :: Spring Boot ::        (v2.0.2.RELEASE)
(...部分内容被省略...)
[           main] o.s.w.r.f.s.s.RouterFunctionMapping       : Mapped (GET && /hello-world) -> thinking.in.spring.boot.config.WebConfiguration$$Lambda$276/1854704090@3e51438d
(...部分内容被省略...)
当前 WebServer 实现类为 org.springframework.boot.web.embedded.tomcat.TomcatWebServer。
[           main] t.i.s.b.f.FirstAppByGuiApplication        : Started FirstAppByGuiApplication in 2.208 seconds (JVM running for 6.633)
```

访问/hello-world URI，检查运行是否正常：

```
$ curl http://127.0.0.1:8080/hello-world
Hello,World
```

程序运行正常，由此证明@SpringBootApplication 并非限定标注于引导类。Spring Boot 官方文档提到@SpringBootApplication 和@EnableAutoConfiguration 均能激活自动装配的特性，假设将 WebConfiguration 的注解@SpringBootApplication 替换为@EnableAutoConfiguration，其类中声明的@Bean 还能正常工作吗？

5.4 @EnableAutoConfiguration 激活自动装配

按照前面的描述，调整 WebConfiguration 类，如下所示。

```
@EnableAutoConfiguration
public class WebConfiguration {

    @Bean
    public RouterFunction<ServerResponse> helloWorld() {
        return route(GET("/hello-world"),
                request -> ok().body(Mono.just("Hello,World"), String.class)
        );
    }

    @EventListener(WebServerInitializedEvent.class)
    public void onWebServerReady(WebServerInitializedEvent event) {
```

```
        System.out.println("当前 WebServer 实现类为: " +
event.getWebServer().getClass().getName());
    }
}
```

发现 WebConfiguration 不再是@Configuration 类，因为@EnableAutoConfiguration 并非@Configuration 类的“派生”注解，继续使用 spring-boot 插件，观察日志的变化：

```
$ mvn spring-boot:run
[INFO] Scanning for projects...
[INFO] ------------------------------------------------------------------------
[INFO] Building first-app-by-gui 0.0.1-SNAPSHOT
[INFO] ------------------------------------------------------------------------
(...部分内容被省略...)
  .   ____          _            __ _ _
 /\\ / ___'_ __ _ _(_)_ __  __ _ \ \ \ \
( ( )\___ | '_ | '_| | '_ \/ _` | \ \ \ \
 \\/  ___)| |_)| | | | | || (_| |  ) ) ) )
  '  |____| .__|_| |_|_| |_\__, | / / / /
 =========|_|==============|___/=/_/_/_/
 :: Spring Boot ::        (v2.0.2.RELEASE)

(...部分内容被省略...)
[           main] o.s.w.r.f.s.s.RouterFunctionMapping      : Mapped (GET && /hello-world) ->
thinking.in.spring.boot.config.WebConfiguration$$Lambda$276/ 510612818@a0ee1af
(...部分内容被省略...)
当前 WebServer 实现类为 org.springframework.boot.web.embedded.tomcat.TomcatWebServer。
[           main] t.i.s.b.f.FirstAppByGuiApplication       : Started
FirstAppByGuiApplication in 2.387 seconds (JVM running for 7.125)
```

再执行 curl 命令检查/hello-world URI：

```
$ curl http://127.0.0.1:8080/hello-world
Hello,World
```

综合两项结果，发现非@Configuration 类 WebConfiguration 在当前项目行为上与改造前无异，说明 SpringApplication#run(Class,String...)方法引导 Spring Boot 应用时，并不强依赖于@Configuration 类作为首参，该参数称为“首要配置源”—— primarySource，更深层次的讨论请参考“**理解 SpringApplication**”部分的“**理解 SpringApplication 主配置类**”一节。不过，

本例的运行结果可能会误导读者，尽管@EnableAutoConfiguration 与@SpringBootApplication 在激活自动装配方面是没有差别的，然而对于被标注类的 Bean 类型则存在差异。

5.5 @SpringBootApplication“继承”@Configuration CGLIB 提升特性

@SpringBootApplication 作为@Configuration 的“派生”注解，同样继承其注解特性，其中最明显的是 CGLIB 提升。Spring Framework 5.0 官方文档在“1.10.5. Defining bean metadata within components”章节中写道：

> The difference is that @Component classes are not enhanced with CGLIB to intercept the invocation of methods and fields. CGLIB proxying is the means by which invoking methods or fields within @Bean methods in @Configuration classes creates bean metadata references to collaborating objects; such methods are *not* invoked with normal Java semantics but rather go through the container in order to provide the usual lifecycle management and proxying of Spring beans even when referring to other beans via programmatic calls to @Bean methods. In contrast, invoking a method or field in an @Bean method within a plain @Component class *has* standard Java semantics, with no special CGLIB processing or other constraints applying.

此段文字描述是对比@Bean 在@Component 类与@Configuration 类中的差异，前者 Bean 的行为与正常的 Java 对象语义相同，不存在 CGLIB 处理，而后者则执行了 CGLIB 提升。同时@Bean 在普通 Java 类被声明后，当该声明类注册到注解驱动的 Spring 应用上下文中（如 AnnotationConfigApplicationContext）时，其 Bean 对象的行为与在@Component 类下的声明一致，这一点在“1.12.1. Basic concepts: @Bean and @Configuration”一节的附录中有相关说明：

> Full @Configuration vs 'lite' @Bean mode?
> When @Bean methods are declared within classes that are *not* annotated with @Configuration they are referred to as being processed in a 'lite' mode. Bean methods declared in a @Component or even in a *plain old class* will be considered 'lite', with a different primary purpose of the containing class and an @Bean method just being a sort of bonus there.

官方称这种@Bean 的声明方式为“轻量模式”(**Lite**)，相反，在@Configuration 下声明的@Bean 则属于“完全模式”(**Full**)，后者会执行 CGLIB 提升的操作，其中实现层面的讨论请查看“**走向自动化配置**”部分的“**装载@Configuration Class**”一节。由此，可为 first-app-by-gui 项目增加 Bean 的类型输出，如下所示。

```
//@Configuration
//@SpringBootApplication
@EnableAutoConfiguration
public class WebConfiguration {

    @Bean
    public RouterFunction<ServerResponse> helloWorld() {
        ...
    }
    ...
    @Bean
    public ApplicationRunner runner(BeanFactory beanFactory) {
        return args -> {
            System.out.println("当前 helloWorld Bean 实现类为："
                    + beanFactory.getBean("helloWorld").getClass().getName());

            System.out.println("当前 WebConfiguration Bean 实现类为："
                    + beanFactory.getBean(WebConfiguration.class).getClass().getName());
        };
    }
}
```

运行 spring-boot Maven 插件：

```
$ mvn spring-boot:run
[INFO] Scanning for projects...
(...部分内容被省略...)
当前 helloWorld Bean 的实现类为 org.springframework.web.reactive.function.server.
RouterFunctions$DefaultRouterFunction。
当前 WebConfiguration Bean 的实现类为 thinking.in.spring.boot.config. WebConfiguration。
```

明显地发现，当前 WebConfiguration 类为非@Configuration 类，因此，两个 Bean 均属于“**Lite**”模式，其类型为原有的 Java 类型。将其重新调回@SpringBootApplication 标注，观察其日志变化：

```
//@Configuration
@SpringBootApplication
//@EnableAutoConfiguration
public class WebConfiguration {
    ...
}
```

执行 spring-boot Maven 插件：

```
$ mvn spring-boot:run
[INFO] Scanning for projects...
(...部分内容被省略...)
```

当前 helloWorld Bean 的实现类为 org.springframework.web.reactive.function.server.RouterFunctions$DefaultRouterFunction。

当前 WebConfiguration Bean 的实现类为 thinking.in.spring.boot.config.WebConfiguration$$EnhancerBySpringCGLIB$$8a57c306。

从此运行结果不难发现，所谓 CGLIB 提升并非是为@Bean 对象提供的，而是为@Configuration 类准备的。

不过 WebConfiguration 属于编码方式的导入式装配，而非自动装配。相反，其他自动装配的 Bean 肯定由某种机制完成。这种机制就是自动配置机制。

5.6 理解自动配置机制

在 Spring Boot 出现之前，Spring Framework 提供 Bean 生命周期管理和 Spring 编程模型。在框架层面，它支持注解的“派生”或扩展，然而无法自动地装配@Configuraion 类。为此，Spring Boot 1.0 在 Spring Framework 4.0 的基础上，添加了约定配置化导入@Configuration 类的方式。

Spring Boot 2.0 官方文档在“46. Creating Your Own Auto-configuration”章节中指导开发人员如何构建自己的自动装配类：

> If you work in a company that develops shared libraries, or if you work on an open-source or commercial library, you might want to develop your own auto-configuration. Auto-configuration classes can be bundled in external jars and still be picked-up by Spring Boot.
> Auto-configuration can be associated to a “starter” that provides the auto-configuration code as well as the typical libraries that you would use with it. We first cover what you need to know to build

> your own auto-configuration and then we move on to the typical steps required to create a custom starter.

文中指出，自动装配类能够打包到外部的 JAR 文件中，并且将被 Spring Boot 装载。同时，自动装配也能被关联到“starter”中，这些“starter”提供自动装配的代码及关联的依赖。关于如何自定义专业化 Spring Boot Starter 的讨论，将安排在“**走向自动装配**”部分的“**自定义 Spring Boot 自动装配**”章节中。

在“46.1 Understanding Auto-configured Beans”中，文档继续介绍 Spring Boot 自动装配底层实现与 Spring Framework 注解`@Configuration`和`@Conditional`的联系：

> Under the hood, auto-configuration is implemented with standard `@Configuration` classes. Additional `@Conditional` annotations are used to constrain when the auto-configuration should apply. Usually, auto-configuration classes use `@ConditionalOnClass` and `@ConditionalOnMissingBean` annotations. This ensures that auto-configuration applies only when relevant classes are found and when you have not declared your own `@Configuration`.

文中提到的`@ConditionalOnClass`和`@ConditionalOnMissingBean`是最常见的注解，顾名思义，当`@ConditionalOnClass`标注在`@Configuration`类上时，当且仅当目标类存在于 Class Path 下时才予以装配，这也就是为什么官方文档提到当 `HSQLDB` 存在于应用的 Class Path 下时，开发人员无须手动配置数据库连接的 Beans。明显地，Spring Boot 框架提供了装配 `HSQLDB` 的逻辑，其中`EmbeddedDatabaseConnection`为嵌入式数据连接枚举：

```
public enum EmbeddedDatabaseConnection {
    ...
  /**
   * HSQL Database Connection.
   */
  HSQL(EmbeddedDatabaseType.HSQL, "org.hsqldb.jdbcDriver", "jdbc:hsqldb:mem:%s");
    ...
    public static EmbeddedDatabaseConnection get(ClassLoader classLoader) {
      for (EmbeddedDatabaseConnection candidate : EmbeddedDatabaseConnection.values()) {
            if (candidate != NONE && ClassUtils.isPresent(candidate.getDriverClassName(),
                    classLoader)) {
                return candidate;
            }
      }
      return NONE;
```

```
    }
    ...
}
```

而 get(ClassLoader)方法用于 DataSourceAutoConfiguration.EmbeddedDatabaseCondition 实现：

```
@Configuration
@ConditionalOnClass({ DataSource.class, EmbeddedDatabaseType.class })
...
public class DataSourceAutoConfiguration {

 @Configuration
 @Conditional(EmbeddedDatabaseCondition.class)
 @ConditionalOnMissingBean({ DataSource.class, XADataSource.class })
 @Import(EmbeddedDataSourceConfiguration.class)
 protected static class EmbeddedDatabaseConfiguration {

 }
    ...
static class EmbeddedDatabaseCondition extends SpringBootCondition {
        ...
     @Override
     public ConditionOutcome getMatchOutcome(ConditionContext context,
             AnnotatedTypeMetadata metadata) {
          ...
         EmbeddedDatabaseType type = EmbeddedDatabaseConnection
                 .get(context.getClassLoader()).getType();
         if (type == null) {
             return ConditionOutcome
                     .noMatch(message.didNotFind("embedded database").atAll());
         }
         return ConditionOutcome.match(message.found("embedded database").items(type));
     }

}

}
```

当 DataSourceAutoConfiguration.EmbeddedDatabaseCondition 匹配后，即 HSQLDB 的 JDBC 驱动类 org.hsqldb.jdbcDriver 存在于 Class Path 下时，@Configuration 类 DataSourceAutoConfiguration.EmbeddedDatabaseConfiguration 将被装配，故 EmbeddedDataSourceConfiguration 将被@Import 注解导入：

```
@Configuration
@EnableConfigurationProperties(DataSourceProperties.class)
public class EmbeddedDataSourceConfiguration implements BeanClassLoaderAware {
    ...
    @Bean
    public EmbeddedDatabase dataSource() {
        EmbeddedDatabaseBuilder builder = new EmbeddedDatabaseBuilder()
                .setType(EmbeddedDatabaseConnection.get(this.classLoader).getType())
                .setName(this.properties.determineDatabaseName());
        this.database = builder.build();
        return this.database;
    }
    ...
}
```

不难看出，DataSourceAutoConfiguration 和嵌套@Configuration 类同样综合了@Configuration 类和@Conditional 的特性，不过嵌套@Configuration 类件随着 DataSourceAutoConfiguration 的装配而装配，那么问题则回到 DataSourceAutoConfiguration 是如何被装配的。部分答案同样在 Spring Boot 官方文档中有描述，如“46.1 Understanding Auto-configured Beans”章节：

> You can browse the source code of spring-boot-autoconfigure to see the @Configuration classes that Spring provides (see theMETA-INF/spring.factories file).

文中的 spring-boot-autoconfigure 是 Spring Boot 核心模块（JAR），其中提供了大量的内建自动装配@Configuration 类，它们统一存放在 org.springframework. boot.autoconfigure 包或子包下，比如 DataSourceAutoConfiguration 的包名为 org.springframework.boot.autoconfigure.jdbc。同时，这些类均配置在 META-INF/spring.factories 资源中：

```
...
# Auto Configure
org.springframework.boot.autoconfigure.EnableAutoConfiguration=\
org.springframework.boot.autoconfigure.admin.SpringApplicationAdminJmxAutoConfiguration,\
```

```
org.springframework.boot.autoconfigure.aop.AopAutoConfiguration,\
org.springframework.boot.autoconfigure.amqp.RabbitAutoConfiguration,\
...
```

如果开发人员需要自定义自动装配类，那么文档在“46.2 Locating Auto-configuration Candidates”章节继续补充：

> Spring Boot checks for the presence of a `META-INF/spring.factories` file within your published jar. The file should list your configuration classes under the `EnableAutoConfiguration` key, as shown in the following example:
>
> ```
> org.springframework.boot.autoconfigure.EnableAutoConfiguration=\
> com.mycorp.libx.autoconfigure.LibXAutoConfiguration,\
> com.mycorp.libx.autoconfigure.LibXWebAutoConfiguration
> ```

不难发现，`META-INF/spring.factories` 属于 Java `Properties` 文件格式，其中激活自动装配注解`@EnableAutoConfiguration` 充当该 `Properties` 的 Key，而自动装配类为 Value。换言之，无论内建自动装配类，还是自定义的，均可采用该模式配置。值得注意的是，它们的类命名均以 `AutoConfiguration` 作为后缀。接下来将予以实践。

5.7 创建自动配置类

当前 `first-app-by-gui` 项目中的引导类 `FirstAppByGuiApplication` 采用的是直接导入 `WebConfiguration` 的方式：

```
public class FirstAppByGuiApplication {

    public static void main(String[] args) {
        SpringApplication.run(WebConfiguration.class, args);
    }
    ...
}
```

此刻需将 `FirstAppByGuiApplication` 标注 `@EnableAutoConfiguration`，并作为 `SpringApplication#run(Class,String...)`方法的首参：

```
@EnableAutoConfiguration
public class FirstAppByGuiApplication {
```

```
    public static void main(String[] args) {
        SpringApplication.run(FirstAppByGuiApplication.class, args);
    }
    ...
}
```

已知标注@SpringBootApplication 的 WebConfiguration 是@Configuration 类，又激活了自动装配，现在让它仅作为@Configuration 类：

```
@Configuration
public class WebConfiguration {
    ...
}
```

创建自动装配类 WebAutoConfiguration，并使用@Import 导入 WebConfiguration：

```
package thinking.in.spring.boot.autoconfigure;
...
@Configuration
@Import(WebConfiguration.class)
public class WebAutoConfiguration {
}
```

在项目 src/main/resources 的目录下新建 META-INF/spring.factories 资源，并配置 WebAutoConfiguration 类：

```
# 自动装配
org.springframework.boot.autoconfigure.EnableAutoConfiguration=\
thinking.in.spring.boot.autoconfigure.WebAutoConfiguration
```

重新执行 spring-boot 插件，从控制台日志观察 WebAutoConfiguration 是否实现自动装配：

```
$ mvn spring-boot:run
[INFO] Scanning for projects...
(...部分内容被省略...)
o.s.w.r.f.s.s.RouterFunctionMapping      : Mapped (GET && /hello-world) ->
thinking.in.spring.boot.config.WebConfiguration$$Lambda$276/1148773630@560a1a59
(...部分内容被省略...)
当前 WebServer 实现类为 org.springframework.boot.web.embedded.tomcat. TomcatWebServer。
t.i.s.b.f.FirstAppByGuiApplication       : Started FirstAppByGuiApplication in 2.153 seconds
(JVM running for 6.076)
```

当前 helloWorld Bean 的实现类为 org.springframework.web.reactive.function.server.RouterFunctions$DefaultRouterFunction。

当前 WebConfiguration Bean 的实现类为 thinking.in.spring.boot.config.WebConfiguration$$EnhancerBySpringCGLIB$$d09b16e8。

不难发现，改造后的 first-app-by-gui 与改造前无异，而且说明 WebAutoConfiguration 已被自动装配。至此，一个简单的自定义自动装配类就实现了，其/hello-world Web 服务也正常工作：

```
$ curl http://127.0.0.1:8080/hello-world
Hello,World
```

在以上 Spring Boot 应用中，不再依赖 XML 配置，完全使用注解驱动进行 Spring Bean 的组装，这样解析注解所带来的时间成本直接影响了应用的启动速度。因此，Spring Framework 5.0 又引入了新注解@Indexed，在代码编译时，向@Component 和“派生”注解增添索引，从而减少运行时性能消耗。

不难发现，Spring Boot 自动装配特性与 Spring Framework 之间存在密切的联系，那么自动装配到底是如何从中发展而来的呢？以及 Spring Framework 为此又付出了哪些努力？其中原委将在“**走向自动装配**”部分“娓娓道来”。不可否认的是，随着自动装配及 Starter 的引入，大量的 Spring Bean 的装配变成黑盒，并且搭配@Conditional 注解之后，情况变得更加扑朔迷离。那么，开发人员需要一套运维的方法，了解 Bean 的组装情况，甚至是其他应用相关的信息。同时，为了支持以配置化的方式调整应用行为，如 Web 服务端口等，Spring Boot 提供了 Production-Ready 特性。

第6章 理解 Production-Ready 特性

回顾官方首页中对 Production-Ready 特性的描述：

> Provide production-ready features such as metrics, health checks and externalized configuration.

从字面意思来翻译“**production-ready features**”仍然不太好理解。什么叫“为生产准备的特性”？请读者注意，后续的文案中有如下说明：**such as metrics, health checks and externalized configuration**。也就是说，metrics（指标）、health checks（健康检查）和 externalized configuration（外部配置）均属于为生产准备的特性。

作为 Spring Boot 核心特性之一，如此扼要的说明必然让人产生疑惑，不过 Production-Ready 概念达到了先入为主的效果。为此，小马哥也查询过不少关于“Production-Ready”名词解释的资料，如从 Google 中搜索关键字“Production-Ready”，惊奇地发现搜索结果的首页并没有出现于任何与 Spring 相关的内容。

其中前三搜索结果分别是：

- Define "production-ready" - Software Engineering Stack Exchange
- What does it mean when code is 'production ready'? - Quora

- What does it mean to be "production ready"? · Issue #3 · mitodl…

6.1 理解 Production-Ready 一般性定义

按照上述次序，首先讨论的是“Define "production-ready" - Software Engineering Stack Exchange”。

1. Production-Ready 在 Stack Exchange 讨论中的定义

Stack Exchange 用户为“Production-Ready”的意思感到困惑：

> “I have been curious about this for a while. What exactly is meant by "production-ready" or its variants?”

其中得票数最高的答复是：

> Programmer's definition of "production-ready":
> - it runs
> - it satisfies the project requirements
> - its design was well thought out
> - it's stable
> - it's maintainable
> - it's scalable
> - it's documented
>
> Management's definition of "production-ready":
> - it runs
> - it'll turn a profit

此处对于“production-ready”的定义不同，程序员强调产品职能和维护，管理人员则关注产品的收益。不过两者均属于软件工程（Software Engineering）的范畴，明显与 Spring Boot 所定义的“Production-Ready”存在较大的差异。因此，再来分析“Quora”所给出的定义。

2. Production-Ready 在“Quora”讨论中的定义

同样，有人在“Quora”上讨论代码何时才算“Prodution Ready”及它的含义：

> What does it mean when code is "production ready"?

Google 软件工程师 Paul K. Young 的回复是：

> “Production ready” means that the author of the code believes it is ready to run in a production environment ...
> 原文地址：https://www.quora.com/What-does-it-mean-when-code-is-production-ready。

他的答复偏重于代码方面，同样与 Spring Boot 所定义的“Production-Ready”相去甚远。于是再来分析排名第三的讨论。

3．Production-Ready 在 mitodl handbook 讨论中的定义

GitHub 用户 justinabrahms 向 bdero 请教关于“production ready”的意思：

> I'm assuming we have standards for what "production ready" entails. I'd further assume that there are things that devs should know about, like taking configuration information in as env vars. Hoping someone could maybe write up some guidelines on what this means.

而 bdero 的回答就与 Spring Boot“Producation Ready”相关了：

> the high level description of what I see as "production ready" from an deployment/ops standpoint is this: http://12factor.net/

原来 Production-Ready 是 DevOps 的立足点，来自 http://12factor.net/，在该 issue 的结尾，bdero 在该项目工程中给出了“Production ready”的定义，其中部分 checklists 与 Spring Boot 的说明重合，比如与外部化配置基本吻合的检查项有：

- Does it support configuration via environment variables?
- Does it support configuration via flatfile (e.g. YAML)?
- Does it have a sane default configuration (e.g. as close to a realistic production configuration as possible)?

与 Health Checks 相关的有：

- Does it have a status route (which requires a secret token) that can be used for health checking by a monitoring service?
- Does the health check validate that each service the app depends on is reachable (e.g. databases, caching, external APIs when necessary)?
- Does the health check return a response body with the availability information above in a machine-readable format in all cases?

- Does the health check return a 500 status code when any component in the health check is detected to be failing?

与日志（Loggers）相关的有：

- Does it log to standard out/error?
- Does it log to syslog?
- Can it send stack trace emails to a configurable email list?
- Does it have configurable log levels (e.g. DEBUG, ERROR, WARNING, INFO)?

与应用监控相关的有：

- Does it have application/resource monitoring in production
- Does it have service availability monitoring in production

bdero 认为当应用满足以上特性和架构时，才可认为是“Production Ready”的应用。

> 以上原文链接：https://github.com/mitodl/handbook/blob/master/production-ready.md。

若将 bdero 的文章与 Spring Boot“Production-Ready”概念直接关联起来似乎有些牵强，该文章首次提交日期为 2015 年 12 月 1 日，而 Spring Boot 在这之前就已提出概念。而该文章开门见山地表达其“Production Ready”的定义遵循 Twelve-Factor App：

> At the application scope, we adhere closely to Heroku's Twelve-Factor App guidelines as a baseline for production readiness.

相信熟悉 Spring Cloud 的读者自然对 12factor 不会感到陌生，因为它也是 Spring Cloud 的理论基础：

> Cloud Native is a style of application development that encourages easy adoption of best practices in the areas of continuous delivery and value-driven development. A related discipline is that of building 12-factor Applications

因此，Spring Boot“Production-Ready”的概念就与 12-factor Applications 产生了某种联结。这也就是为什么几乎 Spring Cloud 组件均扩展 Spring Boot Actuator 特性的原因。尽管在不同的技术领域对于同一名词存在各执一词的情况，Spring Boot Actuator 也可以视为将“Production-Ready”的概念具体化。类似地，Spring `Environment` 抽象与外部化配置（Externalized Configuration）的关系亦是如此。

如果仅从 Spring Boot 官网首页的文字说明来理解“Production-Ready”，那么显然是片面的，Spring Boot 官方从 1.0.0 文档开始为“Production-Ready”特性独立成章，下一节将尝试从官方的文字描述中挖掘更深层次的含义。

6.2 理解 Spring Boot Actuator

从 Spring Boot 1.0.0 开始，“Production-Ready”特性在官方文档“Part V. Spring Boot Actuator: Production-ready features”章节中进行讨论，开篇言简意赅地表达了其作用：

> Spring Boot includes a number of additional features to help you monitor and manage your application when you push it to production. You can choose to manage and monitor your application by using HTTP endpoints or with JMX. Auditing, health, and metrics gathering can also be automatically applied to your application.
> Spring Boot 1.0.x 中使用的标题略有差异，即“Part V. Production-ready features”。

大致上可将以上描述总结为以下几点。

- 使用场景：监视和管理（下文省略为“监管”）投入生产的应用。
- 监管媒介：HTTP 或 JMX 端点（Endpoints）。
- 端点类型：审计（Auditing）、健康（Health）和指标收集（metrics gathering）。
- 基本特点：自动运用（automatically applied）。

由此观之，Spring Boot Production-Ready 特性至少使用了 Java Web 和 JMX 两种技术，并暗示开发人员“自动运用（automatically applied）”是基于 Spring Boot 自动装配实现的。

为此，Spring Boot 引入了 Actuator 实现，并对其做出如下定义：

> **Definition of Actuator**
> An actuator is a manufacturing term that refers to a mechanical device for moving or controlling something. Actuators can generate a large amount of motion from a small change.

“Actuator”的中文意思为“制动器”或“传动装置”，如官方说明的那样，它属于制造业术语，是机器移动装备或某些控制部件，通过细小的调整，能够产生大量的动能。官方特别说明的意图或许让开发人员理解 Spring Boot Actuator 类似于传统 Actuator，能够做到“四两拨千斤”。同时，所谓“兼听则明，偏听则暗”，官方说明在“民间”又是怎么解读的呢？

Spring Boot Actuator 民间解读

https://www.baeldung.com 是一个偏好 Spring 技术的博客网站，其中博主 José Carlos Valero Sánchez 就职于 Spring 母公司 Pivotal，他发表了一篇名为“Spring Boot Actuator”的文章，在文中“**2. What is an Actuator?**”部分写道：

> **Monitoring our app, gathering metrics, understanding traffic or the state of our database becomes trivial with this dependency.**
> The main benefit of this library is that we can get production grade tools without having to actually implement these features ourselves.
> Actuator is mainly used to **expose operational information about the running application** – health, metrics, info, dump, env, etc. It uses HTTP endpoints or JMX beans to enable us to interact with it.
> Once this dependency is on the classpath several endpoints are available for us out of the box. As with most Spring modules, we can easily configure or extend it in many ways.

一言以蔽之，José Carlos Valero Sánchez 认为 Spring Boot Actuator 用于监控和管理 Spring 应用，可通过 HTTP Endpoint 或 JMX Bean 与其交互。下一节将结合代码理解 JMX 与 HTTP 两种 Endpoints。

6.3 Spring Boot Actuator Endpoints

Spring Boot Actuator 并非默认存在，而需要手动地添加依赖至项目 `pom.xml` 文件，如 Spring Boot 2.0 官方文档的“49. Enabling Production-ready Features”章节所示。

> To add the actuator to a Maven based project, add the following ‘Starter’ dependency:
```xml
<dependencies>
   <dependency>
        <groupId>org.springframework.boot</groupId>
        <artifactId>spring-boot-starter-actuator</artifactId>
   </dependency>
</dependencies>
```

同时，“50. Endpoints”章节列举了所有内建的 Endpoints，比如常用的 Endpoints。

- `beans`：显示当前 Spring 应用上下文的 Spring Bean 完整列表（包含所有 `ApplicationContext` 的层次）。
- `conditions`：显示当前应用所有配置类和自动装配类的条件评估结果（包含匹配和非匹配）。

- env：暴露 Spring ConfigurableEnvironment 中的 PropertySource 属性。
- health：显示应用的健康信息。
- info：显示任意的应用信息。

由“50.2 Exposing Endpoints”章节可知，仅有 health 和 info 为默认暴露的 Web Endpoints。如果需要暴露其他 Endpoints，则可增加 management.endpoints.web.exposure.include=*的配置属性到 application.properties 或启动参数中，这类属性配置源称为“外部化配置”，也是后面所讨论的内容。

故先增加 Spring Boot Actuator 依赖到 first-app-by-gui 项目的 pom.xml 文件中：

```
<?xml version="1.0" encoding="UTF-8"?>
<project xmlns="http://maven.apache.org/POM/4.0.0"
xmlns:xsi="http://www.w3.org/2001/XMLSchema-instance"
         xsi:schemaLocation="http://maven.apache.org/POM/4.0.0
http://maven.apache.org/xsd/maven-4.0.0.xsd">
    <modelVersion>4.0.0</modelVersion>

    <groupId>thinking-in-spring-boot</groupId>
    <artifactId>first-app-by-gui</artifactId>
    <version>0.0.1-SNAPSHOT</version>
    <packaging>war</packaging>

    <name>first-app-by-gui</name>
    <description>Demo project for Spring Boot</description>
    ...
    <dependencies>
        ...
        <!-- Spring Boot Actuator 依赖 -->
        <dependency>
            <groupId>org.springframework.boot</groupId>
            <artifactId>spring-boot-starter-actuator</artifactId>
        </dependency>
        ...
    </dependencies>
    ...
</project>
```

再在 spring-boot Maven 插件的启动参数中添加配置属性，暴露 beans 为 Web Endpoint：

```
$ mvn spring-boot:run -Dmanagement.endpoints.web.exposure.include=beans
[INFO] Scanning for projects...
(...部分内容被省略...)
[           main] .b.a.e.w.r.WebFluxEndpointHandlerMapping : Mapped
"{[/actuator/beans],methods=[GET],produces=[application/vnd.spring-boot.actuator.v2+json ||
application/json]}" onto public
org.reactivestreams.Publisher<org.springframework.http.ResponseEntity<java.lang.Object>>
org.springframework.boot.actuate.endpoint.web.reactive.AbstractWebFluxEndpointHandlerMapping$
ReadOperationHandler.handle(org.springframework.web.server.ServerWebExchange)
(...部分内容被省略...)
```

此时 beans 的 Web URI 相对于默认路径/actuator，访问 beans Web Endpoint，即执行$ curl http://127.0.0.1:8080/actuator/beans | json_pp：

```
{
   "contexts" : {
      "application" : {
         "beans" : {
            ...
            "org.springframework.boot.autoconfigure.web.reactive. WebFluxAutoConfiguration" :
{
               "dependencies" : [],
               "aliases" : [],
               "scope" : "singleton",
               "type" :
"org.springframework.boot.autoconfigure.web.reactive.WebFluxAutoConfiguration$$EnhancerBySpri
ngCGLIB$$861ddcdf",
               "resource" : null
            },
            ...
            "firstAppByGuiApplication" : {
               "scope" : "singleton",
               "aliases" : [],
               "dependencies" : [],
               "resource" : null,
               "type" : "thinking.in.spring.boot.firstappbygui.FirstAppByGuiApplication"
            },
            ...
         }
```

```
        }
    }
}
```

其中，自动装配类 WebFluxAutoConfiguration 与引导类 FirstAppByGuiApplication 均被注册为 Spring Bean。同时，向 WebAutoConfiguration 添加 Web 应用的前置条件注解 ConditionalOnWebApplication：

```
@ConditionalOnWebApplication
@Configuration
@Import(WebConfiguration.class)
public class WebAutoConfiguration {
}
```

再次向 spring-boot Maven 插件添加配置属性，暴露 conditions 和 env 的 Web Endpoint：

```
$ mvn spring-boot:run -Dmanagement.endpoints.web.exposure.include=beans,conditions,env
[INFO] Scanning for projects...
(...部分内容被省略...)
[           main] o.s.b.a.e.web.EndpointLinksResolver      : Exposing 3 endpoint(s) beneath
base path '/actuator'
[           main] .b.a.e.w.r.WebFluxEndpointHandlerMapping : Mapped "{[/actuator/beans],...
[           main] .b.a.e.w.r.WebFluxEndpointHandlerMapping : Mapped
"{[/actuator/conditions],...
[           main] .b.a.e.w.r.WebFluxEndpointHandlerMapping : Mapped "{[/actuator/env],...
(...部分内容被省略...)
```

同样地，访问 conditions Web Endpoint——$ curl http://127.0.0.1:8080/actuator/conditions | json_pp：

```
{
   "contexts" : {
      "application" : {
         "positiveMatches" : {
            ...
            "WebFluxAutoConfiguration" : [
               {
                  "message" : "@ConditionalOnClass found required class
'org.springframework.web.reactive.config.WebFluxConfigurer'; @ConditionalOnMissingClass did not
find unwanted class",
```

```
                    "condition" : "OnClassCondition"
                  },
                  {
                    "message" : "found ReactiveWebApplicationContext",
                    "condition" : "OnWebApplicationCondition"
                  },
                  {
                    "condition" : "OnBeanCondition",
                    "message" : "@ConditionalOnMissingBean (types:
org.springframework.web.reactive.config.WebFluxConfigurationSupport; SearchStrategy: all) did
not find any beans"
                  }
                ],
                ...
              }
              ...
            }
          }
        }
```

其中，positiveMatches 表示正向条件匹配，所以 WebFluxAutoConfiguration 能够被装配，/hello-world Web 服务运行如故：

```
$ curl http://127.0.0.1:8080/hello-world
Hello,World
```

同时，属性配置 management.endpoints.web.exposure.include=beans,conditions,env 也出现在 env Web Endpoint 中——$ curl http://127.0.0.1:8080/actuator/env/management.endpoints.web.exposure.include | json_pp：

```
{
  "activeProfiles" : [],
  "property" : {
    "value" : "beans,conditions,env",
    "source" : "systemProperties"
  },
  "propertySources" : [
    {
      "name" : "server.ports"
```

```
        },
        {
            "property" : {
                "value" : "beans,conditions,env"
            },
            "name" : "systemProperties"
        },
        {
            "name" : "systemEnvironment"
        },
        {
            "name" : "random"
        },
        {
            "name" : "Management Server"
        }
    ]
}
```

在该 JSON 响应内容中，`source` 为 `systemProperties`，表示属性配置 `management.endpoints.web.exposure.include` 来源于 Java 系统属性，同样可通过 `spring-boot` Maven 插件命令行 `-D` 参数执行，并且其内容为 `beans,conditions,env`。

当然以上示例仅从现象说明内建 Endpoints 的作用，更多关于 Spring Boot Actuator 的深度探讨将在下一册的**“简化 Spring 应用运维体系”**中展开，届时将全面解读 Spring Boot Actuator 发展历程。

6.4 理解“外部化配置”

在 Spring Boot 官方参考文档的**“24. Externalized Configuration”**章节中，并没有直接解释什么是**“Externalized Configuration”**，而是列出了在 Spring Boot 场景下如何使配置外部化（**to externalize configuration**）的手段：

> Spring Boot allows you to externalize your configuration so you can work with the same application code in different environments. You can use properties files, YAML files, environment variables and command-line arguments to externalize configuration.
>
> 以上文字来源：https://docs.spring.io/spring-boot/docs/2.0.2.RELEASE/reference/htmlsingle/#boot-features-external-config。

结合上文，官方试图告诉开发人员，相同的应用代码可以根据所处的环境，差别化地使用外部的配置来源，可能通过 Properties 文件、YAML 文件、环境变量或命令行参数实现。文档后续补充说明：

> Property values can be injected directly into your beans by using the `@Value annotation`, accessed through Spring's `Environment` abstraction or bound to structured objects through `@ConfigurationProperties`.

“Property values”是指外部化配置属性值，并且能在 Spring Boot 场景中提供三种用途：

- Bean 的 **`@Value`** 注入；
- Spring **`Environment`** 读取；
- **`@ConfigurationProperties`** 绑定到结构化对象。

既然“外部化配置”属性能被消费，那么自然会有它的生产来源。Spring Boot 官方文档继续说明：

> Spring Boot uses a very particular `PropertySource` order that is designed to allow sensible overriding of values. Properties are considered in the following order:
>
> 1. Devtools global settings properties on your home directory (`~/.spring-boot-devtools.properties` when devtools is active).
> 2. @TestPropertySource annotations on your tests.
> 3. @SpringBootTest#properties annotation attribute on your tests.
> 4. Command line arguments.
> 5. Properties from `SPRING_APPLICATION_JSON` (inline JSON embedded in an environment variable or system property).
> 6. `ServletConfig` init parameters.
> 7. `ServletContext` init parameters.
> 8. JNDI attributes from `java:comp/env`.
> 9. Java System properties (`System.getProperties()`).
> 10. OS environment variables.
> 11. A `RandomValuePropertySource` that has properties only in `random.*`.
> 12. Profile-specific application properties outside of your packaged jar (`application-{profile}.properties` and YAML variants).
> 13. Profile-specific application properties packaged inside your jar (`application-{profile}.properties` and YAML variants).

> 14. Application properties outside of your packaged jar (`application.properties` and YAML variants).
> 15. Application properties packaged inside your jar (`application.properties` and YAML variants).
> 16. @PropertySource annotations on your `@Configuration` classes.
> 17. Default properties (specified by setting `SpringApplication.setDefaultProperties`).

从字面意思理解，`PropertySource` 为属性源，从而暗示开发人员它是“外部化配置”属性的生产源，并且 Spring Boot 规定了这 17 种内建的 `PropertySource` 顺序。不过，就目前所了解的 Spring Boot 知识，唯一有印象的恐怕是 `application.properties`。显然，它也是 `PropertySource` 的一种。

于是，Spring Boot 官方文档“‘码’文并茂”，解释它们之间的联系：

```
import org.springframework.stereotype.*;
import org.springframework.beans.factory.annotation.*;

@Component
public class MyBean {

    @Value("${name}")
    private String name;

    // ...

}
```

> On your application classpath (e.g. inside your jar) you can have an `application.properties` that provides a sensible default property value for `name`. When running in a new environment, an `application.properties` can be provided outside of your jar that overrides the `name`; and for one-off testing, you can launch with a specific command line switch (e.g. `java -jar app.jar --name="Spring"`).

以上内容告诉开发人员，名为 `name` 的配置属性在 `application.properties` 中为默认值。当应用运行在新的环境中时，可通过命令行（command line）的方式覆盖该默认值，被 `@Value("${name}")`注入的字段 **name** 的内容也随之变化。由于 Spring Boot 内定 `PropertySource` 的读取顺序，“外部化配置” `4.Command line arguments.`优先于 `15.Applicationproperties packaged inside your jar (application.properties and YAML variants).`，因此

application.properties 为默认值被忽略，从而达到覆盖的目的。

然而，何谓“外部化配置”，官方文档并没有正面解释。根据小马哥的经验，做出如下解释。

通常，对于可扩展性应用，尤其是中间件，它们的功能性组件是可配置化的，如认证信息、端口范围、线程池规模及连接时间等。假设需要设置 Spring 应用的 Profile 为“dev”，则可通过调用 Spring ConfigurableEnvironment 的 setActiveProfiles("dev")方法实现。这种方式是一种显式的代码配置，配置数据来源于应用内部实现，所以称为“内部化配置”。“内部化配置”虽能达成目的，然而配置行为是可以枚举的，必然缺少相应的弹性。

对于同样实现相同的需求，Spring Boot 提供了另外的选择，比如添加配置 spring.profiles.active=dev 到 application.properties 文件，或者追加命令行参数的方式，即--spring.profiles.active=dev。官方文档清楚地告知以上两种配置方式均是“外部化配置”。换句话说，PropertySource（属性源）就是“外部化配置”API 的描述方式。因此，外部配置属性源（PropertySource）是“外部化配置”读取的媒介，它可能是配置文件的形式，如 application.properties 或 application.yml，也可能是环境变量（OS environment variables）或命令行参数（Command line arguments）等。

目前对于“外部化配置”的理解基本上停留在感性认识的层面，不免让人留下一些疑问。首先，列出的三种“外部化配置”的使用场景具体如何做到“因地制宜”？其次，Spring Boot 内建的 17 种“外部化配置”来源何处？最后，如何自定义 Spring Boot 的外部化配置？针对这些疑问，下一册的**“超越外部化配置”**将分别通过应用**“外部化配置”**、追溯**“外部化配置”**和扩展**“外部化配置”**三部分深入探讨。

6.5 理解“规约大于配置”

Spring Boot 官方描述的最后一个特性：

> "Absolutely no code generation and no requirement for XML configuration".

按照字面意思，Spring Boot 中间绝对无代码生成，也无须 XML 配置。前半句是向开发人员做出承诺，Spring Boot 尽管功能强大，然而不会生成中间代码，从而影响 Spring Boot 应用运行时行为。后半句则是耳熟能详的“规约大于配置”。相较而言，后者更重要。

小马哥在日常推广微服务的过程中，发现绝大多数的开发人员非常欢迎 Spring Boot。按照大家的说法，Spring Boot 重回 main 方法的启动方式，让自己找回了初学 Java 时的记忆。尽管不少人的开发模式已习惯以 XML 文件配置驱动，在逐渐转化为注解驱动的过程中，也遇到了一些困惑。大多数人以为注解驱动是 Spring Boot 引入的，这无伤大雅，反而说明了 Spring Boot 与 Spring Framework 的关系是“我中有你，你中有我”。

从技术的角度来叙述，Spring Framework 是 Spring Boot 的“基础设施”，Spring Boot 的基本特性均来自 Spring Framework。Spring Boot 不但无法取代 Spring Framework，而且还实现了 Spring Framework 的“自我救赎”。

不可否认的是，Spring Framework 的注解驱动之路可谓“路漫漫其修远兮，吾将上下而求索”。从 Spring Framework 2.5 开始，Spring Bean 注册方式由 Annotation 驱动逐步替代 XML 文件驱动，通过`@Component` 及“派生”注解（如@Service）与 XML 元素`<context:component-scanbase-package="..."/>`相互配合，将 Spring `@Component` Bean 扫描并注册至 Spring Bean 容器（`BeanFactory`），通过 DI 注解`@Autowired` 获取相应的 Spring 组件 Bean。然而`@Component` Bean 必须在`<context:component-scan>`规定的 `base-package` 集合范围中。

到了 Spring Framework 3.0 时代，新引入的 Annotation @Configuration 是 XML 配置文件的替代物。而 Bean 的定义不再需要在 XML 文件中声明`<bean>`元素，可使用`@Bean` 来代替。同时，框架提供了更细粒度的 Annotation `@Import` 来导入`@Configuration` Class，将其注册为 Spring Bean，并进一步地解析其声明的 Spring Bean 注册相关的注解，如`@Import` 或`@Bean`。尽管在 Bean 的装配方面有明显提升，但仍是以硬编码的方式指定范围。

当 Spring Framework 3.1 发布后，虽说从 3.0 到 3.1 不过是小版本的变化，然而其实际更新动作并不小，此时的 Spring 注解才完全取代 XML 配置文件，引入了 Annotation `@ComponentScan` 来代替 XML 元素`<context:component-scan>`，初看起来并没有多少提升。应用方可以实现 `ImportSelector` 接口（实现 `selectImports(AnnotationMetadata)`方法），程序动态地决定哪些 Spring Bean 需要被导入。可是 `ImportSelector` 实现类必须暴露成 Spring Bean，否则 `selectImports(AnnotationMetadata)`方法不会被 Spring 容器调用。为了简化实现成本，Spring Framework 3.1 内建了不少的功能“模块”激活的注解，如`@EnableCache`。即使如此，被标注`@EnableCache` 的类同样需要通过`@ComponentScan` 或`@Import` 等方式被 Spring 容器感知，但却又成了“先有鸡还是先有蛋”的问题。硬编码的问题得到了缓解，但仍然与自动装配的能力相去甚远。

Spring Framework 到了 4.0 版本增加了条件化的 Spring Bean 装配注解`@Conditional`，其 `value()` 属性可指定 `Condition` 的实现类，而 `Condition` 提供装配条件的实现逻辑，`@Conditional` 的引入更直观表达了 Spring Bean 装载时所需的前置条件。或许 Spring Boot 官方对自动化装配的后半段描述容易被人忽视——**“whenever possible”**，正因为 Spring Framework 条件注解`@Conditional` 的引入，使得条件性装配成为可能。Spring Boot 在此基础上，实现了最显著特性之一——**条件化自动装配**。为此，在“**走向自动装配**”部分的“**走向注解驱动编程（Annotation-Driven）**”章节中，会深度探讨 Spring Framework 从 1.x 到 5.0 注解驱动上的“心路历程”。

6.6 小马哥有话说

至此，Spring Boot 官方描述的六大特性均已初步讨论完毕。严格地说，第三点（Provide opinionated 'starter' dependencies to simplify your build configuration）和第六点（Absolutely no code generation and no requirement for XML configuration）并非 Spring Boot 引入的特性，前者属于 Maven 依赖管理的运用，后者从 Spring Framework 3.1 就已完全支持，而其他特性均有 Spring Boot 的专属名词。综上所述，Spring Boot 主要有五大特性：

- `SpringApplication`；
- 自动装配；
- 外部化配置；
- Spring Boot Actuator；
- 嵌入式 Web 容器。

这五大特性构成了 Spring Boot 作为微服务中间件的基础，又提供了 Spring Cloud 的基础设施。

6.6.1 Spring Boot 作为微服务中间件

小马哥与不少开发人员交换过对微服务开发的看法，部分（尤其是年轻的）开发人员的看法是 Java 开发微服务的首选是 Spring Boot，甚至是唯一选择。

2017 年 7 月 11 日，Spring 技术布道师 Josh Long 在官方博客中发布博文（https://spring.io/blog/2017/07/11/this-week-in-spring-july-11th-2017），其中有一条耐人寻味：

> Oracle have announced as massive-online-only-course on developing REST applications with Spring Boot. Sounds interesting!

Josh Long 援引了 Oracle 的博文 **“Massive Open Online Course: Java Microservices”**：https://blogs.oracle.com/java/java-microservices-mooc。

实际上，微服务架构作为细粒度的 SOA，并未限制实现的技术。因此，传统 Java EE 容器也能实现微服务。除上述框架外，一个名为 KumuluzEE 的轻量级容器赢得了甲骨文“2015Duke 选择奖”。

尽管 KumuluzEE 赢得了 Oracle 大奖，可是在 Java 世界中，基本上默默无闻。赢取人们的信任需要长时间的投入和付出。正因如此，Spring 社区十几年的开源策略和技术演进，使得 Spring Boot 在 Java 微服务的世界里独占鳌头。

6.6.2 Spring Boot 作为 Spring Cloud 基础设施

尽管 Spring Boot 像瑞士军刀那样提供了丰富的功能特性。但仅使用 Spring Boot 可能是不够的，因为微服务架构必须建立在分布式系统中，然而 Spring Boot 天然性地缺少快速构建分布式系统的能力。为此，Spring 官方在 Spring Boot 的基础上研发出 Spring Cloud，致力于为开发人员提供一些快速构建通用的分布式系统，其包含的核心特性如下：

- Distributed/versioned configuration（分布式配置）；
- Service registration and discovery（服务注册和发现）；
- Routing（路由）；
- Service-to-service calls（服务调用）；
- Load balancing（负载均衡）；
- Circuit Breakers（熔断机制）；
- Distributed messaging（分布式消息）。

Spring Cloud 提供的这些功能大多数被大型互联网公司实现。既然均已被实现，为何 Spring 官方还要去做一遍呢？Spring 官方的最大优势在于其强大的 API 设计能力。相比传统 Java EE 阵营，如 SUN（现 Oracle）、Apache 和 Eclipse 等，Spring 的高度抽象及使用简化的能力有过之而无不及。随着云平台的蓬勃发展，在云平台上 Java 语言（以及派生语系）处于垄断地位，这一点基本上在业界达成了共识。Spring Cloud 高度抽象的接口对于应用开发人员而言，可以说是透明的，他们不需要关心底层的实现。当需要更替实现时，按需要配置即可，不需要过多的业务回归测试。

Spring Cloud 的第二大优势是 Spring Cloud Stream 整合，这个整合模式的灵感来源于 Martin Fowler 的名著 *Enterprise Integration Pattern*，中文书名为《企业整合模式》。简单地说，通过 Stream 编程模式，使得不同的通道之间可以自由地切换传输介质，达到数据通信的目的，比如通过消息、文件、网络等。

读者想要了解更多细节，可以参考官方网站：http://projects.spring.io/spring-cloud/。

小马哥同时向读者推荐国内两本非常优秀的 Spring Cloud 的书籍，分别是《Spring Cloud 微服务实战》（作者：翟永超）和《Spring Cloud 与 Docker 微服务架构实战》（作者：周立）。

6.7 下一站：走向自动装配

或许你我共同意识到 Spring Boot 背靠着 Spring Framework 这座大山，Spring 应用开发变得简单了，独立应用的部署模式使其成为微服务中间件的不二选择。Spring Boot 作为 Spring Cloud 的基础设施，让实现 Cloud Native 成为可能。Spring Boot 的出现改变了沉寂已久的 Spring 社区，作为家族中流砥柱的角色，Spring 技术栈焕发了“第二春”。当然 Spring Boot 的成功并非偶然，它必然存在优于 Spring Framework 的特性。其中自动装配作为最显著的特性之一，与 Spring Framework 注解驱动密不可分，接下来共同进入“走向自动装配”部分的讨论。

第 2 部分　走向自动装配

> “改变，人人都会有。可当你真正面对改变时，你也就觉得坦然了。”——《廊桥遗梦》

改变是一个不破不立的过程。“破”不是否定过去，而是去除糟粕；“立”不是无中生有，而是与时俱进。

微服务架构（Micro-Services Architecture）的“春风”吹向了软件行业的“大地”，无意间使得 Spring Boot 在 Java 社区变得出乎意料地风靡。这种趋势的蔓延让开发人员产生了对 Spring Boot 的“狂热”，以及对 Spring Framework 的“忽视”。甚至认为 Spring Boot 是一个新的分水岭，Spring Boot 应用代表了微服务架构，其他的框架支撑着“单体应用”（Monolithic Application）。诚如上述所言，小马哥当初也如此“偏执地”认为 Spring Boot 引领了“新风向”。不过随着系统性地深入研究，才幡然醒悟，殊不知 Spring Framework 是 Spring Boot 核心，Java 规范才是它们的基石。

其中的是非曲直，都反映了一个客观的事实——Java EE 的世界发生了“改变”。

有一种观点认为 Spring Boot 是“Java EE 颠覆者”，另一种观点则认为 Spring Boot 实现了 Java EE 的“自我救赎”，小马哥认为这两种观点无所谓对错优劣，也无意反驳和批判他人的观点。从 Java 技术的发展脉络的角度来看，Spring Boot 既没有也无法“颠覆”Java EE。与其说 Spring Boot 在实现“自我救赎”，还不如认为它在不断地“兼容并包，继往开来”。

优秀的技术架构必然遵循“兼容并包、继往开来”的原则。“兼容并包”更强调软件版本的前后“兼容”、新老技术的“并包”，而“继往开来”则偏重于“承继”前理念、“开辟”新架构。具体而言，Spring Framework 作为 Spring 技术栈最核心的框架，打下了“兼容并包”的基础，而 Spring Boot 完善了“继往开来”的使命。

正如唐太宗所言，“以史为镜，可以知兴替”。小马哥会引证许多相关材料，大多数内容是实践经验的总结，即使在官方参考文档中也无处寻迹，其目的是为后续 Spring Framework 及 Spring Boot 的源码分析做铺陈。

第 7 章 走向注解驱动编程（Annotation-Driven）

小马哥相信大多数资深的开发人员在正式使用 Spring Framework 前，必先理解两大核心思想：IoC（Inversion of Control，控制反转）和 DI（Dependency Inject，依赖注入），其推崇的理念是应用系统不应以 Java 代码的方式直接控制依赖关系，而是通过容器来加以管理。随后 Spring Framework 的作者说明其框架对 IoC 和 DI 支持的天然性。由于当时 Java 5 Annotation 尚未发布，结合 J2EE（Java EE 的前身，当时还称为 J2EE）的传统，通过 XML 文件的方式管理 Bean 之间的依赖关系。

7.1 注解驱动发展史

7.1.1 注解驱动启蒙时代：Spring Framework 1.x

在 Spring Framework 1.x 时代，其中 `1.2.0` 版本是这个时代的分水岭。

> 为了确保论证的严谨性，小马哥比对了 Spring Framework 最早的几个 changelog.txt，发现版本为 `1.2.0` 的 changelog.txt 文件中，记录了 Spring Framework 支持 Annotation 的文字描述(感兴趣的读者请访问 https://docs.spring.io/spring/docs/1.2.0/changelog.txt 一探究竟）：

```
...
* marked JDK 1.5+ annotations (ManagedResource, ManagedAttribute, ManagedOperation, etc)
as @Documented
...
* marked JDK 1.5+ Transactional annotation as "@Inherited" and "@Documented"
...
```

当时，Java 5 刚刚出炉，业界正刮起使用 Annotation 的技术风，Spring Framework 自然也“不甘寂寞”，予以支持。

虽然框架层面均已支持`@ManagedResource`和`@Transactional`等 Annotation，然而被注解的 Spring Bean 的装配仍需要使用 XML 方式。那时还处于大学时代的小马哥，仍记得当时的国内网络论坛及博客几乎没有关注或使用 Annotation 方式，然而时至今日却被广泛地讨论，算不算是一种“朝花夕拾”呢？

由于 Spring Framework 1.x 实现的局限性，XML 配置方式是唯一选择。

7.1.2 注解驱动过渡时代：Spring Framework 2.x

时间回到 2006 年，Spring Framework 2.0 正式发布。在软件兼容性方面，与同行相比较，Spring Framework 可以说得上无出其右，2.x 版本几乎完全兼容 1.x 版本。

虽然 Spring Framework 2.0 在 Annotation 支持方面增添了新的成员，比如 Bean 相关的`@Required`、数据相关的`@Repository`及 AOP 相关的`@Aspect`等，但同时提升了 XML 配置能力，即“**可扩展的 XML 编写（Extensible XML authoring）**”。这种扩展能力的出现，无形中为 XML 配置的价值增加了筹码，比如国内的 Dubbo 等开源框架均扩展了自身的 XML 配置方式及 Schema 文件。

同样作为重要分水岭的 Spring Framework 2.5，新引入了一些骨架式的 Annotation。

- 依赖注入 Annotation：`@Autowired`；
- 依赖查找 Annotation：`@Qualifier`；
- 组件声明 Annotation：`@Component`、`@Service`；
- Spring MVC Annotation：`@Controller`、`@RequestMapping`及`@ModelAttribute`等。

在依赖注入方面，Spring Framework 2.5 开始支持`@Autowired`注解注入的 Spring Bean，这些 Bean 可以来自 XML 元素`<bean>`定义，或者 Spring 模式注解，如`@Component`、`@Repository`、`@Service`等。当然`@Autowired`也允许注入某种类型（Class）Spring Bean 的集合，例如：

```java
@Component("nameRepositoryHolder")
public class NameRepositoryHolder {
```

```
    @Autowired
    private Collection<NameRepository> repositories;
    ...
}
```

> 源码位置：以上示例代码可通过查找 spring-framework-samples/spring-framework-2.5.6-sample 工程获取。

无论@Autowired 注入单个 Spring Bean，还是注入 Spring Bean 集合，其依赖查找的实现均属于限定类型（Class）的方式。假设需要在相同类型中再细粒度地筛选，则需要注解@Qualifier 的配合。也许读者曾使用@Qualifier 依赖查找某个命名的 Bean，例如：

```
@Component("nameRepositoryHolder")
public class NameRepositoryHolder {
    ...
    @Autowired
    @Qualifier("chineseNameRepository")
    private NameRepository nameRepository;
    ...
}
```

> 源码位置：以上示例代码可通过查找 spring-framework-samples/spring-framework-2.5.6-sample 工程获取

除此之外，@Qualifier 也支持“逻辑类型”限定。例如，Spring Boot 中的外部化配置注解@ConfigurationPropertiesBinding：

```
@Qualifier
@Target({ ElementType.TYPE, ElementType.METHOD })
@Retention(RetentionPolicy.RUNTIME)
@Documented
public @interface ConfigurationPropertiesBinding {

}
```

以及 Spring Cloud 中的负载均衡注解@LoadBalanced：

```
@Target({ ElementType.FIELD, ElementType.PARAMETER, ElementType.METHOD })
```

```
@Retention(RetentionPolicy.RUNTIME)
@Documented
@Inherited
@Qualifier
public @interface LoadBalanced {
}
```

两者均标注了@Qualifier，均属于“逻辑类型”限定。在 Spring 应用上下文生命周期中，@ConfigurationPropertiesBinding 或@LoadBalanced Bean 的处理实现先通过@Qualifier 筛选，再对其进行加工处理。

> @ConfigurationPropertiesBinding 的处理实现在 org.springframework.boot.context.properties.ConfigurationPropertiesBindingPostProcessor 中。而@LoadBalanced 的处理则对应 org.springframework.cloud.client.loadbalancer.LoadBalancerAutoConfiguration。

在 Spring Framework 中，@Autowired 和@Qualifier 的处理实现是相对复杂的，在后续的章节中再深入探讨。

除了@Autowired 注入，Spring Framework 2.5 也支持 **JSR-250**（Java 规范）@Resource 注入。Spring Framework 存在此等举措也不难理解，它需要寻求业界的“最大公约数”。包括 **JSR-250** 生命周期回调注解@PostConstruct 和@PreDestroy，前者可替代<bean init-method="..."/>或 Spring InitializingBean 接口回调，后者能替换<bean destroy-method="..." />或 DisposableBean。

在组件声明方面，前文提到@Autowired 能够注入标注 Spring 模式注解的 Bean，如@Component、@Service、@Repository 及@Controller。同时，Spring Framework 2.5 允许自定义 Spring 模式注解，不过该版本仅支持单层次的 Spring 模式注解“派生”，从 Spring Framework 3.0 开始支持多层次 Spring 模式注解“派生”，这方面的详细讨论将安排在后续章节。同时引入@Scope 注解替换 XML 属性<bean scope="..." />。

尽管 Spring Framework 2.x 时代提供了为数不少的注解，然而编程手段却不多，最主要的原因在于框架层面仍未**“直接”**提供驱动注解的 Spring 应用上下文，并且仍需要 XML 配置驱动，即 XML 元素<context:annotation-config>和<context:component-scan>，前者的职责是注册 Annotation 处理器，后者负责扫描相对于 ClassPath 下的指定 Java 根包（Base Packages），寻找被 Spring 模式注解标记的类（Class），将它们注册成 Spring Bean。

在 Spring Framework 1.x 时代，当多个 Spring Bean 需要进行排序时，通常的做法是实现 Ordered 接口。从 Spring Framework 2.0 开始，通过在@Component Class 中标注@Order 的方式进行替代。

值得称赞的是，从 Spring Framework 2.5 开始，Web MVC 方面的开发模式逐步从面向接口编程过渡到注解驱动编程，如@Controller 、@RequestMapping 及@RequestParam 等，更多的细节将在 Web 篇中详细探讨。

Spring Framework 2.x 尚未完全替换 XML 配置驱动，因此，这个时期称为注解驱动的“过渡时代”。不过这种“尴尬”处境从 3.0 版本开始得到了改善。

7.1.3 注解驱动黄金时代：Spring Framework 3.x

Spring Framework 3.x 是一个里程碑式的时代，其功能特性出现了“井喷”现象，包括全面拥抱 Java 5（泛型、变量参数等），以及 Spring Annotation 如雨后春笋般地引入，足以体现 Spring 官方全面替代 XML 配置的决心。

不过，其替代的过程并非一蹴而就，在 Spring Framework 3.0 中，除了提升 Spring 模式注解“派生”的层次性，首要任务是替换 XML 配置方式，所以引入了配置类注解@Configuration，该注解也是内建的@Component“派生”注解：

```java
@Target(ElementType.TYPE)
@Retention(RetentionPolicy.RUNTIME)
@Documented
@Component
public @interface Configuration {
    ...
 String value() default "";

}
```

遗憾的是，Spring Framework3.0 还是没有引入替换 XML 元素<context: component-scan>的注解，而是选择过渡方案——@ImportResource 和@Import。@ImportResource 允许导入遗留的 XML 配置文件，而@Import 则允许导入一个或多个类作为 Spring Bean，这些类无须标注 Spring 模式注解，如@Service。通常标注@ImportResource 或@Import 的类需要再标注@Configuration，例如：

```java
@ImportResource("classpath:/META-INF/spring/others.xml")        // 替代<import>
@Configuration
public class SpringContextConfiguration {
    ...
}
```

> 源码位置：以上示例代码可通查找 spring-framework-samples/spring-framework-3.2.x-sample 工程获取。

然而新的问题出现了，SpringContextConfiguration 又由谁来引导呢？假设仍由 XML 配置文件来装配，无论 XML 元素<bean>还是<context:component-scan>，岂不是“多此一举”吗？所以，Spring Framework 3.0 又引入了新的 Spring 应用上下文实现 AnnotationConfigApplicationContext，作为前时代 ApplicationContext 实现的替代者。即使如此，这种注解驱动开发方式仍让人感觉别扭，先使用 AnnotationConfigApplicationContext 注册@Configuration Class，然后在该 Class 上再标注@ImportResource 或@Import。

不过 Spring Framework 3.0 在替换 XML 元素<bean>上的诚意还是足的，例如：@Bean 替换 XML 元素<bean>、@DependsOn 替代 XML 属性<bean depends-on= "..."/>、@Lazy 替代 XML 属性<bean lazy-init="true|false"/>，以及@Primary 替代 XML 属性<bean primary= "true|false" />。

```
@ImportResource("classpath:/META-INF/spring/others.xml")// 替代<import>
@Configuration("springContextConfiguration")
public class SpringContextConfiguration {

    @Lazy
    @Primary
    @DependsOn("springContextConfiguration") // 依赖 springContextConfiguration
    @Bean(name = "user") // Bean 的名称为 user
    public User user() {
        User user = new User();
        user.setName("小马哥"); // 设置 property
        return user;
    }
}
```

> 源码位置：以上示例代码可通过查找 spring-framework-samples/spring-framework-3.2.x-sample 工程获取。

一向注重细节的 Spring 阵营，当然不会让上述情况影响编程体验，这也促使了 Spring Framework 3.1“脱胎换骨”的改变。新引入的注解@ComponentScan 替换 XML 元素<context:component-scan>。虽然只是提升了一小步，却是 Spring Framework 全面进入注解驱动时代的一大步，例如：

```
...
```

```
    @ComponentScan(basePackages = "thinking.in.spring.boot.samples") // 替代
<context:component-scan>
    public class SpringContextConfiguration {
        ...
    }
```

> 源码位置：以上示例代码可通过查找 spring-framework-samples/spring-framework-3.2.x- sample 工程获取。

在 Bean 定义的声明中，`@Bean` 允许使用注解`@Role` 设置其角色。或许这方面的内容对于大多数开发人员而言相对陌生，不过它却实现了注解驱动替代 `AbstractBeanDefinition#setRole(int)` API 编程方式，例如：

```
    ...
    public class SpringContextConfiguration {

        ...
        @Bean(name = "user") // Bean 名称为“user”
        @Role(BeanDefinition.ROLE_APPLICATION) //应用角色
        public User user() {
            User user = new User();
            user.setName("小马哥"); // 设置 property
            return user;
        }
    }
```

> 源码位置：以上示例代码可通过查找 spring-framework-samples/spring-framework-3.2.x-sample 工程获取。

同时，注解`@Profile` 使得 Spring 应用上下文具备条件化 Bean 定义的能力，该特性的使用场景极为常见，比如开发环境和生产环境采用不同的 Bean 定义，可类比 Maven Profile：

```
    ...
    @Profile("!production") // 非生产环境
    public class SpringContextConfiguration {
        ...
    }
```

> 源码位置：以上示例代码可通过查找 spring-framework-samples/spring-framework-3.2.x-sample 工程获取。

Spring Framework 3.x 在 Web 方面的提升更是突飞猛进。例如，请求处理注解`@RequestHeader`、`@CookieValue` 和`@RequestPart` 的出现，使得 Spring Web MVC `@Controller` 类不必直接使用 Servlet API。除此之外，Spring Framework 3.0 开辟了另外一条开发路线，即 REST 开发。例如，`@PathVariable` 便于 REST 动态路径的开发，`@RequestBody` 能够直接反序列化请求内容。对应的响应方面，`@ResponseBody` 表示将处理方法返回对象序列化为 REST 主体内容，并且 `@ResponseStatus` 可以补充 HTTP 响应的状态信息。当年小马哥接触过 Spring Web MVC 之后，隐约地感觉到 Spring 在尝试解除 Servlet 的“束缚”，形成一股新的技术势力。果然一语成谶，Web Flux 的出现证明了之前的判断。不过，“始作俑者”并非 Spring Web MVC，而是 JAX-RS（Java REST 规范）。其中原委将在 Web 篇中展开讨论。更重要的是，Spring Web 整合了 Servlet 3.0+规范，利用 `javax.servlet.ServletContainerInitializer` API 实现在传统 Servlet 容器中“自动装配”的能力，在本章的“Web 自动装配”一节中再叙。

Spring Framework 3.x 除了提升已有的注解驱动能力，新增的部分自然也不少。

小马哥认为最为精妙的部分是，Spring Framework 3.1 抽象了一套全新并统一的配置属性 API，包括配置属性存储接口 `Environment`，以及配置属性源抽象 `PropertySources`，这两个核心 API 奠定了 Spring Boot 外部化配置的基础，也是 Spring Cloud 分布式配置的基石。不过开发人员想要直接使用 API 编程的难度还是挺大的，开发人员不仅需要合理地掌握 Spring 上下文及 Bean 的生命周期，同时需要理解接口之间的关系。为了降低开发成本，Spring Framework 提供了注解 `@PropertySource` 简化实现。该注解的 `value()`属性方法所关联的 `Properties` 资源类与 Spring Boot 应用配置文件 `application.properties` 并没有本质的区别。关于这方面的内容将在“外部化配置”章节中详细讨论。

其次的部分是缓存（Caching）抽象，Spring Framework 3.1 同样加以抽象了，主要 API 包括缓存 `Cache` 及缓存管理器 `CacheManager`。配套的注解 `Caching` 和 `Cacheable` 等极大地简化了数据缓存的开发。

在异步支持方面，Spring Framework 引入了异步操作注解`@Async`、周期异步执行注解`@Scheduled` 及异步 Web 请求处理 `DeferredResult`。

在检验方面，Spring Framework 3.1 新增了校验注解`@Validated`，不但整合了 JSR-303（Bean Validation 1.0），而且适配了 Spring 早期的 Validator 抽象。

面对如此多的注解驱动提升，Spring 的作者仍觉得诚意不够，再在 Spring Framework 3.1 中加码，引入了小马哥称之为“**`@Enable` 模块驱动**”的特性，该特性将相同职责的功能组件以模块化的方式装配，极大地简化了 Spring Bean 配置。以`@EnableWebMvc` 为例，该注解被`@Configuration`

Class 标注后，RequestMappingHandlerMapping、RequestMappingHandlerAdapter 及 HandlerExceptionResolver 等 Bean 被装配。由于@Enable 模块驱动需要显式地标注在@Configuration Class 上，所以它属于“手动装配”。相关的细节将在“理解@Enable 模块驱动”一节中讨论。

综上所述，小马哥称 Spring Framework 3.x 为注解驱动的“黄金时代”。

“革命尚未成功，同志仍需努力”，Spring Framework 的作者没有停止对注解能力的耕耘，毕竟它不够完善，比如@Profile 条件装配相对简单，以及@PropertySource 资源的来源单一等。Spring Framework 4.x 进入了注解驱动的完善时代。

7.1.4 注解驱动完善时代：Spring Framework 4.x

Spring Framework 4.x 在注解驱动方面趋于完善，注解如“潮水般”涌入的情况不复存在。

首先要完善的任务是提升条件装配能力。从 Spring Framework 3.1 开始，注解@Profile 提供了配置化的条件组装，不过这方面的能力仍旧单薄，无法自定义条件判断。从 Spring Framework 4.0 开始，条件化注解@Conditional 被引入，通过与自定义 Condition 实现配合，弥补了之前版本条件化装配的短板：

```
@Target({ElementType.TYPE, ElementType.METHOD})
@Retention(RetentionPolicy.RUNTIME)
@Documented
public @interface Conditional {

    /**
     * All {@link Condition}s that must {@linkplain Condition#matches match}
     * in order for the component to be registered.
     */
    Class<? extends Condition>[] value();

}
```

以至于@Profile 从 Spring Framework 4.0 开始重新声明，通过@Conditional 实现：

```
@Retention(RetentionPolicy.RUNTIME)
@Target({ElementType.TYPE, ElementType.METHOD})
@Documented
@Conditional(ProfileCondition.class)
public @interface Profile {
```

```
    /**
     * The set of profiles for which the annotated component should be registered.
     */
    String[] value();

}
```

注解@Conditional 的出现自然也对 Spring Boot 的条件装配产生了影响，Spring Boot 的所有@ConditionalOn*注解均基于@Conditional“派生”注解，其抽象类 SpringBootCondition 也是 Condition 的实现，例如：

```
@Target({ ElementType.TYPE, ElementType.METHOD })
@Retention(RetentionPolicy.RUNTIME)
@Documented
@Conditional(OnBeanCondition.class)
public @interface ConditionalOnBean {
    ...
}
```

Spring Framework 4.x 对 Java 语言特性的支持也是与时俱进的，尽管它并不强制使用 Java 8，然而巧妙地兼容了 Java Time API（JSR-310）、@Repeatable 及参数名称发现，援引官方文档的描述：

> Spring Framework 4.0 provides support for several Java 8 features. You can make use of *lambda expressions* and *method references* with Spring's callback interfaces. There is first-class support for `java.time` (JSR-310), and several existing annotations have been retrofitted as `@Repeatable`. You can also use Java 8's parameter name discovery (based on the `-parameters` compiler flag) as an alternative to compiling your code with debug information enabled.

Java 8 @Repeatable 的出现，解决了以往 Annotation 无法重复标注在同一类上的限制，详情请参考 JSR-337（Java Language Specification）“**9.6.3 Repeatable Annotation Types**”章节。

> 资源共享：读者可从小马哥的 JSR GitHub 资源中获取 JSR-337 规范：https://github.com/mercyblitz/jsr/tree/master/JLS。

根据@Repeatable特性，Spring Framework 4.0 将@PropertySource提升为可重复标注的注解：

```
@Target(ElementType.TYPE)
```

```
@Retention(RetentionPolicy.RUNTIME)
@Documented
@Repeatable(PropertySources.class)
public @interface PropertySource {
    ...
}
```

相较于 Spring Framework 3.x 的实现，@PropertySource 新增了@Repeatable (PropertySources.class)，即使 Spring 应用的 Java 版本低于 8 时，也可以使用注解@PropertySources 与@PropertySource 配合，如 Java 8+的写法：

```
@PropertySource("classpath:/config/default.properties")
@PropertySource("classpath:/config/override.properties") // 重复标注@PropertySource
@Configuration
public class RepeatablePropertySourceConfiguration {
}
```

Java 8 之前的等效写法：

```
@PropertySources({
        @PropertySource("classpath:/config/default.properties"),
        @PropertySource("classpath:/config/override.properties")
})
@Configuration
public class PropertySourcesConfiguration {
}
```

> 源码位置：以上示例代码可通过查找 spring-framework-samples/spring-framework-4.3.x- sample 工程获取。

同样@ComponentScan 在 Spring Framework 4.3 中得到了提升并引入了 @ComponentScans。

除了以上注解的完善，Spring Framework 4.2 新增了事件监听器注解@EventListener，作为 ApplicationListener 接口编程的第二选择。关于 Spring Framework 和 Spring Boot 事件/监听器的内容将在“Spring Boot 事件”章节中讨论。

由于 Java 注解之间不存在继承关系，因此 Spring 模式注解“派生”特性从 Spring Framework 2.5 诞生，然而这种“派生”特性需要确保注解之间的属性方法签名完全一致。例如@Component 与@Repository：

```
@Target(ElementType.TYPE)
```

```
@Retention(RetentionPolicy.RUNTIME)
@Documented
public @interface Component {

 ...
 String value() default "";

}
@Target({ElementType.TYPE})
@Retention(RetentionPolicy.RUNTIME)
@Documented
@Component
public @interface Repository {

 ...
 String value() default "";

}
```

@Repository 作为@Component 的“派生”注解，两者均存在相同的属性方法：value()，同样的限制也适用于@Component 与@Configuration 之间。这种限制被 Spring Framework 4.2 新注解@AliasFor 解除，同时它还能在同一注解内实现属性方法的别名，如 Spring Web MVC 中的@RequestMapping：

```
@Target({ElementType.METHOD, ElementType.TYPE})
@Retention(RetentionPolicy.RUNTIME)
@Documented
@Mapping
public @interface RequestMapping {
    ...
 String name() default "";
    ...
 @AliasFor("path")
 String[] value() default {};
    ...
 @AliasFor("value")
 String[] path() default {};
 ...
```

```
    RequestMethod[] method() default {};
    ...
}
```

而 Spring Framework 4.3 引入的@GetMapping 作为@RequestMapping 的“派生”注解，同样利用@AliasFor 实现了不同注解之间的属性方法别名：

```
@Target(ElementType.METHOD)
@Retention(RetentionPolicy.RUNTIME)
@Documented
@RequestMapping(method = RequestMethod.GET)
public @interface GetMapping {

    /**
     * Alias for {@link RequestMapping#name}.
     */
    @AliasFor(annotation = RequestMapping.class)
    String name() default "";
        ...
}
```

用过 Spring Cloud 的读者一定不会对@GetMapping 陌生，实际上它既不是 Spring Cloud 注解，也不属于 Spring Boot。当然，在 Spring Framework 的底层不会对@GetMapping 单独处理，否则兄弟注解@PostMapping、@PutMapping 等也需如此。因此，Spring Framework 必然对@AliasFor 进行特殊处理，详情将在后面的章节中探讨。

Spring Framework 4.x 在 Web 注解驱动编程方面也有小幅的提升，除了上文提到的@GetMapping、@PostMapping，最显著的注解莫过于@RestController，它大量地出现在 Spring Boot 官方示例中，导致大多数人认为它是 Spring Boot 注解。其次值得关注的是@RestControllerAdvice，作为@RestController AOP 拦截通知，它扮演类似于@ControllerAdvice 对于@Controller 的角色。在浏览器跨域资源访问方面，Spring Framework 4.2 开始引入@CrossOrigin，作为 CorsRegistry 替换注解方案，前者更加集中在@Controller 的处理方法上，后者则更关注请求 URL。深入的讨论仍旧在 Web 篇章节中展开。

除此之外，Spring Framework 4.x 新增了依赖查找注解@Lookup，不过它无论在 Spring Framework 中，还是在 Spring Boot 中，均处于边缘化的境地，读者知道其存在即可。

截至本书出版，最新的 Spring Framework 4.x 版本为 4.3.17.RELEASE。按照以往的经验，由于 Spring Framework 5.0 已经发布，Spring 官方应该不会再开发 Spring Framework 4.4 版本，所以尚能定义 Spring Framework 4.x 为注解驱动的“完善时代”。同时，最新的 Spring Framework 版本为

5.0.6.RELEASE，暂时还无法推断未来在 5.x 版本中会出现哪些新特性。因此，只能称之为“当下时代”。

7.1.5　注解驱动当下时代：Spring Framework 5.x

Spring Framework 5.0 作为 Spring Boot 2.0 的底层核心框架，就目前已发布的版本来看，相对于 Spring Framework 4.x 而言，注解驱动的性能提升不是那么明显。然而随着 Spring Framework 注解驱动能力逐渐受到开发人员的关注，尤其在 Spring Boot 应用场景中，大量使用注解@ComponentScan 扫描指定的 package，当扫描的 package 所包含的类越多时，Spring 模式注解解析的耗时就越长。面对这个问题，Spring Framework 5.0 引入了注解@Indexed，为 Spring 模式注解添加索引，以提升应用启动性能，例如：

```
@Indexed
@Configuration
public class AnnotationIndexedConfiguration {
}
```

> 源码位置：以上示例代码可通过查找 spring-framework-samples/spring-framework-5.0.x-sample 工程获取。

注解@Indexed 不能孤立地存在，需要在工程 pom.xml 中增加 org.springframework:spring-context-indexer 依赖：

```
<dependencies>
    <dependency>
        <groupId>org.springframework</groupId>
        <artifactId>spring-context-indexer</artifactId>
        <version>5.0.6.RELEASE</version>
        <optional>true</optional>
    </dependency>
</dependencies>
```

当工程打包为 JAR 或在 IDE 工具中重新构建后，META-INF/spring.components 文件将自动生成。换言之，该文件在编译时生成。当 Spring 应用上下文执行@ComponentScan 扫描时，META-INF/spring.components 将被 CandidateComponentsIndexLoader 读取并加载，转化为 CandidateComponentsIndex 对象，进而@ComponentScan 不再扫描指定的 package，而是读取 CandidateComponentsIndex 对象，从而达到提升性能的目的。不过这种方式存在缺陷，比如 Spring

Framework 官方文档的“1.10.9. Generating an index of candidate components”章节所述：

> In this mode, all modules of the application must use this mechanism as, when the ApplicationContext detects such index, it will automatically use it rather than scanning the classpath.
> 资源地址：https://docs.spring.io/spring/docs/5.0.x/spring-framework-reference/core.html#beans- scanning-index。

假设 Spring 应用存在一个包含 `META-INF/spring.components` 资源的 `a.jar`，`b.jar` 包仅存在模式注解，那么`@ComponentScan` 扫描这两个 JAR 中的 package 时，`b.jar` 中的模式注解不会被识别。因此，读者在后续使用时，请务必注意这样的问题。

> 如果对 `org.springframework:spring-context-indexer` 处理机制感兴趣，则可参考 `org.springframework.context.index.CandidateComponentsIndexer` 的实现，其底层技术为“Java Annotation Processor”：https://docs.oracle.com/javase/7/docs/api/javax/annotation/processing/Processor.html。

Spring Framework 5.0 也引入了 JSR-305 适配注解，如`@NonNull`、`@Nullable` 等，为 Java 与 Kotlin 之间提供技术杠杆。其使用方式相对简单，读者可自行探索。

至此，注解驱动的各个时代均已讨论，下一节总结 Spring Framework 各个版本引入的核心注解。

7.2 Spring 核心注解场景分类

下面将 Spring 核心注解场景进行分类，按照 Spring Framework 版本由低至高，注解重要性由高到低，依次总结如下。此处不会将功能性注解纳入讨论的范畴，如 Web MVC 等，它们将在后续具体章节中回顾。

Spring 模式注解如下表所示。

Spring 注解	场景说明	起始版本
`@Repository`	数据仓储模式注解	2.0
`@Component`	通用组件模式注解	2.5
`@Service`	服务模式注解	2.5
`@Controller`	Web 控制器模式注解	2.5
`@Configuration`	配置类模式注解	3.0

装配注解如下表所示。

Spring 注解	场 景 说 明	起 始 版 本
@ImportResource	替换 XML 元素<import>	2.5
@Import	限定@Autowired 依赖注入范围	3.0
@ComponentScan	扫描指定 package 下标注 Spring 模式注解的类	3.1

依赖注入注解如下表所示。

Spring 注解	场 景 说 明	起 始 版 本
@Autowired	Bean 依赖注入，支持多种依赖查找方式	2.5
@Qualifier	细粒度的@Autowired 依赖查找	2.5
Java 注解	**场 景 说 明**	**起 始 版 本**
@Resource	Bean 依赖注入，仅支持名称依赖查找方式	2.5

Bean 定义注解如下表所示。

<table>
<tr><th>Spring 注解</th><th>场 景 说 明</th><th>起 始 版 本</th></tr>
<tr><td>@Bean</td><td>替换 XML 元素<bean></td><td>3.0</td></tr>
<tr><td>@DependsOn</td><td>替代 XML 属性<bean depends-on="..."/></td><td>3.0</td></tr>
<tr><td>@Lazy</td><td>替代 XML 属性<bean lazy-init="true|false"/></td><td>3.0</td></tr>
<tr><td>@Primary</td><td>替换 XML 元素<bean primary="true|false"/></td><td>3.0</td></tr>
<tr><td>@Role</td><td>替换 XML 元素<bean role="..."/></td><td>3.1</td></tr>
<tr><td>@Lookup</td><td>替代 XML 属性<bean lookup-method="..."></td><td>4.1</td></tr>
</table>

Spring 条件装配注解如下表所示。

Spring 注解	场 景 说 明	起 始 版 本
@Profile	配置化条件装配	3.1
@Conditional	编程条件装配	4.0

配属属性注解如下表所示。

Spring 注解	场 景 说 明	起 始 版 本
@PropertySource	配置属性抽象 PropertySource 注解	3.1
@PropertySources	@PropertySource 集合注解	4.0

生命周期回调注解如下表所示。

Java 注解	场 景 说 明	起 始 版 本
@PostConstruct	替换 XML 元素<bean init-method="..."/>或 InitializingBean	2.5

续表

Java 注解	场 景 说 明	起 始 版 本
@PreDestroy	替换 XML 元素<bean destroy-method="..." />或 DisposableBean	2.5

注解属性注解如下表所示。

Spring 注解	场 景 说 明	起 始 版 本
@AliasFor	别名注解属性，实现复用的目的	4.2

性能注解如下表所示。

Spring 注解	场 景 说 明	起 始 版 本
@Indexed	提升 Spring 模式注解的扫描效率	5.0

7.3 Spring 注解编程模型

前面回顾了 Spring Framework 注解驱动编程从 1.x 版本到 5.x 版本的演进过程，并且将不同职能的注解加以归类。大体掌握了 Spring Framework 注解驱动编程的能力，然而一些相关的术语和概念并未解释，其中疑团尚未理清，距深层次探讨仍有距离。因此，本节将主题定为“Spring Framework 注解编程模型”，主要原因是该主题也在 Spring Framework GitHub 的标题为“Spring Annotation Programming Model”的 Wiki 中出现，按其议题逐一讨论，包括：

- 元注解（Meta-Annotations）；
- Spring 模式注解（Stereotype Annotations）；
- Spring 组合注解（Composed Annotations）；
- Spring 注解属性别名和覆盖（Attribute Aliases and Overrides）。

> https://github.com/spring-projects/spring-framework/wiki/Spring-Annotation-Programming-Model。

理解 Spring 注解编程模型是深入探讨 Spring Boot 注解驱动及自动装配的重要前提。

7.3.1 元注解（Meta-Annotations）

元注解是注解编程模型的首要术语，在 Wiki 中开篇直述其定义：

> A *meta-annotation* is an annotation that is declared on another annotation. An annotation is therefore *meta-annotated* if it is annotated with another annotation. For example, any annotation that is declared to be *documented* is meta-annotated with `@Documented` from the `java.lang.annotation` package.

文中指出，所谓的元注解是指一个能声明在其他注解上的注解，如果一个注解标注在其他注解上，那么它就是元注解。同时举例说明`@Documented`能够作为任何注解的元注解。不难看出，元注解并非仅限定在Spring的使用场景中，而是“放诸Java而皆准”。同理，Java标准注解`@Inherited`、Java 8 `@Repeatable`均属于元注解。在Spring使用场景中，`@Component`就是标准的元注解，它在`@Repository`、`@Service`等注解上均有标注。不过在Spring注解编程模型中，`@Component`及被其标注的注解有一种专属术语描述，也就是下一节的讨论主题——“Spring 模式注解（Stereotype Annotations）”。

7.3.2 Spring模式注解（Stereotype Annotations）

“Spring模式注解”在前面的讨论中反复出现，然而并未对其做出具体的解释，此处引用在上述Wiki中的描述：

> A *stereotype annotation* is an annotation that is used to declare the role that a component plays within the application. For example, the `@Repository` annotation in the Spring Framework is a marker for any class that fulfills the role or *stereotype* of a repository (also known as Data Access Object or DAO).

“Stereotype annotation”中文直译为“模式注解”，Wiki举例说明`@Repository`作为仓储标记注解，管理和存储某种领域对象。它又进一步解释了`@Component`的定义：

> `@Component` is a generic stereotype for any Spring-managed component. Any component annotated with `@Component` is a candidate for component scanning. Similarly, any component annotated with an annotation that is itself meta-annotated with `@Component` is also a candidate for component scanning. For example, `@Service` is meta-annotated with `@Component`.

`@Component`作为一种由Spring容器托管的通用模式组件，任何被`@Component`标注的组件均为组件扫描的候选对象。类似地，凡是被`@Component`元标注（meta-annotated）的注解，如`@Service`，所标注的任何组件，也被视作组件扫描的候选对象：

```
@Target({ElementType.TYPE})
```

```
@Retention(RetentionPolicy.RUNTIME)
@Documented
@Component
public @interface Service {
    ...
 String value() default "";

}
```

由于`@Service`元标注（meta-annotated）`@Component`，当类 A 标注`@Service`注解后，A 也被认为是候选组件。官方继续解释：

> Core Spring provides several stereotype annotations out of the box, including but not limited to: `@Component`, `@Service`, `@Repository`, `@Controller`, `@RestController`, and `@Configuration`. `@Repository`, `@Service`, etc. are specializations of `@Component`.

Spring 核心部分提供了几种内建模式注解，如`@Component`、`@Service`、`@Repository`、`@Controller`、`@RestController`及`@Configuration`。这些注解皆为`@Component`的规范。实际上此处的描述有些含糊，确切地说，Spring Framework 并不限于这些内建模式注解，也能自定义模式注解。

为此，小马哥提出了一个容易理解和便于记忆的概念，Spring 模式注解即`@Component`“派生”注解。由于 Java 语言规范的规定，Annotation 不允许继承，没有类派生子类的能力。因此，Spring Framework 采用元标注方式实现注解之间的“派生”。

> 关于 Annotation 的语言限制，请参考 Java 语言规范（第三版）9.6 节：http://docs.oracle.com/javase/specs/jls/se6/html/interfaces.html#9.6。在 Discussion 部分中：
> No extends clause is permitted.(Annotation types implicitly extend `annotation.Annotation`.)

接下来的内容将结合具体实例来理解`@Component`“派生性”。

1. 理解@Component“派生性”

前文提到，在 Spring Framework 2.0 中，`@Repository`被新引入，被该 Annotation 标注的类被视作“仓储”，管理和存储某种领域对象，其原始设计来源 Evans 在 2003 年所编著的 *Domain-Driven Design*（领域驱动设计，简称 **DDD**）。

> Domain-Driven Design（领域驱动设计）也是目前流行的微服务架构的理论基础。

在 Spring Framework 2.0 中，**@Repository** 注解的定义如下：

```
@Target({ElementType.TYPE})
@Retention(RetentionPolicy.RUNTIME)
@Inherited
@Documented
public @interface Repository {
}
```

援引 Spring 官方**@Repository** JavaDoc 中的文字描述（节选）：

> The annotated class is also clarified as to its role in the overall application architecture for the purpose of tools, aspects, etc.

说明在 Spring Framwork 2.0 中，被**@Repository** 标注的类仅仅起到标记作用。

然而到了 Spring Framework 2.5，**@Repository** 的定义发生了根本性的变化：

```
@Target({ElementType.TYPE})
@Retention(RetentionPolicy.RUNTIME)
@Documented
@Component
public @interface Repository {

 /**
  * The value may indicate a suggestion for a logical component name,
  * to be turned into a Spring bean in case of an autodetected component.
  * @return the suggested component name, if any
  */
 String value() default "";

}
```

在 Spring Framework 2.5 中，官方在 **Repository** JavaDoc 中补充了一些文字：

> As of Spring 2.5, this annotation also serves as a specialization of @Component, allowing for implementation classes to be autodetected through classpath scanning.

简言之，在 Spring Framework 2.5 中，**@Repository** 不但是标记注解，而且作为 Spring Framework 的组件（Component）。值得关注的是，**@Repository** 在注解定义上，也被注解**@Component** 标注，

这不禁让人思考两者之间的关联。

回顾前文所提，Spring Framework 2.5 新引入了 XML 元素<context:component-scan>。小马哥也曾“剧透”Spring 模式注解，如@Component、@Repository、@Service，以及“派生”关系，@Repository 就是@Component 派生 Annotation。

为了检验@Component“派生性”结论的正确性，下面通过自定义@Component“派生”注解的方式加以验证。

> 任何论证过程都离不开其所处的环境，自然也少不了严谨的推导过程。这需要开发人员具备一定的工程意识，包括软件版本、特性范围，以及兼容情况等。因此，论证的过程中从最低版本开始推导，逐步证明不同版本的提升和差异。依照以上原则，这样的认证会在本书中大量地付诸实践，如本例检验@Component“派生性”，将 Spring Framework 2.5 作为起始版本来展开讨论。

2. 自定义@Component“派生”注解

（1）自定义注解@StringRepository。

模仿 Spring Framework @Repository 定义，构造一个类似的 Annotation @StringRepository：

```
@Target({ElementType.TYPE})
@Retention(RetentionPolicy.RUNTIME)
@Documented
@Component
public @interface StringRepository {

    /**
     * 属性方法名称必须与{@link Component#value()}保持一致
     * @return Bean 的名称
     */
    String value() default "";
}
```

（2）标注@StringRepository。

将@StringRepository 标注在一个类名为 NameRepository 的实现上：

```
@StringRepository("chineseNameRepository")
public class NameRepository {

    /**
```

```
     * 查找所有的名字
     *
     * @return non-null List
     */
    public List<String> findAll() {
        return Arrays.asList("张三", "李四", "小马哥");
    }
}
```

（3）部署@StringRepository Bean。

新建 META-INF/spring/context.xml 配置文件，并使用<context:component- scan/>元素扫描 NameRepository 类元信息：

```
<?xml version="1.0" encoding="UTF-8"?>
<beans xmlns="http://www.springframework.org/schema/beans"
       xmlns:xsi="http://www.w3.org/2001/XMLSchema-instance"
       xmlns:context="http://www.springframework.org/schema/context"
       xsi:schemaLocation="http://www.springframework.org/schema/beans
http://www.springframework.org/schema/beans/spring-beans.xsd
     http://www.springframework.org/schema/context
http://www.springframework.org/schema/context/spring-context.xsd">

  <!-- 找寻被@Component 或者其派生 Annotation 标记的类（Class），将它们注册为 Spring Bean -->
  <context:component-scan base-package="thinking.in.spring.boot.samples.spring25" />

</beans>
```

（4）实现@StringRepository 引导类。

实现引导程序，代码如下：

```
public class DerivedComponentAnnotationBootstrap {

    static {
        // 解决 Spring 2.5.x 不兼容 Java 8 的问题
        // 同时，请注意 Java Security 策略，必须具备 PropertyPermission
        System.setProperty("java.version", "1.7.0");
    }

    public static void main(String[] args) {
```

```
        // 构建 XML 配置驱动 Spring 上下文
        ClassPathXmlApplicationContext context = new ClassPathXmlApplicationContext();
        // 设置 XML 配置文件的位置
        context.setConfigLocation("classpath:/META-INF/spring/context.xml");
        // 启动上下文
        context.refresh();
        // 获取名称为“chineseNameRepository”的 Bean 对象
        NameRepository nameRepository = (NameRepository)
context.getBean("chineseNameRepository");
        // 输出用户名称：[张三, 李四, 小马哥]
        System.out.printf("nameRepository.findAll() = %s \n", nameRepository.findAll());
    }
}
```

> 源码位置：以上示例代码可通过查找 spring-framework-samples/spring-framework-2.5.6-sample 工程获取。

执行后，控制台输出如下：

```
(...部分内容被省略...)
nameRepository.findAll() = [张三, 李四, 小马哥]
```

上述示例的运行结果可以清楚地表明@StringRepository作为@Component派生的Annotation，经 Spring Framework 扫描后，其语义与@Repository类似，它们所标注的目标类被 Spring 应用上下文初始化为 Spring Bean，被 Spring 上下文管理。那么就不难理解，即使源码中没有出现@Repository的处理逻辑，同样能承担@Component 的职责。因此，小马哥称此等机制为**@Component“派生性”**。

或许读者会好奇，小马哥是如何发现@Component“派生性”的“秘密”呢？

这还得从一段往事说起。众所周知，HSF 是阿里巴巴的内部使用的 RPC 框架。早期的 HSF 并不支持注解驱动开发，后来小马哥的师傅实现了这种编程模式。通常 RPC 的使用方分为服务提供方和服务消费方。其中@HsfProvider 表示服务提供方注解，标注在服务实现类上。类似的注解在 Dubbo 中也有，即@Service。在 API 设计上，@HsfProvider 标注了 Spring Framework @Component。按照@Component“派生性”，凡是标注@HsfProvider 的服务实现类都能被 XML 元素<context:component-scan>或注解@ComponentScan 识别并注册为 Spring Bean。具备注解驱动能力的 HSF 大受同事的欢迎，再结合 Spring Boot Starter 自动装配后，小伙伴们纷纷升级体验其中开发的乐趣。然而，在一次 HSF 升级后，心急如焚的小伙伴反馈生产环境 HSF 服务不见了。经排查，由于当前 HSF 版本中的@HsfProvider 移除了@Component，导致 Spring 应用上下文无法识别服务实现类。当应用部署后，无法再对外提供服务，下游的服务消费应用自然出现故障。所幸及时发现问题，且微服务化改造后能够快速回滚，所以影响面不大。问题定位后，后续的 HSF 版本将移除的@Component 还原。然而这种修复方案并不是非常妥当。

实际上，我师傅他“老人家”在@HsfProvider 的设计上存在一些瑕疵。@HsfProvider 作为 HSF 独立的 Annotation，不应该依赖于 Spring Framework @Component，尽管它能运用@Component 的“派生性”。因此，小马哥在实现 Dubbo 注解驱动的过程中，就没有在服务实现类注解@Service 上标注@Component，而是利用 ClassPathBeanDefinitionScanner 和 AnnotationTypeFilter 相互配合，实现@Service 服务实现类的识别和注册。关于 Dubbo 注解驱动的相关细节，读者请参考小马哥的博文——Dubbo 注解驱动（Annotation-Driven）：https://mercyblitz.github.io/-注释驱动。

故事的发展并没有因问题的修复而终止，后来一些小伙伴找小马哥讨论其中的缘故，或许读者和他们一样，对技术充满强烈的热情。关于@Component“派生性”的奥秘，会与读者在下一节中一同探索，包括 ClassPathBeanDefinitionScanner 和 AnnotationTypeFilter 之间的关系。

3. @Component“派生性”原理

由于以上示例所依赖的 Spring Framework 版本为 2.5，该版本尚不支持@ComponentScan 注解的方式，扫描并注册@Component 组件，只能使用传统的 XML 配置文件的方式，利用 <context:component-scan>元素扫描@Component 组件。从 Spring Framework 2.0 开始，XML 配置文件的文档结构描述从 DTD（文档类型定义）变成 XML Schema，同时该版本引入了**可扩展的 XML 编写（Extensible XML authoring）**机制。该特性为开发人员提供了一种 XML 元素与 Bean 定义解析器之间的扩展机制。

Spring Framework 2.5 延续了以上特性。框架内建了<context:component-scan>元素，如示例配置文件 META-INF/spring/context.xml。该元素包含两部分信息：元素前缀 context 和 local 元素 component-scan。根据 XML Schema 规范，元素前缀需要显式地关联命名空间（namespace），如 XML 配置文件中的 xmlns:context="http://www.springframework.org/schema/context"。XML Schema 允许开发人员自定义元素前缀，而命名空间则是预先约定的，开发人员无法直接调整其内容。

尽管<context:component-scan>是 Spring Framework 内建元素，然而它同样遵照 Spring **可扩展的 XML 编写（Extensible XML authoring）**扩展机制。按照该机制规定，元素 XML Schema 命名空间需要与其处理类建立映射关系，且配置在相对于 classpath 的规约资源/META-INF/spring.handlers 文件中：

```
http\://www.springframework.org/schema/context=org.springframework.context.config.ContextNamespaceHandler
...
```

> 以上/META-INF/spring.handlers 资源存在于 spring-context-2.5.6.SEC03.jar 文件中。

/META-INF/spring.handlers 资源是以 Key-Value 格式配置的，且元素前缀并未出现在 Key 中。在 Spring 配置文件中，命名空间 http://www.springframework.org/schema/context 不会直接配置在元素上，而是使用元素前缀 context 代替，如上例中的<beans>元素声明：

```
    <beans xmlns="http://www.springframework.org/schema/beans"
         xmlns:xsi="http://www.w3.org/2001/XMLSchema-instance"
         xmlns:context="http://www.springframework.org/schema/context"
         xsi:schemaLocation="...
       http://www.springframework.org/schema/context
http://www.springframework.org/schema/context/spring-context.xsd">
    <beans/>
```

其中默认的命名空间声明为 xmlns="http://www.springframework.org/schema/beans"，所以 <bean> 元素不需要添加前缀。反之，由于 xmlns:context="http://www.springframework.org/schema/context"的作用，component-scan 元素需要增加 context 的前缀。在 XML Schema 校验的过程中，需要定位命名空间所对应的 schema 文件，这部分声明由 xsi:schemaLocation 属性关联，故上例配置中出现 xsi:schemaLocation="http://www.springframework.org/schema/context http://www.springframework.org/schema/context/spring-context.xsd"的声明。

> 关于 Schema 的简明教程，读者可参考 http://www.w3school.com.cn/schema/schema_schema.asp。而 DTD 的教程可参考 http://www.w3school.com.cn/dtd/index.asp。

综上所述，Spring 容器根据/META-INF/spring.handlers 的配置，定位到命名空间 context 所对应的处理器 ContextNamespaceHandler。当 Spring 应用上下文启动时，调用 ContextNamespaceHandler#init()方法，随后注册该命名空间下所有 local 元素的 Bean 定义解析器，包括当前运用的 component-scan 元素：

```
    public class ContextNamespaceHandler extends NamespaceHandlerSupport {

    public void init() {
        ...
        registerJava5DependentParser("component-scan",

    "org.springframework.context.annotation.ComponentScanBeanDefinitionParser");
        ...
```

```
    }
        ...
}
```

> 以上实现源码源于 Spring Framework 2.5.6.SEC03。

因此，<context:component-scan>元素的 Bean 定义解析器为 ComponentScanBeanDefinitionParser。

> 可扩展的 XML 编写（Extensible XML authoring）相关资源可参考官方文档：https://docs.spring.io/spring/docs/2.0.8/reference/extensible-xml.html。

顾名思义，Bean 定义解析器用于解析 Bean 的定义，其 API 为 BeanDefinitionParser，ComponentScanBeanDefinitionParser 为其中一种实现。回到上例的场景，当 Spring 应用上下文加载并解析 XML 配置文件/META-INF/spring/context.xml 后，当解析至<context:component-scan>元素时，ComponentScanBeanDefinitionParser#parse(Element, ParserContext)方法被调用：

```
public class ComponentScanBeanDefinitionParser implements BeanDefinitionParser {

    private static final String BASE_PACKAGE_ATTRIBUTE = "base-package";
        ...
    public BeanDefinition parse(Element element, ParserContext parserContext) {
        String[] basePackages =

    StringUtils.commaDelimitedListToStringArray(element.getAttribute(BASE_PACKAGE_ATTRIBUTE)
);

        // Actually scan for bean definitions and register them.
        ClassPathBeanDefinitionScanner scanner = configureScanner(parserContext, element);
        Set<BeanDefinitionHolder> beanDefinitions = scanner.doScan(basePackages);
        registerComponents(parserContext.getReaderContext(), beanDefinitions, element);

        return null;
    }
        ...
}
```

> 以上实现源码源于 Spring Framework 2.5.6.SEC03。

当 ComponentScanBeanDefinitionParser 读取 base-package 属性后，属性值作为扫描根路径，传入 ClassPathBeanDefinitionScanner#doScan(String...) 方法并返回 BeanDefinitionHolder 集合。而 BeanDefinitionHolder 包含 Bean 定义（BeanDefinition）与其 Bean 名称相关的信息：

```
public class BeanDefinitionHolder implements BeanMetadataElement {

    private final BeanDefinition beanDefinition;

    private final String beanName;

    private final String[] aliases;
        ...
}
```

> 以上实现源码源于 Spring Framework 2.5.6.SEC03。

按照上例的运行结果来推导，说明标注@StringRepository 的 NameRepository 被解析为 BeanDefinitionHolder 集合中的一员，最终 Spring 应用上下文将其初始化为 Spring Bean。那么，@Component“派生性”的奥秘将在 ClassPathBeanDefinitionScanner 中揭晓。

当 ClassPathBeanDefinitionScanner#doScan(String...)方法被 ComponentScanBeanDefinitionParser#parse(Element,ParserContext)方法调用后，它将利用 basePackages 参数迭代地执行 findCandidateComponents(String)方法。每次执行的结果都生成候选的 BeanDefnition 集合，即 candidates 变量：

```
public class ClassPathBeanDefinitionScanner extends
ClassPathScanningCandidateComponentProvider {
        ...
    protected Set<BeanDefinitionHolder> doScan(String... basePackages) {
        Set<BeanDefinitionHolder> beanDefinitions = new LinkedHashSet<BeanDefinitionHolder>();
        for (int i = 0; i < basePackages.length; i++) {
            Set<BeanDefinition> candidates = findCandidateComponents(basePackages[i]);
            ...
        }
        return beanDefinitions;
```

```
    }
        ...
}
```

以上实现源码源于 Spring Framework 2.5.6.SEC03。

而 findCandidateComponents(String) 方法则从父类 ClassPathScanningCandidateComponentProvider 中继承：

```
public class ClassPathScanningCandidateComponentProvider implements ResourceLoaderAware {
    ...
    public Set<BeanDefinition> findCandidateComponents(String basePackage) {
        Set<BeanDefinition> candidates = new LinkedHashSet<BeanDefinition>();
        try {
            String packageSearchPath = ResourcePatternResolver.CLASSPATH_ALL_URL_PREFIX +
                    resolveBasePackage(basePackage) + "/" + this.resourcePattern;
            Resource[] resources =
this.resourcePatternResolver.getResources(packageSearchPath);
            ...
            for (int i = 0; i < resources.length; i++) {
                Resource resource = resources[i];
                ...
                if (resource.isReadable()) {
                    MetadataReader metadataReader =
this.metadataReaderFactory.getMetadataReader(resource);
                    if (isCandidateComponent(metadataReader)) {
                        ScannedGenericBeanDefinition sbd = new
ScannedGenericBeanDefinition(metadataReader);
                        sbd.setResource(resource);
                        sbd.setSource(resource);
                        if (isCandidateComponent(sbd)) {
                            ...
                            candidates.add(sbd);
                        }
                        ...
                    }
                    ...
                }
```

```
                ...
            }
        }
        ...
        return candidates;
    }
    ...
}
```

> 以上实现源码源于 Spring Framework 2.5.6.SEC03。

findCandidateComponents(String)方法首先将 basePackage 参数转化为 ClassLoader 类资源（.class）搜索路径：

```
String packageSearchPath = ResourcePatternResolver.CLASSPATH_ALL_URL_PREFIX +
        resolveBasePackage(basePackage) + "/" + this.resourcePattern;
```

其中，resolveBasePackage(basePackage)方法先处理 backagePage 中的占位符，将${...}替换为实际的配置值，然后将其中的 Java package 路径分隔符“.”替换成资源路径分隔符“/”。假设 basePackage 为“thinking.in.spring.boot”，默认情况下，处理后的 packageSearchPath 为“classpath*:thinking/in/spring/boot/**/*.class”。该值随后作为 PathMatchingResourcePatternResolver#getResources(String)方法的参数，方法执行后得到类资源集合：

```
Resource[] resources = this.resourcePatternResolver.getResources(packageSearchPath);
```

接着，resources 被迭代地执行，当资源可读取时，即 resource.isReadable()为 true，获取该资源的 MetadataReader 对象：

```
MetadataReader metadataReader = this.metadataReaderFactory.getMetadataReader(resource);
```

尽管在日常开发中，开发人员几乎不会直接接触 MetadataReader 接口，不过它包含了类和注解的元信息读取方法：

```
public interface MetadataReader {

/**
 * Read basic class metadata for the underlying class.
 */
ClassMetadata getClassMetadata();
```

```
    /**
     * Read full annotation metadata for the underlying class.
     */
    AnnotationMetadata getAnnotationMetadata();

    }
```

这些元信息是后续 isCandidateComponent(MetadataReader)方法的判断依据，同样是 ScannedGenericBeanDefinition 的元信息来源：

```
    public class ScannedGenericBeanDefinition extends GenericBeanDefinition implements
AnnotatedBeanDefinition {

    private final AnnotationMetadata metadata;


    /**
     * Create a new ScannedGenericBeanDefinition for the class that the
     * given MetadataReader describes.
     * @param metadataReader the MetadataReader for the scanned target class
     */
    public ScannedGenericBeanDefinition(MetadataReader metadataReader) {
         Assert.notNull(metadataReader, "MetadataReader must not be null");
         this.metadata = metadataReader.getAnnotationMetadata();
         setBeanClassName(this.metadata.getClassName());
    }


    public final AnnotationMetadata getMetadata() {
         return this.metadata;
    }

    }
```

而 BeanDefinition 集合 candidates 的候选条件由两个 isCandidateComponent 方法决定。在 isCandidateComponent(MetadataReader)方法中，条件的判断由 excludeFilters 和 includeFilters 字段决定：

```
public class ClassPathScanningCandidateComponentProvider implements ResourceLoaderAware {
    ...
    protected boolean isCandidateComponent(MetadataReader metadataReader) throws IOException
{
        for (TypeFilter tf : this.excludeFilters) {
            if (tf.match(metadataReader, this.metadataReaderFactory)) {
                return false;
            }
        }
        for (TypeFilter tf : this.includeFilters) {
            if (tf.match(metadataReader, this.metadataReaderFactory)) {
                return true;
            }
        }
        return false;
    }
    ...
}
```

而 ClassPathBeanDefinitionScanner 对象在被 ComponentScanBeanDefinitionParser #parse(Element,ParserContext)方法构造时，默认调用以下构造器：

```
public class ClassPathBeanDefinitionScanner extends
ClassPathScanningCandidateComponentProvider {
    ...
    public ClassPathBeanDefinitionScanner(BeanDefinitionRegistry registry, boolean
useDefaultFilters) {
        super(useDefaultFilters);
        ...
    }
    ...
}
```

默认情况下，构造参数 useDefaultFilters 为 true，并且显式地传递给父类构造器：

```
public class ClassPathScanningCandidateComponentProvider implements ResourceLoaderAware {
    ...
    public ClassPathScanningCandidateComponentProvider(boolean useDefaultFilters) {
        if (useDefaultFilters) {
```

```
            registerDefaultFilters();
        }
    }
    ...
    protected void registerDefaultFilters() {
      this.includeFilters.add(new AnnotationTypeFilter(Component.class));
    }
    ...
}
```

构造器执行 registerDefaultFilters()方法，随后给 includeFilters 字段增添一个包含@Component 类型信息的 AnnotationTypeFilter 实例。同时，excludeFilters 字段为空。换言之，new AnnotationTypeFilter(Component.class)决定了 BeanDefinition 在扫描后是否被选。具体而言，AnnotationTypeFilter 能够识别标注@Component、@Repository、@Service 或@Controller 的 BeanDefinition。诚如 ClassPathBeanDefinitionScanner JavaDoc 的描述：

> The default filters include classes that are annotated with Spring's @Component, @Repository, @Service, or @Controller stereotype.

ClassPathBeanDefinitionScanner 的默认过滤器引入标注@Component、@Repository、@Service 或@Controller 的类。同理，它也能够标注所有@Component 的“派生”注解，比如本例的@StringRepository 及 Spring Framework 3.0 引入的@Configuration：

```
@Target(ElementType.TYPE)
@Retention(RetentionPolicy.RUNTIME)
@Documented
@Component
public @interface Configuration {
   ...
   String value() default "";

}
```

同时，请注意 ClassPathBeanDefinitionScanner 的另外一句 JavaDoc 描述：

> Candidate classes are detected through configurable type filters.

ClassPathBeanDefinitionScanner 允许自定义类型过滤规则，因此，Dubbo 的@Service 在

没有标注@Component 的情况下，通过 scanner.addIncludeFilter(new AnnotationTypeFilter (Service.class))的方式达到识别@Service 所标注类的目的。不过，这种方式没有使用@Component“派生性”。

> 关于识别 Dubbo @Service 的相关实现，可参考 com.alibaba.dubbo.config.spring.beans.factory.annotation.ServiceAnnotationBeanPostProcessor#registerServiceBeans 方法，其中 Dubbo 版本必须为 2.5.8 或更高。

4. 多层次@Component“派生性”

构建在 Spring Framework 4.x 及更高版本上的 Spring Boot，自然也继承了 Spring Framework 2.5 以来的@Component“派生性”的能力。比如最常见的@SpringBootApplication，Spring Boot 2.0.0.RELEASE 官方文档中对其有如下描述：

> The @SpringBootApplication annotation is equivalent to using @Configuration, @EnableAutoConfiguration, and @ComponentScan with their default attributes

按照文档的说明，@SpringBootApplication 等价于@Configuration、@EnableAutoConfiguration 和@ComponentScan 的联合注解。再参考@SpringBootApplication 的注解声明：

```
@Target(ElementType.TYPE)
@Retention(RetentionPolicy.RUNTIME)
@Documented
@Inherited
@SpringBootConfiguration
@EnableAutoConfiguration
@ComponentScan(excludeFilters = {
      @Filter(type = FilterType.CUSTOM, classes = TypeExcludeFilter.class),
      @Filter(type = FilterType.CUSTOM, classes = AutoConfigurationExcludeFilter.class) })
public @interface SpringBootApplication {
    ...
}
```

发现@EnableAutoConfiguration 和@ComponentScan 的确标注在@SpringBootApplication 注解上，然而并没有找到@Configuration 的声明。出于职业的嗅觉，@SpringBootConfiguration 看起来与@Configuration 有些神似，不由自主地观察其注解声明：

```
@Target(ElementType.TYPE)
```

```
@Retention(RetentionPolicy.RUNTIME)
@Documented
@Configuration
public @interface SpringBootConfiguration {

}
```

惊奇地发现 SpringBootConfiguration 标注了@Configuration，已知@Configuration 是@Component 的“派生”注解，那是不是意味着@SpringBootApplication 也是@Component 的“派生”注解呢？这样的猜测并非毫无根据，分析@SpringBootApplication 和@Component 之间层次关系，如下所示。

```
@SpringBootApplication
 |- @SpringBootConfiguration
     |- @Configuration
         |- @Component
```

> 在 Spring Boot 1.4 之前，@SpringBootApplication 标注的注解为@Configuration、@EnableAutoConfiguration 和@ComponentScan。从 Spring Boot 1.4 开始，@SpringBootConfiguration 被引入，替换了@Configuration。

假设之前的猜测成立，那么是否意味着@Component“派生性”具备多层次的能力呢？

要证明以上能力，只需证明标注@SpringBootApplication 的引导类也是 Spring Bean 即可，如下例所示。

```
@SpringBootApplication
public class DerivedComponentHierarchyBootstrap {

    public static void main(String[] args) {
        // 当前引导类
        Class<?> bootstrapClass = DerivedComponentHierarchyBootstrap.class;
        // 运行 Spring Boot，并返回 Spring 应用上下文
        ConfigurableApplicationContext context = new
SpringApplicationBuilder(bootstrapClass)
                .web(WebApplicationType.NONE) // 非 Web 类型
                .run();
        System.out.println("当前引导类 Bean : " + context.getBean(bootstrapClass));
        // 关闭 Spring 应用上下文
```

```
        context.close();
    }
}
```

> 源码位置：以上示例代码可通过查找 spring-boot-2.0-samples/auto-configuration-sample 工程获取。

当 `SpringApplicationBuilder#run()`方法执行后，运行 Spring Boot，并返回 Spring 应用上下文 `ConfigurableApplicationContext` 对象。运行以上引导类，观察日志输出：

```
(...部分内容被省略...)
当前引导类 Bean ：
thinking.in.spring.boot.samples.auto.configuration.bootstrap.DerivedComponentHierarchyBootstrap$$EnhancerBySpringCGLIB$$e1e284fb@22875539
(...部分内容被省略...)
```

从结果来看，确实@SpringBootApplication 也是多层次@Component 派生注解。换言之，证明了@Component“派生性”具备多层次。

回到 `DerivedComponentAnnotationBootstrap` 的例子，该引导类依赖于 Spring Framework 2.5，已知该版本也具备**@Component“派生性”**的能力。不过该 Spring Framework 版本是否支持多层次@Component“派生性”尚未可知。因此，将该例中的**@StringRepository** 注解定义调整如下（之前标注为@Component，被替换成@Repository）：

```
@Target({ElementType.TYPE})
@Retention(RetentionPolicy.RUNTIME)
@Documented
@Repository // 之前标注为@Component，被替换成@Repository
public @interface StringRepository {

    /**
     * 属性方法必须与{@link Component#value()}保持一致
     * @return Bean 的名称
     */
    String value() default "";
}
```

其他代码不变，再次运行引导程序 `DerivedComponentAnnotationBootstrap`，观察控制台输出：

```
Exception in thread "main" org.springframework.beans.factory.NoSuchBeanDefinitionException:
No bean named 'chineseNameRepository' is defined at
    (...部分内容被省略...)
    thinking.in.spring.boot.samples.spring25.bootstrap.DerivedComponentAnnotationBootstrap.ma
in(DerivedComponentAnnotationBootstrap.java:30)
```

异常内容提醒 `NameRepository` 不再视作 Spring Bean。这也意味着`@StringRepository` 的`@Component` 派生层次性发生了变化，调整前的层次性如下：

```
@StringRepository
 |- @Component
```

调整后的层次性如下：

```
@StringRepository
 |- @Repository
      |- @Component
```

说明 Spring Framework 2.5 并不支持多层次`@Component` 派生性。而在 Spring Boot 示例中，`DerivedComponentHierarchyBootstrap` 却能支持。换言之，Spring Boot 底层所依赖的 Spring Framework 4.x 弥补了 Spring Framework 2.5 的`@Component` 派生性限制。必然从 Spring Framework 某个版本开始支持多层次`@Component` 派生性，那么具体是哪个版本呢？

这个问题无法直接证明，只能逐个版本尝试。`DerivedComponentAnnotationBootstrap` 所依赖的 Spring Framework 版本是 `2.5.6.SEC03`，也是 `2.5.x` 版本中的最高发行版本。

第一步，将 **spring-framework-samples/spring-framework-2.5.6-sample** 工程中的 `pom.xml` 文件的 Spring Framework 依赖版本临时升级到下一个发行版本：`3.0.0.RELEASE`。

```
<?xml version="1.0" encoding="UTF-8"?>
<project xmlns="http://maven.apache.org/POM/4.0.0"
         xmlns:xsi="http://www.w3.org/2001/XMLSchema-instance"
         xsi:schemaLocation="http://maven.apache.org/POM/4.0.0
http://maven.apache.org/xsd/maven-4.0.0.xsd">
    ...
    <properties>
        <!--<spring.version>2.5.6.SEC03</spring.version>-->
        <!-- 测试 @StringRepository 多层次 @Component 派生时，请将以下内容反注释，升级 Spring
Framework 到 3.0.0.RELEASE  -->
        <spring.version>3.0.0.RELEASE</spring.version>
    </properties>
```

```
    <dependencies>
        <!-- Spring Framework 的版本: ${spring.version} -->
        <dependency>
            <groupId>org.springframework</groupId>
            <artifactId>spring-context</artifactId>
            <version>${spring.version}</version>
        </dependency>
    </dependencies>

</project>
```

第二步，调整 @StringRepository 的@Component “派生”层次性。

```
@Target({ElementType.TYPE})
@Retention(RetentionPolicy.RUNTIME)
@Documented
//@Component // 测试多层次@Component 派生，将当前注释
 @Repository // 测试多层次@Component 派生，将当前反注释，并且将 spring-context 升级到
3.0.0.RELEASE
public @interface StringRepository {

    /**
     * 属性方法必须与{@link Component#value()}保持一致
     * @return Bean 的名称
     */
    String value() default "";
}
```

重启 DerivedComponentAnnotationBootstrap，观察日志的变化：

```
(...部分内容被省略...)
nameRepository.findAll() = [张三, 李四, 小马哥]
```

运行结果说明 Spring Framework 3.0.0.RELEASE 开始支持**多层次@Component“派生性”**。或许 Spring Framework 的作者意识到在 2.5.x 版本中@Component “派生”的局限性，所以在 3.0.0 版本中提升了该特性。

上述的方式仅从表现上说明了 Spring Framework 3.0.0 开始对**多层次@Component“派生性”**的支持，然而实现层面发生了哪些变化呢？由于以下的讨论过程相对复杂，对细节不太敏感的读者可

以跳过该部分，记住“多层次@Component 派生性”的结论即可。

> **建议**：架构师或中间件开发人员可参考以下原理分析，为后续理解 Spring Boot 复杂的注解提供帮助。

5. 多层次@Component 派生性原理

在讨论@Component“派生性”原理一节中，已知 ClassPathScanningCandidateComponentProvider#findCandidateComponents 默认在指定根包路径（basePackage）下，将查找所有标注@Component 及“派生”注解的 BeanDefinition 集合，并且默认的过滤规则由 AnnotationTypeFilter 及@Component 的元信息（ClassMetaData 和 AnnotationMetaData）共同决定。然而由于@Component 派生性在 Spring Framework 2.5 和 3.0 之间的差异，以上实现必然存在变更。下面将使用源码对比的方式，找出 ClassPathScanningCandidateComponentProvider、AnnotationTypeFilter 及 MetadataReader 在 2.5.6.SEC03 和 3.0.0.RELEASE 之间的实现变化。

1）对比 ClassPathScanningCandidateComponentProvider 之间的实现差异

第一处变化是 registerDefaultFilters()方法，Spring Framework 3.0.0.RELEASE 开始新增 JSR-250 @ManagedBean 和 JSR-330 @Named 的 AnnotationTypeFilter：

```
public class ClassPathScanningCandidateComponentProvider implements, ResourceLoaderAware {
    ...
    protected void registerDefaultFilters() {
        this.includeFilters.add(new AnnotationTypeFilter(Component.class));
        ClassLoader cl = ClassPathScanningCandidateComponentProvider.class.getClassLoader();
        try {
            this.includeFilters.add(new AnnotationTypeFilter(
                    ((Class<? extends Annotation>)
ClassUtils.forName("javax.annotation.ManagedBean", cl)), false));
            logger.debug("JSR-250 'javax.annotation.ManagedBean' found and supported for
component scanning");
        }
        catch (ClassNotFoundException ex) {
            // JSR-250 1.1 API (as included in Java EE 6) not available - simply skip.
        }
        try {
            this.includeFilters.add(new AnnotationTypeFilter(
                    ((Class<? extends Annotation>) ClassUtils.forName("javax.inject.Named",
cl)), false));
            logger.debug("JSR-330 'javax.inject.Named' annotation found and supported for
```

```
component scanning");
        }
        catch (ClassNotFoundException ex) {
            // JSR-330 API not available - simply skip.
        }
    }
    ...
}
```

第二处差异在 `findCandidateComponents(String)`方法，如下图所示。

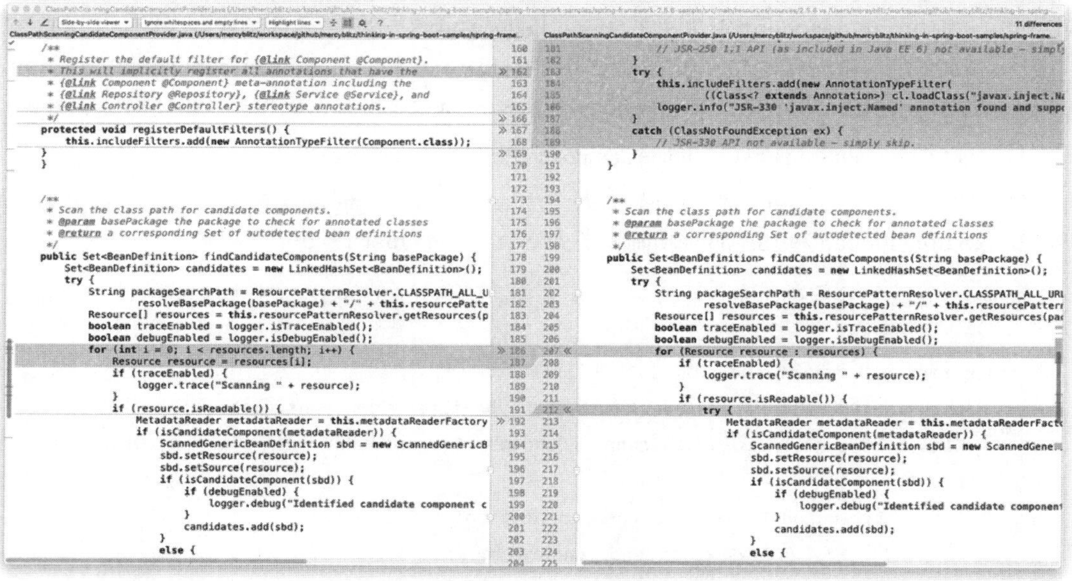
Side-by-side viewer
Ignore whitespaces and empty lines
Highlight lines
11 differences
ClassPathScanningCandidateComponentProvider.java (/Users/mercyblitz/workspace/github/mercyblitz/thinking-in-spring-boot-samples/spring-frame...
/**
* Register the default filter for {@link Component @Component}.
* This will implicitly register all annotations that have the
* {@link Component @Component} meta-annotation including the
* {@link Repository @Repository}, {@link Service @Service}, and
* {@link Controller @Controller} stereotype annotations.
*/
protected void registerDefaultFilters() {
this.includeFilters.add(new AnnotationTypeFilter(Component.class));
}
}
/**
* Scan the class path for candidate components.
* @param basePackage the package to check for annotated classes
* @return a corresponding Set of autodetected bean definitions
*/
public Set<BeanDefinition> findCandidateComponents(String basePackage) {
Set<BeanDefinition> candidates = new LinkedHashSet<BeanDefinition>();
try {
String packageSearchPath = ResourcePatternResolver.CLASSPATH_ALL_U
resolveBasePackage(basePackage) + "/" + this.resourcePatte
Resource[] resources = this.resourcePatternResolver.getResources(p
boolean traceEnabled = logger.isTraceEnabled();
boolean debugEnabled = logger.isDebugEnabled();
for (int i = 0; i < resources.length; i++) {
Resource resource = resources[i];
if (traceEnabled) {
logger.trace("Scanning " + resource);
}
if (resource.isReadable()) {
MetadataReader metadataReader = this.metadataReaderFactory
if (isCandidateComponent(metadataReader)) {
ScannedGenericBeanDefinition sbd = new ScannedGenericB
sbd.setResource(resource);
sbd.setSource(resource);
if (isCandidateComponent(sbd)) {
if (debugEnabled) {
logger.debug("Identified candidate component c
}
candidates.add(sbd);
}
else {
ClassPathScanningCandidateComponentProvider.java (/Users/mercyblitz/workspace/github/mercyblitz/thinking-in-spring-boot-samples/spring-frame...
// JSR-250 1.1 API (as included in Java EE 6) not available - simpl
}
try {
this.includeFilters.add(new AnnotationTypeFilter(
((Class<? extends Annotation>) cl.loadClass("javax.inject.N
logger.info("JSR-330 'javax.inject.Named' annotation found and supp
}
catch (ClassNotFoundException ex) {
// JSR-330 API not available - simply skip.
}
}
/**
* Scan the class path for candidate components.
* @param basePackage the package to check for annotated classes
* @return a corresponding Set of autodetected bean definitions
*/
public Set<BeanDefinition> findCandidateComponents(String basePackage) {
Set<BeanDefinition> candidates = new LinkedHashSet<BeanDefinition>();
try {
String packageSearchPath = ResourcePatternResolver.CLASSPATH_ALL_URL
resolveBasePackage(basePackage) + "/" + this.resourcePatter
Resource[] resources = this.resourcePatternResolver.getResources(pac
boolean traceEnabled = logger.isTraceEnabled();
boolean debugEnabled = logger.isDebugEnabled();
for (Resource resource : resources) {
if (traceEnabled) {
logger.trace("Scanning " + resource);
}
if (resource.isReadable()) {
try {
MetadataReader metadataReader = this.metadataReaderFact
if (isCandidateComponent(metadataReader)) {
ScannedGenericBeanDefinition sbd = new ScannedGene
sbd.setResource(resource);
sbd.setSource(resource);
if (isCandidateComponent(sbd)) {
if (debugEnabled) {
logger.debug("Identified candidate componen
}
candidates.add(sbd);
}
else {

> 以上对比图片的源代码存放在 spring-framework-samples/spring-framework-2.5.6-sample 工程的版本子目录中：
>
> - sources/2.5.6/ClassPathScanningCandidateComponentProvider.java；
> - sources/3.0.0/ClassPathScanningCandidateComponentProvider.java。
>
> 请读者后续自行比较。

图片左边的内容是 `2.5.6.SEC03` 实现，右边则是 `3.0.0.RELEASE` 实现，实际上两个版本并没有本质的变化。将“for”语句调整为 Java 5 “for each”语句，且增加了一个无关判断的“try catch”。

> 由于 Spring Framework 3 最小的 Java 依赖版本为 5，故有此调整。

2）对比 AnnotationTypeFilter 的实现差异

`AnnotationTypeFilter` 唯一的代码差异如下图所示。

```java
    /**
     * Create a new AnnotationTypeFilter for the given annotation type.
     * @param annotationType the annotation type to match
     * @param considerMetaAnnotations whether to also match on meta-annotations
     */
    public AnnotationTypeFilter(Class<? extends Annotation> annotationType, boo
        super(annotationType.isAnnotationPresent(Inherited.class), false);
        this.annotationType = annotationType;
        this.considerMetaAnnotations = considerMetaAnnotations;
    }


    @Override
    protected boolean matchSelf(MetadataReader metadataReader) {
        AnnotationMetadata metadata = metadataReader.getAnnotationMetadata();
        return metadata.hasAnnotation(this.annotationType.getName()) ||
                (this.considerMetaAnnotations && metadata.hasMetaAnnotation(thi
    }

    @Override
    protected Boolean matchSuperClass(String superClassName) {
        if (Object.class.getName().equals(superClassName)) {
            return Boolean.FALSE;
        }
        else if (superClassName.startsWith("java.")) {
            try {
                Class clazz = getClass().getClassLoader().loadClass(superClassN
                return Boolean.valueOf(clazz.getAnnotation(this.annotationType)
            }
            catch (ClassNotFoundException ex) {
                // Class not found - can't determine a match that way.
            }
        }
        return null;
    }

}
```

```java
53
54      /**
55       * Create a new AnnotationTypeFilter for the given annotation type.
56       * @param annotationType the annotation type to match
57       * @param considerMetaAnnotations whether to also match on meta-annotations
58       */
59      public AnnotationTypeFilter(Class<? extends Annotation> annotationType, bool
60          super(annotationType.isAnnotationPresent(Inherited.class), false);
61          this.annotationType = annotationType;
62          this.considerMetaAnnotations = considerMetaAnnotations;
63      }
64
65
66      @Override
67      protected boolean matchSelf(MetadataReader metadataReader) {
68          AnnotationMetadata metadata = metadataReader.getAnnotationMetadata();
69          return metadata.hasAnnotation(this.annotationType.getName()) ||
70                  (this.considerMetaAnnotations && metadata.hasMetaAnnotation(this.
71      }
72
73      @Override
74      protected Boolean matchSuperClass(String superClassName) {
75          if (Object.class.getName().equals(superClassName)) {
76              return Boolean.FALSE;
77          }
78          else if (superClassName.startsWith("java.")) {
79              try {
80                  Class<?> clazz = getClass().getClassLoader().loadClass(superClas
81                  return (clazz.getAnnotation(this.annotationType) != null);
82              }
83              catch (ClassNotFoundException ex) {
84                  // Class not found - can't determine a match that way.
85              }
86          }
87          return null;
88      }
89
90  }
91
```

> 图片左边的内容是 `2.5.6.SEC03` 实现，右边则是 `3.0.0.RELEASE` 实现，对比源代码分别存放在 spring-framework-samples/spring-framework-2.5.6-sample 工程的版本子目录中：
>
> - sources/2.5.6/AnnotationTypeFilter.java；
> - sources/3.0.0/AnnotationTypeFilter.java。
>
> 请读者后续自行比较。

两者同样没有本质上的差异，不过后者使用了 Java 5 的“Boxing”特性，并不影响结果的判断。

> 由于 Spring Framework 3 最小的 Java 依赖版本为 5，故有此调整。

3）对比 MetadataReader 的实现差异

在 Spring Framework 中，`MetadataReader` 接口唯一的实现是非公开类 `SimpleMetadataReader`，该实现类在 3.0.0.RELEAS 与 2.5.6.SEC03 之间并没有本质上的区别，如下图所示。

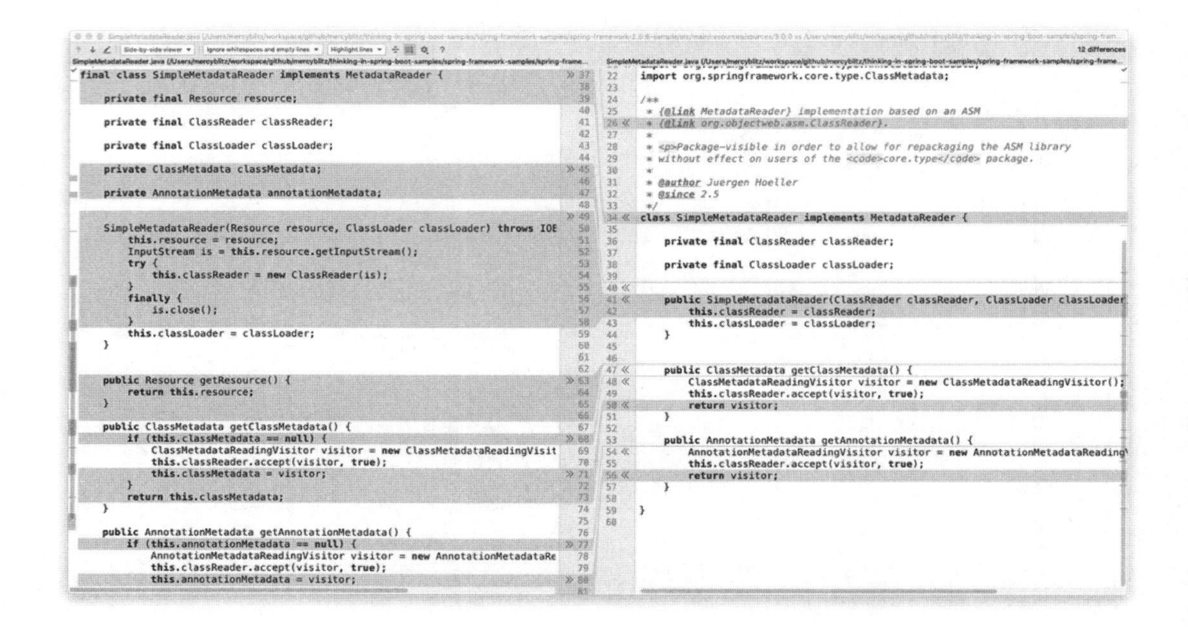

> 图片左边的内容是 `2.5.6.SEC03` 实现，右边则是 `3.0.0.RELEASE` 实现，对比源代码存放在 spring-framework-samples/spring-framework-2.5.6-sample 工程的版本子目录中：
>
> - sources/2.5.6/SimpleMetadataReader.java；
> - sources/3.0.0/SimpleMetadataReader.java。
>
> 请读者后续自行比较。

`SimpleMetadataReader` 在 `3.0.0.RELEASE` 中增加了 `classMetadata` 和 `annotationMetadata` 字段，当方法多次调用时，可以避免重复创建。`ClassPathScanningCandidateComponentProvider`、`AnnotationTypeFilter` 及 `SimpleMetadataReader` 都没有发生根本性的变化。然而代码之中没有奇迹，就算有的话，也不存在稳定复现，因此当前的讨论必然在某个环节存在漏洞。从 `SimpleMetadataReader` 的实现可以看出，`ClassMetadataReadingVisitor` 和 `AnnotationMetadataReadingVisitor` 分别是 `ClassMetadata` 和 `AnnotationMetadata` 的实现类。已知 `ClassPathScanningCandidateComponentProvider` 在寻找候选的 `BeanDefinition` 的过程中，将指定 `basePackage` 参数下的`*.class` 资源进行元信息解析，也就是 `ClassMetadata` 和 `AnnotationMetadata` 的对象。这两个 API 的变更自然也会影响 `BeanDefinition` 候选的结果。因此，仍需将实现类 `ClassMetadataReadingVisitor` 与 `AnnotationMetadataReadingVisitor` 分别展开讨论。

4）对比 ClassMetadataReadingVisitor 的实现差异

此处仍采用前后版本对比的方式，下图左边的是 `2.5.6.SEC03` 的实现，右边是 `3.0.0.RELEASE` 的实现：

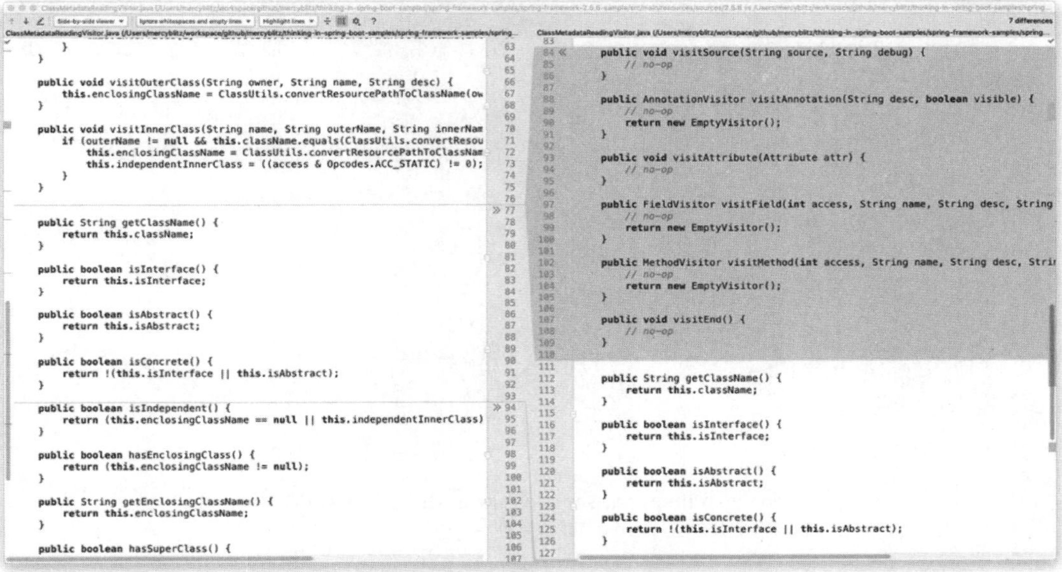

> 对比源代码存放在 spring-framework-samples/spring-framework-2.5.6-sample 工程的版本子目录中：
>
> - sources/2.5.6/ClassMetadataReadingVisitor.java；
> - sources/3.0.0/ClassMetadataReadingVisitor.java。
>
> 请读者后续自行比较。

`3.0.0.RELEASE` 版本的 `ClassMetadataReadingVisitor` 新增了一些 `visit*`方法，并且其中不是空实现，就是 `EmptyVisitor`，`EmptyVisitor` 所有的方法也是空实现。因此，差异性并未在 `ClassMetadataReadingVisitor` 实现层面体现。那么两个版本的差异应该会在 `AnnotationMetadataReadingVisitor` 中出现。

5）对比 AnnotationMetadataReadingVisitor 的实现差异

继续对比 `AnnotationMetadataReadingVisitor` 在 `2.5.6.SEC03`（下图右）和 `3.0.0.RELEASE`（下图左）中的差别。

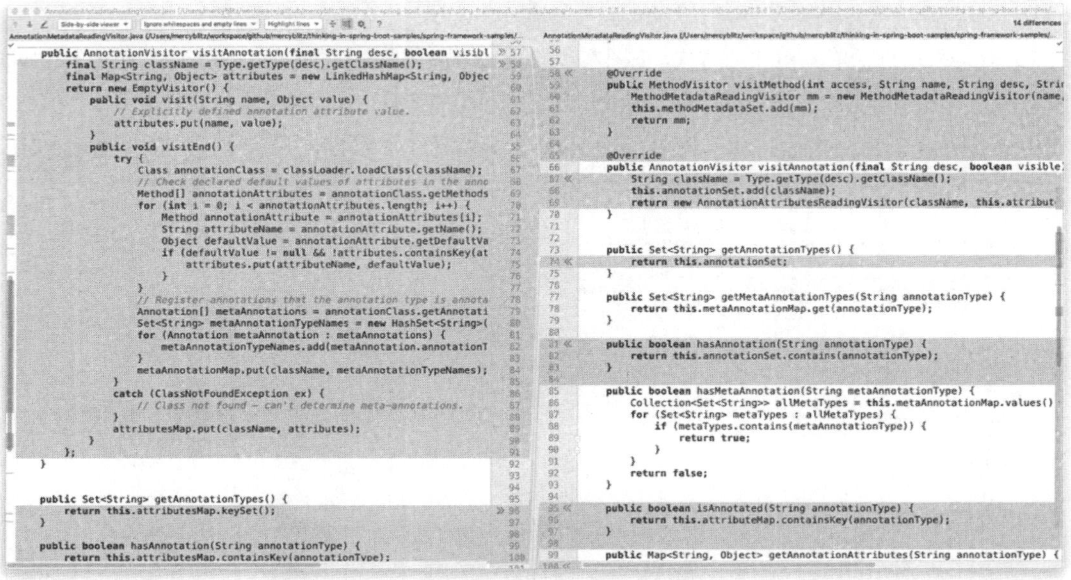

Side-by-side viewer   Ignore whitespaces and empty lines   Highlight lines   14 differences

AnnotationMetadataReadingVisitor.java (/Users/mercyblitz/workspace/github/mercyblitz/thinking-in-spring-boot-samples/spring-framework-samples/...

```
    public AnnotationVisitor visitAnnotation(final String desc, boolean visibl
        final String className = Type.getType(desc).getClassName();
        final Map<String, Object> attributes = new LinkedHashMap<String, Objec
        return new EmptyVisitor() {
            public void visit(String name, Object value) {
                // Explicitly defined annotation attribute value.
                attributes.put(name, value);
            }
            public void visitEnd() {
                try {
                    Class annotationClass = classLoader.loadClass(className);
                    // Check declared default values of attributes in the anno
                    Method[] annotationAttributes = annotationClass.getMethods
                    for (int i = 0; i < annotationAttributes.length; i++) {
                        Method annotationAttribute = annotationAttributes[i];
                        String attributeName = annotationAttribute.getName();
                        Object defaultValue = annotationAttribute.getDefaultVa
                        if (defaultValue != null && !attributes.containsKey(at
                            attributes.put(attributeName, defaultValue);
                        }
                    }
                    // Register annotations that the annotation type is annota
                    Annotation[] metaAnnotations = annotationClass.getAnnotati
                    Set<String> metaAnnotationTypeNames = new HashSet<String>(
                    for (Annotation metaAnnotation : metaAnnotations) {
                        metaAnnotationTypeNames.add(metaAnnotation.annotationT
                    }
                    metaAnnotationMap.put(className, metaAnnotationTypeNames);
                }
                catch (ClassNotFoundException ex) {
                    // Class not found - can't determine meta-annotations.
                }
                attributesMap.put(className, attributes);
            }
        };
    }


    public Set<String> getAnnotationTypes() {
        return this.attributesMap.keySet();
    }

    public boolean hasAnnotation(String annotationType) {
        return this.attributesMap.containsKey(annotationType);
```

AnnotationMetadataReadingVisitor.java (/Users/mercyblitz/workspace/github/mercyblitz/thinking-in-spring-boot-samples/spring-framework-samples/...

```
    @Override
    public MethodVisitor visitMethod(int access, String name, String desc, Strin
        MethodMetadataReadingVisitor mm = new MethodMetadataReadingVisitor(name,
        this.methodMetadataSet.add(mm);
        return mm;
    }

    @Override
    public AnnotationVisitor visitAnnotation(final String desc, boolean visible
        String className = Type.getType(desc).getClassName();
        this.annotationSet.add(className);
        return new AnnotationAttributesReadingVisitor(className, this.attribut
    }


    public Set<String> getAnnotationTypes() {
        return this.annotationSet;
    }

    public Set<String> getMetaAnnotationTypes(String annotationType) {
        return this.metaAnnotationMap.get(annotationType);
    }

    public boolean hasAnnotation(String annotationType) {
        return this.annotationSet.contains(annotationType);
    }

    public boolean hasMetaAnnotation(String metaAnnotationType) {
        Collection<Set<String>> allMetaTypes = this.metaAnnotationMap.values()
        for (Set<String> metaTypes : allMetaTypes) {
            if (metaTypes.contains(metaAnnotationType)) {
                return true;
            }
        }
        return false;
    }

    public boolean isAnnotated(String annotationType) {
        return this.attributeMap.containsKey(annotationType);
    }

    public Map<String, Object> getAnnotationAttributes(String annotationType) {
```

> 对比源代码存放在 spring-framework-samples/spring-framework-2.5.6-sample 工程的版本子目录中：
>
> - sources/2.5.6/AnnotationMetadataReadingVisitor.java；
> - sources/3.0.0/AnnotationMetadataReadingVisitor.java。
>
> 请读者后续自行比较。

两者的实现区别主要集中在 visitAnnotation(String,boolean）方法上，3.0.0.RELEASE 中的 AnnotationMetadataReadingVisitor 实现使用 AnnotationAttributesReadingVisitor 取代 2.5.6.SEC03 中的匿名 EmptyVisitor 类实现，两种方式的实现差别在各自的 visitEnd()方法上。先分析 2.5.6.SEC03 匿名 EmptyVisitor 类的 visitEnd()方法实现：

```
public void visitEnd() {
    try {
        Class annotationClass = classLoader.loadClass(className);
        ...
        // Register annotations that the annotation type is annotated with.
        Annotation[] metaAnnotations = annotationClass.getAnnotations();
        Set<String> metaAnnotationTypeNames = new HashSet<String>();
        for (Annotation metaAnnotation : metaAnnotations) {
            metaAnnotationTypeNames.add(metaAnnotation.annotationType().getName());
```

```
            }
            metaAnnotationMap.put(className, metaAnnotationTypeNames);
        }
        catch (ClassNotFoundException ex) {
            // Class not found - can't determine meta-annotations.
        }
        attributesMap.put(className, attributes);
    }
```

回顾@NameRepository 模式注解层次关系，@NameRepository 元标注了@SpringRepository，而后者又元标注了@Component：

```
@Component
 @SpringRepository
    |- @NameRepository
```

运用到以上 visitEnd() 方法实现，annotationClass 即@NameRepository，由于 metaAnnotations 注解数组仅获取当前注解@NameRepository 所标注的注解，因此只包含了@Repository，而不存在@Component。反之，假设 annotationClass 是 @Repository，那么 metaAnnotations 自然包含@Component。综上所述，2.5.6.SEC03 中的实现由于未采用层次递归获取 Annotation[]，所以仅支持单层次的@Component“派生”，这也就解释了为什么 StringRepository 在 2.5.6.SEC03 中标注@NameRepository 后，最终无法被组件扫描候选对象的原因。再来分析 3.0.0.RELEASE 中的实现：

```
final class AnnotationAttributesReadingVisitor implements AnnotationVisitor {
    ...

    public void visitEnd() {
        this.attributesMap.put(this.annotationType, this.localAttributes);
        try {
            Class<?> annotationClass = this.classLoader.loadClass(this.annotationType);
            ...
            // Register annotations that the annotation type is annotated with.
            Set<String> metaAnnotationTypeNames = new LinkedHashSet<String>();
            for (Annotation metaAnnotation : annotationClass.getAnnotations()) {
                metaAnnotationTypeNames.add(metaAnnotation.annotationType().getName());
                ...
                for (Annotation metaMetaAnnotation :
metaAnnotation.annotationType().getAnnotations()) {
```

```
metaAnnotationTypeNames.add(metaMetaAnnotation.annotationType().getName());
                    }
                }
                ...
            }
            catch (ClassNotFoundException ex) {
                // Class not found - can't determine meta-annotations.
            }
        }
    }
```

其中 registerMetaAnnotations(Class)方法用于注册元注解映射，即 metaAnnotationMap 字段。第一个 for (AnnotationmetaAnnotation : annotationClass.getAnnotations())循环获取的是第一层元注解数组，第二个 for (AnnotationmetaMetaAnnotation : metaAnnotation.annotationType().getAnnotations()) 循环是在第一层元注解 metaAnnotation 的基础上再次获取元注解数组。结合@StringRepository 进行分析，其第一层元注解@Repository 是 metaAnnotation，而第二层元注解@Component 则是 metaMetaAnnotation。不过这个实现仍存在问题，它仅处理了两层@Component“派生”的情况，未能识别多层次@Component“派生”。下面结合示例检验以上推断。

（1）声明@FirstLevelRepository 和@SecondLevelRepository。

- FirstLevelRepository 为两层@Component“派生”注解

```
@Target({ElementType.TYPE})
@Retention(RetentionPolicy.RUNTIME)
@Documented
@Repository
public @interface FirstLevelRepository {

    String value() default "";

}
```

- @SecondLevelRepository 为三层@Component“派生”注解

```
@Target({ElementType.TYPE})
@Retention(RetentionPolicy.RUNTIME)
@Documented
@FirstLevelRepository
```

```java
public @interface SecondLevelRepository {

    String value() default "";

}
```

- 整体@Component“派生”层次性

```
@SecondLevelRepository
 |- @FirstLevelRepository
      |- @Repository
           |- @Component
```

（2）标注@FirstLevelRepository 和@SecondLevelRepository

- 标注@FirstLevelRepository 到 MyFirstLevelRepository

```java
@FirstLevelRepository("myFirstLevelRepository")
public class MyFirstLevelRepository {
}
```

- 标注@SecondLevelRepository 到 MySecondLevelRepository

```java
@SecondLevelRepository("mySecondLevelRepository")
public class MySecondLevelRepository {
}
```

（3）配置@Component 扫描。

由于 Spring Framework 3.0.0.RELEASE 尚未引入@ComponentScan，因此仍采用 XML 元素 <context:component-scan>组件扫描的方式（XML 配置资源位置：META-INF/spring/context.xml）：

```xml
<?xml version="1.0" encoding="UTF-8"?>
<beans xmlns="http://www.springframework.org/schema/beans"
       xmlns:xsi="http://www.w3.org/2001/XMLSchema-instance"
       xmlns:context="http://www.springframework.org/schema/context"
       xsi:schemaLocation="http://www.springframework.org/schema/beans
http://www.springframework.org/schema/beans/spring-beans.xsd
     http://www.springframework.org/schema/context
http://www.springframework.org/schema/context/spring-context.xsd">
```

```
    <!-- 激活注解驱动特性 -->
    <context:annotation-config />

    <!-- 寻找被@Component 或其派生 Annotation 标记的类（Class），将它们注册为 Spring Bean -->
    <context:component-scan base-package="thinking.in.spring.boot.samples.spring3" />

</beans>
```

（4）引导类实现。

加载以上 XML 配置文件（META-INF/spring/context.xml），并检验 Bean myFirstLevelRepository 及 mySecondLevelRepository 是否存在：

```
public class HierarchicalDerivedComponentAnnotationBootstrap {

    public static void main(String[] args) {
        ClassPathXmlApplicationContext context = new
ClassPathXmlApplicationContext("META-INF/spring/context.xml");
        // 检验 Bean myFirstLevelRepository 及 mySecondLevelRepository 是否存在
        System.out.println("myFirstLevelRepository Bean 是否存在：" +
context.containsBean("myFirstLevelRepository"));
        System.out.println("mySecondLevelRepository Bean 是否存在：" +
context.containsBean("mySecondLevelRepository"));
        // 关闭上下文
        context.close();
    }

}
```

> 以上示例代码可通过查找 spring-framework-samples/spring-framework-3.0.x-sample 工程获取。

（5）运行引导类。

万事俱备，运行 HierarchicalDerivedComponentAnnotationBootstrap，观察日志输出：

```
(...部分内容被省略...)
myFirstLevelRepositoryBean 是否存在：true
mySecondLevelRepositoryBean 是否存在：false
```

运行结果证明了之前的推断，即 Spring Framework 3.0.0.RELEASE 仅支持两层 @Component“派生”，因为 AnnotationAttributesReadingVisitor 未递归查找所有层次的元注解。

那么 AnnotationAttributesReadingVisitor 递归查找元注解的实现会出现在哪个 Spring Framework 版本中呢？小马哥尚未找到关于以上实现变化的公开资料，只好逐一尝试 Spring Framework 3 的各个最高 RELEASE 版本。

- 3.0.x 最高 RELEASE：3.0.7.RELEASE；
- 3.1.x 最高 RELEASE：3.1.4.RELEASE；
- 3.2.x 最高 RELEASE：3.2.18.RELEASE。

调整 spring-framework-samples/spring-framework-3.0.x-sample 工程的 pom.xml 依赖如下：

```
<dependencies>
    <dependency>
        <groupId>org.springframework</groupId>
        <artifactId>spring-context</artifactId>
        <version>${spring.version}</version>
    </dependency>
</dependencies>
```

无论${spring.version}调整为 3.0.7.RELEASE、3.1.4.RELEASE 或 3.2.18.RELEASE，HierarchicalDerivedComponentAnnotationBootstrap 的结果都没有变化：

```
(...部分内容被省略...)
myFirstLevelRepositoryBean 是否存在：true
mySecondLevelRepositoryBean 是否存在：false
```

> 请读者对比以上三个版本中的 AnnotationAttributesReadingVisitor 的实现，可以发现三者在解析 metaAnnotationMap 数据方面并不存在明显的变化。

综上所述，Spring Framework 3 仅支持两层@Component 的“派生”。将${spring.version}升级到 Spring Framework 4 的第一个发行版本 4.0.0.RELEASE，再次执行 HierarchicalDerivedComponentAnnotationBootstrap，观察结果变化：

```
(...部分内容被省略...)
myFirstLevelRepositoryBean 是否存在：true
mySecondLevelRepositoryBean 是否存在：true
```

以上运行结果说明 Spring Framework 从 4.0.0.RELEASE 开始支持多层次@Component“派生”。由于 Spring Framework 3 的前车之鉴，不得不再次确认 AnnotationAttributesReadingVisitor 的实现变化：

```
final class AnnotationAttributesReadingVisitor extends RecursiveAnnotationAttributesVisitor
{
    ...
    @Override
    public void doVisitEnd(Class<?> annotationClass) {
        super.doVisitEnd(annotationClass);
        ...
        Set<String> metaAnnotationTypeNames = new LinkedHashSet<String>();
        for (Annotation metaAnnotation : annotationClass.getAnnotations()) {
            recursivelyCollectMetaAnnotations(metaAnnotationTypeNames, metaAnnotation);
        }
        if (this.metaAnnotationMap != null) {
            this.metaAnnotationMap.put(annotationClass.getName(), metaAnnotationTypeNames);
        }
    }

    private void recursivelyCollectMetaAnnotations(Set<String> visited, Annotation annotation)
{
        if (visited.add(annotation.annotationType().getName())) {
            // Only do further scanning for public annotations; we'd run into
IllegalAccessExceptions
            // otherwise, and don't want to mess with accessibility in a SecurityManager
environment.
            if (Modifier.isPublic(annotation.annotationType().getModifiers())) {
                this.attributesMap.add(annotation.annotationType().getName(),
                        AnnotationUtils.getAnnotationAttributes(annotation, true, true));
                for (Annotation metaMetaAnnotation :
annotation.annotationType().getAnnotations()) {
                    recursivelyCollectMetaAnnotations(visited, metaMetaAnnotation);
                }
            }
        }
    }
}
```

终于 AnnotationAttributesReadingVisitor 采用递归的方式查找元注解，此刻才放心地得出结论，即 Spring Framework 从 4.0 版本开始支持多层次@Component“派生性”。由于 Spring Boot 1.x 最低依赖 Spring Framework 4.0，因此多层次@Component“派生性”被 Spring Boot 继承。

掌握 Spring 模式注解（Stereotype Annotations）及`@Component`“派生性”对深度理解并运用 Spring Boot 复杂注解极其重要，故大篇幅讨论之。

7.3.3 Spring 组合注解（Composed Annotations）

Wiki“Spring Annotation Programming Model”中关于“Composed Annotations”的描述：

> A *composed annotation* is an annotation that is *meta-annotated* with one or more annotations with the intent of combining the behavior associated with those meta-annotations into a single custom annotation. For example, an annotation named `@TransactionalService` that is meta-annotated with Spring's `@Transactional` and `@Service` annotations is a composed annotation that combines the semantics of `@Transactional` and `@Service`. `@TransactionalService` is technically also a custom *stereotype annotation.*

所谓的“组合注解”是指某个注解“元标注”一个或多个其他注解，其目的在于将这些关联的注解行为组合成单个自定义注解。其中举例说明`@TransactionalService`标注了`@Transactional`和`@Service` 注解，因此`@TransactionalService` 组合了这两个注解的语义，同时，`@TransactionalService`也是一个自定义 Spring 模式注解（Stereotype Annotation）。后续的 Wiki 并没有给出更多的解释，将`@TransactionalService`通过代码方式描述则是：

```
@Target({ElementType.TYPE})
@Retention(RetentionPolicy.RUNTIME)
@Documented
@Transactional
@Service
public @interface TransactionalService {
}
```

> 以上示例代码可通过查找 spring-framework-samples/spring-framework-5.0.x-sample 工程获取。

值得注意的是，`@TransactionalService` 元标注了`@Transactional` 和`@Service`，其中`@Service`是 Spring 模式注解（Stereotype Annotation），而`@Transactional`则是 Spring 事务注解。言外之意，Spring 组合注解（Composed Annotations）中的元注解允许是 Spring 模式注解（Stereotype Annotation）与其他 Spring 功能性注解的任意组合。

以 Spring Boot 为例，比如 Spring Boot 1.2 之前，`@SpringBootApplication` 没有被引入，通常在引导类上标注`@Configuration`，表明它是 Spring 模式注解（Stereotype Annotation），再标注`@EnableAutoConfiguration`，激活 Spring Boot 自动装配，还可能追加`@ComponentScan` 指定扫描`@Compoent` 的包范围。从 Spring Boot 1.2 开始，引导类直接标注`@SpringBootApplication`，综合以上三项特性，减少了开发人员理解和记忆的成本：

```
@Target(ElementType.TYPE)
@Retention(RetentionPolicy.RUNTIME)
@Documented
@Inherited
@SpringBootConfiguration
@EnableAutoConfiguration
@ComponentScan(excludeFilters = {
      @Filter(type = FilterType.CUSTOM, classes = TypeExcludeFilter.class),
      @Filter(type = FilterType.CUSTOM, classes = AutoConfigurationExcludeFilter.class) })
public @interface SpringBootApplication {
    ...
}
```

明显地，`@SpringBootApplication` 既是 Spring 模式注解（Stereotype Annotation），又是组合注解（Composed Annotation）。类似的注解在 Spring Boot 中比较常见，然而这些注解特性实际上继承了 Spring Framework 的“衣钵”。下面一同进入更深层次的探讨，尝试多方位地理解 Spring 组合注解。

理解 Spring 组合注解

以`@TransactionalService` 为例，其元注解的层次关系如下：

```
@TransactionalService
    |- @Transactional
    |- @Service
        |- @Component
```

一方面，`@Service` 作为 Spring 模式注解。按照`@Component`“派生性”的特点，`@TransactionalService` 也是 Spring 模式注解（或`@Component`“派生”注解），它被 `ClassPathScanningCandidateComponentProvider` 识别，被其子类 `ClassPathBeanDefinitionScanner` 注册为 Spring Bean。另一方面，`@Transactional` 作为 Spring 事务注解，属于原子注解（无法再次拆分）：

```
@Target({ElementType.METHOD, ElementType.TYPE})
@Retention(RetentionPolicy.RUNTIME)
@Inherited
@Documented
public @interface Transactional {
    ...
}
```

> 本节示例代码均在 spring-framework-samples/spring-framework-5.0.x-sample 工程中。

Spring Framework 必然提供@Transactional 的处理实现，问题在于 Spring 容器如何通过@TransactionalService 感知元注解@Transactional 的存在。或许通过 Java 反射获取元注解信息是优先考虑的方案。然而在前文@Component“派生性”原理的讨论中，Spring Framework 并没有考虑使用 Java 反射的手段来解析元注解信息，而是抽象出 AnnotationMetadata 接口，其实现类为 AnnotationMetadataReadingVisitor，并且从 Spring Framework 4.0 开始，AnnotationMetadataReadingVisitor 所关联的 AnnotationAttributesReadingVisitor 采用递归查找元注解，使得多层次元注解信息保存在 AnnotationMetadataReadingVisitor 的 metaAnnotationMap 字段中。

> 表面上 metaAnnotationMap 字段被 AnnotationAttributesReading Visitor 关联并处理，实际上该字段来自 AnnotationMetadataReadingVisitor，通过 AnnotationAttributesReadingVisitor 构造器参数与其关联：
>
> new AnnotationAttributesReadingVisitor(className, this.attributeMap, this.metaAnnotationMap, this.classLoader)

如此表达可能过于抽象，结合 Java 反射原理或许更容易理解。在 Java 中，Class 对象是类的元信息载体，承载了其成员的元信息对象，包括字段（Field）、方法（Method）、构造器（Constructor）及注解 Annotation 等，而 Class 的加载通过 ClassLoader#loadClass(String)方法实现。而 Spring Framework 的类加载则通过 ASM 实现，如 ClassReader。相对于 ClassLoader 体系，Spring ASM 更为底层，读取的是类资源，直接操作其中的字节码，获取相关元信息，同时便于 Spring 相关的字节码提升。在读取元信息方面，Spring 抽象出 MetadataReader 接口：

```
public interface MetadataReader {

 /**
  * Return the resource reference for the class file.
  */
 Resource getResource();
```

```
    /**
     * Read basic class metadata for the underlying class.
     */
    ClassMetadata getClassMetadata();

    /**
     * Read full annotation metadata for the underlying class,
     * including metadata for annotated methods.
     */
    AnnotationMetadata getAnnotationMetadata();

}
```

getClassMetadata()用于读取类的元信息，注解的元信息由 getAnnotationMetadata()方法获取。无论 ClassMetadata，还是 AnnotationMetadata，均没有 Java Class 和 Annotation API 那样丰富的关联属性。

MetadataReader 有明显的资源特性，getResource()方法关联了类资源的 Resource 信息，它在 Spring Framework 中仅存在一个非公开的实现，即 SimpleMetadataReader。其关联的 ClassMetadata 和 AnnotationMetadata 信息在构造的阶段完成初始化：

```
final class SimpleMetadataReader implements MetadataReader {
    ...
    SimpleMetadataReader(Resource resource, @Nullable ClassLoader classLoader) throws
IOException {
        InputStream is = new BufferedInputStream(resource.getInputStream());
        ClassReader classReader;
        try {
            classReader = new ClassReader(is);
        }
        catch (IllegalArgumentException ex) {
            throw new NestedIOException("ASM ClassReader failed to parse class file - " +
                    "probably due to a new Java class file version that isn't supported yet:
" + resource, ex);
        }
        finally {
            is.close();
        }
```

```
        AnnotationMetadataReadingVisitor visitor = new
AnnotationMetadataReadingVisitor(classLoader);
        classReader.accept(visitor, ClassReader.SKIP_DEBUG);

        this.annotationMetadata = visitor;
        // (since AnnotationMetadataReadingVisitor extends ClassMetadataReadingVisitor)
        this.classMetadata = visitor;
        this.resource = resource;
    }
    ...
}
```

然而 AnnotationMetadataReadingVisitor 同时实现了 ClassMetadata 和 AnnotationMetadata 接口。因此，在 Spring 注解编程模型中，元注解（Meta Annotation）的实现仍旧集中在 AnnotationMetadataReadingVisitor 及 AnnotationAttributesReadingVisitor 之中。

下面结合示例加以说明，先回顾 ClassPathScanningCandidateComponentProvider#findCandidateComponents(String)方法中读取 MetadataReader 实例的方式：

```
    public Set<BeanDefinition> findCandidateComponents(String basePackage) {
        Set<BeanDefinition> candidates = new LinkedHashSet<BeanDefinition>();
        try {
            ...
            Resource[] resources =
this.resourcePatternResolver.getResources(packageSearchPath);
            ...
            for (int i = 0; i < resources.length; i++) {
                Resource resource = resources[i];
                ...
                if (resource.isReadable()) {
                    MetadataReader metadataReader =
this.metadataReaderFactory.getMetadataReader(resource);
                    ...
                }
            }
        }
        catch (IOException ex) {
```

```
            throw new BeanDefinitionStoreException("I/O failure during classpath scanning",
ex);
        }
        return candidates;
    }
```

其中MetadataReader对象通过MetadataReaderFactory#getMetadataReader(resource)方法获取，而方法中的 this.metadataReaderFactory 对象默认是 CachingMetadataReaderFactory实例,并且MetadataReaderFactory提供另一个重载方法getMetadataReader(String):

```
public interface MetadataReaderFactory {

    /**
     * Obtain a MetadataReader for the given class name.
     * @param className the class name (to be resolved to a ".class" file)
     * @return a holder for the ClassReader instance (never <code>null</code>)
     * @throws IOException in case of I/O failure
     */
    MetadataReader getMetadataReader(String className) throws IOException;

    /**
     * Obtain a MetadataReader for the given resource.
     * @param resource the resource (pointing to a ".class" file)
     * @return a holder for the ClassReader instance (never <code>null</code>)
     * @throws IOException in case of I/O failure
     */
    MetadataReader getMetadataReader(Resource resource) throws IOException;

}
```

为了简化开发，在示例代码中可利用该方法读取@TransactionalService 注解元信息 AnnotationMetadata，并基于 Spring Framework 5.0 API 实现，代码如下：

```
@TransactionalService
public class TransactionalServiceAnnotationMetadataBootstrap {

    public static void main(String[] args) throws IOException {
        // @TransactionalService 标注在当前类上
TransactionalServiceAnnotationMetadataBootstrap
```

```
        String className = TransactionalServiceAnnotationMetadataBootstrap.class.getName();
        // 构建 MetadataReaderFactory 实例
        MetadataReaderFactory metadataReaderFactory = new CachingMetadataReaderFactory();
        // 读取@TransactionService MetadataReader 信息
        MetadataReader metadataReader = metadataReaderFactory.getMetadataReader(className);
        // 读取@TransactionService AnnotationMetadata 信息
        AnnotationMetadata annotationMetadata = metadataReader.getAnnotationMetadata();

        annotationMetadata.getAnnotationTypes().forEach(annotationType -> {


            Set<String> metaAnnotationTypes =
annotationMetadata.getMetaAnnotationTypes(annotationType);

            metaAnnotationTypes.forEach(metaAnnotationType -> {
                System.out.printf("注解 @%s 元标注 @%s\n", annotationType,
metaAnnotationType);
            });

        });
    }
}
```

> 尽管示例代码构建于 Spring Framework 5.0，不过 MetadataReaderFactory 及实现类在 Spring Framework 2.5 被引入，因此以上实现适应于 2.5+版本。

运行该引导类，观察日志结果：

```
    注解@thinking.in.spring.boot.samples.spring5.annotation.TransactionalService 元标注
@org.springframework.transaction.annotation.Transactional
    注解@thinking.in.spring.boot.samples.spring5.annotation.TransactionalService 元标注
@org.springframework.stereotype.Service
    注解@thinking.in.spring.boot.samples.spring5.annotation.TransactionalService 元标注
@org.springframework.stereotype.Component
    注解@thinking.in.spring.boot.samples.spring5.annotation.TransactionalService 元标注
@org.springframework.stereotype.Indexed
```

将运行结果与前文总结的@TransactionalService 层次结构进行对比：

```
    @TransactionalService
        |- @Transactional
```

```
        |- @Service
            |- @Component
```

@Indexed 没有纳入以上层次结构之中，而该注解则元标注在@Component 之上：

```
@Target(ElementType.TYPE)
@Retention(RetentionPolicy.RUNTIME)
@Documented
@Indexed
public @interface Component {
    ...
}
```

由于@TransactionalService 标注在 TransactionalServiceAnnotationMetadataBootstrap 类上，annotationMetadata.getAnnotationTypes()返回的集合仅包含@TransactionalService 类名的元素，而@TransactionalService 所关联的元注解类名则是四个，即 annotationMetadata.getMetaAnnotationTypes(annotationType)返回值：

```
org.springframework.transaction.annotation.Transactional
org.springframework.stereotype.Service
org.springframework.stereotype.Component
org.springframework.stereotype.Indexed
```

最后分析 AnnotationMetadata#getMetaAnnotationTypes(String)方法的实现，即 AnnotationMetadataReadingVisitor#getMetaAnnotationTypes(String)方法：

```
    public class AnnotationMetadataReadingVisitor extends ClassMetadataReadingVisitor
implements AnnotationMetadata {
        ...
        @Override
    public AnnotationVisitor visitAnnotation(final String desc, boolean visible) {
        String className = Type.getType(desc).getClassName();
        this.annotationSet.add(className);
        return new AnnotationAttributesReadingVisitor(
                className, this.attributesMap, this.metaAnnotationMap, this.classLoader);
    }
        ...
    @Override
    public Set<String> getMetaAnnotationTypes(String annotationName) {
        return this.metaAnnotationMap.get(annotationName);
```

```
    }
        ...
    }
```

AnnotationMetadataReadingVisitor#getMetaAnnotationTypes(String)方法实际上返回的是底层 metaAnnotationMap 字段的结果，而在前文“多层次@Component 派生性原理”一节的讨论中，已知该字段的初始化在 AnnotationAttributesReadingVisitor 中完成。

分析至此，得出以下结论，在抽象层面，无论 Spring 模式注解（多层次@Compoent“派生性”）的元信息，还是 Spring 组合注解的元信息，均由 AnnotationMetadata API 抽象表达，具体某个注解的“元注解”信息则通过 getMetaAnnotationTypes(String)方法查询。

> 之所以不提 AnnotationMetadataReadingVisitor 的原因是该类并非 AnnotationMetadata 的唯一实现。其另外一处实现为 StandardAnnotationMetadata。两者读取注解元信息的手段不同，前者利用 ASM 方式读取，后者的读取方式则是 Java 反射。在后续的源码分析中，StandardAnnotationMetadata 的“身影”将频繁出现。

AnnotationMetadata API 不仅通过 getMetaAnnotationTypes(String)方法暴露“元注解”信息，而且提供 getAnnotationAttributes(String)方法抽象指定注解的属性方法，该方法也是下一节的讨论重点。

7.3.4 Spring 注解属性别名和覆盖(Attribute Aliases and Overrides)

在正式讨论之前，首先理解 Spring 对于注解属性的抽象。在 Java 反射编程模型中，注解之间无法继承，也不能实现接口，不过 Java 语言默认将所有注解实现 Annotation 接口，被标注的对象用 API AnnotatedElement 表达。通过 AnnotatedElement#getAnnotation(Class)方法返回指定类型的注解对象，获取注解属性则需要显式地调用对应的属性方法。

1. 理解 Spring 注解元信息抽象 AnnotationMetadata

仍以@TransactionalService 为例，为其增加属性方法 name()，表示服务 Bean 名称，且默认值为空字符串：

```
@Target({ElementType.TYPE})
@Retention(RetentionPolicy.RUNTIME)
@Documented
@Transactional
@Service
```

```
public @interface TransactionalService {

    /**
     * @return 服务 Bean 名称
     */
    String name() default "";
}
```

> 本节示例代码均在 spring-framework-samples/spring-framework-5.0.x-sample 工程中。

利用 Java 反射 API 实现获取@TransactionalService 属性方法 name()的内容：

```
@TransactionalService(name = "test") // name 属性内容
public class TransactionalServiceAnnotationReflectionBootstrap {

    public static void main(String[] args) {
        // Class 实现了 AnnotatedElement 接口
        AnnotatedElement annotatedElement =
TransactionalServiceAnnotationReflectionBootstrap.class;
        // 从 AnnotatedElement 获取 TransactionalService
        TransactionalService transactionalService =
annotatedElement.getAnnotation(TransactionalService.class);
        // 显式地调用属性方法 TransactionalService#name() 获取属性
        String nameAttribute = transactionalService.name();
        System.out.println("TransactionalService.name() = " + nameAttribute);
    }
}
```

运行 TransactionalServiceAnnotationReflectionBootstrap，并观察输出结果：

```
@TransactionalService.name() = test
```

结果非常简单，即来源于@TransactionalService(name = "test")语句。不过，以上实现并没有将 Java 反射进行到底，name 属性是显式地执行@TransactionalService.name()方法获取的。将以上示例替换为完全 Java 反射实现：

```
@TransactionalService(name = "test") // name 属性内容
public class TransactionalServiceAnnotationReflectionBootstrap {

    public static void main(String[] args) {
```

```
        // Class 实现了 AnnotatedElement 接口
        AnnotatedElement annotatedElement =
TransactionalServiceAnnotationReflectionBootstrap.class;
        // 从 AnnotatedElement 获取 TransactionalService
        TransactionalService transactionalService =
annotatedElement.getAnnotation(TransactionalService.class);
        // 完全 Java 反射实现（ReflectionUtils 为 Spring 反射工具类）
        ReflectionUtils.doWithMethods(TransactionalService.class,
                method -> System.out.printf("@TransactionalService.%s() = %s\n",
method.getName(),          //执行 Method 反射调用
                    ReflectionUtils.invokeMethod(method, transactionalService))
                , method -> method.getParameterCount() == 0); //选择无参数方法
    }
}
```

执行结果如下：

```
    @TransactionalService.name() = test
    @TransactionalService.toString() =
@thinking.in.spring.boot.samples.spring5.annotation.TransactionalService(name=test)
    @TransactionalService.hashCode() = 431984231
    @TransactionalService.annotationType() = interface
thinking.in.spring.boot.samples.spring5.annotation.TransactionalService
```

由于所有的注解实现了 Annotation 接口，因此以上日志包含了无参方法的调用结果：

```
    public interface Annotation {

        ...
        boolean equals(Object obj);

        ...
        int hashCode();

        ...
        String toString();

        ...
        Class<? extends Annotation> annotationType();
```

```
}
```

在具体执行时，需要将 Annotation 接口的方法予以排除，方可得到当前注解的属性方法。具体调整如下：

```
@TransactionalService(name = "test") // name 属性内容
public class TransactionalServiceAnnotationReflectionBootstrap {

    public static void main(String[] args) {
        // Class 实现了 AnnotatedElement 接口
        AnnotatedElement annotatedElement =
TransactionalServiceAnnotationReflectionBootstrap.class;
        // 从 AnnotatedElement 获取 TransactionalService
        TransactionalService transactionalService =
annotatedElement.getAnnotation(TransactionalService.class);
        // 完全 Java 反射实现（ReflectionUtils 为 Spring 反射工具类）
        ReflectionUtils.doWithMethods(TransactionalService.class,
                method -> System.out.printf("@TransactionalService.%s() = %s\n",
method.getName(),         // 执行 Method 反射调用
                        ReflectionUtils.invokeMethod(method, transactionalService))
                        // 选择非 Annotation 方法
               , method -> !method.getDeclaringClass().equals(Annotation.class));
    }
}
```

执行结果如下：

```
@TransactionalService.name() = test
```

以上示例经过复杂的调整后，其运行结果才符合期望。然而@TransactionalService 已元标注@Transactional 和@Service，其中@Transactional 存在不少属性方法声明：

```
@Target({ElementType.METHOD, ElementType.TYPE})
@Retention(RetentionPolicy.RUNTIME)
@Inherited
@Documented
public @interface Transactional {

 ...
 @AliasFor("transactionManager")
```

```
    String value() default "";

    ...
    @AliasFor("value")
    String transactionManager() default "";

    ...
    Propagation propagation() default Propagation.REQUIRED;

    ...
    Isolation isolation() default Isolation.DEFAULT;

    ...
    int timeout() default TransactionDefinition.TIMEOUT_DEFAULT;

    ...
    boolean readOnly() default false;

    ...
    Class<? extends Throwable>[] rollbackFor() default {};

    ...
    String[] rollbackForClassName() default {};

    ...
    Class<? extends Throwable>[] noRollbackFor() default {};

    ...
    String[] noRollbackForClassName() default {};

}
```

此处不关心@Transactional 属性方法的具体意义，用 Java 反射 API 表述，@Transactional 作为 Annotation 对象标注在@TransactionalService 上，因此 TransactionalService.class 也是 AnnotatedElement 对象，需要再次调用其 getAnnotation(Class) 方法，获取@Transactional 对象。将示例代码重构，输出@Transactional 属性方法：

```
    @TransactionalService(name = "test") // name 属性内容
    public class TransactionalServiceAnnotationReflectionBootstrap {

        public static void main(String[] args) {
            // Class 实现了 AnnotatedElement 接口
            AnnotatedElement annotatedElement =
TransactionalServiceAnnotationReflectionBootstrap.class;
            // 从 AnnotatedElement 获取 TransactionalService
            TransactionalService transactionalService =
annotatedElement.getAnnotation(TransactionalService.class);
            // 获取 transactionalService 的所有元注解
            Set<Annotation> metaAnnotations = getAllMetaAnnotations(transactionalService);
             // 输出结果

metaAnnotations.forEach(TransactionalServiceAnnotationReflectionBootstrap::printAnnotationAtt
ribute);
        }

        private static Set<Annotation> getAllMetaAnnotations(Annotation annotation) {

            Annotation[] metaAnnotations = annotation.annotationType().getAnnotations();

            if (ObjectUtils.isEmpty(metaAnnotations)) { // 没有找到，返回空集合
                return Collections.emptySet();
            }
            // 获取所有非 Java 标准元注解集合
            Set<Annotation> metaAnnotationsSet = Stream.of(metaAnnotations)
                    // 排除 Java 标准注解，如@Target、@Documented 等，它们因相互依赖，将
                    // 导致递归不断
                    // 通过 java.lang.annotation 包名排除
                    .filter(metaAnnotation
-> !Target.class.getPackage().equals(metaAnnotation.annotationType().getPackage()))
                    .collect(Collectors.toSet());

            // 递归查找元注解的元注解集合
            Set<Annotation> metaMetaAnnotationsSet = metaAnnotationsSet.stream()
                    .map(TransactionalServiceAnnotationReflectionBootstrap::getAllMetaAnnotati
ons)
```

```
                .collect(HashSet::new, Set::addAll, Set::addAll);

        // 添加递归结果
        metaAnnotationsSet.addAll(metaMetaAnnotationsSet);

        return metaAnnotationsSet;
    }

    private static void printAnnotationAttribute(Annotation annotation) {
        Class<?> annotationType = annotation.annotationType();
        // 完全 Java 反射实现（ReflectionUtils 为 Spring 反射工具类）
        ReflectionUtils.doWithMethods(annotationType,
                method -> System.out.printf("@%s.%s() = %s\n",
annotationType.getSimpleName(),
                        method.getName(), ReflectionUtils.invokeMethod(method, annotation))
// 执行 Method 反射调用
    //, method -> method.getParameterCount() == 0); // 选择无参数方法
                 // 选择非 Annotation 方法
                , method -> !method.getDeclaringClass().equals(Annotation.class));
    }
}
```

运行结果如下：

```
@Service.value() =
@Component.value() =
@Transactional.value() =
@Transactional.timeout() = -1
@Transactional.transactionManager() =
@Transactional.propagation() = REQUIRED
@Transactional.isolation() = DEFAULT
@Transactional.rollbackFor() = [Ljava.lang.Class;@3fa77460
@Transactional.rollbackForClassName() = [Ljava.lang.String;@619a5dff
@Transactional.noRollbackFor() = [Ljava.lang.Class;@1ed6993a
@Transactional.noRollbackForClassName() = [Ljava.lang.String;@7e32c033
@Transactional.readOnly() = false
```

日志内容中出现了@TransactionalService 所有元注解的属性方法值。很明显，基于 Java 反射 API 获取元（嵌套）注解及属性信息的实现是颇为复杂的，需要递归地获取元注解。从 Spring

Framework 4.0 开始，AnnotationMetadata#getMetaAnnotationTypes(String)方法能够获取所有元注解类型集合。再结合 getAnnotationAttributes(String)方法返回注解所关联的属性信息，以 Map 结构存储。同时，AnnotationMetadata 存在两种实现：基于 ASM 的 AnnotationMetadataReadingVisitor 和 Java 反射 API 的 StandardAnnotationMetadata。下面使用 StandardAnnotationMetadata 将示例代码重新实现：

```
@TransactionalService
public class TransactionalServiceStandardAnnotationMetadataBootstrap {

    public static void main(String[] args) throws IOException {

        // 读取@TransactionService AnnotationMetadata 信息
        AnnotationMetadata annotationMetadata = new
StandardAnnotationMetadata(TransactionalServiceStandardAnnotationMetadataBootstrap.class);

        // 获取所有的元注解类型（全类名）集合
        Set<String> metaAnnotationTypes = annotationMetadata.getAnnotationTypes()
                .stream() // TO Stream
                // 读取单注解的元注解类型集合
                .map(annotationMetadata::getMetaAnnotationTypes)
               // 合并元注解类型（全类名）集合
    .collect(LinkedHashSet::new, Set::addAll, Set::addAll);
        metaAnnotationTypes.forEach(metaAnnotation -> { // 读取所有元注解类型
            // 读取元注解属性信息
            Map<String, Object> annotationAttributes =
annotationMetadata.getAnnotationAttributes(metaAnnotation);
            if (!CollectionUtils.isEmpty(annotationAttributes)) {
                annotationAttributes.forEach((name, value) ->
                        System.out.printf("注解 @%s 属性 %s = %s\n",
ClassUtils.getShortName(metaAnnotation), name, value));
            }
        });
    }
}
```

运行结果：

```
注解 @Transactional 属性 value =
注解 @Transactional 属性 transactionManager =
```

```
注解 @Transactional 属性 propagation = REQUIRED
注解 @Transactional 属性 isolation = DEFAULT
注解 @Transactional 属性 rollbackFor = [Ljava.lang.Class;@ea30797
注解 @Transactional 属性 rollbackForClassName = [Ljava.lang.String;@7e774085
注解 @Transactional 属性 noRollbackFor = [Ljava.lang.Class;@3f8f9dd6
注解 @Transactional 属性 noRollbackForClassName = [Ljava.lang.String;@aec6354
注解 @Transactional 属性 timeout = -1
注解 @Transactional 属性 readOnly = false
注解 @Service 属性 value =
注解 @Component 属性 value =
```

对比重构前的示例，经过 AnnotationMetadata API 重构的示例将复杂的递归查找元注解的逻辑“扁平化”，降低了开发的成本。

> 进阶阅读：
> 在 AnnotationMetadata 语义上，基于 Java 反射 API 实现的 StandardAnnotationMetadata 与 AnnotationMetadataReadingVisitor 保持一致。那么，Spring Framework 为什么并存两套实现？对此，小马哥尚未找到任何官方的说明，仅根据个人经验加以推断。首先，两者适用于不同的使用场景。笼统地说，凡是基于 Java 反射 API 的实现必然需要一个大前提，即被反射的 Class 必须被 ClassLoader 加载。当 Spring 应用指定的 Java package 扫描 Spring 模式注解时，StandardAnnotationMetadata 显然不适应了，因为应用不需要、更不应该把这些 package 下的 Class 悉数加载。所以，基于 ASM 实现的 AnnotationMetadataReadingVisitor 更适合这种场景。这也解释了为什么 AnnotationMetadataReadingVisitor 会出现在 ClassPathScanningCandidateComponentProvider 的实现中。反之，当 Class 已被加载并能够被程序获取时，再通过 ASM 读取就显得多此一举了。另一方面，两者的性能表现差异较大，执行以下示例：

```java
public class AnnotationMetadataPerformanceBootstrap {

    public static void main(String[] args) throws IOException {

        // 反射实现
        AnnotationMetadata standardAnnotationMetadata = new
StandardAnnotationMetadata(TransactionalService.class);

        SimpleMetadataReaderFactory factory = new SimpleMetadataReaderFactory();
```

```
        MetadataReader metadataReader =
factory.getMetadataReader(TransactionalService.class.getName());
        // ASM 实现
        AnnotationMetadata annotationMetadata = metadataReader.getAnnotationMetadata();

        int times = 10 * 10000; // 10 万次

        testAnnotationMetadataPerformance(standardAnnotationMetadata, times);
        testAnnotationMetadataPerformance(annotationMetadata, times);

        times = 100 * 10000;    // 100 万次

        testAnnotationMetadataPerformance(standardAnnotationMetadata, times);
        testAnnotationMetadataPerformance(annotationMetadata, times);

        times = 1000 * 10000;   // 1000 万次

        testAnnotationMetadataPerformance(standardAnnotationMetadata, times);
        testAnnotationMetadataPerformance(annotationMetadata, times);

        times = 10000 * 10000; // 1 亿次

        testAnnotationMetadataPerformance(standardAnnotationMetadata, times);
        testAnnotationMetadataPerformance(annotationMetadata, times);
    }

    private static void testAnnotationMetadataPerformance(AnnotationMetadata
annotationMetadata, int times) {
        long startTime = System.currentTimeMillis();
        for (int i = 0; i < times; i++) {
            annotationMetadata.getAnnotationTypes();
        }
        long costTime = System.currentTimeMillis() - startTime;
        System.out.printf("%d 次 %s.getAnnotationTypes() 方法执行消耗 %s ms\n",
                times, annotationMetadata.getClass().getSimpleName(), costTime);
    }
}
```

以上示例代码可通过查找 spring-framework-samples/spring-framework-5.0.x-sample 工程获取。

运行结果：

100000 次 StandardAnnotationMetadata.getAnnotationTypes()方法执行消耗 75ms
100000 次 AnnotationMetadataReadingVisitor.getAnnotationTypes()方法执行消耗 1ms
1000000 次 StandardAnnotationMetadata.getAnnotationTypes()方法执行消耗 300ms
1000000 次 AnnotationMetadataReadingVisitor.getAnnotationTypes()方法执行消耗 0ms
10000000 次 StandardAnnotationMetadata.getAnnotationTypes()方法执行消耗 2317ms
10000000 次 AnnotationMetadataReadingVisitor.getAnnotationTypes()方法执行消耗 0ms
100000000 次 StandardAnnotationMetadata.getAnnotationTypes()方法执行消耗 17404ms
100000000 次 AnnotationMetadataReadingVisitor.getAnnotationTypes()方法执行消耗 0ms

排除预热的干扰，不考虑第一次运算结果，在 getAnnotationTypes()的执行性能上，AnnotationMetadataReadingVisitor 明显要优于 StandardAnnotationMetadata。以上结论仅供参考，或许在设计方面存在更多的考量。

可预料的是，StandardAnnotationMetadata#getMetaAnnotationTypes 方法的调用链路较为复杂，如下图所示。

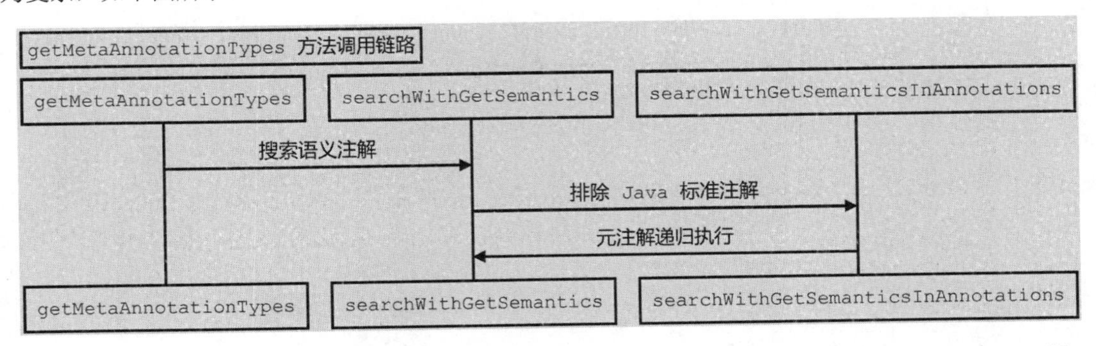

其中 searchWithGetSemanticsInAnnotations 方法通过 AnnotationUtils.isInJavaLangAnnotationPackage(Class)方法过滤 Java 标准注解，同时递归调用 hasSearchableMetaAnnotations 方法获取所有的元注解：

```
    public class StandardAnnotationMetadata extends StandardClassMetadata implements
AnnotationMetadata {
        ...
        @Nullable
    private static <T> T searchWithGetSemanticsInAnnotations(@Nullable AnnotatedElement
element,
                List<Annotation> annotations, @Nullable Class<? extends Annotation>
annotationType,
                @Nullable String annotationName, @Nullable Class<? extends Annotation>
```

```
containerType,
            Processor<T> processor, Set<AnnotatedElement> visited, int metaDepth) {

        // Search in annotations
        for (Annotation annotation : annotations) {
            Class<? extends Annotation> currentAnnotationType = annotation.annotationType();
            if (!AnnotationUtils.isInJavaLangAnnotationPackage(currentAnnotationType)) {
                ...
            }
        }

        // Recursively search in meta-annotations
        for (Annotation annotation : annotations) {
            Class<? extends Annotation> currentAnnotationType = annotation.annotationType();
            if (hasSearchableMetaAnnotations(currentAnnotationType, annotationType,
annotationName)) {
                T result = searchWithGetSemantics(currentAnnotationType, annotationType,
                        annotationName, containerType, processor, visited, metaDepth + 1);
                ...
            }
        }

        return null;
    }
    ...
}
```

> 以上实现来源于 Spring Framework 5.0.6.RELEASE。

同样地，`StandardAnnotationMetadata#getAnnotationAttributes(java.lang.String)` 实际调用了 `AnnotatedElementUtils.getMergedAnnotationAttributes` 方法：

```
public class StandardAnnotationMetadata extends StandardClassMetadata implements
AnnotationMetadata {
    ...
    @Override
    public Map<String, Object> getAnnotationAttributes(String annotationName) {
        return getAnnotationAttributes(annotationName, false);
```

```
    }

    @Override
    @Nullable
    public Map<String, Object> getAnnotationAttributes(String annotationName, boolean
classValuesAsString) {
        return (this.annotations.length > 0 ?
AnnotatedElementUtils.getMergedAnnotationAttributes(
                getIntrospectedClass(), annotationName, classValuesAsString,
this.nestedAnnotationsAsMap) : null);
    }
    ...
}
```

引用 `AnnotatedElementUtils.getMergedAnnotationAttributes` JavaDoc 中的描述：

> Get the first annotation of the specified annotationName within the annotation hierarchy above the supplied element and merge that annotation's attributes with matching attributes from annotations in lower levels of the annotation hierarchy.

再结合该方法的声明一同理解：

```
public static AnnotationAttributes getMergedAnnotationAttributes(AnnotatedElement element,

        String annotationName, boolean classValuesAsString, boolean nestedAnnotationsAsMap){

    ...
}
```

大意是指在指定 `element` 的元注解层次中查找第一个类型名为 `annotationName` 的注解，将它及更底层的元注解属性合并到 `AnnotationAttributes` 之中。

2. 理解 Spring 注解属性抽象 AnnotationAttributes

以`@TransactionalService` 为例，假设当 `annotationName` 为 `org.springframework.stereotype.Service` 时，按照层次性分析，`@Service` 更底层的元注解为`@Component`，`@Component.value()`与`@Service.value()`将发生合并，所以最终的 `AnnotationAttributes` 对象仅包含一个元素，即 `value:""`。同理，当 `annotationName` 为 `org.springframework.transaction.annotation.Transactional` 时，由于它更底层的元注解仅为 Java 标准注解，所以 `AnnotationAttributes` 只包含`@Transactional` 自身的注解。为此，可以稍作示例检验之：

```
public class AnnotationAttributesBootstrap {

    public static void main(String[] args) {

        AnnotatedElement annotatedElement = TransactionalService.class;

        // 获取@Service 注解属性独享
        AnnotationAttributes serviceAttributes =
                AnnotatedElementUtils.getMergedAnnotationAttributes(annotatedElement,
Service.class);

        // 获取@Transactional 注解属性独享
        AnnotationAttributes transactionalAttributes =
                AnnotatedElementUtils.getMergedAnnotationAttributes(annotatedElement,
Transactional.class);

        // 输出
        print(serviceAttributes);

        print(transactionalAttributes);
    }

    private static void print(AnnotationAttributes annotationAttributes) {

        System.out.printf("注解 @%s 属性集合 : \n",
annotationAttributes.annotationType().getName());

        annotationAttributes.forEach((name, value) ->
                System.out.printf("\t 属性 %s : %s \n", name, value)
        );
    }
}
```

运行结果：

```
注解 @org.springframework.stereotype.Service 属性集合 :
        属性 value :
注解 @org.springframework.transaction.annotation.Transactional 属性集合 :
        属性 value :
```

```
        属性 transactionManager : 
        属性 propagation : REQUIRED
        属性 isolation : DEFAULT
        属性 rollbackFor : [Ljava.lang.Class;@1cf4f579
        属性 rollbackForClassName : [Ljava.lang.String;@18769467
        属性 noRollbackFor : [Ljava.lang.Class;@46ee7fe8
        属性 noRollbackForClassName : [Ljava.lang.String;@7506e922
        属性 readOnly : false
        属性 timeout : -1
```

运行结果和预期一致，说明之前的理解是无误的。值得注意的是，Spring Framework 将注解属性抽象为 `AnnotationAttributes` 类，它直接扩展了 `LinkedHashMap`，目的不言而喻，既要使用 Key-Value 的数据结构，又要确保其顺序保持与属性方法声明一致。不过，在多层次元注解的场景中存在一个限制，其根本原因在于 Java 注解的静态性。按照目前`@TransactionalService` 的声明，其元注解分别为`@Transactional` 和`@Service`，而`@Service` 的元注解是`@Component`，并且两者均有属性方法 value()，以及默认值均为""。`@TransactionalService` 无法修改它们的值，那么 `AnnotationAttributes` 中的"value"属性到底是`@Component.value()`覆盖 `@Service.value()`，还是反过来呢？由于两者同值，所以在前例的运行结果中显得不那么重要。如果是复杂的场景呢？所以 `AnnotatedElementUtils.getMergedAnnotationAttributes` 方法的 JavaDoc 还有补充：

> Attributes from lower levels in the annotation hierarchy override attributes of the same name from higher levels.

按照它的说法，较低层注解属性将覆盖较高层，不过问题在于所谓的"较低层注解"到底是`@Component` 还是`@Service`？因此尝试修改`@TransactionalService` 中`@Service` 的声明：

```java
@Target({ElementType.TYPE})
@Retention(RetentionPolicy.RUNTIME)
@Documented
@Transactional
@Service(value = "transactionalService")
public @interface TransactionalService {

    /**
     * @return 服务 Bean 名称
     */
    String name() default "";
}
```

将 `@Service.value()` 显式地设置为“transactionalService”，再次运行 `AnnotationAttributesBootstrap`，观察日志的变化：

```
注解 @org.springframework.stereotype.Service 属性集合 :
属性 value : transactionalService
(...部分内容被省略...)
```

由于 `AnnotationAttributes` 采用 Map 的存储结构，若不同层次元注解之间出现同名属性，则必然出现重叠的情况。

3. 理解 Spring 注解属性覆盖（Overrides）

结合“Spring Annotation Programming Model”中“Attribute Aliases and Overrides”一节给出的 *attribute override* 定义：

> An attribute override is an annotation attribute that overrides (or shadows) an annotation attribute in a meta-annotation.

综上所述，较低层注解能够覆盖其元注解的同名属性，并且 `AnnotationAttributes` 采用注解就近覆盖的设计原则，如下所示。

```
@Compoent
    |- @Service
        |- @TransactionalService
```

`@Service` 较`@Compoent` 而言，距`@TransactionalService` 更近，所以它是较低层注解。反之，较高层注解则是`@Compoent`。同理，`@TransactionalService` 也可以覆盖`@Service` 的同名属性。

由于元注解的层次高低关系，从而衍生出“Spring 注解属性覆盖（Overrides）”的规则。在“Attribute Aliases and Overrides”一节中称其为“隐性覆盖（Implicit Overrides）”：

> **1. Implicit Overrides**: given attribute `A` in annotation `@One` and attribute `A` in annotation `@Two`, if `@One` is meta-annotated with `@Two`, then attribute `A` in annotation `@One` is an *implicit override* for attribute `A` in annotation `@Two` based solely on a naming convention (i.e., both attributes are named `A`).

与此相反的是“显性覆盖（Explicit Overrides）”：

> **2. Explicit Overrides**: if attribute `A` is declared as an alias for attribute `B` in a meta-annotation via `@AliasFor`, then `A` is an *explicit override* for `B`.

文档中并没有明确指出属性 A 和 B 是否在同一注解中，仅说明当 A @AliasFor B 时，属性 A 显性覆盖了属性 B 的内容。文档继续补充“传递的显性覆盖（Transitive Explicit Overrides）”：

> **3. Transitive Explicit Overrides**: if attribute A in annotation @One is an explicit override for attribute B in annotation @Two and B is an explicit override for attribute C in annotation @Three, then A is a *transitive explicit override* for C following the law of transitivity.

此处结合“显性覆盖（Explicit Overrides）”的内容，间接地说明@AliasFor 可建立在不同注解层次的属性之间。一言以蔽之，“隐性覆盖（Implicit Overrides）”属于低层次注解属性覆盖高层次元注解同名属性的特性，而“显性覆盖（Explicit Overrides）”是@AliasFor 提供的属性覆盖能力。因此，如何深入理解“注解属性别名（Alias）”成为本节讨论的关键。

下面仍以@TransactionalService 为例进行说明，@TransactionalService 的元注解为@Service 和@Transactional，其中@Transactional 有不少属性方法，如 transactionManager()和 propagation()等。由于 Java 注解的静态性，@TransactionalService 无法调整其元注解@Transactional 属性方法的默认值。因此，AnnotationAttributes 只能读取这些默认值。假设@TransactionalService 需要调整@Transactional.transactionManager()的默认值，可结合“隐性覆盖”原则进行设计，即@TransactionalService 增加 transactionManager()属性方法，覆盖默认值：

```
@Target({ElementType.TYPE})
@Retention(RetentionPolicy.RUNTIME)
@Documented
@Transactional
@Service(value = "transactionalService")
public @interface TransactionalService {

    /**
     * @return 服务 Bean 名称
     */
    String name() default "";

    /**
     * 覆盖{@link Transactional#transactionManager()}默认值
     *
     * @return {@link PlatformTransactionManager}Bean 名称，默认关联 "txManager" Bean
     */
```

```
    String transactionManager() default "txManager";

}
```

将@TransactionalService 标注到具体类上：

```
@TransactionalService
public class TransactionalServiceBean {

    public void save() {
        System.out.println("保存操作...");
    }
}
```

实现引导类：

```
@Configuration
// 扫描 TransactionalServiceBean 所在 package
@ComponentScan(basePackageClasses = TransactionalServiceBean.class)
@EnableTransactionManagement // 激活事务管理
public class TransactionalServiceBeanBootstrap {

    public static void main(String[] args) {
        // 注册当前引导类作为 Configuration Class
        ConfigurableApplicationContext context = new
AnnotationConfigApplicationContext(TransactionalServiceBeanBootstrap.class);
        // 获取所有 TransactionalServiceBean 类型 Bean，其中 Key 为 Bean 名称
        Map<String, TransactionalServiceBean> beansMap =
context.getBeansOfType(TransactionalServiceBean.class);
        beansMap.forEach((beanName, bean) -> {
            System.out.printf("Bean 名称 : %s , 对象 : %s\n", beanName, bean);
            bean.save();
        });
        context.close();
    }
}
```

运行并观察输出：

```
(...部分内容被省略...)
```

```
    Bean 名称 : transactionalServiceBean , 对象 :
thinking.in.spring.boot.samples.spring5.bean.TransactionalServiceBean@402bba4f
    Exception in thread "main" org.springframework.beans.factory.NoSuchBeanDefinitionException:
No bean named 'txManager' available: No matching PlatformTransactionManager bean found for
qualifier 'txManager' - neither qualifier match nor bean name match!
```

示例程序读取 `TransactionalServiceBean` Bean 无误，当执行 `save()`方法时，由于 Spring 应用上下文不存在名为“txManager”的 `PlatformTransactionManager` Bean，故有此异常，符合预期。同时，该结果证明了`@TransactionalService.transactionManager()`确实覆盖了`@Transactional.transactionManager()`默认值。故再次调整引导类，配置两个 `PlatformTransactionManager` Bean，分别命名为“txManager”和“txManager2”。此处定义两处 `PlatformTransactionManager` Bean 的目的是为了排除单一 Bean 被`@Transactional.transactionManager()`默认值筛选的可能。同时，将各自的 `commit(TransactionStatus)`方法差异化实现：

```
    @Configuration
    // 扫描 TransactionalServiceBean 所在 package
    @ComponentScan(basePackageClasses = TransactionalServiceBean.class)
    @EnableTransactionManagement // 激活事务管理
    public class TransactionalServiceBeanBootstrap {
        ...

        @Bean("txManager")
        public PlatformTransactionManager txManager() {
            return new PlatformTransactionManager() {
                @Override
                public TransactionStatus getTransaction(TransactionDefinition definition) throws
TransactionException {
                    return new SimpleTransactionStatus();
                }

                @Override
                public void commit(TransactionStatus status) throws TransactionException {
                    System.out.println("txManager : 事务提交...");
                }
                ...
            };
        }
```

```
    @Bean("txManager2")
    public PlatformTransactionManager txManger2() {
        return new PlatformTransactionManager() {
            @Override
            public TransactionStatus getTransaction(TransactionDefinition definition) throws
TransactionException {
                return new SimpleTransactionStatus();
            }

            @Override
            public void commit(TransactionStatus status) throws TransactionException {
                System.out.println("txManger2 : 事务提交...");
            }
            ...
        };
    }
}
```

再次运行，观察日志变化：

```
(...部分内容被省略...)
Bean 名称 : transactionalServiceBean , 对象 :
thinking.in.spring.boot.samples.spring5.bean.TransactionalServiceBean@14dd9eb7
保存操作...
txManager : 事务提交...
(...部分内容被省略...)
```

运行结果显示“txManager”Bean 被选择，更严谨地证明了`@Transactional.transactionManager()`默认值确实被覆盖。

4．理解 Spring 注解属性别名（Aliases）

属性方法`@Transactional.transactionManager()`是从 Spring Framework 4.2 开始被引入的，早期的版本则使用 `value()`：

```
@Target({ElementType.METHOD, ElementType.TYPE})
@Retention(RetentionPolicy.RUNTIME)
@Inherited
@Documented
```

```
public @interface Transactional {
    ...
    @AliasFor("transactionManager")
    String value() default "";

    /**
     * ...
     * @since 4.2
     * @see #value
     */
    @AliasFor("value")
    String transactionManager() default "";

}
```

观察其属性方法定义，两者相互“@AliasFor”，说明@AliasFor可用于同一注解属性方法之间相互别名。明显地，transactionManager()相对于value()更具有语义。因此，再次重构@TransactionalService注解，暂时将transactionManager()属性方法注释，替换为value()，默认值仍旧为“txManager”，进而覆盖@Transactional.value()的内容：

```
@Target({ElementType.TYPE})
@Retention(RetentionPolicy.RUNTIME)
@Documented
@Transactional
@Service(value = "transactionalService")
public @interface TransactionalService {

    /**
     * @return 服务 Bean 名称
     */
    String name() default "";


//String transactionManager() default "txManager";

    /**
     * 覆盖 {@link Transactional#value()} 默认值
     *
```

```
     * @return {@link PlatformTransactionManager} Bean 名称，默认关联 "txManager" Bean
     */
    String value() default "txManager";
}
```

重启引导类 TransactionalServiceBeanBootstrap，观察日志输出是否变化：

```
(...部分内容被省略...)
Bean 名称 : txManager , 对象 :
thinking.in.spring.boot.samples.spring5.bean.TransactionalServiceBean@52e6fdee
保存操作...
txManager2 : 事务提交...
(...部分内容被省略...)
```

理论上，运行结果不应该发生变化，然而实际上却又是“txManager2”事务提交，并且 Bean 名称从“transactionalServiceBean”调整为“txManager2”。更重要的一点是，当前 Spring 应用上下文存在两个 PlatformTransactionManager Bean，然而事务在执行时，上下文却选择了“txManager2”，为什么没有报错呢？运行结果已给出了解释，当前 TransactionalServiceBean 的 Bean 名称为“txManager”，同名“txManager”的 PlatformTransactionManager Bean 被覆盖，所以当前上下文仅存在一个 PlatformTransactionManager Bean，即“txManager2”。换言之，@TransactionalService.value() 的默认值“txManager”作为 TransactionalServiceBean Bean 的名称。这也是能够解释的，因为 @TransactionalService 作为 Spring 模式注解，其元注解@Service 和@Compoent 均定义了 value() 属性方法，根据“隐性覆盖”原则，value()属性被 Spring Framework 视作标注类的 Bean 名称。按照目前对@AliasFor 的理解，它能够构建同注解属性间的别名，所以将 @TransactionalService 的 name()单向“@AliasFor”到 value()：

```
@Target({ElementType.TYPE})
@Retention(RetentionPolicy.RUNTIME)
@Documented
@Transactional
@Service(value = "transactionalService")
public @interface TransactionalService {

    /**
     * @return 服务 Bean 名称
     */
    @AliasFor("value")
    String name() default "";
```

```
    /**
     * 覆盖 {@link Transactional#value()} 默认值
     *
     * @return {@link PlatformTransactionManager} Bean 名称，默认关联 "txManager" Bean
     */
    String value() default "txManager";

}
```

重启引导类 TransactionalServiceBeanBootstrap，观察日志输出：

```
(...部分内容被省略...)
Exception in thread "main" org.springframework.beans.factory.BeanDefinitionStoreException:
Failed to parse configuration class
[thinking.in.spring.boot.samples.spring5.bootstrap.TransactionalServiceBeanBootstrap]; nested
exception is org.springframework.core.annotation.AnnotationConfigurationException: Attribute
'value' in annotation [thinking.in.spring.boot.samples.spring5.annotation.TransactionalService]
must be declared as an @AliasFor [name].
(...部分内容被省略...)
```

异常信息提示 value()也必须“@AliasFor”到 name()，于是继续调整@TransactionalService.name()的声明：

```
@Target({ElementType.TYPE})
@Retention(RetentionPolicy.RUNTIME)
@Documented
@Transactional
@Service(value = "transactionalService")
public @interface TransactionalService {

    /**
     * @return 服务 Bean 名称
     */
    @AliasFor("value")
    String name() default "";

    /**
     * 覆盖 {@link Transactional#value()} 默认值
```

```
     *
     * @return {@link PlatformTransactionManager} Bean 名称，默认关联 "txManager" Bean
     */
    @AliasFor("name")
    String value() default "txManager";

}
```

再次运行引导类 TransactionalServiceBeanBootstrap，观察日志输出：

```
    (...部分内容被省略...)
    Misconfigured aliases: attribute 'name' in annotation
[thinking.in.spring.boot.samples.spring5.annotation.TransactionalService] and attribute 'value'
in annotation [thinking.in.spring.boot.samples.spring5.annotation.TransactionalService] must
declare the same default value.
    (...部分内容被省略...)
```

异常再次提示 name 属性与 value 属性的默认值必须相等。不过这些内容并没有在“Attribute Aliases and Overrides”一节中体现，与之对应地提到“显性别名（Explicit Aliases）”的概念：

> 1. **Explicit Aliases**: if two attributes in one annotation are declared as aliases for each other via @AliasFor, they are *explicit aliases*.

文中仅提到相同注解中的两个属性方法需要相互“@AliasFor”，并没有说明默认值必须相等。故下面再次调整@TransactionalService.name()的默认值声明：

```
@Target({ElementType.TYPE})
@Retention(RetentionPolicy.RUNTIME)
@Documented
@Transactional
@Service(value = "transactionalService")
public @interface TransactionalService {

    /**
     * @return 服务 Bean 名称
     */
    @AliasFor("value")
    String name() default "txManager";
```

```
    /**
     * 覆盖 {@link Transactional#value()} 默认值
     *
     * @return {@link PlatformTransactionManager} Bean 名称，默认关联 "txManager" Bean
     */
    @AliasFor("name")
    String value() default "txManager";

}
```

再次运行：

```
(...部分内容被省略...)
Bean 名称 : txManager , 对象 :
thinking.in.spring.boot.samples.spring5.bean.TransactionalServiceBean@1a451d4d
保存操作...
txManger2 : 事务提交...
(...部分内容被省略...)
```

运行结果正常，不过这里仍然存在一个疑惑，既然 name 和 value 的默认值都是“txManager”，那么到底哪个属性方法被 Spring Framework 框架选择了呢？为此调整 TransactionalServiceBean 中的注解声明：

```
@TransactionalService(name = "transactionalServiceBean")
public class TransactionalServiceBean {

    public void save() {
        System.out.println("保存操作...");
    }
}
```

显式地设置 Bean 的名称为“transactionalServiceBean”，再次运行示例程序：

```
(...部分内容被省略...)
Bean 名称 : transactionalServiceBean , 对象 :
thinking.in.spring.boot.samples.spring5.bean.TransactionalServiceBean@62150f9e
No qualifying bean of type 'org.springframework.transaction.PlatformTransactionManager'
available: expected single matching bean but found 2: txManager,txManager2
(...部分内容被省略...)
```

本次运行结果说明@TransactionalService(name = "transactionalServic eBean")语句驱动 TransactionalServiceBean 的 Bean 名称调整成功，不过导致 PlatformTransactionManager Bean“txManager”和“txManager2”同时存在于 Spring 应用上下文中，所以无法精确匹配。尽管如此，结果并不影响@AliasFor 在同注解属性中的使用，只要属性方法之间需要相互“@AliasFor”，则它们的默认值就必须相等。如同@TransactionalService 这样的注解，通过注解属性方法 name()与 value()相互“@AliasFor”，并且 value()又覆盖其元注解@Service.value()属性，因此@TransactionalService.name()与@Service.value()之间的关系被“Attribute Aliases and Overrides”一节定义为“隐性别名”：

> 2. **Implicit Aliases**: if two or more attributes in one annotation are declared as explicit overrides for the same attribute in a meta-annotation via @AliasFor, they are *implicit aliases*.

以上说明结合下图来加深理解。

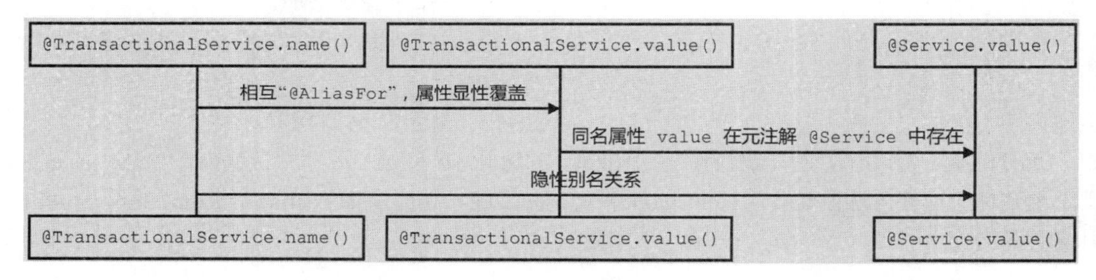

除此之外，文档还定义了更进一步的关系——传递的隐性别名（Transitive Implicit Aliases）：

> 3. **Transitive Implicit Aliases**: given two or more attributes in one annotation that are declared as explicit overrides for attributes in meta-annotations via @AliasFor, if the attributes *effectively* override the same attribute in a meta-annotation following the law of transitivity, they are *transitive implicit aliases*.

实际上，传递的隐性别名（Transitive Implicit Aliases）关系在@TransactionalService.name()属性中也是存在的，如下图所示。

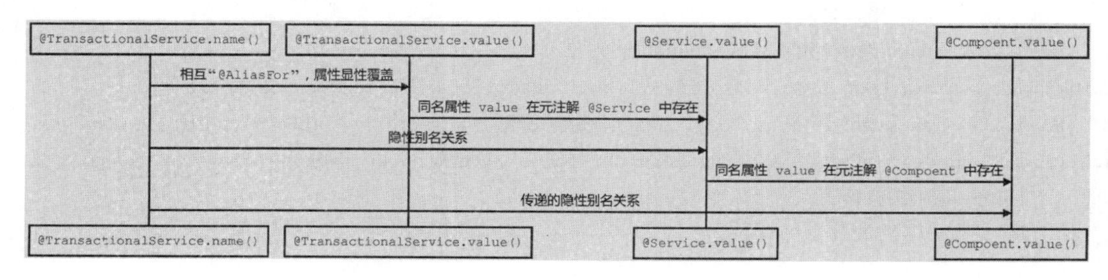

前面的示例主要讨论同注解属性之间“@AliasFor”的关系，那么不同层次注解属性之间相互“@AliasFor”需要哪些步骤呢？Spring Boot 中最常见的注解@SpringBootApplication 提供了参考：

```
@Target(ElementType.TYPE)
@Retention(RetentionPolicy.RUNTIME)
@Documented
@Inherited
@SpringBootConfiguration
@EnableAutoConfiguration
@ComponentScan(excludeFilters = {
        @Filter(type = FilterType.CUSTOM, classes = TypeExcludeFilter.class),
        @Filter(type = FilterType.CUSTOM, classes = AutoConfigurationExcludeFilter.class) })
public @interface SpringBootApplication {
    ...
    @AliasFor(annotation = EnableAutoConfiguration.class)
    Class<?>[] exclude() default {};
    ...
}
```

@SpringBootApplication 在 exclude()的属性上与其元注解@EnableAutoConfiguration 建立了单向@AliasFor 关系。如果反向建立关系，那么必然是不合理的，Spring 元注解之间不应该相互元标注，当然 Java 标准注解是一个例外，如@Documented 与@Target 之间。简言之，多层次注解属性之间的@AliasFor 关系只能由较低层向较高层建立。按此逻辑，@TransactionalService 新增 manager()属性定义，将其“@AliasFor”到@Transactional.transactionManager()，并设置其默认值为“txManager”：

```
@Target({ElementType.TYPE})
@Retention(RetentionPolicy.RUNTIME)
@Documented
@Transactional
@Service(value = "transactionalService")
public @interface TransactionalService {
    ...
    /**
     * 建立 {@link Transactional#transactionManager()} 别名
     *
     * @return {@link PlatformTransactionManager} Bean 名称，默认关联 "txManager" Bean
```

```
     */
    @AliasFor(attribute = "transactionManager", annotation = Transactional.class)
    String manager() default "txManager";

}
```

重启引导类 TransactionalServiceBeanBootstrap，观察日志输出：

```
(...部分内容被省略...)
Bean 名称 : transactionalServiceBean , 对象 :
thinking.in.spring.boot.samples.spring5.bean.TransactionalServiceBean@1a451d4d
保存操作...
txManager : 事务提交...
(...部分内容被省略...)
```

结果证明当前示例所构建的多层次注解@AliasFor 关系是无误的。此处需要说明的是，AnnotatedElementUtils.getMergedAnnotationAttributes 方法也是符合属性别名完整语义的，故在其 JavaDoc 中补充：

> and @AliasFor semantics are fully supported, both within a single annotation and within the annotation hierarchy.

据此说明，对 AnnotationAttributesBootstrap 做出微调，将 TransactionalService 替换为 TransactionalServiceBean：

```
public class AnnotationAttributesBootstrap {

    public static void main(String[] args) {

    //    AnnotatedElement annotatedElement = TransactionalService.class;

        AnnotatedElement annotatedElement = TransactionalServiceBean.class;

        // 获取@Service 注解属性独享
        AnnotationAttributes serviceAttributes =
                AnnotatedElementUtils.getMergedAnnotationAttributes(annotatedElement,
Service.class);

        // 获取@Transactional 注解属性独享
```

```
        AnnotationAttributes transactionalAttributes =
                AnnotatedElementUtils.getMergedAnnotationAttributes(annotatedElement,
Transactional.class);

        // 输出
        print(serviceAttributes);

        print(transactionalAttributes);
    }
    ...
}
```

运行并且观察 TransactionalServiceBean 元注解属性的变化：

```
注解 @org.springframework.stereotype.Service 属性集合 :
        属性 value : transactionalService
注解 @org.springframework.transaction.annotation.Transactional 属性集合 :
        属性 value : txManager
        属性 transactionManager : txManager
        ...
```

尽管在 Java 编程语言级别上，Java 注解是绝对静态性的，且属性方法是表达注解状态的唯一途径。而 Spring Framework 为 Spring 元注解和@AliasFor 提供了属性覆盖和别名的特性，最终由 AnnotationAttributes 对象来表达语义。它的重要程度在后续的 Spring 注解驱动编程中将频繁地体现。

> 进阶阅读：若觉得意犹未尽，则不妨参考小马哥为 Dubbo 2.5.8 提供的激活 Dubbo 注解驱动特性注解@EnableDubbo，其中的特性包括本节的主题“Spring 注解属性别名和覆盖（Attribute Aliases and Overrides）”，以及讨论的“@Enable 模块驱动”内容。
>
> ```
> @Target({ElementType.TYPE})
> @Retention(RetentionPolicy.RUNTIME)
> @Inherited
> @Documented
> @EnableDubboConfig
> @DubboComponentScan
> public @interface EnableDubbo {
>
>     /**
>      * Base packages to scan for annotated @Service classes.
> ```

```
     * <p>
     * Use {@link #scanBasePackageClasses()} for a type-safe alternative to String-based
     * package names.
     *
     * @return the base packages to scan
     * @see DubboComponentScan#basePackages()
     */
    @AliasFor(annotation = DubboComponentScan.class, attribute = "basePackages")
    String[] scanBasePackages() default {};

    /**
     * Type-safe alternative to {@link #scanBasePackages()} for specifying the packages to
     * scan for annotated @Service classes. The package of each class specified will be
     * scanned.
     *
     * @return classes from the base packages to scan
     * @see DubboComponentScan#basePackageClasses
     */
    @AliasFor(annotation = DubboComponentScan.class, attribute = "basePackageClasses")
    Class<?>[] scanBasePackageClasses() default {};


    /**
     * It indicates whether {@link AbstractConfig} binding to multiple Spring Beans.
     *
     * @return the default value is <code>false</code>
     * @see EnableDubboConfig#multiple()
     */
    @AliasFor(annotation = EnableDubboConfig.class, attribute = "multiple")
    boolean multipleConfig() default false;

}
```

第 8 章 Spring 注解驱动设计模式

前文的讨论主要集中在单一 Annotation 配置和少数组件组合上。接下来进入“**@Enable 模块驱动**”部分，系统性地介绍这部分内容，理解模式的意义。

8.1 Spring @Enable 模块驱动

再次将目光投向 Spring Framework 3.x，其中 Spring Framework 3.1 是一个具有里程碑意义的发行版本，从此版本开始，Spring Framework 开始支持“**@Enable 模块驱动**”。所谓“模块”是指具备相同领域的功能组件集合，组合所形成的一个独立的单元，比如 Web MVC 模块、AspectJ 代理模块、Caching（缓存）模块、JMX（Java 管理扩展）模块、Async（异步处理）模块等。

8.1.1 理解@Enable 模块驱动

在 Spring Framework 3.1 中，框架实现者有意识地形成了一种新“设计模式”，然而在官方的文档中着墨不多。这种设计模式有别于传统的面向对象 GoF 23 设计模式（23 种设计模式），小马哥将其定义为“**@Enable 模块驱动**”。

@Enable 模块驱动在后续 Spring Framework、Spring Boot 和 Spring Cloud 中一以贯之，这种模块化的 Annotation 均以@Enable 作为前缀，如下表所示。

框 架 实 现	@Enable 注解模块	激 活 模 块
Spring Framework	@EnableWebMvc	Web MVC 模块
	@EnableTransactionManagement	事务管理模块
	@EnableCaching	Caching 模块
	@EnableMBeanExport	JMX 模块
	@EnableAsync	异步处理模块
	@EnableWebFlux	Web Flux 模块
	@EnableAspectJAutoProxy	AspectJ 代理模块
Spring Boot	@EnableAutoConfiguration	自动装配模块
	@EnableManagementContext	Actuator 管理模块
	@EnableConfigurationProperties	配置属性绑定模块
	@EnableOAuth2Sso	OAuth 2 单点登录模块
Spring Cloud	@EnableEurekaServer	Eureka 服务器模块
	@EnableConfigServer	配置服务器模块
	@EnableFeignClients	Feign 客户端模块
	@EnableZuulProxy	服务网关 Zuul 模块
	@EnableCircuitBreaker	服务熔断模块

引入“**@Enable 模块驱动**”的意义在于能够简化装配步骤，实现了“**按需装配**”，同时屏蔽组件集合装配的细节。然而凡有利必有弊，该模式必须手动触发，也就是说实现 Annotation 必须标注在某个配置 Bean 中，同时实现该模式的成本相对较高，尤其是在理解其中原理和加载机制及单元测试方面。

理解了“@Enable 模块驱动”，对于 Spring Boot 甚至是 Spring Cloud 的源码解读必然有所裨益。

8.1.2 自定义@Enable 模块驱动

根据小马哥的经验，Spring Framework 大致分为两类实现，按照实现难易程度，从易到难来排序，分为“**注解驱动**”和“**接口编程**”。

无论哪种实现方式，均需要依赖 Spring Framework 3.0 新引入的@Import。前文中探讨过一个类似的 Annotation @ImportResource，@ImportResource 用于导入 XML 配置文件。而 Spring Framework 3.0 中的@Import 则用于导入一个或多个 ConfigurationClass，将其注册为 Spring Bean，不过在 3.0 版本中存在一定的限制，即仅支持被@Configuration 标注的类。

> Spring Framework 3.0 中对@Import JavaDoc 描述：
> Indicates one or more Configuration classes to import.
> Provides functionality equivalent to the element in Spring XML. Only supported for actual @Configuration-annotated classes.

然而在 Spring Framework 3.1 中，@Import 的职责范围有所扩大，还可以用于声明至少一个@Bean 方法的类，以及 ImportSelector 或 ImportBeanDefinitionRegistrar 的实现类。

> Spring Framework 3.1 中对@Import JavaDoc 描述：
> Indicates one or more @Configuration classes to import.
> Provides functionality equivalent to the `<import/>` element in Spring XML. Only supported for classes annotated with `@Configuration` or declaring at least one `@Bean` method, as well as ImportSelector and ImportBeanDefinitionRegistrarimplementations.

因此，将@Configuration 类和@Bean 方法声明类归类为“注解驱动”，而 ImportSelector 或 ImportBeanDefinitionRegistrar 的实现类则归于“接口编程”。

1. 基于“注解驱动”实现@Enable 模块

小马哥和朋友们分享和交流技术时，嘴边常常会挂着一句话——“处处留心皆学问”。当某个技术点被使用时，万万不可得过且过，而是应将其进行系统性研究。因此，在正式进入编码实现之前，先参考和学习 Spring Framework 中已有的实现，比如@EnableWebMvc。

首先，参考@EnableWebMvc 的注解定义：

```java
@Retention(RetentionPolicy.RUNTIME)
@Target(ElementType.TYPE)
@Documented
@Import(DelegatingWebMvcConfiguration.class)
public @interface EnableWebMvc {
}
```

请读者留意第 4 行@Import(DelegatingWebMvcConfiguration.class)，前文已提及@Import 的作用，自然需要一窥 DelegatingWebMvcConfiguration 的源码：

```java
@Configuration
public class DelegatingWebMvcConfiguration extends WebMvcConfigurationSupport {
...
}
```

实现源码说明 DelegatingWebMvcConfiguration 是一个@Configuration 类，这种实现方式提供了样板。下面根据以上经验，实现“注解驱动实现”的“@Enable 模块驱动”。

（1）实现 Configuration 类：HelloWorldConfiguration。

```
@Configuration
public class HelloWorldConfiguration {

    @Bean
    public String helloWorld() { // 创建名为“helloWorld”String 类型的 Bean
        return "Hello,World";
    }
}
```

（2）实现“@Enable 模块驱动”Annotation：@EnableHelloWorld。

```
@Target(ElementType.TYPE)
@Retention(RetentionPolicy.RUNTIME)
@Documented
@Import(HelloWorldConfiguration.class) // 导入 HelloWorldConfiguration
public @interface EnableHelloWorld {
}
```

（3）标注@EnableHelloWorld 到引导类 EnableHelloWorldBootstrap。

```
@EnableHelloWorld
public class EnableHelloWorldBootstrap {

    public static void main(String[] args) {
        // 构建 Annotation 配置驱动 Spring 上下文
        AnnotationConfigApplicationContext context = new
AnnotationConfigApplicationContext();
        // 注册当前引导类（被@Configuration 标注）到 Spring 上下文
        context.register(EnableHelloWorldBootstrap.class);
        // 启动上下文
        context.refresh();
        // 获取名称为“helloWorld”的 Bean 对象
        String helloWorld = context.getBean("helloWorld", String.class);
        // 输出用户名称："Hello,World"
```

```
        System.out.printf("helloWorld = %s \n", helloWorld);
        // 关闭上下文
        context.close();
    }
}
```

> 以上示例代码可通过查找 spring-framework-samples/spring-framework-3.2.x-sample 工程获取。

引导类 EnableHelloWorldBootstrap 启动后，helloWorld Bean 能够正常获取，并且成功地输出到控制台：

```
(...部分内容被省略...)
helloWorld = Hello,World
(...部分内容被省略...)
```

注解驱动实现“@Enable 模块驱动”相对比较容易，接下来的接口编程实现稍显复杂。

2. 基于“接口编程”实现@Enable 模块

前文提到，接口编程需要实现 ImportSelector 或 ImportBeanDefinitionRegistrar 接口：

- ImportSelector 接口相对简单，使用 Spring 注解元信息抽象 AnnotationMetadata 作为方法参数，该参数的内容为导入 ImportSelector 实现的@Configuration 类元信息，进而动态地选择一个或多个其他@Configuration 类进行导入。故在 ImportSelector JavaDoc 中有如下描述：

> Interface to be implemented by types that determine which @Configuration class(es) should be imported based on a given selection criteria, usually one or more annotation attributes.

- ImportBeanDefinitionRegistrar 相对于 ImportSelector 而言，其编程复杂度更高，除注解元信息 AnnotationMetadata 作为入参外，接口将 Bean 定义（BeanDefinition）的注册交给开发人员决定。故在 ImportBeanDefinitionRegistrar JavaDoc 中有如下描述：

> Interface to be implemented by types that register additional bean definitions when processing @Configuration classes. Useful when operating at the bean definition level (as opposed to @Bean method/instance level) is desired or necessary.
> Along with @Configuration and ImportSelector, classes of this type may be provided to the @Import annotation (or may also be returned from an ImportSelector).

采用与“注解驱动实现”章节中相同的经验，参考@EnableCaching 的注解定义：

```
@Target(ElementType.TYPE)
@Retention(RetentionPolicy.RUNTIME)
@Documented
@Import(CachingConfigurationSelector.class)
public @interface EnableCaching {
  ...
}
```

再进入 CachingConfigurationSelector 的实现：

```
public class CachingConfigurationSelector extends AdviceModeImportSelector<EnableCaching> {

    /**
     * {@inheritDoc}
     * @return {@link ProxyCachingConfiguration} or {@code AspectJCacheConfiguration} for
     * {@code PROXY} and {@code ASPECTJ} values of {@link EnableCaching#mode()}, respectively
     */
    public String[] selectImports(AdviceMode adviceMode) {
        switch (adviceMode) {
            case PROXY:
                return new String[] { AutoProxyRegistrar.class.getName(),
ProxyCachingConfiguration.class.getName() };
            case ASPECTJ:
                return new String[]
{ AnnotationConfigUtils.CACHE_ASPECT_CONFIGURATION_CLASS_NAME };
            default:
                return null;
        }
    }

}
```

虽然 CachingConfigurationSelector 没有注解直接实现 ImportSelector#selectImports (AnnotationMetadata)方法，然而 AdviceModeImportSelector final 实现了该方法：

```
public abstract class AdviceModeImportSelector<A extends Annotation> implements
ImportSelector {
    ...
    public final String[] selectImports(AnnotationMetadata importingClassMetadata) {
```

```
        Class<?> annoType = GenericTypeResolver.resolveTypeArgument(getClass(),
AdviceModeImportSelector.class);
        AnnotationAttributes attributes = MetadataUtils.attributesFor(importingClassMetadata,
annoType);
        Assert.notNull(attributes, String.format(
              "@%s is not present on importing class '%s' as expected",
              annoType.getSimpleName(), importingClassMetadata.getClassName()));

        AdviceMode adviceMode = attributes.getEnum(this.getAdviceModeAttributeName());
        String[] imports = selectImports(adviceMode);
        Assert.notNull(imports, String.format("Unknown AdviceMode: '%s'", adviceMode));
        return imports;
    }
    ...
    protected abstract String[] selectImports(AdviceMode adviceMode);

}
```

由于实现 ImportSelector 的弹性相对于“注解驱动”实现更大，因此，ImportSelector 实现类在 Spring Framework 中也屈指可数。尽管如此，也并不影响实现的动力，希望以下的实现能起到抛砖引玉的作用。

- 基于 ImportSelector 接口实现

假设当前应用支持两种服务类型：HTTP 和 FTP，通过@EnableServer 设置服务器类型（type）提供对应的服务。

（1）定义服务器接口——Server 及服务类型：Server.Type。

```
public interface Server {

    /**
     * 启动服务器
     */
    void start();

    /**
     * 关闭服务器
     */
    void stop();
```

```
    /**
     * 服务器类型
     */
    enum Type {

        HTTP, // HTTP 服务器
        FTP   // FTP 服务器
    }
}
```

（2）实现 HTTP 和 FTP 服务器——HttpServer 和 FtpServer。

```
@Component // 根据 ImportSelector 的契约，请确保实现为 Spring 组件
public class HttpServer implements Server {

    @Override
    public void start() {
        System.out.println("HTTP 服务器启动中...");
    }

    @Override
    public void stop() {
        System.out.println("HTTP 服务器关闭中...");
    }
}
@Component // 根据 ImportSelector 的契约，请确保实现为 Spring 组件
public class FtpServer implements Server {

    @Override
    public void start() {
        System.out.println("FTP 服务器启动中...");
    }

    @Override
    public void stop() {
        System.out.println("FTP 服务器关闭中...");
    }
}
```

（3）实现“@Enable 模块驱动”Annotation——@EnableServer。

```
@Target(ElementType.TYPE)
@Retention(RetentionPolicy.RUNTIME)
@Documented
@Import(ServerImportSelector.class) // 导入 ServerImportSelector
public @interface EnableServer {

    /**
     * 设置服务器类型
     * @return non-null
     */
    Server.Type type();
}
```

（4）实现 Server ImportSelector——ServerImportSelector。

```
public class ServerImportSelector implements ImportSelector {

    @Override
    public String[] selectImports(AnnotationMetadata importingClassMetadata) {
        // 读取 EnableServer 中所有的属性方法，本例中仅有 type()属性方法
        // 其中 key 为属性方法的名称，value 为属性方法的返回对象
        Map<String, Object> annotationAttributes =
importingClassMetadata.getAnnotationAttributes(EnableServer.class.getName());
        // 获取名为“type”的属性方法，并且强制转化成 Server.Type 类型
        Server.Type type = (Server.Type) annotationAttributes.get("type");
        // 导入的类名称数组
        String[] importClassNames = new String[0];
        switch (type) {
            case HTTP: // 当设置 HTTP 服务器类型时，返回 HttpServer 组件
                importClassNames = new String[]{HttpServer.class.getName()};
                break;
            case FTP: //  当设置 FTP 服务器类型时，返回 FtpServer 组件
                importClassNames = new String[]{FtpServer.class.getName()};
                break;
        }
        return importClassNames;
    }
```

```
}
```

（5）标注@EnableServer 到引导类 EnableServerBootstrap。

```
@Configuration
@EnableServer(type = Server.Type.HTTP) // 设置 HTTP 服务器
public class EnableServerBootstrap {

    public static void main(String[] args) {
        // 构建 Annotation 配置驱动 Spring 上下文
        AnnotationConfigApplicationContext context = new
AnnotationConfigApplicationContext();
        // 注册当前引导类（被 @Configuration 标注）到 Spring 上下文
        context.register(EnableServerBootstrap.class);
        // 启动上下文
        context.refresh();
        // 获取 Server Bean 对象，实际为 HttpServer
        Server server = context.getBean(Server.class);
        // 启动服务器
        server.start();
        // 关闭服务器
        server.stop();
    }
}
```

> 以上示例代码可通过查找 spring-framework-samples/spring-framework-3.2.x-sample 工程获取。

引导类 EnableServerBootstrap 启动后，Server Bean 能够正常获取，并且成功地输出到控制台：

```
...(部分内容被忽略)...
HTTP 服务器启动中...
HTTP 服务器关闭中...
```

再将以上示例调整为 ImportBeanDefinitionRegistrar 实现。

- 基于 ImportBeanDefinitionRegistrar 接口实现

本示例中待调整的部分较少，只需将@EnableServer 导入（@Import）的 ServerImportSelector 替换为 ImportBeanDefinitionRegistrar 实现即可，并复用 ServerImportSelector 的实现。

（1）实现 ServerImportBeanDefinitionRegistrar:ServerImportBeanDefinitionRegistrar。

```
public class ServerImportBeanDefinitionRegistrar implements ImportBeanDefinitionRegistrar {

    @Override
    public void registerBeanDefinitions(AnnotationMetadata importingClassMetadata,
BeanDefinitionRegistry registry) {
        // 复用 {@link ServerImportSelector} 实现，避免重复劳动
        ImportSelector importSelector = new ServerImportSelector();
        // 筛选 Class 名称集合
        String[] selectedClassNames = importSelector.selectImports(importingClassMetadata);
        // 创建 Bean 定义
        Stream.of(selectedClassNames)
                // 转化为 BeanDefinitionBuilder 对象
                .map(BeanDefinitionBuilder::genericBeanDefinition)
                // 转化为 BeanDefinition
                .map(BeanDefinitionBuilder::getBeanDefinition)
                .forEach(beanDefinition ->
                        // 注册 BeanDefinition 到 BeanDefinitionRegistry
                        BeanDefinitionReaderUtils.registerWithGeneratedName(beanDefinition,
registry)
                );
    }
}
```

（2）替换@EnableServer @Import。

```
@Target(ElementType.TYPE)
@Retention(RetentionPolicy.RUNTIME)
@Documented
//@Import(ServerImportSelector.class) // 导入 ServerImportSelector
@Import(ServerImportBeanDefinitionRegistrar.class) // 替换 ServerImportSelector
public @interface EnableServer {

    /**
     * 设置服务器类型
     * @return non-null
     */
    Server.Type type();
```

```
}
```

重启引导类 EnableServerBootstrap，观察日志变化：

```
...(部分内容被忽略)...
HTTP 服务器启动中...
HTTP 服务器关闭中...
```

运行结果证明当前重构是成功的。

> 进阶阅读：如果读者仍不觉得过瘾，请参考小马哥为 Dubbo 2.5.8+提供的@Enable 模块实现@EnableDubboConfig，该实现用于简化 Dubbo 外部化配置开发。
>
> ```
> @Target({ElementType.TYPE})
> @Retention(RetentionPolicy.RUNTIME)
> @Inherited
> @Documented
> @Import(DubboConfigConfigurationSelector.class)
> public @interface EnableDubboConfig {
>
>     /**
>      * It indicates whether binding to multiple Spring Beans.
>      *
>      * @return the default value is <code>false</code>
>      * @revised 2.5.9
>      */
>     boolean multiple() default false;
>
> }
> ```
>
> 如果读者在参考 Dubbo 外部化配置的实现中遇到了困难，请提前翻阅“外部化配置”章节内容，里面有非常详尽的讲解。

8.1.3 @Enable 模块驱动原理

在前文的讨论中，@Enable 模块驱动模块无论来自 Spring 内建，还是自定义，均使用@Import 实现，并且@Import 的职责在于装载导入类（Importing Class），将其定义为 Spring Bean。结合当前场景，导入类主要为@Configuration Class、ImportSelector 实现及 ImportBeanDefinitionRegistrar 实现，它们也是接下来讨论的重点。

1．装载@Configuration Class

@Configuration 从 Spring Framework 3.0 开始引入，该版本还未引入@ComponentScan，因此配套的导入注解是@Import。尽管 Spring Framework 3.0 提供了注解驱动上下文实现 AnnotationConfigApplicationContext，然而与@Import 配合仍比较烦琐，仅支持 Spring 组件类的逐个导入，如@Import({A.class,B.class,...})，因此当时无法完全替代 XML 元素 <context:component-scan/>。即使 Spring 应用上下文与<context:component-scan/>结合使用，@Import 的处理也无法执行。因此，开发人员经常看到 XML 元素<context:component-scan/>与<context:annotation-config/>同时存在。根据 Spring Framework“可扩展的 XML 编写（Extensible XML authoring）”的特性，可知<context:annotation-config/>所对应的 BeanDefinitionParser 实现为 AnnotationConfigBeanDefinitionParser。该实现在解析过程中，将注册注解配置处理器 Bean：

```
public class AnnotationConfigBeanDefinitionParser implements BeanDefinitionParser {

    public BeanDefinition parse(Element element, ParserContext parserContext) {
        Object source = parserContext.extractSource(element);

        // Obtain bean definitions for all relevant BeanPostProcessors.
        Set<BeanDefinitionHolder> processorDefinitions =

    AnnotationConfigUtils.registerAnnotationConfigProcessors(parserContext.getRegistry(),
source);
            ...
        return null;
    }

    }
```

> 以上实现源码的 Maven GAV 坐标为：org.springframework:spring-context:3.0.0.RELEASE。

这里可以看出，parse(Element,ParserContext)并没有直接解析 BeanDefinition 的实例，而是调用 AnnotationConfigUtils#registerAnnotationConfigProcessors(BeanDefinitionRegistry, Object)方法实现的。该方法从 Spring Framework 3.0 开始，新增了@Configuration Class

的处理实现 ConfigurationClassPostProcessor：

```
public class AnnotationConfigUtils {
    ...
    /**
  * The bean name of the internally managed Configuration annotation processor.
  */
 public static final String CONFIGURATION_ANNOTATION_PROCESSOR_BEAN_NAME =

 "org.springframework.context.annotation.internalConfigurationAnnotationProcessor";
    ...
 public static Set<BeanDefinitionHolder> registerAnnotationConfigProcessors(
         BeanDefinitionRegistry registry, Object source) {

     Set<BeanDefinitionHolder> beanDefs = new LinkedHashSet<BeanDefinitionHolder>(4);

     if (!registry.containsBeanDefinition(CONFIGURATION_ANNOTATION_PROCESSOR_BEAN_NAME)) {
         RootBeanDefinition def = new
RootBeanDefinition(ConfigurationClassPostProcessor.class);
         def.setSource(source);
         beanDefs.add(registerPostProcessor(registry, def,
CONFIGURATION_ANNOTATION_PROCESSOR_BEAN_NAME));
     }
     ...
     return beanDefs;
 }
    ...
}
```

> 以上实现源码的 Maven GAV 坐标为：org.springframework:spring-context:3.0.0.RELEASE。

ConfigurationClassPostProcessor 被封装成 Spring Bean 定义（BeanDefinition），后续注册为 Spring Bean，且其 Bean 名称为“org.springframework. context.annotation. internalConfiguration AnnotationProcessor”。按其 JavaDoc 中的描述：

> BeanFactoryPostProcessor used for bootstrapping processing of @Configuration classes. Registered by default when using context:annotation-config/ or context:component-scan/. Otherwise, may be declared manually as with any other BeanFactoryPostProcessor.

ConfigurationClassPostProcessor 默认被 XML 元素 <context:annotation-config/> 或 <context:component-scan/>注册，所以在 Spring Framework 中，<context:annotation-config/>的底层实现类 AnnotationConfigBeanDefinitionParser 和 <context:component-scan/>的底层实现类 ComponentScanBeanDefinitionParser 均调用了 AnnotationConfigUtils#registerAnnotationConfigProcessors(BeanDefinitionRegistry,Object)方法注册 ConfigurationClassPostProcessor Bean。同时文档提到，除此之外，可能需要手动使其被其他 BeanFactoryPostProcessor 装配。这种说法挑不出毛病，不过有些“偏颇”，似乎在暗示开发人员 ConfigurationClassPostProcessor 仅在 XML 配置驱动场景和手动方式下才可装配。文档忽略了 Spring Framework 3.0 注解驱动上下文实现 AnnotationConfigApplicationContext 的存在，其 AnnotatedBeanDefinitionReader 类型成员 reader 在构造时：

```
public class AnnotationConfigApplicationContext extends GenericApplicationContext {

    private final AnnotatedBeanDefinitionReader reader = new
AnnotatedBeanDefinitionReader(this);
        ...
}
```

也显式地调用了 AnnotationConfigUtils#registerAnnotationConfigProcessors(BeanDefinitionRegistry, Object)方法：

```
public class AnnotatedBeanDefinitionReader {
    ...
    public AnnotatedBeanDefinitionReader(BeanDefinitionRegistry registry) {
        this.registry = registry;
        AnnotationConfigUtils.registerAnnotationConfigProcessors(this.registry);
    }
    ...
}
```

> 以上实现源码的 Maven GAV 坐标为：org.springframework:spring-context:3.0.0.RELEASE。

简言之，ConfigurationClassPostProcessor 无论是在 XML 配置驱动还是在注解驱动的使用场景下，均通过 AnnotationConfigUtils#registerAnnotationConfigProcessors(BeanDefinitionRegistry, Object)方法的执行得到了装载。JavaDoc 继续向开发人员说明，ConfigurationClassPostProcessor 为最高优先级（Ordered.HIGHEST_PRECEDENCE）的 BeanFactoryPostProcessor 实现，不但处理@Configuration Class，也负责@Bean 方法的 Bean 定义：

> This post processor is Ordered.HIGHEST_PRECEDENCE as it is important that any Bean methods declared in Configuration classes have their respective bean definitions registered before any other BeanFactoryPostProcessor executes.

ConfigurationClassPostProcessor 所赋予的能力远超出 JavaDoc 的描述。在 Spring 应用上下文启动过程中（AbstractApplicationContext#refresh()方法被调用时），Spring 容器（BeanFactory）将 ConfigurationClassPostProcessor 初始化为 Spring Bean。它作为 BeanFactoryPostProcessor 的实现，随后其 postProcessBeanFactory(ConfigurableListableBeanFactory)方法被调用：

```
    public abstract class AbstractApplicationContext extends DefaultResourceLoader
          implements ConfigurableApplicationContext, DisposableBean {
        ...
    protected void invokeBeanFactoryPostProcessors(ConfigurableListableBeanFactory beanFactory)
{
          ...
          String[] postProcessorNames =
                    beanFactory.getBeanNamesForType(BeanFactoryPostProcessor.class, true,
false);
          ...
          List<String> orderedPostProcessorNames = new ArrayList<String>();
          ...
          for (String ppName : postProcessorNames) {
               ...
               else if (isTypeMatch(ppName, Ordered.class)) {
                    orderedPostProcessorNames.add(ppName);
               }
               ...
          }
          ...
```

```
        List<BeanFactoryPostProcessor> orderedPostProcessors = new
ArrayList<BeanFactoryPostProcessor>();
        for (String postProcessorName : orderedPostProcessorNames) {
            orderedPostProcessors.add(getBean(postProcessorName,
BeanFactoryPostProcessor.class));
        }
        OrderComparator.sort(orderedPostProcessors);
        invokeBeanFactoryPostProcessors(orderedPostProcessors, beanFactory);
        ...
    }
    ...
    private void invokeBeanFactoryPostProcessors(
            List<BeanFactoryPostProcessor> postProcessors, ConfigurableListableBeanFactory
beanFactory) {

        for (BeanFactoryPostProcessor postProcessor : postProcessors) {
            postProcessor.postProcessBeanFactory(beanFactory);
        }
    }
    ...
}
```

ConfigurationClassPostProcessor#postProcessBeanFactory(ConfigurableListable BeanFactory)方法被调用后，随之处理@Configuration 类和@Bean 方法：

```
    public class ConfigurationClassPostProcessor implements BeanFactoryPostProcessor,
BeanClassLoaderAware {
        ...
        public void postProcessBeanFactory(ConfigurableListableBeanFactory beanFactory) {
          if (!(beanFactory instanceof BeanDefinitionRegistry)) {
              throw new IllegalStateException(
                      "ConfigurationClassPostProcessor expects a BeanFactory that implements
BeanDefinitionRegistry");
          }
          processConfigBeanDefinitions((BeanDefinitionRegistry) beanFactory);
          enhanceConfigurationClasses(beanFactory);
    }
        ...
        public void processConfigBeanDefinitions(BeanDefinitionRegistry registry) {
```

```
        Set<BeanDefinitionHolder> configCandidates = new
LinkedHashSet<BeanDefinitionHolder>();
        for (String beanName : registry.getBeanDefinitionNames()) {
            BeanDefinition beanDef = registry.getBeanDefinition(beanName);
            if (checkConfigurationClassCandidate(beanDef)) {
                configCandidates.add(new BeanDefinitionHolder(beanDef, beanName));
            }
        }

        // Return immediately if no @Configuration classes were found
        if (configCandidates.isEmpty()) {
            return;
        }

        // Populate a new configuration model by parsing each @Configuration classes
        ConfigurationClassParser parser = new
ConfigurationClassParser(this.metadataReaderFactory, this.problemReporter);
        for (BeanDefinitionHolder holder : configCandidates) {
            BeanDefinition bd = holder.getBeanDefinition();
            try {
                if (bd instanceof AbstractBeanDefinition && ((AbstractBeanDefinition)
bd).hasBeanClass()) {
                    parser.parse(((AbstractBeanDefinition) bd).getBeanClass(),
holder.getBeanName());
                }
                else {
                    parser.parse(bd.getBeanClassName(), holder.getBeanName());
                }
            }
            catch (IOException ex) {
                throw new BeanDefinitionStoreException("Failed to load bean class: " +
bd.getBeanClassName(), ex);
            }
        }
        parser.validate();

        // Read the model and create bean definitions based on its content
        new ConfigurationClassBeanDefinitionReader(registry,
```

```
this.sourceExtractor).loadBeanDefinitions(parser.getConfigurationClasses());
    }
        ...
    }
```

执行期间，最重要的组件莫过于 ConfigurationClassParser，它将已注册的 Spring BeanDefinition 进行注解元信息解析，其中两个 parse 重载方法分别采用了基于 ASM 实现的 AnnotationMetadataReadingVisitor 和 Java 反射实现的 StandardAnnotationMetadata：

```
class ConfigurationClassParser {
    ...
    public void parse(String className, String beanName) throws IOException {
        MetadataReader reader = this.metadataReaderFactory.getMetadataReader(className);
        processConfigurationClass(new ConfigurationClass(reader, beanName));
    }
    ...
    public void parse(Class<?> clazz, String beanName) throws IOException {
        processConfigurationClass(new ConfigurationClass(clazz, beanName));
    }
    ...
}
```

其中，metadataReaderFactory 默认是熟悉的 CachingMetadataReaderFactory 实例。同时，@Configuration Class 被 ConfigurationClass 类所抽象，并且当前处理方法为 processConfigurationClass(ConfigurationClass)：

```
class ConfigurationClassParser {
    ...
    protected void processConfigurationClass(ConfigurationClass configClass) throws
IOException {
        AnnotationMetadata metadata = configClass.getMetadata();
        while (metadata != null) {
            doProcessConfigurationClass(configClass, metadata);
            ...
        }
        ...
    }

    protected void doProcessConfigurationClass(ConfigurationClass configClass,
```

```
AnnotationMetadata metadata) throws IOException {
        if (metadata.isAnnotated(Import.class.getName())) {
            processImport(configClass, (String[])
metadata.getAnnotationAttributes(Import.class.getName(), true).get("value"));
        }
        if (metadata.isAnnotated(ImportResource.class.getName())) {
            ...
        }
        Set<MethodMetadata> methods = metadata.getAnnotatedMethods(Bean.class.getName());
        for (MethodMetadata methodMetadata : methods) {
            ...
        }
    }

    private void processImport(ConfigurationClass configClass, String[] classesToImport)
throws IOException {
        ...
        else {
            this.importStack.push(configClass);
            for (String classToImport : classesToImport) {
                processClassToImport(classToImport);
            }
            this.importStack.pop();
        }
    }

    private void processClassToImport(String classToImport) throws IOException {
        ...
        else {
            processConfigurationClass(new ConfigurationClass(reader, null));
        }
    }

    ...
}
```

在 `processConfigurationClass(ConfigurationClass)`方法中，Spring 注解元数据抽象 `AnnotationMetadata` 同样是“关键先生”，在其被递归执行 `doProcessConfigurationClass`

(ConfigurationClass,AnnotationMetadata) 方法时，不但@Import 注解被处理了，@ImportResource 和@Bean 也被处理了，共同点是三者处理过程中的元信息均存储在 ConfigurationClass 对象中。其中，@Import 处理方法 processImport(ConfigurationClass,String[])与 processConfigurationClass(ConfigurationClass)方法形成了递归调用，实现多层次@Import 元标注 ConfigurationClass 的解析并压栈（importStack 字段）。解析后的 ConfigurationClass 集合将被 ConfigurationClassBeanDefinitionReader 再次注册为 Spring Bean：

```
class ConfigurationClassBeanDefinitionReader {
    ...
    public void loadBeanDefinitions(Set<ConfigurationClass> configurationModel) {
     for (ConfigurationClass configClass : configurationModel) {
         loadBeanDefinitionsForConfigurationClass(configClass);
     }
}
    ...
    private void loadBeanDefinitionsForConfigurationClass(ConfigurationClass configClass) {
     doLoadBeanDefinitionForConfigurationClass(configClass);

     for (ConfigurationClassMethod method : configClass.getMethods()) {
         loadBeanDefinitionsForModelMethod(method);
     }

     loadBeanDefinitionsFromImportedResources(configClass.getImportedResources());
 }
    ...
}
```

当然，ConfigurationClassBeanDefinitionReader 将@Import、@ImportResource 和@Bean 所关联的 Bean 定义一并注册了。

值得一提的是，ConfigurationClassPostProcessor 并非仅检查@Configuration 和@Bean 两种 Bean 定义方式，也处理@Component 的情况：

```
    public class ConfigurationClassPostProcessor implements BeanFactoryPostProcessor,
BeanClassLoaderAware {
        private static final String CONFIGURATION_CLASS_ATTRIBUTE =
            Conventions.getQualifiedAttributeName(ConfigurationClassPostProcessor.class,
"configurationClass");
```

```
    private static final String CONFIGURATION_CLASS_FULL = "full";

    private static final String CONFIGURATION_CLASS_LITE = "lite";
      ...
      protected boolean checkConfigurationClassCandidate(BeanDefinition beanDef) {
        AnnotationMetadata metadata = null;
        ...
        if (metadata != null) {
             if (metadata.isAnnotated(Configuration.class.getName())) {
                  beanDef.setAttribute(CONFIGURATION_CLASS_ATTRIBUTE,
CONFIGURATION_CLASS_FULL);
                  return true;
             }
             else if (metadata.isAnnotated(Component.class.getName()) ||
                       metadata.hasAnnotatedMethods(Bean.class.getName())) {
                  beanDef.setAttribute(CONFIGURATION_CLASS_ATTRIBUTE,
CONFIGURATION_CLASS_LITE);
                  return true;
             }
        }
        return false;
    }
      ...
    }
```

当目标 BeanDefinition 的 AnnotationMetadata 信息包含元注解@Component 时，同样认为该 BeanDefinition 所关联的 Class 也是候选的 ConfigurationClass。不过该 BeanDefinition 的 CONFIGURATION_CLASS_ATTRIBUTE 属性有所调整，其中@Configuration Class 是“完全模式”（CONFIGURATION_CLASS_FULL ="full"），而@Component Class 和@Bean 方法则是“轻量模式”（CONFIGURATION_CLASS_LITE = "lite"）。最后，ConfigurationClassPostProcessor 使用 CGLib 实现 ConfigurationClassEnhancer，用于提升 @Configuration Class：

```
   public class ConfigurationClassPostProcessor implements BeanFactoryPostProcessor,
BeanClassLoaderAware {
      ...
   public void enhanceConfigurationClasses(ConfigurableListableBeanFactory beanFactory) {
        Map<String, AbstractBeanDefinition> configBeanDefs = new LinkedHashMap<String,
```

```
AbstractBeanDefinition>();
        for (String beanName : beanFactory.getBeanDefinitionNames()) {
            BeanDefinition beanDef = beanFactory.getBeanDefinition(beanName);
            if
(CONFIGURATION_CLASS_FULL.equals(beanDef.getAttribute(CONFIGURATION_CLASS_ATTRIBUTE))) {
                ...
                configBeanDefs.put(beanName, (AbstractBeanDefinition) beanDef);
            }
        }
        ...
        ConfigurationClassEnhancer enhancer = new ConfigurationClassEnhancer(beanFactory);
        for (Map.Entry<String, AbstractBeanDefinition> entry : configBeanDefs.entrySet()) {
            AbstractBeanDefinition beanDef = entry.getValue();
            try {
                Class configClass = beanDef.resolveBeanClass(this.beanClassLoader);
                Class enhancedClass = enhancer.enhance(configClass);
                ...
                beanDef.setBeanClass(enhancedClass);
            }
            ...
        }
    }
    ...
}
```

> 以上实现源码的 Maven GAV 坐标为：org.springframework:spring-context:3.0.0.RELEASE。

2. 装载 ImportSelector 和 ImportBeanDefinitionRegistrar 实现

由于 ImportSelector 和 ImportBeanDefinitionRegistrar 从 Spring Framework 3.1 才开始引入，所以 3.0 版本中不会出现两者的实现。由于 BeanDefinitionRegistryPostProcessor 从 Spring Framework 3.0.1 开始引入，ConfigurationClassPostProcessor 的实现也随之发生变化，其实现接口从 BeanFactoryPostProcessor 替换为 BeanDefinitionRegistryPostProcessor，且 BeanDefinitionRegistryPostProcessor 扩展了 BeanFactoryPostProcessor 接口，所以 ConfigurationClassPostProcessor 存在两阶段实现：

```
public class ConfigurationClassPostProcessor implements
```

```
BeanDefinitionRegistryPostProcessor,
        Ordered, ResourceLoaderAware, BeanClassLoaderAware, EnvironmentAware {
    ...
    public void postProcessBeanDefinitionRegistry(BeanDefinitionRegistry registry) {
        ...
    }

    public void postProcessBeanFactory(ConfigurableListableBeanFactory beanFactory) {
        ...
    }
    ...
}
```

除两阶段实现外，ConfigurationClassPostProcessor 在 Spring Framework 3.1 中并没有太多变化，该版本的主要变化还是集中在 ConfigurationClassParser 的实现上，在其 doProcessConfigurationClass(ConfigurationClass,AnnotationMetadata)方法中，增加了 @PropertySource 和@ComponentScan 注解处理，并且更新了 processImport(ConfigurationClass, String[], boolean)方法的实现：

```
class ConfigurationClassParser {
    ...
    private void processImport(ConfigurationClass configClass, String[] classesToImport,
boolean checkForCircularImports) throws IOException {
        ...
        else {
            this.importStack.push(configClass);
            AnnotationMetadata importingClassMetadata = configClass.getMetadata();
            for (String candidate : classesToImport) {
                MetadataReader reader =
this.metadataReaderFactory.getMetadataReader(candidate);
                if (new AssignableTypeFilter(ImportSelector.class).match(reader,
metadataReaderFactory)) {
                    // the candidate class is an ImportSelector -> delegate to it to determine
imports
                    try {
                        ImportSelector selector =
BeanUtils.instantiateClass(Class.forName(candidate), ImportSelector.class);
                        processImport(configClass,
```

```
selector.selectImports(importingClassMetadata), false);
                    } catch (ClassNotFoundException ex) {
                        throw new IllegalStateException(ex);
                    }
                }
                else if (new
AssignableTypeFilter(ImportBeanDefinitionRegistrar.class).match(reader, metadataReaderFactory))
{
                    // the candidate class is an ImportBeanDefinitionRegistrar -> delegate
to it to register additional bean definitions
                    try {
                        ImportBeanDefinitionRegistrar registrar =
BeanUtils.instantiateClass(Class.forName(candidate), ImportBeanDefinitionRegistrar.class);
                        registrar.registerBeanDefinitions(importingClassMetadata,
registry);
                    } catch (ClassNotFoundException ex) {
                        throw new IllegalStateException(ex);
                    }
                }
                ...
            }
            this.importStack.pop();
        }
    }
```

通过 AssignableTypeFilter 判断当前候选 Class 元注解@Import 是否赋值 ImportSelector 或 ImportBeanDefinitionRegistrar 实现，从而决定是否执行 ImportSelector 或 ImportBeanDefinitionRegistrar 的处理，其中 importingClassMetadata 就是当前元注解@Import 的 AnnotationMetadata 对象。

综上所述，ConfigurationClassPostProcessor 负责筛选@Component Class、@Configuration Class 及@Bean 方法的 Bean 定义（BeanDefinition），ConfigurationClassParser 则从候选的 Bean 定义中解析出 ConfigurationClass 集合，随后被 ConfigurationClassBeanDefinitionReader 转化并注册 BeanDefinition。

通过参考和学习 Spring Framework“@Enable 模块驱动”的实现，并且加以实践，相信读者对该模式已有一定的认识。

虽然“模块装配”需要手动触发，没有 Spring Boot 所提供的“自动装配”的能力，但如果将“模块装配”和“自动装配”比作汽车驾驶模式，则前者属于“手动挡”，后者属于“自动挡”。两者并不完全排斥，而是相互依存。不过可以看得出来，Spring Framework 在走向“自动装配”上迈出了一大步。

反观“自动装配”，似乎已成为 Spring Boot 的特有标签，实际情况是怎样的呢？接下来一同探讨 Spring Framework 中的自动装配。

8.2 Spring Web 自动装配

Spring Framework 3.1 里程碑的意义不仅在于提供“模块装配”的能力，还有一项更具有象征性意义的能力——“Web 自动装配”。

“自动装配”是 Spring Boot 的三大特征之一，不过 Spring Boot 的“自动装配”大致可以分为两种应用场景，即 Web 应用和非 Web 应用。而 Spring Framework 3.1 及更高版本支持的“自动装配”仅限于 Web 应用场景，同时依赖于 Servlet 3.0+容器，如 Tomcat 7.x 或 Jetty 7.x。

8.2.1 理解 Web 自动装配

引述 Spring Framework `3.1.0.RELEASE` 官方文档的“3.1.10 Support for Servlet 3 code-based configuration of Servlet Container”（https://docs.spring.io/spring/docs/3.1.0.RELEASE/spring-framework-reference/htmlsingle/）章节中的描述:

> The new `WebApplicationInitializer` builds atop Servlet 3.0's `ServletContainerInitializer` support to provide a programmatic alternative to the traditional web.xml.
>
> - See org.springframework.web.WebApplicationInitializer Javadoc
> - Diff from Spring's Greenhouse reference application demonstrating migration from web.xml to `WebApplicationInitializer`

文档告知开发人员，新引入的 `WebApplicationInitializer` 构建在 Servlet 3.0 `ServletContainerInitializer` 之上，后者支持以编程的方式替换传统的 `web.xml` 文件。随后，文档又引导开发人员参考 `WebApplicationInitializer` 的 JavaDoc 及官方示例。无论 JavaDoc，还是官方示例，均较为复杂。也许出于这样的原因，从 Spring Framework `3.2.0.RELEASE` 开始，官方文档对 `WebApplicationInitializer` 做出了示例说明，如“17.2 The DispatcherServlet”章节：

> the following example shows such a `DispatcherServlet` declaration and mapping:
>
> ```
> <web-app>
> ```

```
  <servlet>
      <servlet-name>example</servlet-name>
      <servlet-class>org.springframework.web.servlet.DispatcherServlet</servlet-class>
      <load-on-startup>1</load-on-startup>
  </servlet>

  <servlet-mapping>
      <servlet-name>example</servlet-name>
      <url-pattern>/example/*</url-pattern>
  </servlet-mapping>

</web-app>
```

以上示例为传统 Servlet 容器 `web.xml` 部署 `DispatcherServlet` 的示例。文档告诉开发人员，应用在 Servlet 3.0+的环境中也可以采用编程的手段实现：

> In a Servlet 3.0+ environment, you also have the option of configuring the Servlet container programmatically. Below is the code based equivalent of the above `web.xml` example:

```java
public class MyWebApplicationInitializer implements WebApplicationInitializer {

   @Override
   public void onStartup(ServletContext container) {
       ServletRegistration.Dynamic registration = container.addServlet("dispatcher", new
DispatcherServlet());
       registration.setLoadOnStartup(1);
       registration.addMapping("/example/*");
   }

}
```

同时，文档也给出了 `WebApplicationInitializer` 的定义：

> WebApplicationInitializer is an interface provided by Spring MVC that ensures your code-based configuration is detected and automatically used to initialize any Servlet 3 container. An abstract base class implementation of WebApplicationInitializer named AbstractDispatcherServletInitializer makes it even easier to register the DispatcherServlet by simply specifying its servlet mapping.

WebApplicationInitializer 属于 Spring MVC 提供的接口，确保 WebApplicationInitializer 自定义实现（如上例 MyWebApplicationInitializer）能够被任何 Servlet 3 容器侦测并自动地初始化。如果实现 WebApplicationInitializer 接口较为困难，那么文档也透漏了一种简化实现方案，即 AbstractDispatcherServletInitializer。除此之外，在“4.6 Abstract base class for code-based Servlet 3+ container initialization”一节中，文档描述了另外一种实现，即 AbstractAnnotationConfigDispatcherServletInitializer：

> The new class is named AbstractDispatcherServletInitializer and its sub-class AbstractAnnotationConfigDispatcherServletInitializer can be used with Java-based Spring configuration. For more details see Section 17.14, “Code-based Servlet container initialization”.

AbstractAnnotationConfigDispatcherServletInitializer 是 AbstractDispatcherServletInitializer 的子类。文档再将读者引导到 17.14 节“Code-based Servlet container initialization”，依次提供了 AbstractAnnotationConfigDispatcherServletInitializer 与 AbstractDispatcherServletInitializer 注册 DispatcherServlet 的示例：

```java
public class MyWebAppInitializer extends AbstractAnnotationConfigDispatcherServletInitializer
{

    @Override
    protected Class<?>[] getRootConfigClasses() {
        return null;
    }

    @Override
    protected Class<?>[] getServletConfigClasses() {
        return new Class[] { MyWebConfig.class };
    }

    @Override
    protected String[] getServletMappings() {
        return new String[] { "/" };
    }

}
```

> The above example is for an application that uses Java-based Spring configuration. If using XML-based Spring configuration, extend directly from AbstractDispatcherServletInitializer:
>
> ```
> public class MyWebAppInitializer extends AbstractDispatcherServletInitializer {
>
>     @Override
>     protected WebApplicationContext createRootApplicationContext() {
>         return null;
>     }
>
>     @Override
>     protected WebApplicationContext createServletApplicationContext() {
>         XmlWebApplicationContext cxt = new XmlWebApplicationContext();
>         cxt.setConfigLocation("/WEB-INF/spring/dispatcher-config.xml");
>         return cxt;
>     }
>
>     @Override
>     protected String[] getServletMappings() {
>         return new String[] { "/" };
>     }
>
> }
> ```

按照文档的说法，第一种基于 `AbstractAnnotationConfigDispatcherServletInitializer` 的实现属于 Spring Java 代码配置驱动，第二种基于 `AbstractDispatcherServletInitializer` 的实现是 Spring XML 配置驱动。到此，对 `AbstractDispatcherServletInitializer` 和 `AbstractAnnotationConfigDispatcherServletInitializer` 的使用场景及层次关系有了一定的认识。出于注解驱动编程的需要，下面将基于抽象类 `AbstractAnnotationConfigDispatcherServletInitializer` 自定义 Web 自动装配。

> 在实践过程中要注意 Spring Framework 版本间的差异。`WebApplicationInitializer` 是 Spring Framework 3.1 提供的接口，它的抽象实现 `AbstractDispatcherServletInitializer` 和 `AbstractAnnotationConfigDispatcherServletInitializer` 是从 Spring Framework 3.2 才予以支持的。

8.2.2 自定义 Web 自动装配

（1）新增@Controller——HelloWorldController。

```
@Controller
public class HelloWorldController {

    @RequestMapping
    @ResponseBody
    public String hellloWorld() {
        return "Hello,World!!!";
    }
}
```

（2）新增 Spring Web MVC 配置——SpringWebMvcConfiguration。

```
@EnableWebMvc
@Configuration
@ComponentScan(basePackageClasses = SpringWebMvcConfiguration.class)
public class SpringWebMvcConfiguration {
}
```

（3）实现 AbstractAnnotationConfigDispatcherServletInitializer—SpringWebMvcServletInitializer。

```
    public class SpringWebMvcServletInitializer extends
AbstractAnnotationConfigDispatcherServletInitializer {

        @Override
        protected Class<?>[] getRootConfigClasses() {
            return new Class[0];
        }

        @Override
    // DispatcherServlet 配置 Bean
        protected Class<?>[] getServletConfigClasses() {
            return of(SpringWebMvcConfiguration.class);
        }
```

```
    @Override
    protected String[] getServletMappings() { // DispatcherServlet URL Pattern 映射
        return of("/*");
    }

    private static <T> T[] of(T... values) {  // 便利 API ，减少 new T[] 代码
        return values;
    }
}
```

（4）pom.xml 依赖和插件配置。

此处引入了 Tomcat 7 Maven 插件（tomcat7-maven-plugin），目的在于实现以 java -jar 的方式执行传统的 Servlet war 文件。

```
(...部分内容被省略...)
<dependencies>
        <!-- Servlet 3.0 API -->
        <dependency>
            <groupId>javax.servlet</groupId>
            <artifactId>javax.servlet-api</artifactId>
            <version>3.0.1</version>
            <scope>provided</scope>
        </dependency>
        <!-- Spring 3.x 最新发布版本 -->
        <dependency>
            <groupId>org.springframework</groupId>
            <artifactId>spring-webmvc</artifactId>
        </dependency>
    </dependencies>

    <build>
        <plugins>

            <!-- Maven war 插件 -->
            <plugin>
                <groupId>org.apache.maven.plugins</groupId>
                <artifactId>maven-war-plugin</artifactId>
                <configuration>
```

```
                <!-- 忽略错误，当 web.xml 不存在时 -->
                <failOnMissingWebXml>false</failOnMissingWebXml>
            </configuration>
        </plugin>

        <!-- Tomcat Maven 插件用于构建可执行 WAR -->
        <plugin>
            <groupId>org.apache.tomcat.maven</groupId>
            <artifactId>tomcat7-maven-plugin</artifactId>
            <version>2.1</version>
            <executions>
                <execution>
                    <id>tomcat-run</id>
                    <goals>
                        此处应为：<! 最终打包成可执行的 WAR 包>
                        <goal>exec-war-only</goal>
                    </goals>
                    <phase>package</phase>
                    <configuration>
                        <!-- ServletContext 路径 -->
                        <path>/</path>
                    </configuration>
                </execution>
            </executions>
        </plugin>

    </plugins>
</build>
(...部分内容被省略...)
```

> 源码位置：以上示例代码可通过查找 spring-framework-samples/spring-webmvc-3.2.x- sample 工程获取。

（5）执行 Maven 构建操作。

```
$ mvn clean package
[INFO] Scanning for projects...
[INFO]
```

```
[INFO] ------------------------------------------------------------------------
[INFO] Building spring-webmvc-3.2.x-sample 1.0.0-SNAPSHOT
[INFO] ------------------------------------------------------------------------
(...部分内容被省略...)
[INFO] --- tomcat7-maven-plugin:2.1:exec-war-only (tomcat-run) @ spring-webmvc-3.2.x-sample
---
[INFO] ------------------------------------------------------------------------
[INFO] BUILD SUCCESS
[INFO] ------------------------------------------------------------------------
(...部分内容被省略...)
```

（6）执行 `java -jar`，运行可执行 `war` 文件。

```
$ java -jar target/spring-webmvc-3.2.x-sample-1.0.0-SNAPSHOT-war-exec.jar
(...部分内容被省略...)
org.apache.catalina.core.ApplicationContext log
信息: Spring WebApplicationInitializers detected on classpath: [thinking.in.spring.boot.
samples.chapter3.spring3.web.SpringWebMvcServletInitializer@53f45428]
org.apache.catalina.core.ApplicationContext log
信息: Initializing Spring FrameworkServlet 'dispatcher'
org.springframework.web.servlet.DispatcherServlet initServletBean
信息: FrameworkServlet 'dispatcher': initialization started
(...部分内容被省略...)
信息: Starting ProtocolHandler ["http-bio-8080"]
```

（7）测试结果。

```
$ curl http://localhost:8080
Hello,World!!!
```

测试结果为“Hello,World!!!”，说明它是由 `HelloWorldController#helloWorld()`方法输出的。更重要的是，仅通过 Spring Framework 和 Servlet 容器也能实现 Spring Web MVC 的自动装配，这不就是期盼已久的“Web 自动装配”吗？

原来，Web 应用即使不使用 Spring Boot 框架，也能实现“自动装配”。无独有偶，Spring Security 也有类似的实现——AbstractSecurityWebApplicationInitializer。不过这些特性并没有受到广泛的关注，小马哥认为其根本原因在于广大开发人员（尤其是国内的）对 Servlet 3.0 规范的陌生。

当 Spring Boot 成为技术的潮流时，开发人员很少了解 Spring Framework 在 Web 自动装配上那些鲜为人知的耕耘。此时，当你我熟悉了 `WebApplicationInitializer` 的使用方法及子类的运用场景后，又有什么理由拒绝对 Web 自动装配原理的探索呢？

8.2.3 Web 自动装配原理

实际上，Spring Framework 并不具备“Web 自动装配”原生能力，而是站在“巨人”的肩膀上构建的。这个“巨人”正是 **Servlet 3.0** 技术，其中“**ServletContext 配置方法**”和“**运行时插拔**”两大特性是“Web 自动化装配”的技术保障。为了更好地掌握相关技术，下面的内容将参考 Spring Framework 官方文档和 **Servlet 3.0（JSR 315）**规范，从而加深对 Spring Web 自动装配的理解。

1. ServletContext 配置方法

在传统的 Java Web 应用中，Servlet 技术的运用占比绝对领先。基于 Servlet 编程习惯，当装配 `Servlet`、`Filter` 及各种 `Listener` 时，离不开 `web.xml` 文件（Deployment Descriptor）的配置，如`<servlet>`、`<filter>`和`<listener>`元素。而 `web.xml` 文件一旦配置，运行时就无法调整，显然这种方式的灵活度不够，既不支持占位符，也无法支持条件、循环等逻辑。从 Servlet 3.0 开始，这种限制被打破，其中 `ServletContext` 配置方法是 Servlet 3.0 API 的新特性。Servlet 3.0 规范 **4.4 Configuration methods** 章节中写道：

> The following methods are added to ServletContext since Servlet 3.0 to enable programmatic definition of servlets, filters and the url pattern that they map to.

从 **4.4.1** 章节到 **4.4.3** 章节，依次介绍了通过编程的方式动态地装配 `Servlet`、`Filter` 及各种 `Listener`，增加了运行时配置的弹性。相关配置方法如下表所示。

配置组件	配置方法	配置对象
Servlet	`ServletContext#addServlet`	`ServletRegistration` 或 `ServletRegistration.Dynamic`
Filter	`ServletContext#addFilter`	`FilterRegistration` 或 `FilterRegistration.Dynamic`
Listener	`ServletContext#addListener`	无

以上配置方法均存在重载方法的情况，如 Servlet 部分：

- `addServlet(String,Class)`；
- `addServlet(String,Servlet)`；
- `addServlet(String,className)`。

不难看出 `ServletContext` 配置方法的学习成本不高。

虽然 ServletContext 配置方法为 Web 应用提供了运行时装配的能力，但要达到“Web 自动装配”的目的，还需要在适当的时机加以装配。接下来将讨论如何把握装配时机。

2. 运行时插拔

无论 Servlet、Filter，还是某种 Listener，在动态装配前，都需要在某个时间点调用 ServletContext 配置方法。

Servlet 3.0 规范的“4.4 Configuration methods”章节中写道：

> These methods can only be called during the initialization of the application either from the contexInitialized method of a ServletContextListener implementation or from the onStartup method of a ServletContainerInitializer implementation.

文中的“These methods”是指 ServletContext 配置方法，它们仅能在 ServletContextListener#contexInitialized 或 ServletContainerInitializer#onStartup 方法中被调用。

规范也定义了 ServletContextListener 的职责，它用于监听 Servlet 上下文（ServletContext）的生命周期事件，包括“**初始化**”和“**销毁**”两个事件。其中“初始化”事件由 ServletContextListener#contextInitialized 方法监听。不难理解，Servlet 和 Filter 对外提供服务前，必然经过 Servlet 上下文（ServletContext）初始化事件。

至于 ServletContainerInitializer，则需要参考“**8.2.4 Shared libraries / runtimes pluggability**”章节的内容：

> An instance of the ServletContainerInitializer is looked up via the jar services API by the container at container / application startup time. The framework providing an implementation of the ServletContainerInitializer MUST bundle in the META-INF/services directory of the jar file a file called javax.servlet.ServletContainerInitializer, as per the jar services API, that points to the implementation class of the ServletContainerInitializer.
> In addition to the ServletContainerInitializer we also have an annotation - HandlesTypes. The annotation will be applied on the implementation of ServletContainerInitializer to express interest in classes that are either annotated with the classes specified in the value or if a class extends / implements one of those classes anywhere in the classes super types.

当容器或应用启动时，ServletContainerInitializer#onStartup(Set<Class<?>>, ServletContext)方法将被回调，同时为了选择关心的类型，通过@HandlesTypes 来进行过滤，即关心类型通过@HandlesTypes#value()属性方法来指定。该类型的子类（包括抽象类）候选为类集合（Set<Class<?>>），作为 onStartup 方法的第一个入参。不过 ServletContainerInitializer

的一个或多个实现类需要存放在一个名为“**javax.servlet.ServletContainerInitializer**”的文本文件中，该文件存放在独立 JAR 包中的“**META-INF/services**”目录下。

> 以上存储和加载的技术为“ServiceLoader”，详情请参考：http://docs.oracle.com/javase/6/docs/api/java/util/ServiceLoader.html。

综上分析，`ServletContainerInitializer#onStartup(Set<Class<?>>, ServletContext)` 方法的调用早于 `ServletContextListener#contextInitialized(ServletContextEvent)` 方法。

回顾当前的主要问题——如何实现“Web 自动装配”？按照目前已知的部分，假设一个 `Servlet` 需要装配，并且提供 Web 服务，首先想到的是通过 `ServletContext` 配置方法 `addServlet`，动态地为其装配。随后，在 `ServletContainerInitializer#onStartup` 实现方法中加以实现。以此类推，如果需要装配 *N* 个 `Servlet` 或 `Filter`，那么 `Servlet` 或 `Filter` 及 `ServletContainerInitializer` 的实现打包在若干 JAR 包中。当 Servlet 应用依赖这些 JAR 包后，这些 `Servlet` 或 `Filter` 不就自动装配到 Web 应用中了吗？

目前，实现方案大致理清，或许仍然感到困难重重，无从下手。请不要担忧，有此感受纯属正常。面对这种情况，为何不参考 Spring Framework 是如何运用 Servlet 3.0 技术的呢？

回到具有里程碑意义的 Spring Framework 3.1，在该版本中新增了 `ServletContainerInitializer` 的实现类 `SpringServletContainerInitializer`：

```
    @HandlesTypes(WebApplicationInitializer.class)
    public class SpringServletContainerInitializer implements ServletContainerInitializer {
    ...
    public void onStartup(Set<Class<?>> webAppInitializerClasses, ServletContext 
servletContext)
            throws ServletException {

        List<WebApplicationInitializer> initializers = new 
LinkedList<WebApplicationInitializer>();

        if (webAppInitializerClasses != null) {
            for (Class<?> waiClass : webAppInitializerClasses) {
                // Be defensive: Some servlet containers provide us with invalid classes,
                // no matter what @HandlesTypes says...
                if (!waiClass.isInterface() && !Modifier.isAbstract(waiClass.getModifiers()) 
&&
                        WebApplicationInitializer.class.isAssignableFrom(waiClass)) {
```

```
                    try {
                        initializers.add((WebApplicationInitializer)
waiClass.newInstance());
                    }
                    catch (Throwable ex) {
                        throw new ServletException("Failed to instantiate
WebApplicationInitializer class", ex);
                    }
                }
            }
        }

        if (initializers.isEmpty()) {
            servletContext.log("No Spring WebApplicationInitializer types detected on
classpath");
            return;
        }

        AnnotationAwareOrderComparator.sort(initializers);
        servletContext.log("Spring WebApplicationInitializers detected on classpath: " +
initializers);

        for (WebApplicationInitializer initializer : initializers) {
            initializer.onStartup(servletContext);
        }
    }

}
```

注意第 1 行`@HandlesTypes(WebApplicationInitializer.class)`，结合规范，可以得出`WebApplicationInitializer`的子类（包括抽象类）集合将会作为第一个入参，即“webAppInitializerClasses”。

Spring Framework 3.1 没有提供具体实现，而是将这种弹性能力提供给开发者。随着 Spring Framework 3.2 的发布，框架内部提供了三种抽象实现：

```
AbstractContextLoaderInitializer
 |- AbstractDispatcherServletInitializer
     |- AbstractAnnotationConfigDispatcherServletInitializer
```

为什么这三种实现均为抽象类呢？有两个原因。其一，如果它们是 WebApplicationInitializer 实现类，那么这三个类均会被 SpringServletContainerInitializer 作为具体实现添加到 WebApplicationInitializer 集合 initializers 中，随后顺序迭代执行 onStartup(ServletContext)方法；其二，抽象实现提供模板化的配置接口，最终将相关配置的决策交给开发人员。

简单地介绍这三个抽象类的使用场景。

- AbstractContextLoaderInitializer：如果构建 Web Root 应用上下文（WebApplicationContext）成功则替代 web.xml 注册 ContextLoaderListener；
- AbstractDispatcherServletInitializer：替代 web.xml 注册 DispatcherServlet，并且如果必要的话，创建 Web Root 应用上下文（WebApplicationContext）；
- AbstractAnnotationConfigDispatcherServletInitializer：具备 Annotation 配置驱动能力的 AbstractDispatcherServletInitializer。

接下来将这三种实现结合实现源码逐一分析。

3. AbstractContextLoaderInitializer 装配原理

在 Spring Framework 3.2.0.RELEASE“17.2 The DispatcherServlet”章节中有一段关于 DispatcherServlet WebApplicationContext 与 Root WebApplicationContext 层次关系的描述：

> In the Web MVC framework, each DispatcherServlet has its own WebApplicationContext, which inherits all the beans already defined in the root WebApplicationContext.

在 Spring Web MVC 中，DispatcherServlet 有专属的 WebApplicationContext，它继承了来自 Root WebApplicationContext 的所有 Bean，以便@Controller 等组件依赖注入。两者的层次关系如下图所示。

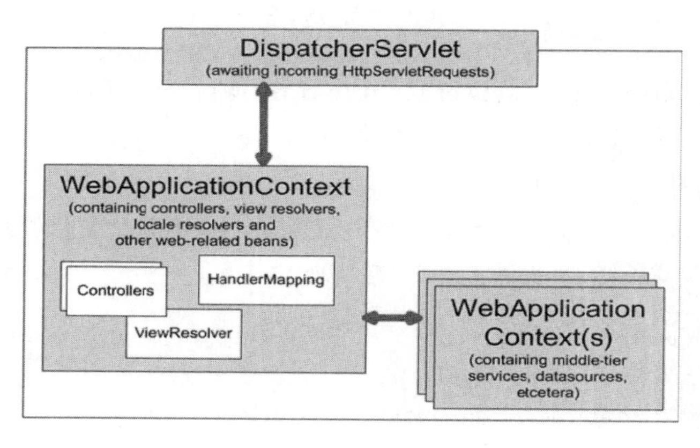

在传统的 Serlvet 应用场景下，Spring Web MVC 的 Root `WebApplicationContext` 由 `ContextLoaderListener` 装载，后者通常配置在 `web.xml` 文件中：

```
<?xml version="1.0" encoding="UTF-8"?>
<web-app xmlns:xsi="http://www.w3.org/2001/XMLSchema-instance"
xmlns="http://java.sun.com/xml/ns/javaee"
         xsi:schemaLocation="http://java.sun.com/xml/ns/javaee
http://java.sun.com/xml/ns/javaee/web-app_3_0.xsd"
         version="3.0">

    <listener>

<listener-class>org.springframework.web.context.ContextLoaderListener</listener-class>
    </listener>

    <!-- All Spring Configuration (both MVC and Security) are in /WEB-INF/spring/ -->
    <context-param>
        <param-name>contextConfigLocation</param-name>
        <param-value>/WEB-INF/spring/*.xml</param-value>
    </context-param>

</web-app>
```

`ContextLoaderListener` 是标准的 `ServletContextListener` 实现，监听 `ServletContext` 生命周期。当 Web 应用启动时，首先，Servlet 容器调用 `ServletContextListener` 实现类的默认构造器，随后 `contextInitialized(ServletContextEvent)`方法被调用。反之，当 Web 应用关闭时，Servlet 容器调用其 `contextDestroyed(ServletContextEvent)`方法：

```
public class ContextLoaderListener extends ContextLoader implements ServletContextListener {

    ...
    public ContextLoaderListener() {
    }

    ...
    public ContextLoaderListener(WebApplicationContext context) {
        super(context);
    }
```

```
    /**
     * Initialize the root web application context.
     */
    @Override
    public void contextInitialized(ServletContextEvent event) {
        initWebApplicationContext(event.getServletContext());
    }


    /**
     * Close the root web application context.
     */
    @Override
    public void contextDestroyed(ServletContextEvent event) {
        closeWebApplicationContext(event.getServletContext());
        ContextCleanupListener.cleanupAttributes(event.getServletContext());
    }

}
```

当 Web 应用运行在 Serlvet 3.0+环境中时，以上 `web.xml` 部署 `ContextLoaderListener` 的方式可替换为实现抽象类 `AbstractContextLoaderInitializer` 来完成。通常情况下，子类只需实现 `createRootApplicationContext()`方法：

```
public abstract class AbstractContextLoaderInitializer implements WebApplicationInitializer {
    ...
    public void onStartup(ServletContext servletContext) throws ServletException {
        registerContextLoaderListener(servletContext);
    }
    ...
    protected void registerContextLoaderListener(ServletContext servletContext) {
        WebApplicationContext rootAppContext = createRootApplicationContext();
        if (rootAppContext != null) {
            servletContext.addListener(new ContextLoaderListener(rootAppContext));
        }
        ...
    }
    ...
```

```
    protected abstract WebApplicationContext createRootApplicationContext();

}
```

在 Spring Web MVC 使用场景中，直接实现 AbstractContextLoaderInitializer 的方式是不推荐的，因此，在 Spring Framework 官方文档中也找不到实例说明。同时，ContextLoaderListener 不允许执行重复注册到 ServletContext，这并非 Servlet 的限制，而是当多个 ContextLoaderListener 监听 contextInitialized 时，其父类 ContextLoader 禁止 Root WebApplicationContext 重复关联 ServletContext：

```
public class ContextLoader {
    ...
    public WebApplicationContext initWebApplicationContext(ServletContext servletContext) {
        if
(servletContext.getAttribute(WebApplicationContext.ROOT_WEB_APPLICATION_CONTEXT_ATTRIBUTE) !=
null) {
            throw new IllegalStateException(
                    "Cannot initialize context because there is already a root application
context present - " +
                    "check whether you have multiple ContextLoader* definitions in your
web.xml!");
        }
        ...

servletContext.setAttribute(WebApplicationContext.ROOT_WEB_APPLICATION_CONTEXT_ATTRIBUTE,
this.context);
        ...
    }
    ...
}
```

如果当前 Web 应用的 ClassPath 下同时存在多个 AbstractContextLoaderInitializer 实现类，则 IllegalStateException 异常将会抛出。同时，核心前端控制器 DispatcherServlet 的注册完全交给开发人员实现。既然如此，何必直接扩展 AbstractContextLoaderInitializer 呢？还有哪些方法值得尝试呢？Spring Framework 官方文档推荐扩展 AbstractDispatcherServletInitializer 或 AbstractAnnotationConfigDispatcherServletInitializer，能够简化配置：

> To simplify all this configuration, consider extending AbstractDispatcherServletInitializer or AbstractAnnotationConfigDispatcherServletInitializer, which automatically set those options and make it very easy to register Filter instances.

接下来将揭开 AbstractDispatcherServletInitializer“神秘的面纱”。

4. AbstractDispatcherServletInitializer 装配原理

AbstractDispatcherServletInitializer 作为 AbstractContextLoaderInitializer 的子类，弥补了父类中没有注册 DispatcherServlet 的遗憾，不过它并没有实现父类 createRootApplicationContext()方法：

```
    public abstract class AbstractDispatcherServletInitializer extends
AbstractContextLoaderInitializer {
        ...
        @Override
    public void onStartup(ServletContext servletContext) throws ServletException {
          super.onStartup(servletContext);

          registerDispatcherServlet(servletContext);
    }
        ...
        protected void registerDispatcherServlet(ServletContext servletContext) {
          String servletName = getServletName();
          ...

          WebApplicationContext servletAppContext = createServletApplicationContext();
          ...

          DispatcherServlet dispatcherServlet = new DispatcherServlet(servletAppContext);
          ServletRegistration.Dynamic registration = servletContext.addServlet(servletName,
dispatcherServlet);
          ...
          registration.addMapping(getServletMappings());
          registration.setAsyncSupported(isAsyncSupported());

          Filter[] filters = getServletFilters();
          if (!ObjectUtils.isEmpty(filters)) {
              for (Filter filter : filters) {
```

```
                registerServletFilter(servletContext, filter);
            }
        }

        customizeRegistration(registration);
    }
    ...
}
```

在执行注册 DispatcherServlet 的 registerDispatcherServlet(ServletContext)方法中，DispatcherServlet 注册信息相关方法的访问修饰符为 protected，例如：

- DispatcherServlet 名称

```
protected String getServletName() {
    return DEFAULT_SERVLET_NAME;
}
```

- DispatcherServlet 映射

```
protected abstract String[] getServletMappings();
```

- DispatcherServlet 异步支持

```
protected boolean isAsyncSupported() {
    return true;
}
```

因此，AbstractDispatcherServletInitializer 子类可通过实现或覆盖这些方法，实现内容调整的目的。

同时，AbstractDispatcherServletInitializer 也提供了注册 Filter 的模板方法：

```
public abstract class AbstractDispatcherServletInitializer extends
AbstractContextLoaderInitializer {
    ...
    protected Filter[] getServletFilters() {
        return null;
    }
    ...
    protected FilterRegistration.Dynamic registerServletFilter(ServletContext servletContext,
Filter filter) {
```

```
        String filterName = Conventions.getVariableName(filter);
        Dynamic registration = servletContext.addFilter(filterName, filter);
        registration.setAsyncSupported(isAsyncSupported());
        registration.addMappingForServletNames(getDispatcherTypes(), false,
getServletName());
        return registration;
    }
    ...
}
```

尽管 AbstractDispatcherServletInitializer 提供了功能相对完善的编程接口，然而实现成本仍旧不低。因此，其子类 AbstractAnnotationConfigDispatcherServletInitializer 进一步简化配置成本，并且更加面向注解驱动开发模式。故在“定义 Web 自动装配”一节中，通过扩展 AbstractAnnotationConfigDispatcherServletInitializer 来实现 Web 自动装配，不过并没有解释子类为什么需要扩展 getRootConfigClasses()等方法，下面将一探究竟。

5. AbstractAnnotationConfigDispatcherServletInitializer 装配原理

前面提到，AbstractDispatcherServletInitializer 既没有实现其父类的 createRootApplicationContext() 方法，又需要子类实现其 createServletApplicationContext() 方法。客观而言，对于 Spring Web MVC 上下文层次性理解不深的开发人员，实现以上方法的成本是较高的。所以，Spring Framework 3.2 提供了更为简单的抽象实现，即 AbstractAnnotationConfigDispatcherServletInitializer。它直接为开发人员实现了这两个方法：

```
public abstract class AbstractAnnotationConfigDispatcherServletInitializer
        extends AbstractDispatcherServletInitializer {
    ...
    @Override
    protected WebApplicationContext createRootApplicationContext() {
        Class<?>[] configClasses = getRootConfigClasses();
        if (!ObjectUtils.isEmpty(configClasses)) {
            AnnotationConfigWebApplicationContext rootAppContext = new
AnnotationConfigWebApplicationContext();
            rootAppContext.register(configClasses);
            return rootAppContext;
        }
        else {
            return null;
```

```
        }
    }
        ...
    @Override
    protected WebApplicationContext createServletApplicationContext() {
        AnnotationConfigWebApplicationContext servletAppContext = new
AnnotationConfigWebApplicationContext();
        Class<?>[] configClasses = getServletConfigClasses();
        if (!ObjectUtils.isEmpty(configClasses)) {
            servletAppContext.register(configClasses);
        }
        return servletAppContext;
    }
    ...
    protected abstract Class<?>[] getRootConfigClasses();
        ...
    protected abstract Class<?>[] getServletConfigClasses();

    }
```

无论 Root `WebApplicationContext`，还是 `DispatcherServlet WebApplicationContext`，均采用注解驱动 Web 应用上下文实现：`AnnotationConfigWebApplicationContext`，这两个 Web 应用上下文需要配置类（Configuration Class），从而驱动其他 Bean 的初始化。因此，这两个配置类分别来自 `AbstractAnnotationConfigDispatcherServletInitializer` 子类的 `getRootConfigClasses()`和 `getServletConfigClasses()`方法。虽然这两个配置类是可选的，不过子类必须覆盖这两个方法，从而让开发人员感知该过程的存在。至于是否装配，则取决于应用的需要。例如，前例 `SpringWebMvcServletInitializer` 就没有提供 Root 配置类（Configuration Class）的来源：

```
    public class SpringWebMvcServletInitializer extends
AbstractAnnotationConfigDispatcherServletInitializer {


        @Override
        protected Class<?>[] getRootConfigClasses() {
            return new Class[0];
        }

        @Override
        // DispatcherServlet 配置 Bean protected Class<?>[] getServletConfigClasses() {
```

```
        return of(SpringWebMvcConfiguration.class);
    }

    @Override
    protected String[] getServletMappings() { // DispatcherServlet URL Pattern 映射
        return of("/*");
    }

    private static <T> T[] of(T... values) {  // 便利 API ，减少 new T[] 代码
        return values;
    }
}
```

至此终于明白了 `getRootConfigClasses()`和 `getServletConfigClasses()`的语义及接口设计意图。`getServletMappings()`方法继承于 `AbstractDispatcherServletInitializer`。

从全局的视角来看，`SpringServletContainerInitializer` 通过实现 Servlet 3.0 SPI 接口 `ServletContainerInitializer`，与`@HandlesTypes` 配合过滤出 `WebApplicationInitializer` 具体实现类集合，随后顺序迭代地执行该集合元素，进而利用 Servlet 3.0 配置 API 实现 Web 自动装配的目的。同时，结合 Spring Framework 3.2 抽象实现 `AbstractAnnotationConfigDispatcherServletInitializer`，极大地简化了注解驱动开发的成本。以上就是 Spring Framework 基于 Servlet 3.0 特性而构建的 Web 自动装配的原理。

> 进阶阅读：通过本章的介绍，相信读者已经意识到 **JSR** 的重要性，既然感兴趣，不妨“**star**”小马哥多年收集的 JSR 资源吧！
> GitHub 地址：https://github.com/mercyblitz/jsr/，其中 Servlet 规范存放在“Servlet”目录下。

接下来进入对“条件装配”的讨论。

8.3 Spring 条件装配

通常企业级的 Spring 应用会提供几套部署环境，用于各个阶段的功能检验，比如：“dev”代表开发环境，“staging” 表示预发环境，“prod”表示生产环境。对于同一个应用，在不同环境中所依赖的资源或表现的行为可能存在差异。大致的手段有两类，其一为编译时差异化，其二是运行时配置化。前者偏资源处理，构建不同环境的归档文件，通常需要依赖外部工具，比如使用 Maven Profile 构建。后者则倾向于利用不同环境的配置控制统一归档文件的应用行为，如设置环境变量或 Java 系统属性。两种方式各有优劣，Spring Framework 则选择后者。在“条件装配注解”一节曾提到，从

Spring Framework 3.1 开始逐步引入条件装配注解@Profile 和@Conditional，不过当时并没有展开讨论，比如两种条件装配方式存在哪些差异，以及在 Spring Framework 3.1 前的时代能否实现条件装配？

8.3.1　理解配置条件装配

在 Spring Framework 3.1 之前，并不存在 XML 属性<beans profile="...">和注解@Profile，也未提供@ComponentScan 注解。即使在这些低版本中，Spring Framework 同样能实现有限的条件装配。假设 Spring 应用在开发环境（dev）中读取 META-INF/dev-context.xml：

```
<?xml version="1.0" encoding="UTF-8"?>
<beans xmlns="http://www.springframework.org/schema/beans"
       xmlns:xsi="http://www.w3.org/2001/XMLSchema-instance"
       xsi:schemaLocation="http://www.springframework.org/schema/beans
http://www.springframework.org/schema/beans/spring-beans.xsd">

    <!-- dev 环境 value Bean 定义-->
    <bean id="name" class="java.lang.String">
        <constructor-arg>
            <value>mercyblitz</value>
        </constructor-arg>
    </bean>

</beans>
```

而在生产环境中需要 META-INF/prod-context.xml：

```
<?xml version="1.0" encoding="UTF-8"?>
<beans xmlns="http://www.springframework.org/schema/beans"
       xmlns:xsi="http://www.w3.org/2001/XMLSchema-instance"
       xsi:schemaLocation="http://www.springframework.org/schema/beans
http://www.springframework.org/schema/beans/spring-beans.xsd">

    <!-- prod 环境 name Bean 定义-->
    <bean id="name" class="java.lang.String">
        <constructor-arg>
            <value>小马哥</value>
        </constructor-arg>
```

```
    </bean>

</beans>
```

当应用不采用代码方式条件化加载时，可以通过增加桥接 XML 上下文配置文件的方式实现，如 `META-INF/configurable-context.xml` 文件：

```xml
<?xml version="1.0" encoding="UTF-8"?>
<beans xmlns="http://www.springframework.org/schema/beans"
       xmlns:xsi="http://www.w3.org/2001/XMLSchema-instance"
       xsi:schemaLocation="http://www.springframework.org/schema/beans
       http://www.springframework.org/schema/beans/spring-beans.xsd">

    <!-- 通过 placeholder "env" 加载所在环境的上下文 XML 文件 -->
    <!-- 如： 开发环境 "env" = "dev"，即加载 "/META-INF/dev-context.xml"文件-->
    <import resource="classpath:/META-INF/${env}-context.xml"/>

</beans>
```

此时，Spring 上下文需要替换占位符变量 `env`，该值可以来自外部化配置，如 Java 系统属性或操作系统环境变量。

> Java 系统属性或操作系统环境变量作为 Spring 外部化配置，贯穿 Spring Framework 和 Spring Boot 时代。

因此，其引导程序实现如下：

```java
public class ConfigurableApplicationContextBootstrap {

    static {
        // 调整系统属性“env”，实现“name”bean 的定义切换
        // envValue 可能来自“-D” 命令行启动参数
        // 参数当不存在时，使用“prod” 作为默认值
        String envValue = System.getProperty("env", "prod");
        System.setProperty("env", envValue);
    }

    public static void main(String[] args) {
        // 定义 XML ApplicationContext
```

```
        ClassPathXmlApplicationContext context = new
ClassPathXmlApplicationContext("META-INF/configurable-context.xml");
        // “name” bean 对象
        String value = (String) context.getBean("name");
        // “name” bean 内容输出
        System.out.println("Bean 'name' 的内容为: " + value);
        // 关闭上下文
        context.close();
    }
}
```

> 源码位置：以上示例代码可通过查找 spring-framework-samples/spring-framework-2.0.x- sample 工程获取。

由于 `envValue` 可能来自“-D”命令行启动参数，当 `String envValue = System.getProperty("env", "prod")`执行后，如果参数存在则使用参数值，如果不存在则使用默认值“prod”。直接运行该引导类，`/META-INF/${env}-context.xml` 被默认值“prod”替换为`/META-INF/prod-context.xml`，所以运行结果为：

```
Bean name 的内容为：小马哥
```

本示例使用中间文件 `META-INF/configurable-context.xml` 代替了编码动态加载所需的 Spring 上下文配置文件。从实现成本而言，两者差别不大。不过在 Spring Framework 3.1 之前的时代，无法避免任意一种装配方式。因此，从 Spring Framework 3.1 开始，引入了 XML 属性`<beans profile="...">`和注解`@Profile`，这类中间桥梁实现不再是必需选项。

Spring Profile 在 Java 世界中并非新鲜事物，比如 Maven 早已支持（Profile），两者的语义类似。

从字面意思来解读，“Profile”通译为“侧面”，既然是侧面，自然是通过某个角度去观察，无法展示“全貌”。假定 Spring 上下文管理了两个 Bean，其名称分别为“bean1”和“bean2”，在不限定任何条件时，Spring 应用上下文的“全貌”是该两个 Bean。如果将这两个 Bean 分别由两个 Profile 来管理，则每个 Profile 管理一个 Bean。Profile 的分治管理通常采用静态配置的方式（XML 文件或 Annotation），因此，Profile 又可称为“配置”。这也意味着 Profile 与某种场景关联，在 Spring 上下文的管理上呈现多样性，故定义这种装配方式为**“配置条件装配”**。

在前例中，开发环境“dev”与生产环境“prod”属于互斥条件，Spring Profile 不但很好地支持，而且还能允许“或”条件，如`@Profile({"dev","prod"})`，以及“非”条件如 `@Profile("!dev")`。

Spring Framework 允许设置两种类型：

- 有效（Active）Profile；

- 默认（Default）Profile。

当有效 Profile 不存在时，采用默认 Profile。同时，Spring 应用有两种 Spring Profile 配置的选择：

- `ConfigurableEnvironment` API 编码配置；
- Java 系统属性配置（见下表）。

设置类型	`ConfigurableEnvironment` API 编码配置	Java 系统属性配置
设置 Active Profile	`#setActiveProfiles(String...`	`spring.profiles.active`
添加 Active Profile	`#addActiveProfile(String)`	
设置 Default Profile	`#setDefaultProfiles(String...`	`spring.profiles.default`

其中，Java 系统属性配置 `spring.profiles.active` 来源于 `AbstractEnvironment#ACTIVE_PROFILES_PROPERTY_NAME` 常量定义，而 `spring.profiles.default` 则由 `AbstractEnvironment#DEFAULT_PROFILES_PROPERTY_NAME` 定义。当应用需要多个 Profile 时，以上两种属性均允许使用“,”连接多值配置数据。

接下来，根据以上理解付诸实践，自定义“配置条件装配”。

8.3.2 自定义配置条件装配

假设有一个计算服务提供多个整数累加，在 Java 8 环境下，使用 Lambda 表达式实现。若是 Java 7 环境，则使用迭代实现。

（1）定义计算服务接口。

```java
public interface CalculatingService {

    /**
     * 累加求和
     * @param values 多个累加值
     * @return 累加结果
     */
    Integer sum(Integer... values);

}
```

（2）实现 Java 8 计算服务接口。

```java
@Service
```

```java
    @Profile("Java8")
    public class LambdaCalculatingService implements CalculatingService {

        @Override
        public Integer sum(Integer... values) {

            int sum =  Stream.of(values).reduce(0, Integer::sum);

            System.out.printf("[Java 8 Lambda 实现] %s 累加结果 : %d\n", Arrays.asList(values),
sum);

            return sum;
        }
    }
```

> LambdaCalculatingService 与 CalculatingService 存放在相同的 package 中。

（3）实现 Java 7 计算服务接口。

```java
    @Service
    @Profile("Java7")
    public class IterationCalculatingService implements CalculatingService {

        @Override
        public Integer sum(Integer... values) {
            int sum = 0;
            for (Integer value : values) {
                sum += value;
            }
            System.out.printf("[Java 7 迭代实现] %s 累加结果 : %d\n", Arrays.asList(values), sum);
            return sum;
        }
    }
```

> IterationCalculatingService 与 CalculatingService 存放在相同的 package 中。

（4）运行“Java8”Profile。

```java
    @Configuration
```

```
@ComponentScan(basePackageClasses = CalculatingService.class)
public class CalculatingServiceBootstrap {

    static {
        // 通过 Java 系统属性设置 Spring Profile
        // 以下语句等效于 ConfigurableEnvironment.setActiveProfiles("Java8")
        System.setProperty(AbstractEnvironment.ACTIVE_PROFILES_PROPERTY_NAME, "Java8");
        // 以下语句等效于 ConfigurableEnvironment.setDefaultProfiles ("Java7")
        System.setProperty(AbstractEnvironment.DEFAULT_PROFILES_PROPERTY_NAME, "Java7");
    }

    public static void main(String[] args) {
        // 构建 Annotation 配置驱动 Spring 上下文
        AnnotationConfigApplicationContext context = new
AnnotationConfigApplicationContext();
        // 注册当前配置 Bean 到 Spring 上下文
        context.register(CalculatingServiceBootstrap.class);
        // 启动上下文
        context.refresh();
        // 获取 CalculatingService Bean
        CalculatingService calculatingService = context.getBean(CalculatingService.class);
        // 输出累加结果
        calculatingService.sum(1, 2, 3, 4, 5);
        // 关闭上下文
        context.close();
    }
}
```

> 源码位置：以上示例代码可通过查找 spring-framework-samples/spring-framework-3.2. x-sample 工程获取。

（5）执行结果。

```
(...部分内容被省略...)
[Java 8 Lambda 实现] [1, 2, 3, 4, 5] 累加结果 : 15
(...部分内容被省略...)
```

代码结合运行结果，不难发现，当激活“Java8”Profile 时，Spring 上下文并没有注册两个 CalculatingService Bean，仅存在 LambdaCalculatingService。在 Spring Boot 场景中，Java

系统属性 spring.profiles.active 和 spring.profiles.default 同样能产生功效，并且两者还能定义其他属性配置源（PropertySource），如 Spring Boot 命令行参数，详细的讨论会在“外部化配置”章节中展开。

8.3.3 配置条件装配原理

回顾上例，在 CalculatingServiceBootstrap 注解声明中，该类标注了@ComponentScan(basePackageClasses = CalculatingService.class)的信息，并且已知 Profile“Java7”的 IterationCalculatingService 及 Profile“Java8”的 LambdaCalculatingService 与 CalculatingService 存放在相同的 package 下。由于 CalculatingServiceBootstrap 受到 Java 系统属性 spring.profiles.active=Java8 的作用，@Profile("Java7") 所标注的 IterationCalculatingService 并未注册在 Spring 应用上下文中，否则 context.getBean (CalculatingService.class)会抛出非唯一 CalculatingService 的异常。换言之，Spring 应用上下文是有选择性地注册的。目前，已知的 Bean 注册方式有两大类：注解驱动和传统 XML 配置驱动，对应的配置进行方式是@Profile 和 XML 属性<beans profile="...">。

1. @Profile 条件装配原理

根据前文的积累，注解驱动 Bean 注册途径大致如下表所示。

注解驱动 Bean 注册方式	使用场景说明	Bean 注解元信息处理类
@ComponentScan	扫描 Spring 模式注解	ClassPathScanningCandidateComponentProvider
@Component 或@Configuration Class	@Import 导入	ConfigurationClassPostProcessor
@Bean	@Bean 方法定义	ConfigurationClassParser
AnnotationConfigApplicationContext	注册 Bean Class	AnnotatedBeanDefinitionReader

从 Spring Framework 3.1 开始，以上三种 Bean 注解元信息处理类均增加了@Profile 的处理。

- ClassPathScanningCandidateComponentProvider#isCandidateComponent(MetadataReader)方法

```
    public class ClassPathScanningCandidateComponentProvider implements EnvironmentCapable,
ResourceLoaderAware {
        ...
    protected boolean isCandidateComponent(MetadataReader metadataReader) throws IOException {
        ...
        for (TypeFilter tf : this.includeFilters) {
```

```
            if (tf.match(metadataReader, this.metadataReaderFactory)) {
                AnnotationMetadata metadata = metadataReader.getAnnotationMetadata();
                if
                (!metadata.isAnnotated(Profile.class.getName())) {
                    return true;
                }
                AnnotationAttributes profile = MetadataUtils.attributesFor(metadata,
Profile.class);
                return this.environment.acceptsProfiles(profile.getStringArray("value"));
            }
        }
        return false;
    }
    ...
}
```

首先，判断 AnnotationMetadata 是否标注@Profile，如果未标注，则认为当前类的元信息匹配。否则，再通过 Environment 判断当前@Profile value 属性值是否能够接受。

- ConfigurationClassParser#processConfigurationClass(ConfigurationClass)方法

```
class ConfigurationClassParser {
    ...
    protected void processConfigurationClass(ConfigurationClass configClass) throws
IOException {
        AnnotationMetadata metadata = configClass.getMetadata();
        if (this.environment != null && metadata.isAnnotated(Profile.class.getName())) {
            AnnotationAttributes profile = MetadataUtils.attributesFor(metadata,
Profile.class);
            if (!this.environment.acceptsProfiles(profile.getStringArray("value"))) {
                return;
            }
        }
        ...
    }
    ...
}
```

如果当前 ConfigurationClass 的元信息 AnnotationMetadata 标注了@Profile，并且与实

际的 Profile 无法匹配，则 ConfigurationClass 未来将不会注册为 Spring Bean。

- AnnotatedBeanDefinitionReader#registerBean(Class,String,Class...)方法

```
public class AnnotatedBeanDefinitionReader {
    ...
    public void registerBean(Class<?> annotatedClass, String name, Class<? extends
Annotation>... qualifiers) {
        AnnotatedGenericBeanDefinition abd = new
AnnotatedGenericBeanDefinition(annotatedClass);
        AnnotationMetadata metadata = abd.getMetadata();
        if
        (metadata.isAnnotated(Profile.class.getName())) {
            AnnotationAttributes profile = MetadataUtils.attributesFor(metadata,
Profile.class);
            if (!this.environment.acceptsProfiles(profile.getStringArray("value"))) {
                return;
            }
        }
        ...
        BeanDefinitionHolder definitionHolder = new BeanDefinitionHolder(abd, beanName);
        definitionHolder = AnnotationConfigUtils.applyScopedProxyMode(scopeMetadata,
definitionHolder, this.registry);
        BeanDefinitionReaderUtils.registerBeanDefinition(definitionHolder, this.registry);
    }
    ...
}
```

注解类Bean注册的前置判断逻辑与前两者一样，均判断Class注解元信息AnnotationMetadata是否包含@Profile 属性元信息 AnnotationAttributes，并且其属性值能够被 Environment Profile 接受：

```
public abstract class AbstractEnvironment implements ConfigurableEnvironment {
    ...
    public boolean acceptsProfiles(String... profiles) {
        Assert.notEmpty(profiles, "Must specify at least one profile");
        for (String profile : profiles) {
            if (profile != null && profile.length() > 0 && profile.charAt(0) == '!') {
                return !isProfileActive(profile.substring(1));
```

```
            }
            if (isProfileActive(profile)) {
                return true;
            }
        }
        return false;
    }
    ...
    protected boolean isProfileActive(String profile) {
        validateProfile(profile);
        return doGetActiveProfiles().contains(profile) ||
                (doGetActiveProfiles().isEmpty() &&
doGetDefaultProfiles().contains(profile));
    }
    ...
}
```

当@Profile 属性值首字母为“!”时，获取 isProfileActive(String)相反的结果，属于“非”条件逻辑。如果 profiles 存在多个 Profile 值，第一个 isProfileActive(String)匹配则表示可接受，此为“或”条件逻辑。同时，isProfileActive(String)首先比较有效 Profile 集合，当该集合不为空并且无法匹配时，认为不可接受，否则返回默认 Profile 集合匹配结果。

> 请读者高度关注 Spring Framework 注解元信息 AnnotationMetadata 及属性元信息 AnnotationAttributes 在 Spring 注解驱动中的运用，它对后续的深入讨论及未来扩展是非常关键的。

2. <beans profile="...">条件装配原理

XML 元素<beans profile="...">条件装配的处理方式相对单一，仅在解析 XML 应用上下文配置文件时进行判断：

```
public class DefaultBeanDefinitionDocumentReader implements BeanDefinitionDocumentReader {
    ...
    /** @see org.springframework.context.annotation.Profile */
public static final String PROFILE_ATTRIBUTE = "profile";
    ...
protected void doRegisterBeanDefinitions(Element root) {
    String profileSpec = root.getAttribute(PROFILE_ATTRIBUTE);
    if (StringUtils.hasText(profileSpec)) {
```

```
            Assert.state(this.environment != null, "environment property must not be null");
            String[] specifiedProfiles = StringUtils.tokenizeToStringArray(profileSpec,
BeanDefinitionParserDelegate.MULTI_VALUE_ATTRIBUTE_DELIMITERS);
            if (!this.environment.acceptsProfiles(specifiedProfiles)) {
                return;
            }
        }
        ...
    }
}
```

`DefaultBeanDefinitionDocumentReader#doRegisterBeanDefinitions(Element)` 方法参数是 XML 元素`<beans>`，同样 `profile="..."`属性也能配置多值，并且也以“,”作为连接符。在实现上，当前方法与`@Profile` 的处理方法没有本质上的差异。简言之，当 Profile 不匹配时，当前 Class 不会被注册为 Spring Bean，以上就是“配置条件装配”的原理。

3. @Conditional 条件装配

在进入实质性讨论之前，回顾 Spring Boot 官方首页中的一句特征文字描述：

> **Features**
> - Automatically configure Spring whenever possible

请注意文字的后半部分：**whenever possible**，表明自动装配实际上存在前置条件，因此这种装配方式也可称为“**条件装配**”。

“`@Conditional` 条件装配”是 Spring Framework 4.0 引入的新特性，它与“配置条件装配”的职责相似，都是加载匹配的 Bean。不同点是，“`@Conditional` 条件装配”具备更大的弹性：

> The @Profile annotation is actually implemented using a much more flexible annotation called @Conditional.

“配置条件装配”偏向于“静态激活和配置”，而“`@Conditional` 条件装配”则更关注运行时动态选择。从 API 层面来描述：

```
@Target({ElementType.TYPE, ElementType.METHOD})
@Retention(RetentionPolicy.RUNTIME)
@Documented
public @interface Conditional {
```

```
    /**
     * All {@link Condition}s that must {@linkplain Condition#matches match}
     * in order for the component to be registered.
     */
    Class<? extends Condition>[] value();

}
```

@Conditional 允许指定一个或多个 Condition，当所有的 Condition 均匹配时，说明当前条件成立。其中，Condition 接口定义如下：

```
public interface Condition {

    /**
     * Determine if the condition matches.
     * @param context the condition context
     * @param metadata metadata of the {@link org.springframework.core.type.AnnotationMetadata
class}
     * or {@link org.springframework.core.type.MethodMetadata method} being checked.
     * @return {@code true} if the condition matches and the component can be registered
     * or {@code false} to veto registration.
     */
    boolean matches(ConditionContext context, AnnotatedTypeMetadata metadata);

}
```

前文中所说的“匹配”是指 Condition#matches(ConditionContext,AnnotatedTypeMetadata)方法执行后返回 true。其中方法参数 ConditionContext 包含 Spring 应用上下文相关：BeanDefinitionRegistry、ConfigurableListableBeanFactory、Environment、ResourceLoader 和 ClassLoader。AnnotatedTypeMetadata 虽然是 Spring Framework 4.0 的新接口，但是其中大部分方法是从 AnnotationMetadata 接口中抽出的，因此，AnnotationMetadata 在 4.0 版本之后成为 AnnotatedTypeMetadata 的子接口。盖因如此，实现@Conditional 和 Condition 的成本并不算高，这部分内容在官方的参考手册中一带而过：

> The @Conditional annotation indicates specific org.springframework. context.annotation. Condition implementations that should be consulted before a @Bean is registered.

> @Conditional 官方文档：https://docs.spring.io/spring/docs/4.0.x/spring-framework-reference/htmlsingle/#beans-java-conditional。

Spring Boot 内建了不少的条件注解，如@ConditionalOnClass、@ConditionalOnBean 和@ConditionalOnProperty 等。@ConditionalOnClass 的注解定义：

```
@Target({ ElementType.TYPE, ElementType.METHOD })
@Retention(RetentionPolicy.RUNTIME)
@Documented
@Conditional(OnClassCondition.class)
public @interface ConditionalOnClass {

    /**
     * The classes that must be present. Since this annotation is parsed by loading class
     * bytecode, it is safe to specify classes here that may ultimately not be on the
     * classpath, only if this annotation is directly on the affected component and
     * <b>not</b> if this annotation is used as a composed, meta-annotation.In order to
     * use this annotation as a meta-annotation, only use the {@link #name} attribute.
     * @return the classes that must be present
     */
    Class<?>[] value() default {};

    /**
     * The classes names that must be present.
     * @return the class names that must be present.
     */
    String[] name() default {};

}
```

Spring Boot 条件注解@ConditionalOnClass 采用元标注（meta-annotated）@Conditional(OnClassCondition.class)的方式定义，所以它包含@Conditional 的属性元信息。结合以上理论，自定义@Conditional 实现，以加深理解。

4. 自定义@Conditional 条件装配

假设通过 Java 系统属性（Properties）设置当前应用的语言（language），通过该语言属性来装配对应 String 类型的语种消息（message）。

（1）定义指定系统属性名称与属性值匹配条件注解——ConditionalOnSystemProperty。

```
@Target({ElementType.METHOD}) // 只能标注在方法上面
@Retention(RetentionPolicy.RUNTIME)
@Documented
@Conditional(OnSystemPropertyCondition.class)
public @interface ConditionalOnSystemProperty {

    /**
     * @return System 属性名称
     */
    String name();

    /**
     * @return System 属性值
     */
    String value();

}
```

（2）实现“系统属性名称与值匹配条件”——OnSystemPropertyCondition。

```
public class OnSystemPropertyCondition implements Condition {

    @Override
    public boolean matches(ConditionContext context, AnnotatedTypeMetadata metadata) {
        // 获取 ConditionalOnSystemProperty 所有的属性方法值
        MultiValueMap<String, Object> attributes =
metadata.getAllAnnotationAttributes(ConditionalOnSystemProperty.class.getName());
        // 获取 ConditionalOnSystemProperty#name()方法值（单值）
        String propertyName = (String) attributes.getFirst("name");
        // 获取 ConditionalOnSystemProperty#value()方法值（单值）
        String propertyValue = (String) attributes.getFirst("value");
        // 获取系统属性值
        String systemPropertyValue = System.getProperty(propertyName);
        // 比较系统属性值与 ConditionalOnSystemProperty#value()方法值是否相等
        if (Objects.equals(systemPropertyValue, propertyValue)) {
            System.out.printf("系统属性[名称 : %s] 找到匹配
值 : %s\n",propertyName,propertyValue);
            return true;
        }
```

```
        return false;
    }
}
```

（3）实现“条件消息配置”——ConditionalMessageConfiguration。

```
@Configuration
public class ConditionalMessageConfiguration {

    @ConditionalOnSystemProperty(name = "language", value = "Chinese")
    @Bean("message") // Bean 名称“message”的中文消息
    public String chineseMessage() {
        return "你好，世界";
    }

    @ConditionalOnSystemProperty(name = "language", value = "English")
    @Bean("message") // Bean 名称“message”的英文消息
    public String englishMessage() {
        return "Hello,World";
    }
}
```

（4）编写 OnSystemPropertyCondition 引导程序。

```
public class ConditionalOnSystemPropertyBootstrap {

    public static void main(String[] args) {
        // 设置 System Property  language = Chinese
        System.setProperty("language", "Chinese");
        // 构建 Annotation 配置驱动 Spring 上下文
        AnnotationConfigApplicationContext context = new
AnnotationConfigApplicationContext();
        // 注册配置 Bean ConditionalMessageConfiguration 到 Spring 上下文
        context.register(ConditionalMessageConfiguration.class);
        // 启动上下文
        context.refresh();
        // 获取名称为“message”的 Bean 对象
        String message = context.getBean("message", String.class);
        // 输出 message 内容
```

```
        System.out.printf("\"message\" Bean 对象 : %s\n", message);
    }
}
```

（5）执行结果。

```
(...部分内容被省略...)
系统属性[名称 : language] 找到匹配值 Chinese
"message" Bean 对象 : 你好，世界
```

明显地发现，系统属性设置 `System.setProperty("language", "Chinese")`被条件注解`@ConditionalOnSystemProperty(name = "language", value = "Chinese")`匹配，因此“message” Bean 的内容为“你好，世界”。`@Conditional` 作为条件装配的元注解，它的语义与`@Profile` 类似，所以`@Profile` 从 Spring Framework 4.0 开始重构为`@Conditional` 实现：

```
@Retention(RetentionPolicy.RUNTIME)
@Target({ElementType.TYPE, ElementType.METHOD})
@Documented
@Conditional(ProfileCondition.class)
public @interface Profile {

/**
 * The set of profiles for which the annotated component should be registered.
 */
String[] value();

}
```

总之，条件装配均采用`@Conditional` 实现，具体原理稍后分析。

5. @Conditional 条件装配原理

在 Spring Framework 3.1～3.2 中，`@Profile` 利用 `Environment#acceptsProfiles(String...)` API 实现配置条件化 Bean 注册。从 4.0 版本开始则替换 `ProfileCondition` 实现：

```
class ProfileCondition implements Condition {

@Override
public boolean matches(ConditionContext context, AnnotatedTypeMetadata metadata) {
    if (context.getEnvironment() != null) {
        MultiValueMap<String, Object> attrs =
```

```
metadata.getAllAnnotationAttributes(Profile.class.getName());
            if (attrs != null) {
                for (Object value : attrs.get("value")) {
                    if
                    (context.getEnvironment().acceptsProfiles(((String[]) value))) {
                        return true;
                    }
                }
                return false;
            }
        }
        return true;
    }

    }
```

在判断逻辑上，仍采用 Environment#acceptsProfiles(String...)方法执行结果。因此，注解驱动的实现类 ClassPathScanningCandidateComponentProvider、ConfigurationClassParser 和 AnnotatedBeanDefinitionReader 中不再出现@Profile 元信息处理，转而抽象出@Conditional 的统一处理实现——ConditionEvaluator，其中 shouldSkip(AnnotatedTypeMetadata, ConfigurationPhase)方法判断当前标注类是否应该被跳过：

```
    class ConditionEvaluator {
        ...
    public boolean shouldSkip(AnnotatedTypeMetadata metadata, ConfigurationPhase phase) {
        if (metadata == null || !metadata.isAnnotated(Conditional.class.getName())) {
            return false;
        }

        if (phase == null) {
            if (metadata instanceof AnnotationMetadata &&
                    ConfigurationClassUtils.isConfigurationCandidate((AnnotationMetadata)
metadata)) {
                return shouldSkip(metadata, ConfigurationPhase.PARSE_CONFIGURATION);
            }
            return shouldSkip(metadata, ConfigurationPhase.REGISTER_BEAN);
        }
```

```
        for (String[] conditionClasses : getConditionClasses(metadata)) {
            for (String conditionClass : conditionClasses) {
                Condition condition = getCondition(conditionClass,
context.getClassLoader());
                ConfigurationPhase requiredPhase = null;
                if (condition instanceof ConfigurationCondition) {
                    requiredPhase = ((ConfigurationCondition)
condition).getConfigurationPhase();
                }
                if (requiredPhase == null || requiredPhase == phase) {
                    if (!condition.matches(context, metadata)) {
                        return true;
                    }
                }
            }
        }

        return false;
    }
    ...
}
```

> 以上实现源码的 Maven GAV 坐标为：org.springframework:spring-context:4.0.3.RELEASE。

该方法将 AnnotatedTypeMetadata 标注的所有 Condition 实例逐一匹配，遇到不匹配的实例时，方法返回 true，表示该 AnnotatedTypeMetadata 的 Bean 注册应被跳过。

> 该方法在 Spring Framework 4.1.1.RELEASE 中增加了 Condition 集合的排序：
>
> ```
> public boolean shouldSkip(AnnotatedTypeMetadata metadata, ConfigurationPhase phase) {
>     ...
>     List<Condition> conditions = new ArrayList<Condition>();
>     ...
>
>     Collections.sort(conditions, AnnotationAwareOrderComparator.INSTANCE);
>
>     for (Condition condition : conditions) {
> ```

```
        ...
    }

    return false;
}
```

后续的版本也保留了这个逻辑，不过在 Condition JavaDoc 中并没有告知开发人员，当多个 Condition 实现类并存时，能够通过 Ordered 或@Order 的方法加以排序。故请读者朋友尤其关注实现和文档出现出入的情况。

所以，之前在 ClassPathScanningCandidateComponentProvider、ConfigurationClassParser 和 AnnotatedBeanDefinitionReader 中移除的@Profile 处理替换为 ConditionEvaluator#shouldSkip 方法的调动：

- ClassPathScanningCandidateComponentProvider#isConditionMatch(MetadataReader)方法

```
public class ClassPathScanningCandidateComponentProvider implements EnvironmentCapable, ResourceLoaderAware {
    ...
    protected boolean isCandidateComponent(MetadataReader metadataReader) throws IOException {
        for (TypeFilter tf : this.excludeFilters) {
            if (tf.match(metadataReader, this.metadataReaderFactory)) {
                return false;
            }
        }
        for (TypeFilter tf : this.includeFilters) {
            if (tf.match(metadataReader, this.metadataReaderFactory)) {
                return isConditionMatch(metadataReader);
            }
        }
        return false;
    }
    ...
    private boolean isConditionMatch(MetadataReader metadataReader) {
        if (this.conditionEvaluator == null) {
            this.conditionEvaluator = new ConditionEvaluator(getRegistry(), getEnvironment(), getResourceLoader());
        }
```

```
        return !this.conditionEvaluator.shouldSkip(metadataReader.getAnnotationMetadata());
    }
        ...
    }
```

- `ConfigurationClassParser#processConfigurationClass(ConfigurationClass)` 方法

```
    protected void processConfigurationClass(ConfigurationClass configClass) throws IOException {
        if (this.conditionEvaluator.shouldSkip(configClass.getMetadata(),
ConfigurationPhase.PARSE_CONFIGURATION)) {
            return;
        }
        ...
    }
```

- `AnnotatedBeanDefinitionReader#registerBean(Class,String,Class...)`方法

```
    public void registerBean(Class<?> annotatedClass, String name,
            @SuppressWarnings("unchecked") Class<? extends Annotation>... qualifiers) {

        AnnotatedGenericBeanDefinition abd = new
AnnotatedGenericBeanDefinition(annotatedClass);
        if (this.conditionEvaluator.shouldSkip(abd.getMetadata())) {
            return;
        }
          ...
        BeanDefinitionHolder definitionHolder = new
        BeanDefinitionHolder(abd, beanName);
        definitionHolder =
        AnnotationConfigUtils.applyScopedProxyMode(scopeMetadata, definitionHolder,
this.registry);
        BeanDefinitionReaderUtils.registerBeanDefinition(definitionHolder, this.registry);
    }
```

> 以上实现源码的 Maven GAV 坐标为：`org.springframework:spring-context:4.0.3.RELEASE`。

不难看出，Spring Framework 4 引入的 `ConditionEvaluator` 完全覆盖了该版本以前`@Profile`

的实现，成为条件装配的唯一“评估者”，并且其影响范围扩大到 ConfigurationClassBeanDefinitionReader 之中：

```
class ConfigurationClassBeanDefinitionReader {
    ...
    private void loadBeanDefinitionsForBeanMethod(BeanMethod beanMethod) {
        if (this.conditionEvaluator.shouldSkip(beanMethod.getMetadata(),
ConfigurationPhase.REGISTER_BEAN)) {
            return;
        }
        ...
        this.registry.registerBeanDefinition(beanName, beanDefToRegister);
    }
    ...
}
```

细心的读者可能发现，ConditionEvaluator 的评估存在两个阶段：“Bean 注册阶段”和“Configuration Class 解析阶段”。定义在枚举 ConfigurationPhase 中：ConfigurationPhase.REGISTER_BEAN 和 ConfigurationPhase.PARSE_CONFIGURATION，依次在 ConfigurationClassBeanDefinitionReader 和 ConfigurationClassParser 中评估条件。

至此，关于 Spring Framework 走向注解驱动编程（Annotation-Driven）的讨论告一段落。深刻地感受到在此过程中，Spring Framework 曾付出的努力。尽管如此，Spring Framework 组件装配的自动化程度仍不是特别理想，比如@Enable 模块驱动不仅需要将@Enable 注解显式地标注在配置类上，而且该类还依赖@Import 或@ComponentScan 的配合。同时，Web 自动装配必须部署在外部 Servlet 3.0+容器中，无法做到 Spring Web 应用自我驱动。或许正因为 Spring Framework 中诸如此类的特性限制，直接或间接地促使 Spring Boot 的出现，这些限制终被 Spring Boot 自动装配和嵌入式 Web 容器等特性“各个击破”。

虽然自动装配是 Spring Boot 的高级特性，但经过本章对 Spring Framework 注解驱动编程的深入探讨，后续的讨论将会变得很顺利。

第9章 Spring Boot自动装配

在 Spring Framework 时代，当 Spring 应用的`@Component`或`@Configuration` Class 需要被装配时，应用需要借助`@Import`或`@ComponentScan`的能力。由于应用依赖 JAR 存在变化的可能，因此其中的`@Component`或`@Configuration` Class 所在的包路径也随之不确定，如要实现当前应用所有组件自动装配，则`@Import`显然是无能为力的，开发人员自然会想到使用`@ComponentScan`扫描应用默认包路径，代码如下所示。

> 默认包（Defaut Package）又称为根包。而 Java 语言规范中的术语为“Unnamed Package”。后续继续保持“默认包”的称谓，主要是为了与 Spring Boot 官方文档保持一致。

```
package thinking.in.spring.boot.samples.spring4.bootstrap;
...
@ComponentScan(basePackages = "") // 扫描默认包
public class ComponentScanDefaultPackageBootstrap {

    public static void main(String[] args) {
        // 注册当前引导类作为配置 Class，并启动当前上下文
        AnnotationConfigApplicationContext context = new
AnnotationConfigApplicationContext(ComponentScanDefaultPackageBootstrap.class);
        // 输出当前 Spring 应用上下文中所有注册的 Bean 名称
```

```
        System.out.println("当前 Spring 应用上下文中所有注册的 Bean 名称：");
        Stream.of(context.getBeanDefinitionNames())
                .map(name -> "\t"+name) // 增加格式缩进
                .forEach(System.out::println);
        // 关闭上下文
        context.close();
    }
}
```

> 源码位置：以上示例代码可通过查找 spring-framework-samples/spring-framework-4.3.x-sample 工程获取。

理论上，默认包及其子包下的所有`@Component`或`@Configuration` Class 均会被`@ComponentScan`注册为 Spring Bean，从而达到组件自动装配的目的，然而实际效果却事与愿违：

```
(...部分内容被省略...)
当前 Spring 应用上下文中所有注册的 Bean 名称：
org.springframework.context.annotation.internalConfigurationAnnotationProcessor
org.springframework.context.annotation.internalAutowiredAnnotationProcessor
org.springframework.context.annotation.internalRequiredAnnotationProcessor
org.springframework.context.annotation.internalCommonAnnotationProcessor
org.springframework.context.event.internalEventListenerProcessor
org.springframework.context.event.internalEventListenerFactory
componentScanDefaultPackageBootstrap
annotationLookupBootstrap
helloWorld
(...部分内容被省略...)
```

除 Spring Framework 内建的 Bean 外，其他 Bean 则来自`ComponentScanDefaultPackageBootstrap`所在包下的其他配置 Class。这种设计方案似是而非，因为`@ComponentScan(basePackages = "")`仅扫描`ComponentScanDefaultPackageBootstrap`所在包，而不是所有包。

在 Spring Boot 官方文档“14.1 Using the ‘default’ Package”一节中，以下文字给出了一点提示：

> When a class does not include a `package` declaration, it is considered to be in the “default package”. The use of the “default package” is generally discouraged and should be avoided. It can cause particular problems for Spring Boot applications that use the `@ComponentScan`, `@EntityScan`, or `@SpringBootApplication` annotations, since every class from every jar is read.

大意是不鼓励开发人员通过`@ComponentScan`或`@SpringBootApplication`注解的方式扫描默认包，因为它将读取所有 JAR 中的类，并且它可能会造成某些 Spring Boot 应用的错误。前例证明使用`@ComponentScan(basePackages = "")`的方式是徒劳之举，所以可以考虑替换为`@ComponentScan(basePackageClasses = ...)`，其中`basePackageClasses`指向默认包中的类即可。不过，按照 Java SE 6 的语言规范“7.5 Import Declartions”章节中的规定：

> A type-import-on-demand declaration (§7.5.2) imports all the accessible (§6.6) types of a named type or package as needed. It is a compile time error to import a type from the unnamed package.

当命名类型（named type）导入非命名包（unnamed package）中的类时，编译时会出现错误。因此，需要在默认包下新建一个引导类，实现自我“`@ComponentScan`”：

```
@ComponentScan(basePackageClasses = DefaultPackageBootstrap.class)
public class DefaultPackageBootstrap {

    public static void main(String[] args) {
        // 注册当前引导类作为配置 Class，并启动当前上下文
        AnnotationConfigApplicationContext context = new
AnnotationConfigApplicationContext(DefaultPackageBootstrap.class);
        // 输出当前 Spring 应用上下文中所有注册的 Bean 名称
        System.out.println("当前 Spring 应用上下文中所有注册的 Bean 名称：");
        Stream.of(context.getBeanDefinitionNames())
                .map(name -> "\t" + name) // 增加格式缩进
                .forEach(System.out::println);
        // 关闭上下文
        context.close();
    }
}
```

> 源码位置：以上示例代码可通过查找 spring-framework-samples/spring-framework-4.3.x-sample 工程获取。

由于读取所有 JAR 中的类的缘故，该引导类在启动秒后数，抛出以下异常：

```
(...部分内容被省略...)
Exception in thread "main"
org.springframework.beans.factory.BeanCreationException: Error creating bean with name
```

```
'org.springframework.context.annotation.internalAsyncAnnotationProcessor' defined in class path
resource [org/springframework/scheduling/annotation/ProxyAsyncConfiguration.class]: Bean
instantiation via factory method failed; nested exception is
org.springframework.beans.BeanInstantiationException: Failed to instantiate
[org.springframework.scheduling.annotation.AsyncAnnotationBeanPostProcessor]: Factory method
'asyncAdvisor' threw exception; nested exception is java.lang.IllegalArgumentException:
@EnableAsync annotation metadata was not injected
    (...部分内容被省略...)
```

提示开发人员 Spring Framework 内建 `ProxyAsyncConfiguration` Bean 创建失败，这是由于没有发现元标注`@EnableAsync` 信息的缘故：

> java.lang.IllegalArgumentException: @EnableAsync annotation metadata was not injected.

明显地，在自动装配 `ProxyAsyncConfiguration` 前，需要对`@EnableAsync` 的元注解情况进行条件判断。透过现象看本质，当 Spring 应用自动装配某些组件时，它需要一种综合性技术手段，重新深度整合 Spring 注解编程模型、`@Enable` 模块驱动及条件装配等 Spring Framework 原生特性，这种技术就是“Spring Boot 自动装配”。

9.1 理解 Spring Boot 自动装配

在 Spring Boot 官方文档的“18. Using the @SpringBootApplication Annotation”一节对`@SpringBootApplication` 和`@EnableAutoConfiguration` 做出了解释：

> A single `@SpringBootApplication` annotation can be used to enable those three features, that is:
> - `@EnableAutoConfiguration`: enable Spring Boot's auto-configuration mechanism.
> - `@ComponentScan`: enable `@Component` scan on the package where the application is located (see the best practices).
> - `@Configuration`: allow to register extra beans in the context or import additional configuration classes.

其中，`@EnableAutoConfiguration` 用于激活 Spring Boot 自动装配的特性。按照命名规范和实现特点，`@EnableAutoConfiguration` 也属于 Spring Boot `@Enable` 模块装配的实现：

```
@Target(ElementType.TYPE)
@Retention(RetentionPolicy.RUNTIME)
@Documented
```

```java
@Inherited
@AutoConfigurationPackage
@Import(AutoConfigurationImportSelector.class)
public @interface EnableAutoConfiguration {
    ...
}
```

通常，Spring Boot 应用引导类不会直接标注@EnableAutoConfiguration，而是选择@SpringBootApplication 来激活@EnableAutoConfiguration、@ComponentScan 和@SpringBootConfiguration。作为 Spring Boot 最核心注解的@SpringBootApplication，其组合注解成员@SpringBootConfiguration 与@Configuration 无异：

```java
@Target(ElementType.TYPE)
@Retention(RetentionPolicy.RUNTIME)
@Documented
@Configuration
public @interface SpringBootConfiguration {

}
```

尽管@SpringBootConfiguration 从 Spring Boot 1.4 开始被引入，但官方文档并没有更新，不过也无妨。其元标注在@SpringBootApplication 上更多的意义在于@SpringBootApplication 的配置类能够被@ComponentScan 识别。

9.1.1 理解@EnableAutoConfiguration

按照官方“Spring Boot's auto-configuration mechanism”的指引，移步至“16. Auto-configuration”一节：

> Spring Boot auto-configuration attempts to automatically configure your Spring application based on the jar dependencies that you have added. For example, if HSQLDB is on your classpath, and you have not manually configured any database connection beans, then Spring Boot auto-configures an in-memory database.

文档告诉开发人员，Spring Boot 自动装配尝试自动配置应用的组件，这些组件来自应用依赖的 JAR，并且举例说明这样的装配是有前提条件的。随后文档告知开发人员，可在@EnableAutoConfiguration 和@SpringBootApplication 中选择其一，激活自动装配特性：

> You need to opt-in to auto-configuration by adding the @EnableAutoConfiguration or @SpringBootApplication annotations to one of your @Configuration classes.

实际上，SpringApplication#run(Object,String...)方法入参的配置类并不依赖@SpringBootConfiguration或@Configuration的标注，比如Spring Boot官方首页中的例子：

```
package hello;

import org.springframework.boot.*;
import org.springframework.boot.autoconfigure.*;
import org.springframework.stereotype.*;
import org.springframework.web.bind.annotation.*;

@Controller
@EnableAutoConfiguration
public class SampleController {

    @RequestMapping("/")
    @ResponseBody
    String home() {
        return "Hello World!";
    }

    public static void main(String[] args) throws Exception {
        SpringApplication.run(SampleController.class, args);
    }
}
```

SampleController仅需标注@EnableAutoConfiguration即可激活Spring Boot的自动装配特性。所以@SpringBootApplication并非Spring Boot自动装配的必需注解，不过它能够减少多注解所带来的配置成本，如配置@SpringBootApplication#scanBasePackages属性，等同于配置@ComponentScan#basePackages属性等。

9.1.2 优雅地替换自动装配

Spring Boot 官方文档的“16.1 Gradually Replacing Auto-configuration”一节中提到：

> Auto-configuration is non-invasive. At any point, you can start to define your own configuration to replace specific parts of the auto-configuration. For example, if you add your own `DataSource` bean, the default embedded database support backs away.

以上文字大意是，Spring Boot 自动装配并非是侵占性的，开发人员可在任意一处定义配置类，从而覆盖那些被自动装配的组件。文档又举例说明，自定义的 `DataSource` Bean 能够覆盖默认的嵌入式数据库的数据源。结合目前已有的技术积累，可将其解读为 Spring Boot 优先解析自定义配置类，并且内建的自动装配配置类实际上为默认的条件配置，即一旦应用存在自定义实现，则不再将它们装配。其中的缘由将在“Spring Boot 自动装配原理”一节中深入讨论。

9.1.3 失效自动装配

Spring Boot 官方文档中的“16.2 Disabling Specific Auto-configuration Classes”继续介绍如何失效自动装配：

> If you find that specific auto-configuration classes that you do not want are being applied, you can use the exclude attribute of `@EnableAutoConfiguration` to disable them, as shown in the following example:
>
> ```
> import org.springframework.boot.autoconfigure.*;
> import org.springframework.boot.autoconfigure.jdbc.*;
> import org.springframework.context.annotation.*;
>
> @Configuration
> @EnableAutoConfiguration(exclude={DataSourceAutoConfiguration.class})
> public class MyConfiguration {
> }
> ```
>
> If the class is not on the classpath, you can use the `excludeName` attribute of the annotation and specify the fully qualified name instead. Finally, you can also control the list of auto-configuration classes to exclude by using the `spring.autoconfigure.exclude` property.

简言之，Spring Boot 提供两种失效手段：

- 代码配置方式
    - 配置类型安全的属性方法：@EnableAutoConfiguration.exclude()；
    - 配置排除类名的属性方法：@EnableAutoConfiguration.excludeName()。
- 外部化配置方式
    - 配置属性：spring.autoconfigure.exclude。

本质上，Spring Boot 失效自动装配是一种类似于 Spring Framework 黑名单方式的条件装配，然而想要在 Spring Framework 时代实现该功能，要么阻断@Configuration Class BeanDefinition 的注册，要么通过@Conditonal 实现条件控制。前者的实现成本较高，后者对@Configuration Class 存在注解侵入性。这种复杂的逻辑由官方来实现是皆大欢喜的。不过，可以大胆地猜想其中实现，@EnableAutoConfiguration 可通过 ImportSelector 实现，且 exclude()和 excludeName()属性可从其 AnnotationMetadata 对象中获取，那么@EnableAutoConfiguration 所装配组件又从哪里获取呢？这个问题的答案将在下一节揭晓。

9.2 Spring Boot 自动装配原理

依照@Enable 模块驱动设计模式，@EnableAutoConfiguration 必然“@Import” ImportSelector 或 ImportBeanDefinitionRegistrar 的实现类。于是参考其注解定义：

```
@Target(ElementType.TYPE)
@Retention(RetentionPolicy.RUNTIME)
@Documented
@Inherited
@AutoConfigurationPackage
@Import(AutoConfigurationImportSelector.class)
public @interface EnableAutoConfiguration {
    ...
}
```

其中，AutoConfigurationImportSelector 就是@EnableAutoConfiguration 所“@Import”的 DeferredImportSelector 实现类，由于 DeferredImportSelector 作为 ImportSelector 的子接口，所以组件自动装配逻辑均在 selectImports(AnnotationMetadata)方法中实现：

```
public class AutoConfigurationImportSelector
        implements DeferredImportSelector, BeanClassLoaderAware, ResourceLoaderAware,
        BeanFactoryAware, EnvironmentAware, Ordered {
```

```
    ...
    @Override
    public String[] selectImports(AnnotationMetadata annotationMetadata) {
        if (!isEnabled(annotationMetadata)) {
            return NO_IMPORTS;
        }
        AutoConfigurationMetadata autoConfigurationMetadata = AutoConfigurationMetadataLoader
                .loadMetadata(this.beanClassLoader);
        AnnotationAttributes attributes = getAttributes(annotationMetadata);
        List<String> configurations = getCandidateConfigurations(annotationMetadata,
                attributes);
        configurations = removeDuplicates(configurations);
        Set<String> exclusions = getExclusions(annotationMetadata, attributes);
        checkExcludedClasses(configurations, exclusions);
        configurations.removeAll(exclusions);
        configurations = filter(configurations, autoConfigurationMetadata);
        fireAutoConfigurationImportEvents(configurations, exclusions);
        return StringUtils.toStringArray(configurations);
    }
    ...
}
```

面对 `selectImports(AnnotationMetadata)`方法复杂的实现，按照执行的顺序，结合类和方法的字面意思，此处不妨大胆地猜测其中的含义：

（1）`AutoConfigurationMetadata autoConfigurationMetadata = AutoConfigurationMetadataLoader.loadMetadata(this.beanClassLoader)`：似乎是加载自动装配的元信息，暂且不求甚解，存疑。

（2）`AnnotationAttributes attributes = getAttributes(annotationMetadata)`：应该是获取`@EnableAutoConfiguration` 标注类的元信息。

（3）`List<String> configurations = getCandidateConfigurations(annotationMetadata,attributes)`：由于 `configurations` 作为 `selectImports(AnnotationMetadata)`方法的返回对象，而该方法返回的是导入类名集合，所以该对象应该是自动装配的候选类名集合。

（4）`configurations = removeDuplicates(configurations)`：方法字面意思是移除重复对象，说明 `configurations` 存在重复的可能，至于为什么会出现重复？这是一个很好的问题。

（5）`Set<String> exclusions = getExclusions(annotationMetadata, attributes)`：由于后续 `configurations` 将移除 `exclusions`，所以 `exclusions` 应该是自动装配组件的排除

名单。

（6）configurations = filter(configurations, autoConfigurationMetadata)：经过去重和排除后的 configurations 再执行过滤操作，该步骤依赖于步骤“1”获取的 AutoConfigurationMetadata 对象，换言之，AutoConfigurationMetadata 对象充当过滤的条件。

（7）fireAutoConfigurationImportEvents(configurations, exclusions)：在 configurations 对象返回前，貌似触发了一个自动装配的导入事件，事件可能包括候选的装配组件类名单和排除名单。

尽管目前无法证明以上猜测是否属实，不过可以基本定位@EnableAutoConfiguration 中的疑惑，例如：

（1）@EnableAutoConfiguration 如何装配组件，以及装配哪些组件？可由 getCandidateConfigurations(AnnotationMetadata,AnnotationAttributes)方法解答。

（2）@EnableAutoConfiguration 如何排除某些组件的自动装配，以及与其配置手段是如何交互的？getExclusions(AnnotationMetadata,AnnotationAttributes)方法可以解释。

不过还有一个更奇怪的问题，为什么应用自定义配置 Class 能够覆盖其自动装配 Class？这个问题留到最后说明。

9.2.1 @EnableAutoConfiguration 读取候选装配组件

首先分析 getCandidateConfigurations(AnnotationMetadata,AnnotationAttributes)方法：

```
public class AutoConfigurationImportSelector
    implements DeferredImportSelector, BeanClassLoaderAware, ResourceLoaderAware,
    BeanFactoryAware, EnvironmentAware, Ordered {
   ...
   protected List<String> getCandidateConfigurations(AnnotationMetadata metadata,
         AnnotationAttributes attributes) {
    List<String> configurations = SpringFactoriesLoader.loadFactoryNames(
            getSpringFactoriesLoaderFactoryClass(), getBeanClassLoader());
    ...
    return configurations;
 }
   ...
 protected Class<?> getSpringFactoriesLoaderFactoryClass() {
    return EnableAutoConfiguration.class;
```

```
    }
    ...
}
```

该方法实际执行的是 SpringFactoriesLoader#loadFactoryNames(Class,ClassLoader) 方法：

```
    public abstract class SpringFactoriesLoader {
        ...
        public static final String FACTORIES_RESOURCE_LOCATION = "META-INF/spring.factories";
        ...
        public static List<String> loadFactoryNames(Class<?> factoryClass, @Nullable ClassLoader
classLoader) {
            String factoryClassName = factoryClass.getName();
            return
            loadSpringFactories(classLoader).getOrDefault(factoryClassName,
Collections.emptyList());
    }

    private static Map<String, List<String>> loadSpringFactories(@Nullable ClassLoader
classLoader) {
        MultiValueMap<String, String> result = cache.get(classLoader);
        if (result != null) {
            return result;
        }

        try {
            Enumeration<URL> urls = (classLoader != null ?
                    classLoader.getResources(FACTORIES_RESOURCE_LOCATION) :
                    ClassLoader.getSystemResources(FACTORIES_RESOURCE_LOCATION));
            result = new LinkedMultiValueMap<>();
            while (urls.hasMoreElements()) {
                URL url = urls.nextElement();
                UrlResource resource = new UrlResource(url);
                Properties properties = PropertiesLoaderUtils.loadProperties(resource);
                for (Map.Entry<?, ?> entry : properties.entrySet()) {
                    List<String> factoryClassNames = Arrays.asList(
                            StringUtils.commaDelimitedListToStringArray((String)
entry.getValue()));
```

```
                    result.addAll((String) entry.getKey(), factoryClassNames);
                }
            }
            cache.put(classLoader, result);
            return result;
        }
        catch (IOException ex) {
            throw new IllegalArgumentException("Unable to load factories from location [" +
                    FACTORIES_RESOURCE_LOCATION + "]", ex);
        }
    }
    ...
}
```

SpringFactoriesLoader 是 Spring Framework 工厂机制的加载器，loadFactoryNames(Class,ClassLoader)方法加载原理如下：

（1）搜索指定 ClassLoader 下所有的 META-INF/spring.factories 资源内容（可能存在多个）。

（2）将一个或多个 META-INF/spring.factories 资源内容作为 Properties 文件读取，合并为一个 Key 为接口的全类名，Value 是实现类全类名列表的 Map，作为 loadSpringFactories(ClassLoader)方法的返回值。

（3）再从上一步返回的 Map 中查找并返回方法指定类名所映射的实现类全类名列表。

> 尽管 SpringFactoriesLoader 来源于 Spring Framework 3.2，但鲜有使用，不过它被 Spring Boot“委以重用”，频繁出现在各处，尤其在 SpringApplication 中。后续的章节将常与其“打交道”，请读者务必加深理解。

以上加载机制的描述在 Spring Boot 官方文档的“46.2 Locating Auto-configuration Candidates”一节中：

> Spring Boot checks for the presence of a META-INF/spring.factories file within your published jar. The file should list your configuration classes under the EnableAutoConfiguration key, as shown in the following example:
>
> ```
> org.springframework.boot.autoconfigure.EnableAutoConfiguration=\
> com.mycorp.libx.autoconfigure.LibXAutoConfiguration,\
> com.mycorp.libx.autoconfigure.LibXWebAutoConfiguration
> ```

结合当前的场景，以 Spring Boot 2.0.2.RELEASE 为例，框架内部.jar 文件默认引入多个META-INF/spring.factories 资源，其中包含@EnableAutoConfiguration 配置信息的.jar 文件有：

- spring-boot-autoconfigure-2.0.2.RELEASE.jar；
- spring-boot-actuator-autoconfigure-2.0.2.RELEASE.jar；
- spring-boot-devtools-2.0.2.RELEASE.jar（可选）。

例如，spring-boot-autoconfigure-2.0.2.RELEASE.jar 中的 META-INF/spring.factories：

```
...
# Auto Configure
org.springframework.boot.autoconfigure.EnableAutoConfiguration=\
org.springframework.boot.autoconfigure.admin.SpringApplicationAdminJmxAutoConfiguration,\
org.springframework.boot.autoconfigure.aop.AopAutoConfiguration,\
...
```

> Spring Boot 内建的@EnableAutoConfiguration 配置项众多，此处予以省略。

由于@EnableAutoConfiguration 配置可能存在自动装配组件类名重复定义的情况，当getCandidateConfigurations(AnnotationMetadata,AnnotationAttributes)方法获取所有的候选类集合名后，立即执行 removeDuplicates(List)方法，利用 Set 不可重复性达到去重的目的：

```
public class AutoConfigurationImportSelector
      implements DeferredImportSelector, BeanClassLoaderAware, ResourceLoaderAware,
      BeanFactoryAware, EnvironmentAware, Ordered {
    ...
    protected final <T> List<T> removeDuplicates(List<T> list) {
      return new ArrayList<>(new LinkedHashSet<>(list));
}
    ...
}
```

当自动装配组件类被 SpringFactoriesLoader 加载并去重后，接下来执行排除操作。

9.2.2　@EnableAutoConfiguration 排除自动装配组件

当 getExclusions(AnnotationMetadata,AnnotationAttributes)方法执行后，程序将获得一个自动装配 Class 的排除名单：

```
    public class AutoConfigurationImportSelector
        implements DeferredImportSelector, BeanClassLoaderAware, ResourceLoaderAware,
        BeanFactoryAware, EnvironmentAware, Ordered {
      ...
      private static final String PROPERTY_NAME_AUTOCONFIGURE_EXCLUDE =
"spring.autoconfigure.exclude";
      ...
      protected Set<String> getExclusions(AnnotationMetadata metadata,
            AnnotationAttributes attributes) {
        Set<String> excluded = new LinkedHashSet<>();
        excluded.addAll(asList(attributes, "exclude"));
        excluded.addAll(Arrays.asList(attributes.getStringArray("excludeName")));
        excluded.addAll(getExcludeAutoConfigurationsProperty());
        return excluded;
    }

    private List<String> getExcludeAutoConfigurationsProperty() {
        if (getEnvironment() instanceof ConfigurableEnvironment) {
            Binder binder = Binder.get(getEnvironment());
            return binder.bind(PROPERTY_NAME_AUTOCONFIGURE_EXCLUDE, String[].class)
                    .map(Arrays::asList).orElse(Collections.emptyList());
        }
        String[] excludes = getEnvironment()
                .getProperty(PROPERTY_NAME_AUTOCONFIGURE_EXCLUDE, String[].class);
        return (excludes != null ? Arrays.asList(excludes) : Collections.emptyList());
    }
      ...
    }
```

该方法的实现相对简单，将标注@EnableAutoConfiguration 配置类的注解属性（AnnotationAttributes）exclude 和 excludeName，以及将 spring.autoconfigure.exclude 配置值累加至排除集合 excluded。随后，检查排除类名集合是否合法：

```
public class AutoConfigurationImportSelector
    implements DeferredImportSelector, BeanClassLoaderAware, ResourceLoaderAware,
    BeanFactoryAware, EnvironmentAware, Ordered {
   ...
 private void checkExcludedClasses(List<String> configurations,
        Set<String> exclusions) {
    List<String> invalidExcludes = new ArrayList<>(exclusions.size());
    for (String exclusion : exclusions) {
        if (ClassUtils.isPresent(exclusion, getClass().getClassLoader())
                && !configurations.contains(exclusion)) {
            invalidExcludes.add(exclusion);
        }
    }
    if (!invalidExcludes.isEmpty()) {
        handleInvalidExcludes(invalidExcludes);
    }
 }
   ...
   protected void handleInvalidExcludes(List<String> invalidExcludes) {
    StringBuilder message = new StringBuilder();
    for (String exclude : invalidExcludes) {
        message.append("\t- ").append(exclude).append(String.format("%n"));
    }
    throw new IllegalStateException(String
            .format("The following classes could not be excluded because they are"
                    + " not auto-configuration classes:%n%s", message));
 }
   ...
}
```

当排除类存在于当前 `ClassLoader` 且不在自动装配候选类名单中时，`handleInvalidExcludes(List)`方法被执行，触发排除类非法异常。

接着该排除集合 `exclusions` 从候选自动装配 Class 名单 `configurations` 中移除：`configurations.removeAll(exclusions)`。计算后的 `configurations` 并非最终自动装配 Class 名单，还需再次过滤。

9.2.3 @EnableAutoConfiguration 过滤自动装配组件

移除排除类名单后的 configurations 配合 AutoConfigurationMetadata 对象执行过滤操作：

```
public class AutoConfigurationImportSelector
    implements DeferredImportSelector, BeanClassLoaderAware, ResourceLoaderAware,
    BeanFactoryAware, EnvironmentAware, Ordered {
   ...
   private List<String> filter(List<String> configurations,
         AutoConfigurationMetadata autoConfigurationMetadata) {
    long startTime = System.nanoTime();
    String[] candidates = StringUtils.toStringArray(configurations);
    boolean[] skip = new boolean[candidates.length];
    boolean skipped = false;
    for (AutoConfigurationImportFilter filter : getAutoConfigurationImportFilters()) {
         invokeAwareMethods(filter);
         boolean[] match = filter.match(candidates, autoConfigurationMetadata);
         ...
    }
    ...
    List<String> result = new ArrayList<>(candidates.length);
    for (int i = 0; i < candidates.length; i++) {
         if (!skip[i]) {
              result.add(candidates[i]);
         }
    }
    ...
    return new ArrayList<>(result);
 }

 protected List<AutoConfigurationImportFilter> getAutoConfigurationImportFilters() {
    return
    SpringFactoriesLoader.loadFactories(AutoConfigurationImportFilter.class,
              this.beanClassLoader);
 }
   ...
}
```

其中 AutoConfigurationImportFilter 对象集合同样被 SpringFactoriesLoader 加载，故查找 AutoConfigurationImportFilter 在所有 META-INF/spring.factories 资源中的配置。Spring Boot 框架默认仅有一处声明，即在 org.springframework.boot:spring-boot-autoconfigure 中：

```
# Auto Configuration Import Filters
org.springframework.boot.autoconfigure.AutoConfigurationImportFilter=\
org.springframework.boot.autoconfigure.condition.OnClassCondition
```

不过此处的 SpringFactoriesLoader#loadFactories(Class,ClassLoader) 方法与之前讨论的 loadFactoryNames(Class,ClassLoader)方法存在细微的差别，由于前者调用了后者，所以它们在类名单加载部分的逻辑是相同的：

```
public abstract class SpringFactoriesLoader {
    ...
    public static <T> List<T> loadFactories(Class<T> factoryClass, @Nullable ClassLoader
classLoader) {
        ...
        List<String> factoryNames = loadFactoryNames(factoryClass, classLoaderToUse);
        ...
        List<T> result = new ArrayList<>(factoryNames.size());
        for (String factoryName : factoryNames) {
            result.add(instantiateFactory(factoryName, factoryClass, classLoaderToUse));
        }
        AnnotationAwareOrderComparator.sort(result);
        return result;
    }
    ...
    private static <T> T instantiateFactory(String instanceClassName, Class<T> factoryClass,
ClassLoader classLoader) {
        try {
            Class<?> instanceClass = ClassUtils.forName(instanceClassName, classLoader);
            if (!factoryClass.isAssignableFrom(instanceClass)) {
                throw new IllegalArgumentException(
                        "Class [" + instanceClassName + "] is not assignable to [" +
factoryClass.getName() + "]");
            }
            return (T) ReflectionUtils.accessibleConstructor(instanceClass).newInstance();
```

```
        }
        catch (Throwable ex) {
            throw new IllegalArgumentException("Unable to instantiate factory class: " +
factoryClass.getName(), ex);
        }
    }
    ...
}
```

两者的区别在于，前者获取工厂类名单 factoryNames 后，将它们逐一进行类加载，这些类必须是参数 factoryClass 的子类，并被实例化且予以排序。换言之，在 META-INF/spring.factories 资源声明中的 OnClassCondition 也是 AutoConfigurationImportFilter 的实现类，进而说明 **AutoConfigurationImportSelector#filter(List,AutoConfigurationMetadata) 方法的实际作用是过滤 META-INF/spring.factories 资源中那些当前 ClassLoader 不存在的 Class** 。

不过，方法参数所依赖的 AutoConfigurationMetadata 对象又是如何而来的呢？需要再次分析 AutoConfigurationMetadataLoader#loadMetadata(ClassLoader)方法：

```
final class AutoConfigurationMetadataLoader {

    protected static final String PATH = "META-INF/"
            + "spring-autoconfigure-metadata.properties";
    ...
    public static AutoConfigurationMetadata loadMetadata(ClassLoader classLoader) {
        return loadMetadata(classLoader, PATH);
    }

    static AutoConfigurationMetadata loadMetadata(ClassLoader classLoader, String path) {
        try {
            Enumeration<URL> urls = (classLoader != null ? classLoader.getResources(path)
                    : ClassLoader.getSystemResources(path));
            Properties properties = new Properties();
            while (urls.hasMoreElements()) {
                properties.putAll(PropertiesLoaderUtils
                        .loadProperties(new UrlResource(urls.nextElement())));
            }
            return loadMetadata(properties);
        }
        catch (IOException ex) {
```

```
            throw new IllegalArgumentException(
                    "Unable to load @ConditionalOnClass location [" + path + "]", ex);
        }
    }
```

该类的 JavaDoc 简要地说明了其用意：

> Internal utility used to load AutoConfigurationMetadata.

换言之，`AutoConfigurationMetadataLoader` 是 `AutoConfigurationMetadata` 的加载器。`AutoConfigurationMetadata` 则是 Spring Boot 1.5 开始引入的自动装配元信息接口，这些信息配置于 `Properties` 格式的资源 `META-INF/spring-autoconfigure-metadata.properties` 中，框架内部仅存在基于 `Properties` 文件格式的实现 `PropertiesAutoConfigurationMetadata`，被 `AutoConfigurationMetadataLoader` 初始化。`AutoConfigurationMetadata` 接口支持多种数据类型的方法，`OnClassCondition` 作为 `AutoConfigurationImportFilter` 的实现类，它依赖于 `AutoConfigurationMetadata#getSet` 方法获取自动装配 Class 的"ConditionalOnClass"元信息：

```
class OnClassCondition extends SpringBootCondition
        implements AutoConfigurationImportFilter, BeanFactoryAware, BeanClassLoaderAware {
    ...
    private final class StandardOutcomesResolver implements OutcomesResolver {
        ...
        private ConditionOutcome[] getOutcomes(String[] autoConfigurationClasses,
                int start, int end, AutoConfigurationMetadata autoConfigurationMetadata) {
            ConditionOutcome[] outcomes = new ConditionOutcome[end - start];
            for (int i = start; i < end; i++) {
                String autoConfigurationClass = autoConfigurationClasses[i];
                Set<String> candidates = autoConfigurationMetadata
                        .getSet(autoConfigurationClass, "ConditionalOnClass");
                if (candidates != null) {
                    outcomes[i - start] = getOutcome(candidates);
                }
            }
            return outcomes;
        }
        ...
    }
    ...
```

```
}
```

根据以上方法的逻辑，自动装配 Class 的集合 autoConfigurationClasses 迭代地调用 AutoConfigurationMetadata#getSet 方法获取它们的“ConditionalOnClass”元信息，以 JmxAutoConfiguration 为例，其“ConditionalOnClass”配置如下：

```
...
org.springframework.boot.autoconfigure.jmx.JmxAutoConfiguration.ConditionalOnClass=org.sp
ringframework.jmx.export.MBeanExporter
...
```

> 以上 META-INF/spring-autoconfigure-metadata.properties 资源存放于 org.springframework.boot:spring-boot-autoconfigure jar 文件中。

当 org.springframework.jmx.export.MBeanExporter 作为方法返回值时，再直接使用 getOutcome(Set)方法计算匹配结果：

```
class OnClassCondition extends SpringBootCondition
        implements AutoConfigurationImportFilter, BeanFactoryAware, BeanClassLoaderAware {
    ...
    private List<String> getMatches(Collection<String> candidates, MatchType matchType,
            ClassLoader classLoader) {
        List<String> matches = new ArrayList<>(candidates.size());
        for (String candidate : candidates) {
            if (matchType.matches(candidate, classLoader)) {
                matches.add(candidate);
            }
        }
        return matches;
    }
    ...
        private ConditionOutcome getOutcome(Set<String> candidates) {
            try {
                List<String> missing = getMatches(candidates, MatchType.MISSING,
                        this.beanClassLoader);
                if (!missing.isEmpty()) {
                    return ConditionOutcome.noMatch(
                            ConditionMessage.forCondition(ConditionalOnClass.class)
                                    .didNotFind("required class", "required classes")
```

```
                                .items(Style.QUOTE, missing));
                }
            }
            catch (Exception ex) {
                // We'll get another chance later
            }
            return null;
        }
    ...
}
```

最终，判断的标准由 MatchType.MISSING#matches 方法决定：

```
private enum MatchType {
    ...
    MISSING {

        @Override
        public boolean matches(String className, ClassLoader classLoader) {
            return !isPresent(className, classLoader);
        }

    };

    private static boolean isPresent(String className, ClassLoader classLoader) {
        if (classLoader == null) {
            classLoader = ClassUtils.getDefaultClassLoader();
        }
        try {
            forName(className, classLoader);
            return true;
        }
        catch (Throwable ex) {
            return false;
        }
    }

    private static Class<?> forName(String className, ClassLoader classLoader)
            throws ClassNotFoundException {
```

```
        if (classLoader != null) {
            return classLoader.loadClass(className);
        }
        return Class.forName(className);
    }
    ...
}
```

总之，JmxAutoConfiguration 是否能够自动装配取决于其“ConditionalOnClass”关联的 org.springframework.jmx.export.MBeanExporter 类是否存在，从而帮助 AutoConfigurationImportSelector#filter 方法过滤那些类依赖不满足的自动装配 Class。如此设计相对复杂，当然也有好处，常规的@ConditionalOnClass 判断需要依赖自动装配 Class 必须被 ClassLoader 提前装载，然后解析其注解元信息，从而根据依赖类是否存在来判断装配与否，同时，Spring 应用上下文处理@Conditional 的时机较晚。然而通过读取 META-INF/spring-autoconfigure-metadata.properties 资源的“ConditionalOnClass”配置元信息，并判断其依赖类的存在性，不但实现的逻辑直接，而且减少了自动装配的计算时间。

总而言之，AutoConfigurationImportSelector 读取自动装配 Class 的流程为：

（1）通过 SpringFactoriesLoader#loadFactoryNames(Class,ClassLoader)方法读取所有 META-INF/spring.factories 资源中@EnableAutoConfiguration 所关联的自动装配 Class 集合。

（2）读取当前配置类所标注的@EnableAutoConfiguration 属性 exclude 和 excludeName，并与 spring.autoconfigure.exclude 配置属性合并为自动装配 Class 排除集合。

（3）检查自动装配 Class 排除集合是否合法。

（4）排除候选自动装配 Class 集合中的排除名单。

（5）再次过滤候选自动装配 Class 集合中 Class 不存在的成员。

当自动装配 Class 读取完毕后，fireAutoConfigurationImportEvents(List, Set)方法被执行，可能触发了一个自动装配的导入事件，具体情况究竟如何，下一节接着讨论。

9.2.4 @EnableAutoConfiguration 自动装配事件

继续探讨 fireAutoConfigurationImportEvents(List, Set)方法的实现：

```
public class AutoConfigurationImportSelector
        implements DeferredImportSelector, BeanClassLoaderAware, ResourceLoaderAware,
        BeanFactoryAware, EnvironmentAware, Ordered {
    ...
```

```
    private void fireAutoConfigurationImportEvents(List<String> configurations,
            Set<String> exclusions) {
        List<AutoConfigurationImportListener> listeners =
getAutoConfigurationImportListeners();
        if (!listeners.isEmpty()) {
            AutoConfigurationImportEvent event = new
            AutoConfigurationImportEvent(this,
                    configurations, exclusions);
            for (AutoConfigurationImportListener listener : listeners) {
                invokeAwareMethods(listener);
                listener.onAutoConfigurationImportEvent(event);
            }
        }
    }

    protected List<AutoConfigurationImportListener> getAutoConfigurationImportListeners() {
        return SpringFactoriesLoader.loadFactories(AutoConfigurationImportListener.class,
                this.beanClassLoader);
    }
    ...
    }
```

Spring Boot 1.5 开始引入 AutoConfigurationImportListener 接口，它有别于传统的 Spring ApplicationListener 的实现。ApplicationListener 与 Spring 应用上下文 ConfigurableApplicationContext 紧密关联，监听 Spring ApplicationEvent。而 AutoConfigurationImportListener 则是自定义 Java EventListener 实现，仅监听 AutoConfigurationImportEvent，然而其实例同样被 SpringFactoriesLoader 加载，因此 Spring Boot 框架层面为开发人员提供了扩展的途径。其中，ConditionEvaluationReportAutoConfigurationImportListener 就是内建实现，用于记录自动装配的条件评估详情，配置在 META-INF/spring.factories 资源中：

```
# Auto Configuration Import Listeners
org.springframework.boot.autoconfigure.AutoConfigurationImportListener=\
org.springframework.boot.autoconfigure.condition.ConditionEvaluationReportAutoConfigurati
onImportListener
```

> 更多关于 Spring 和 Spring Boot 事件的讨论将在“Spring Boot 事件”中展开。

根据以上原理，自定义 AutoConfigurationImportListener 实现。

自定义 AutoConfigurationImportListener

（1）实现 AutoConfigurationImportListener——DefaultAutoConfigurationImportListener。

```
    public class DefaultAutoConfigurationImportListener implements
AutoConfigurationImportListener {

        @Override
        public void
        onAutoConfigurationImportEvent(AutoConfigurationImportEvent event) {
            // 获取当前 ClassLoader
            ClassLoader classLoader = event.getClass().getClassLoader();
            // 候选的自动装配 Class 名单
            List<String> candidates =
                    SpringFactoriesLoader.loadFactoryNames(EnableAutoConfiguration.class,
classLoader);
            // 实际的自动装配 Class 名单
            List<String> configurations = event.getCandidateConfigurations();
            // 排除的自动装配 Class 名单
            Set<String> exclusions = event.getExclusions();
            // 输出各自数量
            System.out.printf("自动装配 Class 名单 - 候选数量：%d，实际数量：%d，排除数量：%s\n",
                    candidates.size(), configurations.size(), exclusions.size());
            // 输出实际和排除的自动装配 Class 名单
            System.out.println("实际的自动装配 Class 名单：");
            event.getCandidateConfigurations().forEach(System.out::println);
            System.out.println("排除的自动装配 Class 名单：");
            event.getExclusions().forEach(System.out::println);
        }
    }
```

（2）新建 META-INF/spring.factories 并配置 DefaultAutoConfigurationImportListener。

```
    # org.springframework.boot.autoconfigure.AutoConfigurationImportListener 配置
    org.springframework.boot.autoconfigure.AutoConfigurationImportListener=\
    thinking.in.spring.boot.samples.auto.configuration.listener.DefaultAutoConfigurationImpor
tListener
```

（3）实现引导类 EnableAutoConfigurationBootstrap。

```
@EnableAutoConfiguration(exclude = SpringApplicationAdminJmxAutoConfiguration.class)
public class EnableAutoConfigurationBootstrap {

    public static void main(String[] args) {
        new
        SpringApplicationBuilder(EnableAutoConfigurationBootstrap.class)
                .web(WebApplicationType.NONE) // 非 Web 应用
                .run(args)                    // 运行
                .close();                     // 关闭当前上下文
    }
}
```

> 源码位置：以上示例代码可通过查找 spring-boot-2.0-samples/auto-configuration-sample 工程获取，该工程依赖 `org.springframework.boot:spring-boot-starter`。

（4）运行引导类 `EnableAutoConfigurationBootstrap`。

```
(...部分内容被省略...)
自动装配类名单 - 候选数量：109，实际数量：9，排除数量：1
实际的自动装配类名单：
org.springframework.boot.autoconfigure.cache.CacheAutoConfiguration
org.springframework.boot.autoconfigure.context.ConfigurationPropertiesAutoConfiguration
org.springframework.boot.autoconfigure.context.MessageSourceAutoConfiguration
org.springframework.boot.autoconfigure.context.PropertyPlaceholderAutoConfiguration
org.springframework.boot.autoconfigure.info.ProjectInfoAutoConfiguration
org.springframework.boot.autoconfigure.jmx.JmxAutoConfiguration
org.springframework.boot.autoconfigure.mail.MailSenderValidatorAutoConfiguration
org.springframework.boot.autoconfigure.security.reactive.ReactiveSecurityAutoConfiguration
org.springframework.boot.autoconfigure.web.embedded.EmbeddedWebServerFactoryCustomizerAut
oConfiguration
排除的自动装配类名单：
org.springframework.boot.autoconfigure.admin.SpringApplicationAdminJmxAutoConfiguration
(...部分内容被省略...)
```

控制台输出了实际和排除的自动装配类名单。从数量上来看，`META-INF/spring.factories` 资源配置的自动装配 Class 名单要远多于实际装载，其中部分被过滤的类名单是 `AutoConfigurationImportSelector#filter` 方法执行的结果。

不过上述方法无法解释为什么应用自定义配置 Class 能够覆盖其自动装配 Class，谜底即将在下一节中揭晓。

9.2.5　@EnableAutoConfiguration 自动装配生命周期

在前面章节中，讨论的议题有选择性地忽略了 AutoConfigurationImportSelector 的接口层次性，而是直奔@EnableAutoConfiguration 自动装配的实现逻辑。前文曾提到：

> AutoConfigurationImportSelector 就是@EnableAutoConfiguration 所“@Import”的 DeferredImportSelector 实现类，由于 DeferredImport Selector 作为 ImportSelector 的子接口，所以组件自动装配逻辑均在 selectImports(AnnotationMetadata)方法中实现。

这样的说法尽管结论没有问题，然而逻辑却不够严谨。Spring Framework 4.0 开始引入 DeferredImportSelector 接口，从字面意义上分析，DeferredImportSelector 可理解为 Deferred（延迟的）ImportSelector，其 JavaDoc 提供了更翔实的说明：

> A variation of ImportSelector that runs after all @Configuration beans have been processed. This type of selector can be particularly useful when the selected imports are @Conditional. Implementations can also extend the org.springframework.core.Ordered interface or use the org.springframework.core.annotation.Order annotation to indicate a precedence against other DeferredImportSelectors.

DeferredImportSelector 作为 ImportSelector 的变种，它在@Configuration Bean 处理完毕后才运作。它在@Conditional 场景中尤其有用，同时该实现类可通过实现 Ordered 接口或标注@Order 的方式调整其优先执行顺序，以 AutoConfigurationImportSelector 为例，其优先级接近最低：

```
public class AutoConfigurationImportSelector
        implements DeferredImportSelector, BeanClassLoaderAware, ResourceLoaderAware,
        BeanFactoryAware, EnvironmentAware, Ordered {
...
public int getOrder() {
        return Ordered.LOWEST_PRECEDENCE - 1;
}
...
}
```

回顾前文“装载 ImportSelector 和 ImportBeanDefinitionRegistrar 实现”一节，ImportSelector 的处理实际上在 ConfigurationClassParser#processImports 方法中执行，

不过从 Spring Framework 4.0 开始，其方法实现发生了一些变化：

```
class ConfigurationClassParser {
...
private void processImports(ConfigurationClass configClass, SourceClass currentSourceClass,
        Collection<SourceClass> importCandidates, boolean checkForCircularImports) {
    ...
    else {
        this.importStack.push(configClass);
        try {
            for (SourceClass candidate : importCandidates) {
                if (candidate.isAssignable(ImportSelector.class)) {
                    // Candidate class is an ImportSelector -> delegate to it to determine
imports
                    Class<?> candidateClass = candidate.loadClass();
                    ImportSelector selector =
BeanUtils.instantiateClass(candidateClass, ImportSelector.class);
                    ParserStrategyUtils.invokeAwareMethods(
                            selector, this.environment, this.resourceLoader,
this.registry);
                    if (this.deferredImportSelectors != null && selector instanceof
DeferredImportSelector) {
                        this.deferredImportSelectors.add(
                                new
                                DeferredImportSelectorHolder(configClass,
(DeferredImportSelector) selector));
                    }
                    else {
                        String[] importClassNames =
selector.selectImports(currentSourceClass.getMetadata());
                        Collection<SourceClass> importSourceClasses =
asSourceClasses(importClassNames);
                        processImports(configClass, currentSourceClass,
importSourceClasses, false);
                    }
                }
                ...
            }
```

```
            }
            ...
        }
    }
    ...
}
```

其方法实现增加了 selector instanceof DeferredImportSelector 的逻辑判断，当条件成立时，将 DeferredImportSelector 包装为 DeferredImportSelectorHolder 对象，添加至待处理队列 deferredImportSelectors 中。否则，继续延续 Spring Framework 4.0 前的方式，调用 ImportSelector#selectImports 方法，并递归执行 processImports 方法。当所有 Configuration Class 处理完毕后，DeferredImportSelectorHolder 待处理队列在 processDeferredImportSelectors()方法中执行：

```
class ConfigurationClassParser {
    ...
    public void parse(Set<BeanDefinitionHolder> configCandidates) {
        this.deferredImportSelectors = new LinkedList<>();

        for (BeanDefinitionHolder holder : configCandidates) {
            BeanDefinition bd = holder.getBeanDefinition();
            try {
                if (bd instanceof AnnotatedBeanDefinition) {
                    parse(((AnnotatedBeanDefinition) bd).getMetadata(),
holder.getBeanName());
                }
                else if (bd instanceof AbstractBeanDefinition && ((AbstractBeanDefinition)
bd).hasBeanClass()) {
                    parse(((AbstractBeanDefinition) bd).getBeanClass(),
holder.getBeanName());
                }
                else {
                    parse(bd.getBeanClassName(), holder.getBeanName());
                }
            }
            ...
        }
```

```
        processDeferredImportSelectors();
    }
    ...
    private void processDeferredImportSelectors() {
        List<DeferredImportSelectorHolder> deferredImports = this.deferredImportSelectors;
        ...
        deferredImports.sort(DEFERRED_IMPORT_COMPARATOR);
        Map<Object, DeferredImportSelectorGrouping> groupings = new LinkedHashMap<>();
        ...
        for (DeferredImportSelectorGrouping grouping : groupings.values()) {
            grouping.getImports().forEach((entry) -> {
                ConfigurationClass configurationClass = configurationClasses.get(
                        entry.getMetadata());
                try {
                    processImports(configurationClass, asSourceClass(configurationClass),
                            asSourceClasses(entry.getImportClassName()), false);
                }
                ...
            });
        }
    }
    ...
}
```

而在 Spring Framework 5.0 中，`DeferredImportSelector` 接口新增了 `getImportGroup()`方法：

```
public interface DeferredImportSelector extends ImportSelector {
    ...
    @Nullable
    default Class<? extends Group> getImportGroup() {
        return null;
    }
    ...
}
```

其中 `DeferredImportSelector.Group` 接口辅助处理 `DeferredImportSelector` 导入的 Configuration Class：

```
public interface DeferredImportSelector extends ImportSelector {
  ...
  interface Group {

      /**
       * Process the {@link AnnotationMetadata} of the importing @{@link Configuration}
       * class using the specified {@link DeferredImportSelector}.
       */
      void process(AnnotationMetadata metadata, DeferredImportSelector selector);

      /**
       * Return the {@link Entry entries} of which class(es) should be imported for this
       * group.
       */
      Iterable<Entry> selectImports();
      ...
  }
  ...
}
```

该接口提供两类方法：`process()`和`selectImports()`。前者二次处理`DeferredImportSelector#selectImports(AnnotationMetadata)`方法返回的结果，而后者负责决定本组应该导入的 Configuration Class 作为实际导入的结果。

所以，在`processDeferredImportSelectors()`方法中，实际迭代的集合是`DeferredImportSelectorGrouping`，而非`DeferredImportSelectorHolder`。相关`ConfigurationClassParser`方法的调用链路和顺序如下图所示。

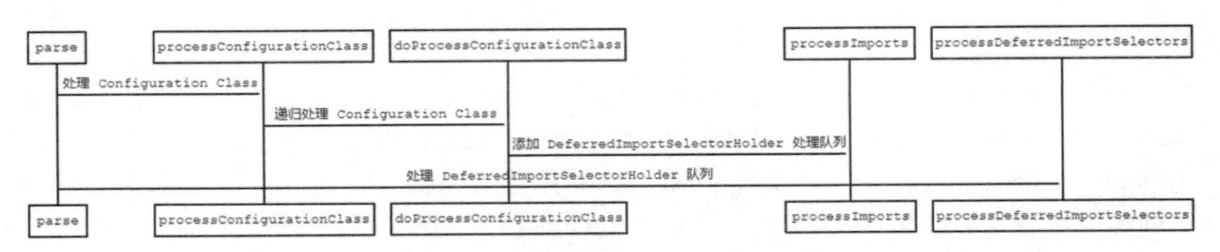

回到 Spring Boot 自动装配场景，在 Spring Boot 2.0 中，`AutoConfigurationImportSelector`并没有继承`DeferredImportSelector#getImportGroup()`方法的默认实现：

```
public class AutoConfigurationImportSelector
    implements DeferredImportSelector, BeanClassLoaderAware, ResourceLoaderAware,
```

```
        BeanFactoryAware, EnvironmentAware, Ordered {
    ...
    @Override
    public Class<? extends Group> getImportGroup() {
        return AutoConfigurationGroup.class;
    }
    ...
    private static class AutoConfigurationGroup implements DeferredImportSelector.Group,
            BeanClassLoaderAware, BeanFactoryAware, ResourceLoaderAware {
                @Override
        public void process(AnnotationMetadata annotationMetadata,
                DeferredImportSelector deferredImportSelector) {
            String[] imports = deferredImportSelector.selectImports(annotationMetadata);
            for (String importClassName : imports) {
                this.entries.put(importClassName, annotationMetadata);
            }
        }

        @Override
        public Iterable<Entry> selectImports() {
            return sortAutoConfigurations().stream()
                    .map((importClassName) -> new Entry(this.entries.get(importClassName),
                            importClassName))
                    .collect(Collectors.toList());
        }
    }
    ...
}
```

AutoConfigurationImportSelector.AutoConfigurationGroup（下面简写为 AutoConfigurationGroup）作为 DeferredImportSelector.Group 的实现，其 process (AnnotationMetadata,DeferredImportSelector) 方法负责缓存导入类名和 AnnotationMetadata 的键值对象，而 selectImports()方法则用于将自动装配 Class 排序，然后被 ConfigurationClassParser#processDeferredImportSelectors()方法用于最终自动装配 Class 的导入：

```
class ConfigurationClassParser {
    ...
```

```
private void processDeferredImportSelectors() {
    ...
    for (DeferredImportSelectorGrouping grouping : groupings.values()) {
        grouping.getImports().forEach((entry) -> {
            ConfigurationClass configurationClass = configurationClasses.get(
                    entry.getMetadata());
            try {
                processImports(configurationClass, asSourceClass(configurationClass),
                        asSourceClasses(entry.getImportClassName()), false);
            }
            ...
        });
    }
}
...
private static class DeferredImportSelectorGrouping {

    private final DeferredImportSelector.Group group;
    ...
    public Iterable<Group.Entry> getImports() {
        for (DeferredImportSelectorHolder deferredImport : this.deferredImports) {
            this.group.process(deferredImport.getConfigurationClass().getMetadata(),
                    deferredImport.getImportSelector());
        }
        return this.group.selectImports();
    }
}
...
}
```

ConfigurationClassParser 方法的调用链路如下图所示。

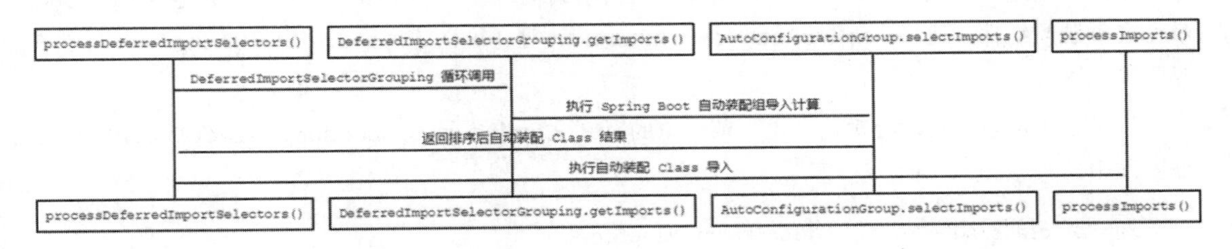

至此，关于@EnableAutoConfiguration生命周期的讨论接近尾声，然而在AutoConfigurationGroup.selectImports()方法返回值上卖了一个关子，该方法返回的是排序后自动装配 Class 的结果，而已知 SpringFactoriesLoader#loadFactoryNames(Class, ClassLoader) API 仅读取自动装配 Class 名单，并没有管理它们的顺序，自动装配 Class 排序的议题将在下一节中讨论。

9.2.6 @EnableAutoConfiguration 排序自动装配组件

在正式讨论之前，首先了解一下 Spring Boot 官方文档对于自动装配组件的顺序控制所提供的说明，在“46.2 Locating Auto-configuration Candidates”章节中有如下描述：

> You can use the @AutoConfigureAfter or @AutoConfigureBefore annotations if your configuration needs to be applied in a specific order. For example, if you provide web-specific configuration, your class may need to be applied after `WebMvcAutoConfiguration`.
> If you want to order certain auto-configurations that should not have any direct knowledge of each other, you can also use `@AutoConfigureOrder`. That annotation has the same semantic as the regular `@Order` annotation but provides a dedicated order for auto-configuration classes.

总结以上说明，Spring Boot 提供了两种自动装配组件的排序手段：

- 绝对自动装配顺序——@AutoConfigureOrder；
- 相对自动装配顺序——@AutoConfigureBefore 和@AutoConfigureAfter。

其中，@AutoConfigureOrder 与 Spring Framework @Order 的语义相同。@AutoConfigureBefore 和@AutoConfigureAfter 提供相对于其他自动装配组件的顺序控制。从以上注解的使用率分析，就 Spring Boot 内建自动装配组件而言，@AutoConfigureOrder 的比重低于@AutoConfigureBefore 或@AutoConfigureAfter 的比重。而在实际的使用场景中，绝对自动装配顺序的@AutoConfigureOrder 更是难以驾驭，因为开发人员想要了解所有的自动装配组件的顺序是不现实的。因此，建议读者尽可能使用注解@AutoConfigureBefore 或@AutoConfigureAfter 来控制，不过这也要求开发人员对应用所涉及的自动装配组件的顺序相当清楚，尤其是 Spring Boot 核心自动装配组件，比如 WebMvcAutoConfiguration 等（这也是为什么常说“Spring Boot 易学难精”的原因）。

以上三个注解的排序处理均在上一节讨论的 AutoConfigurationGroup.selectImports()方法实现中：

```
public class AutoConfigurationImportSelector
```

```
    implements DeferredImportSelector, BeanClassLoaderAware, ResourceLoaderAware,
    BeanFactoryAware, EnvironmentAware, Ordered {
  ...
  private static class AutoConfigurationGroup implements DeferredImportSelector.Group,
          BeanClassLoaderAware, BeanFactoryAware, ResourceLoaderAware {
    ...
    @Override
    public Iterable<Entry> selectImports() {
        return sortAutoConfigurations().stream()
                .map((importClassName) -> new Entry(this.entries.get(importClassName),
                        importClassName))
                .collect(Collectors.toList());
    }

    private List<String> sortAutoConfigurations() {
        ...
        AutoConfigurationMetadata autoConfigurationMetadata =
AutoConfigurationMetadataLoader
                .loadMetadata(this.beanClassLoader);
        return new
        AutoConfigurationSorter(getMetadataReaderFactory(),
                autoConfigurationMetadata).getInPriorityOrder(autoConfigurations);
    }
    ...
  }
  ...
  }
```

其中核心排序处理落在 `sortAutoConfigurations()`方法中，包括 `AutoConfigurationMetadata` 和 `AutoConfigurationSorter` 的交互。在前文“`@EnableAutoConfiguration` 过滤自动装配组件”一节中，曾讨论 `AutoConfigurationMetadata` 是 Spring Boot 1.5 开始引入的自动装配元信息接口，这些信息配置于 `Properties` 格式的资源 `META-INF/spring-autoconfigure-metadata.properties` 中，并且读取该资源的“ConditionalOnClass”配置元信息，判断依赖类存在性的方式，不但实现逻辑直接，而且减少了自动装配的计算时间。当然，该资源文件还包括自动装配 Class 的 AutoConfigureOrder、AutoConfigureBefore 和 AutoConfigureAfter 的配置信息，如 `WebMvcAutoConfiguration`：

```
org.springframework.boot.autoconfigure.web.servlet.WebMvcAutoConfiguration.AutoConfigureO
rder=-2147483638
```

```
org.springframework.boot.autoconfigure.web.servlet.WebMvcAutoConfiguration.AutoConfigureA
fter=org.springframework.boot.autoconfigure.web.servlet.DispatcherServletAutoConfiguration,or
g.springframework.boot.autoconfigure.validation.ValidationAutoConfiguration
```

因此，`META-INF/spring-autoconfigure-metadata.properties` 可认为是自动装配 Class 预处理元信息配置的资源。当该资源文件存在自动装配 Class 的注解元信息配置时，自动装配 Class 无须 ClassLoader 加载，即可得到所需的元信息，减少了运行时的计算消耗。所以，在自动装配 Class 速度方面，Spring Boot 1.5 相较于之前版本是有提升的。

`AutoConfigurationMetadata` 作为 `META-INF/spring-autoconfigure-metadata.properties` 资源的封装对象，再次在 `sortAutoConfigurations()`方法中加载，它与 `MetadataReaderFactory` 对象同时作为 `AutoConfigurationSorter` 构造参数，辅助 `AutoConfigurationSorter#getInPriorityOrder(Collection)`方法对自动装配 Class 集合进行排序：

```
class AutoConfigurationSorter {
...
public List<String> getInPriorityOrder(Collection<String> classNames) {
    AutoConfigurationClasses classes = new AutoConfigurationClasses(
            this.metadataReaderFactory, this.autoConfigurationMetadata, classNames);
    List<String> orderedClassNames = new ArrayList<>(classNames);
    // Initially sort alphabetically
    Collections.sort(orderedClassNames);
    // Then sort by order
    orderedClassNames.sort((o1, o2) -> {
        int i1 = classes.get(o1).getOrder();
        int i2 = classes.get(o2).getOrder();
        return Integer.compare(i1, i2);
    });
    // Then respect @AutoConfigureBefore @AutoConfigureAfter
    orderedClassNames = sortByAnnotation(classes, orderedClassNames);
    return orderedClassNames;
}
...
}
```

该方法排序自动装配 Class 的方式是按照字母顺序依此加载的，换言之，如果自动装配 Class 集合中未包含`@AutoConfigureOrder` 等顺序注解，则它们是按照字母顺序依次加载的。随后进行的是`@AutoConfigureOrder` 排序。当 `AutoConfigurationClasses` 构造时，将指定的自动装配 Class 集合逐一转换为 `AutoConfigurationClass` 对象，并形成与 `AutoConfigurationClasses` 多对一

的关联关系，

```
class AutoConfigurationSorter {
...
private static class AutoConfigurationClasses {

    private final Map<String, AutoConfigurationClass> classes = new HashMap<>();

    AutoConfigurationClasses(MetadataReaderFactory metadataReaderFactory,
            AutoConfigurationMetadata autoConfigurationMetadata,
            Collection<String> classNames) {
        addToClasses(metadataReaderFactory, autoConfigurationMetadata, classNames,
                true);
    }
    ...
    private void addToClasses(MetadataReaderFactory metadataReaderFactory,
            AutoConfigurationMetadata autoConfigurationMetadata,
            Collection<String> classNames, boolean required) {
        for (String className : classNames) {
            if (!this.classes.containsKey(className)) {
                AutoConfigurationClass autoConfigurationClass = new
AutoConfigurationClass(
                        className, metadataReaderFactory, autoConfigurationMetadata);
                boolean available = autoConfigurationClass.isAvailable();
                if (required || available) {
                    this.classes.put(className, autoConfigurationClass);
                }
                ...
            }
        }
    }
    ...
}
...
}
```

当 AutoConfigurationClass 与 AutoConfigurationClasses 建立映射关系后，具体的 @AutoConfigureOrder 排序规则由 AutoConfigurationClass#getOder()方法决定：

```
class AutoConfigurationSorter {
...
private static class AutoConfigurationClasses {
...
    private int getOrder() {
        if (wasProcessed()) {
            return
            this.autoConfigurationMetadata.getInteger(this.className,
                    "AutoConfigureOrder", AutoConfigureOrder.DEFAULT_ORDER);
        }
        Map<String, Object> attributes = getAnnotationMetadata()
                .getAnnotationAttributes(AutoConfigureOrder.class.getName());
        return (attributes != null ? (Integer) attributes.get("value")
                : AutoConfigureOrder.DEFAULT_ORDER);
    }

    private boolean wasProcessed() {
        return (this.autoConfigurationMetadata != null
                && this.autoConfigurationMetadata.wasProcessed(this.className));
    }
...
}
...
}
```

当 getOrder() 方法执行时，首先判断该 wasProcessed() 的结果，而该方法依赖于 AutoConfigurationMetadata#wasProcessed(String) 的结果，已知 AutoConfigurationMetadata#wasProcessed(String) 方法取决于指定的自动装配 Class 是否在 META-INF/spring-autoconfigure-metadata.properties 资源中配置为属性名。假设当前自动装配类为 WebMvcAutoConfiguration，那么 wasProcessed()方法返回 true，因为该类的配置存在：

```
org.springframework.boot.autoconfigure.web.servlet.WebMvcAutoConfiguration=
```

随后，getOrder()方法将读取自动装配类的 AutoConfigureOrder 的配置值，如果不存在，则使用默认值 AutoConfigureOrder.DEFAULT_ORDER，此处仍以 WebMvcAutoConfiguration 为例，它的配置值为：

```
org.springframework.boot.autoconfigure.web.servlet.WebMvcAutoConfiguration.AutoConfigureO
```

```
rder=-2147483638
```

否则，当自动装配类未出现在 META-INF/spring-autoconfigure-metadata.properties 资源中时，仍旧走 ASM 读取类元信息的老路。

在@AutoConfigureOrder 绝对顺序排序之后，再进入对@AutoConfigureBefore 和@AutoConfigureAfter 的排序过程，即 sortByAnnotation(AutoConfigurationClasses,List)方法，这部分的探讨请读者独立完成，其基本原理与@AutoConfigureOrder 类似，仍是读取注解元信息：

```
class AutoConfigurationSorter {
...
private static class AutoConfigurationClasses {
    ...
    public Set<String> getBefore() {
        if (this.before == null) {
            this.before = (wasProcessed()
                    ? this.autoConfigurationMetadata.getSet(this.className,
                            "AutoConfigureBefore", Collections.emptySet())
                    : getAnnotationValue(AutoConfigureBefore.class));
        }
        return this.before;
    }

    public Set<String> getAfter() {
        if (this.after == null) {
            this.after = (wasProcessed()
                    ? this.autoConfigurationMetadata.getSet(this.className,
                            "AutoConfigureAfter", Collections.emptySet())
                    : getAnnotationValue(AutoConfigureAfter.class));
        }
        return this.after;
    }
    ...
}
...
}
```

值得一提的是，自动装配排序实现在 Spring Boot 1.x 中是相对简单的，以 Spring Boot

1.5.10.RELEASE 为例：

```
public class AutoConfigurationImportSelector
     implements DeferredImportSelector, BeanClassLoaderAware, ResourceLoaderAware,
     BeanFactoryAware, EnvironmentAware, Ordered {
...
@Override
public String[] selectImports(AnnotationMetadata annotationMetadata) {
     if (!isEnabled(annotationMetadata)) {
          return NO_IMPORTS;
     }
     try {
          ...
          configurations = sort(configurations, autoConfigurationMetadata);
          ...
          return configurations.toArray(new String[configurations.size()]);
     }
     catch (IOException ex) {
          throw new IllegalStateException(ex);
     }
}
...
private List<String> sort(List<String> configurations,
          AutoConfigurationMetadata autoConfigurationMetadata) throws IOException {
     configurations = new AutoConfigurationSorter(getMetadataReaderFactory(),
               autoConfigurationMetadata).getInPriorityOrder(configurations);
     return configurations;
}
...
}
```

其 selectImports(AnnotationMetadata)方法直接调用了 sort(List,AutoConfigurationMetadata)，该方法的实现与 Spring Boot 2.0 基本一致，不过它没有 Spring Boot 2.0 那样复杂的 AutoConfigurationGroup 生命周期处理。

最后，再分享一些关于@AutoConfigureBefore 和@AutoConfigureAfter 的实践经验。大多数情况下，建议尽可能地使用@AutoConfigureBefore 或@AutoConfigureAftername()属性方法，不推荐使用 value()方法。关于这一点，小马哥是有切身体会的。在阿里巴巴集团 Spring Boot 基础设施的升级过程中，每次多少都有些战战兢兢。基本上，Spring Boot 的大小版本升级，均有不同程度的破坏性 API 变更。无法确认下一次 Spring Boot 升级后当前 Starter 能否兼容。尽管这样的错误不难被发现，不过容易让使用方对我们的专业性产生质疑，当然这种“黑锅”不能总背。又由于 `@AutoConfigureBefore` 和 `@AutoConfigureAfter` 在 Spring Boot 自动装配实现中属于常态化注解，所以尽可能地使用属性方法 `name()`，避免使用 `value()`。以最新开源的 `velocity-spring-boot-stater` 为例，为了兼容 Spring Boot 1.x 和 2.0，小马哥对 Spring Boot 官方实现做出了微调：

```
    @Configuration
    @ConditionalOnClass({VelocityEngine.class, VelocityEngineFactory.class})
    @AutoConfigureAfter(
            name = {
                    "org.springframework.boot.autoconfigure.web.WebMvcAutoConfiguration", //
Compatible with Spring Boot 1.x

"org.springframework.boot.autoconfigure.web.servlet.WebMvcAutoConfiguration"
            }
    )
    @EnableConfigurationProperties(VelocityProperties.class)
    @Deprecated
    public class VelocityAutoConfiguration {
    ...
    }
```

> 由于新模板引擎 Thymeleaf 在性能上不尽如人意，认为它不适合用在前台 Web 应用中。出于性能和迁移成本上的考虑，故“接棒”了 Spring Boot 1.5 开始移除的 Velocity 功能，并扩展了官方未实现的特性。详情请参考：https://github.com/alibaba/velocity-spring-boot-project。

官方实现使用属性方法是 `@AutoConfigureAfter#value()`，当该应用的 Spring Boot 版本为 2.0 时，`org.springframework.boot.autoconfigure.web.WebMvcAutoConfiguration` 类被调整为 `org.springframework.boot.autoconfigure.web.servlet.WebMvcAutoConfiguration`，因此应用启动过程中会抛出 `ArrayStoreException`。

或许读者要问：“为什么在 `@ConditionalOnClass` 中就使用了类型安全的属性方法 `value()` 呢”？原因非常简单，因为 `VelocityEngine` 和 `VelocityEngineFactory` 能被确认不会发生变化。

至此，之前的重重疑云被一一拨开。不过 `@EnableAutoConfiguration` 除了利用 `AutoConfigurationImportSelector` 自动装配 Class，它还将标注类所在的 package 添加至 `BasePackages` 中，为后续扫描组件提供 `BasePackages` 数据来源，如 JPA Entity 扫描，这将是下一节所讨论的范围。

9.2.7 @EnableAutoConfiguration 自动装配 BasePackages

从 Spring Boot 1.3 开始，@EnableAutoConfiguration 元标注新注解@AutoConfigurationPackage：

```
@Target(ElementType.TYPE)
@Retention(RetentionPolicy.RUNTIME)
@Documented
@Inherited
@AutoConfigurationPackage
@Import(AutoConfigurationImportSelector.class)
public @interface EnableAutoConfiguration {
 ...
}
```

在 Spring Boot 1.3 之前，则是直接“@Import”实现类 AutoConfigurationPackages.Registrar：

```
@Target(ElementType.TYPE)
@Retention(RetentionPolicy.RUNTIME)
@Documented
@Inherited
@Import({ EnableAutoConfigurationImportSelector.class,
     AutoConfigurationPackages.Registrar.class })
public @interface EnableAutoConfiguration {

 /**
  * Exclude specific auto-configuration classes such that they will never be applied.
  * @return the classes to exclude
  */
 Class<?>[] exclude() default {};

}
```

> 在 Spring Boot 1.3 之前，@EnableAutoConfiguration “@Import”的是 EnableAutoConfigurationImportSelector，其实现与 AutoConfigurationImportSelector 类似，请读者自行对比。

结合源码分析，元标注@AutoConfigurationPackage 同样“@Import” AutoConfigurationPackages.Registrar：

```
@Target(ElementType.TYPE)
@Retention(RetentionPolicy.RUNTIME)
@Documented
@Inherited
@Import(AutoConfigurationPackages.Registrar.class)
public @interface AutoConfigurationPackage {

}
```

同时，ConfigurationClassPostProcessor 提供递归处理配置 Class 和元注解的能力。换言之，@EnableAutoConfiguration 在 Spring Boot 1.3 前后并没有本质区别。所以，嵌套类 AutoConfigurationPackages.Registrar 是自动装配 BasePackages 的核心实现：

```
public abstract class AutoConfigurationPackages {
 ...
 private static final String BEAN = AutoConfigurationPackages.class.getName();
 ...
 @Order(Ordered.HIGHEST_PRECEDENCE)
 static class Registrar implements ImportBeanDefinitionRegistrar {

     @Override
     public void registerBeanDefinitions(AnnotationMetadata metadata,
             BeanDefinitionRegistry registry) {
         register(registry, ClassUtils.getPackageName(metadata.getClassName()));
     }

 }
 ...
 public static void register(BeanDefinitionRegistry registry, String... packageNames) {
     if (registry.containsBeanDefinition(BEAN)) {
         BeanDefinition beanDefinition = registry.getBeanDefinition(BEAN);
         ConstructorArgumentValues constructorArguments = beanDefinition
                 .getConstructorArgumentValues();
         constructorArguments.addIndexedArgumentValue(0,
                 addBasePackages(constructorArguments, packageNames));
     }
```

```
        else {
            GenericBeanDefinition beanDefinition = new GenericBeanDefinition();
            beanDefinition.setBeanClass(BasePackages.class);
            beanDefinition.getConstructorArgumentValues().addIndexedArgumentValue(0,
                    packageNames);
            beanDefinition.setRole(BeanDefinition.ROLE_INFRASTRUCTURE);
            registry.registerBeanDefinition(BEAN, beanDefinition);
        }
    }

    private static String[] addBasePackages(
            ConstructorArgumentValues constructorArguments, String[] packageNames) {
        String[] existing = (String[]) constructorArguments
                .getIndexedArgumentValue(0, String[].class).getValue();
        Set<String> merged = new LinkedHashSet<>();
        merged.addAll(Arrays.asList(existing));
        merged.addAll(Arrays.asList(packageNames));
        return StringUtils.toStringArray(merged);
    }
    ...
}
```

AutoConfigurationPackages.Registrar 作为 ImportBeanDefinitionRegistrar 的实现，通过 AutoConfigurationPackages#register(BeanDefinitionRegistry,String...) 方法注册当前标注类所在 package。尽管@AutoConfigurationPackage 鲜有直接标注在配置 Class 上的场景，不过不排除@EnableAutoConfiguration 被不同的配置 Class 标注的场景，所以 AutoConfigurationPackages.Registrar 可能存在多次被“@Import”的可能，当第一个被“@Import”时，BasePackagesBeanDefinition 注册到 Spring 应用上下文中，后续的@Import 操作将调整该 BeanDefinition 的构造器参数元信息 ConstructorArgumentValues，即添加当前标注类所在的 package 到已有集合中。此处相对抽象，需要结合 BasePackages 构造器定义加以说明：

```
public abstract class AutoConfigurationPackages {
...
    static final class BasePackages {

    private final List<String> packages;

    private boolean loggedBasePackageInfo;
```

```
        BasePackages(String... names) {
            List<String> packages = new ArrayList<>();
            for (String name : names) {
                if (StringUtils.hasText(name)) {
                    packages.add(name);
                }
            }
            this.packages = packages;
        }

        public List<String> get() {
            ...
            return this.packages;
        }

    }
    ...
}
```

当第一个 ConfigurationClass 被 AutoConfigurationPackages.Registrar 首次执行时，BasePackages 被注册为 Spring Bean，其名称为“org.springframework.boot.autoconfigure.AutoConfigurationPackages”。

> BasePackages Bean 的名称来源于 AutoConfigurationPackages.BEAN 常量定义：AutoConfigurationPackages.class.getName()，故为“org.springframework.boot.autoconfigure.AutoConfigurationPackages”。

随后，它将当前标注类所在的 package 作为 BasePackages 构造器的首参。由于 AutoConfigurationPackages.Registrar#registerBeanDefinitions(AnnotationMetadata, BeanDefinitionRegistry)方法的执行处在 Bean 注册阶段，其 BeanDefinition 拥有调整的机会。当方法再度执行时，先读取 BeanDefinition，然后获取其构造器参数元信息 ConstructorArgumentValues，将本次的标注类所在的 package 追加至其中。当 BasePackages Bean 初始化后，关联的 packages 可由其 get()方法获得，为工具方法 AutoConfigurationPackages#get(BeanFactory)提供数据来源：

```
public abstract class AutoConfigurationPackages {
  ...
```

```
    private static final String BEAN = AutoConfigurationPackages.class.getName();
    ...
    public static List<String> get(BeanFactory beanFactory) {
        try {
            return beanFactory.getBean(BEAN, BasePackages.class).get();
        }
        catch (NoSuchBeanDefinitionException ex) {
            throw new IllegalStateException(
                    "Unable to retrieve @EnableAutoConfiguration base packages");
        }
    }
    ...
}
```

该方法被多处使用，如 JPA 中：

```
@EnableConfigurationProperties(JpaProperties.class)
@Import(DataSourceInitializedPublisher.Registrar.class)
public abstract class JpaBaseConfiguration implements BeanFactoryAware {
...
protected String[] getPackagesToScan() {
    List<String> packages = EntityScanPackages.get(this.beanFactory)
            .getPackageNames();
    if (packages.isEmpty() && AutoConfigurationPackages.has(this.beanFactory)) {
        packages = AutoConfigurationPackages.get(this.beanFactory);
    }
    return StringUtils.toStringArray(packages);
}
...
}
```

值得关注的是，AutoConfigurationPackages 和 AutoConfigurationPackages. Registrar 贯穿于 Spring Boot 1.x 到 2.0。因此，开发人员可以放心地使用 AutoConfigurationPackages，也可以从 Spring Boot 1.3 开始，使用注解@AutoConfigurationPackage 扩大 BasePackages 的搜索范围。

按照@EnableAutoConfiguration 规约的特性，它将自动装配 META-INF/spring. factories 资源中所声明的配置类，这也意味着开发人员能够自定义自动装配实现。那么，如何专业化地实现自动装配不失为一个大学问。同时，不经意地发现 Spring Boot 内建的自动装配组件大量地使用 Spring

Boot `@Conditional` 扩展注解，那么开发人员应该如何合理地复用已有实现则又是一大学问。

由于 `@EnableAutoConfiguration` 基于 `SpringFactoriesLoader#loadFactoryNames(Class,ClassLoader)` API 实现自动装配 Class 名的导入，未对其配置 Class 做优先级排序，如果加入的配置 Class 之间存在初始化前后依赖关系，那么又应该如何实现呢？带着这些疑问，进入“自定义 Spring Boot 自动装配”的讨论。

9.3 自定义 Spring Boot 自动装配

通常开发人员将通用的逻辑打包到独立 `JAR` 文件中，供上层应用使用。不同的运行环境或框架，有其独特的 SPI（Service Provider Interface）机制，如“Web 自动装配”一节中曾提及 `javax.servlet.ServletContainerInitializer` 接口。

> `ServletContainerInitializer` 详情请参考 Servlet 3.0 规范中的“8.2.4 Shared libraries / runtimes pluggability”章节描述。

作为通用框架的 Spring Boot 自然也不例外，当注解`@EnableAutoConfiguration` 激活自动装配后，`META-INF/spring.factories` 资源中声明的配置 Class 随即被装配。这种 SPI 机制不仅在 Spring Boot 框架内部奏效，也允许开发人员自定义 Spring Boot 自动装配，正如 Spring Boot 官方文档“46. Creating Your Own Auto-configuration”一节中的描述：

> If you work in a company that develops shared libraries, or if you work on an open-source or commercial library, you might want to develop your own auto-configuration. Auto-configuration classes can be bundled in external jars and still be picked-up by Spring Boot.

文档继续解释“自动装配”和共享类库的关系：

> Auto-configuration can be associated to a “starter” that provides the auto-configuration code as well as the typical libraries that you would use with it.

此处的“自动装配”包括配置资源 `META-INF/spring.factories` 和配置 Class，两者可被关联到名为“starter”的共享类库。稍微留意最简单的 Spring Boot 应用，它所依赖类库就包括官方的“starter”，如 `spring-boot-starter`。请读者注意上述官方文档中的潜台词，其中“can be associated”传达给开发人员的意思是，“starter”只是其中一种自动装配的实现方式。暂且不论自动装配 Class 是否以 `starter` 方式承载，如何将它们及其所在的 package 合理地命名应该是更迫切的提议。实际上，Spring Boot 官方文档并没有给出自动装配 Class 和 package 的命名规则，因此，后面两节将讨论两者命名的

“潜规则”。

9.3.1 自动装配 Class 命名的潜规则

援引 Spring Boot 官方文档附录“C.1 From the ‘spring-boot-autoconfigure’ module”的内容，寻找 Spring Boot 内建的自动装配 Class 的命名规律：

> The following auto-configuration classes are from the `spring-boot-autoconfigure` module:
>
> | Configuration Class | Links |
> | --- | --- |
> | ActiveMQAutoConfiguration | JavaDoc |
> | AopAutoConfiguration | JavaDoc |
> | ArtemisAutoConfiguration | JavaDoc |
> | BatchAutoConfiguration | JavaDoc |
> | CacheAutoConfiguration | JavaDoc |
> | CassandraAutoConfiguration | JavaDoc |
> | ... | |

不难发现，所有 Spring Boot 内建的自动装配 Class 的名称模式为`*AutoConfiguration`，由于以上规则并不成文，故称之为“自动装配 Class 命名的潜规则”。该规则无论在 Spring Cloud 中，还是第三方整合（如 MyBatis 或 Dubbo）中均得到沿用，因此，建议读者在实现 Spring Boot 自动装配 Class 的过程中也依此惯例。接下来继续讨论自动装配 package 命名的潜规则。

9.3.2 自动装配 package 命名的潜规则

或许继续引用“C.1 From the ‘spring-boot-autoconfigure’ module”的内容无法寻找自动化装配 package 命名的规律。不过，稍微调整即可达到目的。重温`@EnableAutoConfiguration`在`/META-INF/spring.factories`资源中的配置声明，以 Spring Boot `2.0.2.RELEASE`为例，挑选几个具有代表性的自动装配 Class：

```
org.springframework.boot.autoconfigure.EnableAutoConfiguration=\
...
org.springframework.boot.autoconfigure.cache.CacheAutoConfiguration,\
...
org.springframework.boot.autoconfigure.data.jpa.JpaRepositoriesAutoConfiguration,\
```

```
...
org.springframework.boot.autoconfigure.web.servlet.WebMvcAutoConfiguration,\
```

仔细观察以上三例的 package 名称，其中：

- CacheAutoConfiguration 的 pacakge 为 org.springframework.boot.autoconfigure.cache。
- JpaRepositoriesAutoConfiguration 的 pacakge 为 org.springframework.boot.autoconfigure.data.jpa。
- WebMvcAutoConfiguration 的 pacakge 为 org.springframework.boot.autoconfigure.web.servlet。

明显地发现，三者 base package 为 org.springframework.boot.autoconfigure，不过三种实现类位于不同的模块功能子 package 下。将讨论的范围扩大至所有 Spring Boot 内建的自动装配 Class，它们的 package 基本模式为：

```
org.springframework.boot.autoconfigure
 |- ${module-package}
     |- *AutoConfiguration
     |- ${sub-module-package}
         |- ...
```

其中，${module-package}是功能模块 package，如 web.servlet，而${sub-mod ule-package}是子模块 package，如 error。

按照此规律，自动装配 package 命名模式应该是：

```
${root-package}
 |- autoconfigure
     |- ${module-package}
         |- *AutoConfiguration
         |- ${sub-module-package}
             |- ...
```

${root-package}是根 package，如 com.acme。

目前，已知@EnableAutoConfiguration 的配置声明方式及实现原理，同时了解了自动装配 Class 及 package 的命名潜规则，为接下来的自定义 Spring Boot Starter 扫清了技术障碍。

9.3.3 自定义 Spring Boot Starter

Spring Boot 官方文档的“46.5 Creating Your Own Starter”一节开门见山地告诉开发人员完整的 Spring Boot Starter 包含的组件：

> A full Spring Boot starter for a library may contain the following components:
> - The `autoconfigure` module that contains the auto-configuration code.
> - The `starter` module that provides a dependency to the `autoconfigure` module as well as the library and any additional dependencies that are typically useful. In a nutshell, adding the starter should provide everything needed to start using that library.

官方建议将自动装配代码存放在 `autoconfigure` 模块中，`starter` 模块依赖该模块，并且附加其他需要的依赖。换言之，`starter` 是“麻雀虽小，五脏俱全”，需要提供完整的功能和配套资源。以上描述给人的直接感受是，需要将 `autoconfigure` 和 `starter` 各自发布在独立的 `jar` 中，然而官方并没有执意如此，并补充道：

> You may combine the auto-configuration code and the dependency management in a single module if you do not need to separate those two concerns.

简言之，虽然两者是不同的概念，但开发人员完全可以将两者合并到单模块（单 `jar`）中。当 Starter 部署结构确定后，需要为其取一佳名。

1. Spring Boot Starter 命名规则

官方文档在“46.5.1 Naming”中继续说明 Spring Boot Starter 的命名规则：

> As a rule of thumb, you should name a combined module after the starter. For example, assume that you are creating a starter for "acme" and that you name the auto-configure module `acme-spring-boot-autoconfigure` and the starter `acme-spring-boot-starter`. If you only have one module that combines the two, name it `acme-spring-boot-starter`.

官方推荐开发人员使用 `${module}-spring-boot-starter` 的 Starter 命名模式，其中 `${module}`为功能模块的名称。采用如此命名的 Spring Boot Starter 属于“民间 Starter”或“第三方 Starter”，而 Spring Boot 官方的 Starter 则采用 `spring-boot-starter-${module}`的命名模式，比如 `spring-boot-starter-logging`。同时，开发人员可以将 Starter 发布为`${module}-spring-boot-autoconfigure` 和`${module}-spring-boot-starter` 两个 `jar` 文件。比如小马哥维护的

Apache Dubbo 子工程“Dubbo Spring Boot Project”就采用了“双模块制”：dubbo-spring-boot-autoconfigure dubbo-spring-boot-starter。

> 截至本章编写时，Apache Dubbo 正处于 Apache 孵化期，子工程 Dubbo Spring Boot Project 的 GitHub 地址为：https://github.com/apache/incubator-dubbo-spring-boot-project，内有中英文双语文档。

官方文档继续补充道：

> Also, if your starter provides configuration keys, use a unique namespace for them. In particular, do not include your keys in the namespaces that Spring Boot uses (such as `server`, `management`, `spring`, and so on). If you use the same namespace, we may modify these namespaces in the future in ways that break your modules.

除了 Starter 命名，官方语“带威胁”地告诉开发人员不要使用 `server`、`management`、`spring` 等作为配置 Key 命名空间。实际上，这些命名空间是外部化配置`@ConfigurationProperties` 前缀属性`prefix`，同时，`spring-boot-configuration-processor` 能够帮助`@ConfigurationProperties` Bean 生成 IDE 辅助元信息：

> Make sure to trigger meta-data generation so that IDE assistance is available for your keys as well. You may want to review the generated meta-data (`META-INF/spring-configuration-metadata.json`) to make sure your keys are properly documented.

不过，外部化配置 Key 的命名空间规则是从 Spring Boot 1.3 官方文档开始引入的，对于早期版本 Spring Boot Starter 而言，并不一定完全遵照该规则，如阿里巴巴内部大量的 Spring Boot Starter 仍采用 `Spring` 的命名空间，而它们又装配在许多线上 Spring Boot 应用中，多少有些“积重难返”。即便如此，内部的 Starter 可采用二级命名空间的方式，与 Spring Boot 官方 Starter 区分，如 `spring.acme`。不过这样又对维护人员提出了挑战，他们不得不经常关注 Spring Boot 版本间的变化，既有小版本间的变化，如 `MultipartProperties` Spring Boot 1.x 之间的变化，在 Spring Boot 1.4 之前，该类的命名空间为 `multipart`，从 1.4 版本开始，则为 `spring.http.multipart`；又存在大版本的调整，如 `ManagementServerProperties` 在 Spring Boot 1.x 的命名空间是 `management`，到了 Spring Boot 2.0 则变为 `management.server`。所以，开发人员最好遵照官方的建议，不要低估 Spring Boot “we may modify these namespaces in the future”的“决心”。比如 Dubbo 和 Dubbo Spring Boot Starter 就不敢“造次”，采用独立的命名空间：`dubbo`。关于更多的外部化配置细节，请读者翻阅本书的“外部化配置”章节，也是本书的重要讨论议题。

> 读者如果想了解 Dubbo 外部化配置使用和设计的细节，请参考小马哥的博客：https://mercyblitz.github.io/2018/01/18/Dubbo-外部化配置。

根据以上讨论内容，实现一个 Spring Boot Starter，其需求为格式化 Object 对象为文本内容。

2. 实现 Spring Boot Starter

（1）新建 Spring Boot Starter 工程——`formatter-spring-boot-starter`。

构建一个名为 `formatter-spring-boot-starter` 的 Maven 模块工程，增加 Spring Boot Starter 依赖，完整的 `pom.xml` 文件如下：

```xml
<?xml version="1.0" encoding="UTF-8"?>
<project xmlns="http://maven.apache.org/POM/4.0.0"
         xmlns:xsi="http://www.w3.org/2001/XMLSchema-instance"
         xsi:schemaLocation="http://maven.apache.org/POM/4.0.0
http://maven.apache.org/xsd/maven-4.0.0.xsd">
    <parent>
        <artifactId>spring-boot-2.x-samples</artifactId>
        <groupId>thinking-in-spring-boot</groupId>
        <version>1.0.0-SNAPSHOT</version>
        <relativePath>../pom.xml</relativePath>
    </parent>
    <modelVersion>4.0.0</modelVersion>
    <groupId>thinking-in-spring-boot</groupId>
    <artifactId>formatter-spring-boot-starter</artifactId>
    <dependencies>
        <!-- Spring Boot Starter 基础依赖 -->
        <dependency>
            <groupId>org.springframework.boot</groupId>
            <artifactId>spring-boot-starter</artifactId>
           <optional>true</optional>
        </dependency>
    </dependencies>
</project>
```

（2）新建格式化接口——`Formatter`。

```java
public interface Formatter {
```

```
    /**
     * 格式化操作
     *
     * @param object 待格式化对象
     * @return 返回格式化后的内容
     */
    String format(Object object);

}
```

依此自动装配 package 的命名潜规则，Formatter 的 package 应该为 thinking.in.spring.boot.samples.autoconfigure.formatter，后续其他相关实现也存放于此。

下一步则是实现 Formatter 接口。

（3）实现 Formatter 接口——DefaultFormatter。

```
public class DefaultFormatter implements Formatter {

    @Override
    public String format(Object object) {
        return String.valueOf(object); // null 安全实现
    }
}
```

此处 DefaultFormatter 仅简单地利用 String.valueOf(Object)方法输出格式化内容。

（4）实现 DefaultFormatter 自动装配——FormatterAutoConfiguration。

FormatterAutoConfiguration 简单地通过@Bean 方式声明 DefaultFormatter 为 Spring Bean：

```
@Configuration
public class FormatterAutoConfiguration {

    /**
     * 构建 {@link DefaultFormatter} Bean
     *
     * @return {@link DefaultFormatter}
     */
    @Bean
    public Formatter defaultFormatter() {
```

```
        return new DefaultFormatter();
    }
}
```

（5）META-INF/spring.factories 资源声明 FormatterAutoConfiguration。

META-INF/spring.factories 新增 FormatterAutoConfiguration 的配置声明：

```
# FormatterAutoConfiguration 自动装配声明
org.springframework.boot.autoconfigure.EnableAutoConfiguration=\
thinking.in.spring.boot.samples.autoconfigure.formatter.FormatterAutoConfiguration
```

> 源码位置：以上示例代码可通过查找 spring-boot-2.0-samples/formatter-spring-boot-starter 工程获取。

（6）构建 Spring Boot Starter。

将当前目录切换至 spring-boot-2.0-samples/formatter-spring-boot-starter 工程，并执行构建 Maven package 的命令：

```
formatter-spring-boot-starter mercyblitz$ mvn -Dmaven.test.skip -U clean install
[INFO] Scanning for projects...
(...部分内容被省略...)
[INFO] Building formatter-spring-boot-starter 1.0.0-SNAPSHOT
(...部分内容被省略...)
[INFO] BUILD SUCCESS
(...部分内容被省略...)
```

（7）添加 formatter-spring-boot-starter 依赖。

添加 formatter-spring-boot-starter 依赖到 spring-boot-2.0-samples/auto-configuration-sample 工程，pom.xml 文件完整的依赖如下：

```
    <dependencies>
        <!-- Spring Boot 基础依赖 -->
        <dependency>
            <groupId>org.springframework.boot</groupId>
            <artifactId>spring-boot-starter</artifactId>
        </dependency>
        <!-- 添加 formatter-spring-boot-starter -->
        <dependency>
            <groupId>thinking-in-spring-boot</groupId>
```

```
        <artifactId>formatter-spring-boot-starter</artifactId>
        <version>1.0.0-SNAPSHOT</version>
    </dependency>
</dependencies>
```

（8）新建引导类——FormatterBootstrap。

```
@EnableAutoConfiguration
public class FormatterBootstrap {

    public static void main(String[] args) {
        ConfigurableApplicationContext context = new
SpringApplicationBuilder(FormatterBootstrap.class)
                .web(WebApplicationType.NONE)  // 非 Web 应用
                .run(args);                    // 运行
        // 待格式化对象
        Map<String, Object> data = new HashMap<>();
        data.put("name", "小马哥");
        // 获取 Formatter，来自 FormatterAutoConfiguration
        Formatter formatter = context.getBean(Formatter.class);
        System.out.printf("formatter.format(data) : %s\n", formatter.format(data));
        // 关闭当前上下文
        context.close();
    }
}
```

（9）运行引导类——FormatterBootstrap。

观察日志变化：

```
(...部分内容被省略...)
formatter.format(data) : {name=小马哥}
(...部分内容被省略...)
```

Formatter Bean 成功地格式化 Map 对象内容，说明以上 formatter-spring-boot-starter 运作正常。不过在该 Starter pom.xml 声明中，有一处细节有待说明：

```
<dependency>
    <groupId>org.springframework.boot</groupId>
    <artifactId>spring-boot-starter</artifactId>
    <optional>true</optional>
```

```
</dependency>
```

其中 org.springframework.boot:org.spring-boot-starter Maven 依赖声明为 <optional>true</optional>，其目的在于说明 formatter-spring-boot-starter 不应该传递 spring-boot-starter 的依赖。一旦 Starter 发布为 jar，spring-boot-starter 的版本也随之固定。当应用使用该 Starter 后，很有可能与当前应用所依赖的 spring-boot-starter 版本发生冲突，由于不同环境下 ClassLoader 加载的不确定性，最终导致 Class 文件二进制不兼容的情况，可能表现在 IDE 中工作正常，而部署在线上却发生故障，或者相反现象。常见的异常有 ClassNotFoundException、NoClassDefFoundError 和 NoSuchMethodException 等。由于 Spring Boot 应用直接或间接地引入 spring-boot-starter 相关依赖，因此这些依赖不需要也不应该由自定义 Spring Boot Starter 传递引入。

> 读者在开发 Spring Boot Starter 的过程中，请保持 spring-boot-starter 等相关依赖声明为 <optional>true</optional>的良好编程习惯。

值得注意的是，即使自定义 Starter tx-spring-boot-starter 将 spring-boot-starter 相关依赖声明为 <optional>true</optional>，假设它依赖于 Spring Boot 1.5 API PlatformTransactionManagerCustomizer，当它被 Spring Boot 1.4 应用依赖时，如果 tx-spring-boot-starter 中的自动装配 Class 贸然声明 PlatformTransactionManager Customizer @Bean，则当前 Spring Boot 1.4 应用运行时必然会抛出 ClassNotFoundException 的异常。因此，合理地运用条件装配是至关重要的。

不过 formatter-spring-boot-starter 还不够完善，在接下来的章节中，它将深度整合 Spring Boot 条件化装配，逐渐演变为专业化程度较高的 Spring Boot Starter。

9.4 Spring Boot 条件化自动装配

在 Spring Boot 官方文档的“46.1 Understanding Auto-configured Beans”一节中，给出了实现 Spring Boot 自动装配的基本策略：

> Under the hood, auto-configuration is implemented with standard @Configuration classes. Additional @Conditional annotations are used to constrain when the auto-configuration should apply. Usually, auto-configuration classes use @ConditionalOnClass and @ConditionalOnMissingBean annotations. This ensures that auto-configuration applies only when relevant classes are found and when you have not declared your own @Configuration.

简言之，标准@Configuration 类是自动装配的底层实现，并且搭配 Spring Framework @Conditional 注解，使其能够合理地在不同环境中运作。在“Spring Boot 自动装配原理”一节的讨论中，@EnableAutoConfiguration 利用 AutoConfigurationImportFilter 实现类 OnClassCondition 过滤非法自动装配 Class，从而间接地接触条件注解@ConditoinalOnClass。

在“@Conditional 条件装配”一节有如下描述：

> 条件注解@ConditionalOnClass 采用元标注@Conditional(OnClassCondition.class)的方式定义，所以它包含@Conditional 的属性元信息。

实际上，所有 Spring Boot 条件注解@Conditional*均采用元标注@Conditional 的方式实现。

为了让开发人员系统性地掌握 Spring Boot 条件装配，Spring Boot 官方将所有条件注解总结在“46.3 Condition Annotations”一节中：

> You almost always want to include one or more @Conditional annotations on your auto-configuration class. The @ConditionalOnMissingBean annotation is one common example that is used to allow developers to override auto-configuration if they are not happy with your defaults.
> Spring Boot includes a number of @Conditional annotations that you can reuse in your own code by annotating @Configuration classes or individual @Bean methods. These annotations include:
> - Section 46.3.1, “Class Conditions”
> - Section 46.3.2, “Bean Conditions”
> - Section 46.3.3, “Property Conditions”
> - Section 46.3.4, “Resource Conditions”
> - Section 46.3.5, “Web Application Conditions”
> - Section 46.3.6, “SpEL Expression Conditions”

尽管 Spring Boot 条件注解的学习成本不高，然而合理地运用却需要较高的专业化程度。在阿里数以千计的 Spring Boot 应用迁移过程中，小马哥对此深有体会。接下来，依照官方文档给定的顺序，结合个人经验，逐一讨论。

9.4.1 Class 条件注解

Class 条件注解是一对语义相反的@ConditionalOnClass 和@ConditionalOnMissingClass，分别表达“当指定类存在时”和“当指定类缺失时”的语义。在注解声明方面，@ConditionalOnClass

的定义在 Spring Boot 1.0～2.0 中稳定为如下内容：

```
@Target({ ElementType.TYPE, ElementType.METHOD })
@Retention(RetentionPolicy.RUNTIME)
@Documented
@Conditional(OnClassCondition.class)
public @interface ConditionalOnClass {

 /**
  * The classes that must be present. Since this annotation is parsed by loading class
  * bytecode, it is safe to specify classes here that may ultimately not be on the
  * classpath, only if this annotation is directly on the affected component and
  * <b>not</b> if this annotation is used as a composed, meta-annotation. In order to
  * use this annotation as a meta-annotation, only use the {@link #name} attribute.
  * @return the classes that must be present
  */
 Class<?>[] value() default {};

 /**
  * The classes names that must be present.
  * @return the class names that must be present.
  */
 String[] name() default {};

}
```

@ConditionalOnMissingClass 的定义从 Spring Boot 1.4 开始才保持稳定：

```
@Target({ ElementType.TYPE, ElementType.METHOD })
@Retention(RetentionPolicy.RUNTIME)
@Documented
@Conditional(OnClassCondition.class)
public @interface ConditionalOnMissingClass {

 /**
  * The names of the classes that must not be present.
  * @return the names of the classes that must not be present
  */
 String[] value() default {};
```

```
}
```

@ConditionalOnMissingClass 在 Spring Boot 1.1 到 1.2 中声明为：

```
@Target({ ElementType.TYPE, ElementType.METHOD })
@Retention(RetentionPolicy.RUNTIME)
@Documented
@Conditional(OnClassCondition.class)
public @interface ConditionalOnMissingClass {

 /**
  * The classes that must not be present. Since this annotation parsed by loading class
  * bytecode it is safe to specify classes here that may ultimately not be on the
  * classpath.
  * @return the classes that must be present
  * @deprecated Since 1.1.0 due to the fact that the reflection errors can occur when
  * beans containing the annotation remain in the context. Use {@link #name()} instead.
  */
 @Deprecated
 public Class<?>[] value() default {};

 /**
  * The classes names that must not be present.
  * @return the class names that must be present.
  */
 public String[] name() default {};

}
```

当时属性方法 value()的类型为 Class[]，被标记为@Deprecated，并且建议使用 name()。而该定义到了 Spring Boot 1.3 中则是：

```
@Target({ ElementType.TYPE, ElementType.METHOD })
@Retention(RetentionPolicy.RUNTIME)
@Documented
@Conditional(OnClassCondition.class)
public @interface ConditionalOnMissingClass {
```

```
    /**
     * The names of the classes that must not be present.
     * @return the names of the classes that must not be present
     */
    String[] value() default {};

    /**
     * An alias for {@link #value} specifying the names of the classes that must not be
     * present.
     * @return the class names that must not be present.
     * @deprecated since 1.3.0 in favor of {@link #value}.
     */
    String[] name() default {};

}
```

当时属性方法 `name()`还暂时存在，不过其 JavaDoc 中通知开发人员不推荐使用该属性，而使用 `value()`替代：

> @deprecated since 1.3.0 in favor of {@link #value}.

而此时 `value()`的属性方法类型从 `Class[]`调整为 `String[]`。

`@ConditionalOnMissingClass` 从 Spring Boot 1.4 开始进行破坏性升级，说明之前的设计没有思考周全。当指定类不存在时，并不需要该类显式地依赖到当前工程或 Starter。然而其结果则让构建于 Spring Boot 1.4 之前的 Starter 非常麻烦，让开发人员感到 `value()`类型的变化莫测，以及 `name()`存在的不确定性。这类破坏性升级在 Spring Boot 中较为常见，尤其在 Spring Boot 2.0 中的表现可谓“罄竹难书”。因此，开发人员合理地使用`@ConditionalOnClass` 和`@ConditionalOnMissingClass` 特别关键，比如在 Spring Boot 1.5 之前，`ServletRegistrationBean` 所在的 package 为 `org.springframework.boot.context.embedded`，从 Spring Boot 1.4 开始，该类的 package 调整为`org.springframework.boot.web.servlet`。由于 Spring Boot 1.4 采取 `org.springframework.boot.context.embedded.ServletRegistrationBean` 和 `org.springframework.boot.web.servlet.ServletRegistrationBean` 并存的措施，所以前者作为 Servlet 注册 Bean 是没有问题的。不过当 Spring Boot 应用升级到 1.5 或更高版本后，应用在自动装配过程中会抛出 `ClassNotFoundException` 的异常，出于保险起见，可以考虑`@ConditionalOnClass` 和 `@ConditionalOnMissingClass` 配合使用，比如以下实现：

```
@Configuration
```

```
    public class ServletRegistrationAutoConfiguration {

        @ConditionalOnClass(name =
"org.springframework.boot.context.embedded.ServletRegistrationBean")
        @Bean
        public org.springframework.boot.context.embedded.ServletRegistrationBean
servletRegistrationBean() {
            return new org.springframework.boot.context.embedded.ServletRegistrationBean(new
NoOpServlet(), "/no-op");
        }


@ConditionalOnMissingClass("org.springframework.boot.context.embedded.ServletRegistrationBean
")
        @Bean
        public ServletRegistrationBean defaultServletRegistrationBean() {
            return new ServletRegistrationBean(new NoOpServlet(), "/no-op");
        }

        private static class NoOpServlet extends HttpServlet {
        }
    }
```

> 源码位置：以上示例代码可通过查找 spring-boot-1.x-samples/spring-boot-1.4.x-project 工程获取。

如果以上代码出现在目标应用工程中，那么当 org.springframework.boot.context.embedded.ServletRegistrationBean 不存在时，当前工程编译无法通过，开发人员感知错误的存在，可以将错误定义删除。如果代码出现在 Starter 中，则同样当 org.springframework.boot.context.embedded.ServletRegistrationBean 不存在时，其关联的@Bean 定义不再工作，并且也不会遇到编译问题。同理，在参考 Spring Boot 内建 Starter 源码时，IDE 环境会提示一些类无法找到等错误。所以，合理地运用@ConditionalOnClass 和@ConditionalOnMissingClass 至关重要。

有趣的地方在于，@ConditionalOnClass 和@ConditionalOnMissingClass 均使用@Conditional(OnClassCondition.class)实现：

```
class OnClassCondition extends SpringBootCondition {

    @Override
    public ConditionOutcome getMatchOutcome(ConditionContext context,
            AnnotatedTypeMetadata metadata) {

        StringBuffer matchMessage = new StringBuffer();

        MultiValueMap<String, Object> onClasses = getAttributes(metadata,
                ConditionalOnClass.class);
        if (onClasses != null) {
            List<String> missing = getMatchingClasses(onClasses, MatchType.MISSING,
                    context);
            if (!missing.isEmpty()) {
                return ConditionOutcome
                        .noMatch("required @ConditionalOnClass classes not found: "
                                +
StringUtils.collectionToCommaDelimitedString(missing));
            }
            ...
        }

        MultiValueMap<String, Object> onMissingClasses = getAttributes(metadata,
                ConditionalOnMissingClass.class);
        if (onMissingClasses != null) {
            List<String> present = getMatchingClasses(onMissingClasses,
                    MatchType.PRESENT, context);
            if (!present.isEmpty()) {
                return ConditionOutcome
                        .noMatch("required @ConditionalOnMissing classes found: "
                                +
StringUtils.collectionToCommaDelimitedString(present));
            }
            ...
        }

        return ConditionOutcome.match(matchMessage.toString());
    }
```

```
    ...
}
```

这种实现方式可能是作者个人的偏好，后面讨论的 Bean 条件注解@ConditionalOnBean 和@ConditionalOnMissingBean 也是如此。

整合 Class 条件注解——重构 formatter-spring-boot-starter

在 formatter-spring-boot-starter 的实现中，FormatterAutoConfiguration 声明了一个名为“defaultFormatter”的 DefaultFormatter Bean，该 Bean 运用 String#valueOf(Object) 方法实现对象的格式化。然而该实现已不再适合复杂对象格式化的需求，FormatterAutoConfiguration 需要增加一个输出 JSON 格式的 Formatter 实现——JsonFormatter，并需要依赖三方库 Jackson 作为 JSON 序列化的工具。故实现如下。

- 增加 Jackson 依赖到 pom.xml，保持<optional>为 true

```
...
    <dependencies>
      ...
        <!-- Jackson 依赖 -->
        <dependency>
            <groupId>com.fasterxml.jackson.core</groupId>
            <artifactId>jackson-databind</artifactId>
          <optional>true</optional>
        </dependency>

    </dependencies>
...
```

- 新增 JSON Formatter 实现——JsonFormatter

```
public class JsonFormatter implements Formatter {

    private final ObjectMapper objectMapper;

    public JsonFormatter() {
        this.objectMapper = new ObjectMapper();
    }

    @Override
    public String format(Object object) {
```

```
        try {
            return objectMapper.writeValueAsString(object);
        } catch (JsonProcessingException e) {
            // 解析失败返回非法参数异常
            throw new IllegalArgumentException(e);
        }
    }
}
```

> JsonFormatter 的单元测试 JsonFormatterTest 存放在 spring-boot-2.0-samples/formatter-spring-boot-starter 工程下。

- 新增 JsonFormatter Bean 声明到 FormatterAutoConfiguration

```
public class FormatterAutoConfiguration {

    /**
     * 构建 {@link DefaultFormatter} Bean
     *
     * @return {@link DefaultFormatter}
     */
    @Bean
    public Formatter defaultFormatter() {
        return new DefaultFormatter();
    }

    /**
     * JSON 格式 {@link Formatter} Bean
     *
     * @return {@link JsonFormatter}
     */
    @Bean
    public Formatter jsonFormatter() {
        return new JsonFormatter();
    }
}
```

明显地，在同一个 Spring 应用上下文中出现了两个 Formatter Bean，即 DefaultFormatter 和 JsonFormatter，为避免出现 Formatter Bean 唯一性冲突，故有以下步骤。

- Formatter Bean 整合 Class 条件注解

两个 Formatter Bean 可使用互斥的 Class 条件注解@ConditionalOnMissingClass 和 @ConditionalOnClass：

```
public class FormatterAutoConfiguration {

    /**
     * 构建 {@link DefaultFormatter} Bean
     *
     * @return {@link DefaultFormatter}
     */
    @Bean
    @ConditionalOnMissingClass(value = "com.fasterxml.jackson.databind.ObjectMapper")
    public Formatter defaultFormatter() {
        return new DefaultFormatter();
    }

    /**
     * JSON 格式 {@link Formatter} Bean
     *
     * @return {@link JsonFormatter}
     */
    @Bean
    @ConditionalOnClass(name = "com.fasterxml.jackson.databind.ObjectMapper")
    public Formatter jsonFormatter() {
        return new JsonFormatter();
    }
}
```

> 源码位置：以上示例代码可通过查找 spring-boot-2.0-samples/formatter-spring-boot-starter 工程获取。

由于@ConditionalOnMissingClass 在 Spring Boot 2.0 中仅保留 value()属性方法，为提高代码的可读性，建议使用@ConditionalOnClass#name()与其对应。

> @ConditionalOnClass#value()属性方法提供“类型安全”的保障，避免在开发过程中出现全类名拼写的低级失误。通常，@ConditionalOnClass#value()属性方法用于类的物理位置非常稳定的情况，如 Servlet API javax.servlet.Servlet，以及 Java 9 模块化使用场景，如模块“java.management”中的 java.lang.management.ManagementFactory。而

> @ConditionalOnClass#name() 则多用于三方库或高低版本兼容的场景，如 WebMvcAutoConfiguration 在 Spring Boot 1.x 的全类名为 org.springframework.boot.autoconfigure.web.WebMvcAutoConfiguration，而在 2.0 版本中则变为 org.springframework.boot.autoconfigure.web.servlet.WebMvcAutoConfiguration，所以 @ConditionalOnClass#name()更合适。

目前尚不清楚引入 formatter-spring-boot-starter 的 Spring Boot 应用的 Class Path 是否存在 com.fasterxml.jackson.databind.ObjectMapper。以引导类 FormatterBootstrap 为例，需要将其微调，输出 Formatter Bean 的类名，如下所示。

- 微调 FormatterBootstrap

```
@EnableAutoConfiguration
public class FormatterBootstrap {

    public static void main(String[] args) {
        ConfigurableApplicationContext context = new
SpringApplicationBuilder(FormatterBootstrap.class)
                .web(WebApplicationType.NONE)  // 非 Web 应用
                .run(args);                    // 运行
        // 待格式化对象
        Map<String, Object> data = new HashMap<>();
        data.put("name", "小马哥");
        // 获取 Formatter，来自 FormatterAutoConfiguration
        Formatter formatter = context.getBean(Formatter.class);
        System.out.printf("%s.format(data) : %s\n", formatter.getClass().getSimpleName(),
formatter.format(data));
        //  关闭当前上下文
        context.close();
    }
}
```

> 源码位置：以上示例代码可通过查找 spring-boot-2.0-samples/auto-configuration-sample 工程获取。

运行该引导类，观察输出结果：

```
(...部分内容被省略...)
DefaultFormatter.format(data) : {name=小马哥}
(...部分内容被省略...)
```

从运行日志分析，DefaultFormatter 仍旧选为格式化组件 Bean，说明 FormatterBootstrap 所在的 **auto-configuration-sample** 工程并没有依赖 Jackson 的第三方库。换言之，由于 formatter-spring-boot-starter 声明<optional>true</optional>的缘故，com.fasterxml.jackson.core:jackson-databind 的 Maven 依赖并没有传递到当前工程，故需要将该依赖增加至 pom.xml 文件中：

```
...
<dependencies>
    ...
        <!-- Jackson 依赖 -->
        <dependency>
            <groupId>com.fasterxml.jackson.core</groupId>
            <artifactId>jackson-databind</artifactId>
        </dependency>
</dependencies>
...
```

由于 **auto-configuration-sample** 是应用工程，无须再声明<optional>true</optional>，重启 FormatterBootstrap，观察日志输出的变化：

```
(...部分内容被省略...)
org.springframework.boot.autoconfigure.jackson.JacksonAutoConfiguration
(...部分内容被省略...)
JsonFormatter.format(data) : {"name":"小马哥"}
(...部分内容被省略...)
```

格式化组件 Bean 由 DefaultFormatter 替换为 JsonFormatter，输出格式也随之调整。除此之外，由于工程引入 com.fasterxml.jackson.core:jackson-databind 的缘故，所以 JacksonAutoConfiguration 也被自动装配，因为它也是基于@ConditionalOnClass 的条件化自动装配实现的：

```
@Configuration
@ConditionalOnClass(ObjectMapper.class)
public class JacksonAutoConfiguration {
 ...
 @Configuration
 @ConditionalOnClass(Jackson2ObjectMapperBuilder.class)
 static class JacksonObjectMapperConfiguration {

     @Bean
```

```
        @Primary
        @ConditionalOnMissingBean
        public ObjectMapper jacksonObjectMapper(Jackson2ObjectMapperBuilder builder) {
            return builder.createXmlMapper(false).build();
        }

    }
    ...
}
```

同时 ObjectMapper Bean 也被自动装配，那么它是否能够被 JsonFormatter Bean 复用呢？下一节将继续讨论“Bean 条件注解”的相关细节，以及如何与 formatter-spring-boot-starter 整合的内容。

9.4.2 Bean 条件注解

Bean 条件注解也是成对出现的。相对于 Class 注解，@ConditionalOnBean 和 @ConditionalOnMissingBean 的复杂程度不可等同视之。其中，@ConditionalOnBean 的 JavaDoc 部分说明了该条件注解的判断标准：

> Conditional that only matches when the specified bean classes and/or names are already contained in the BeanFactory. The condition can only match the bean definitions that have been processed by the application context so far and, as such, it is strongly recommended to use this condition on auto-configuration classes only

以上文字翻译成程序语言是，@ConditionalOnBean 仅匹配 BeanFactory 中 Bean 的类型和名字，而在实现层面的 Condition 仅匹配应用上下文中已处理的 BeanDefinition。因此 ConditionalOnMissingBean 是相反的逻辑。所以两者属性方法签名基本保持同步，如下表所示。

属性方法	属性类型	语义说明	使用场景	起始版本
value()	Class[]	Bean 类型集合	类型安全的属性设置	1.0
type()	String[]	Bean 类名集合	当类型不存在时的属性设置	1.3
annotation()	Class[]	Bean 声明注解类型集合	当 Bean 标注了某种注解类型时	1.0
name()	String[]	Bean 名称集合	指定具体 Bean 名称集合	1.0
search()	SearchStrategy	层次性应用上下文搜索策略	三种应用上下文搜索策略：当前、父及所有	1.0

从 Spring Boot 1.2.5 开始，@ConditionalOnMissingBean 引入了两个新的属性方法：

```
@Target({ ElementType.TYPE, ElementType.METHOD })
@Retention(RetentionPolicy.RUNTIME)
@Documented
@Conditional(OnBeanCondition.class)
public @interface ConditionalOnMissingBean {
...
/**
 * The class type of beans that should be ignored when identifying matching beans.
 * @return the class types of beans to ignore
 * @since 1.2.5
 */
Class<?>[] ignored() default {};

/**
 * The class type names of beans that should be ignored when identifying matching
 * beans.
 * @return the class type names of beans to ignore
 * @since 1.2.5
 */
String[] ignoredType() default {};
...
}
```

其中，属性方法 ignored()用于 Bean 类型的忽略或排除，而 ignoredType()则用于忽略或排除指定 Bean 名称。显然，两者不会孤立存在，需要配合如属性方法 value()或 annotation()提供细粒度忽略匹配。

与 Class 条件注解类似，@ConditionalOnBean 和 ConditionalOnMissingBean 同样采用单 Condition 实现处理语义对立条件注解。其 Condition 实现类为 OnBeanCondition，该类同样扩展了抽象类 SpringBootCondition：

```
@Order(Ordered.LOWEST_PRECEDENCE)
class OnBeanCondition extends SpringBootCondition implements ConfigurationCondition {
...
    @Override
public ConditionOutcome getMatchOutcome(ConditionContext context,
```

```
            AnnotatedTypeMetadata metadata) {
        ConditionMessage matchMessage = ConditionMessage.empty();
        if (metadata.isAnnotated(ConditionalOnBean.class.getName())) {
            BeanSearchSpec spec = new BeanSearchSpec(context, metadata,
                    ConditionalOnBean.class);
            MatchResult matchResult = getMatchingBeans(context, spec);
            if (!matchResult.isAllMatched()) {
                String reason = createOnBeanNoMatchReason(matchResult);
                return ConditionOutcome.noMatch(ConditionMessage
                        .forCondition(ConditionalOnBean.class, spec).because(reason));
            }
            matchMessage =
                    matchMessage.andCondition(ConditionalOnBean.class, spec)
                    .found("bean", "beans")
                    .items(Style.QUOTE,
                    matchResult.getNamesOfAllMatches());
        }
        ...
        return ConditionOutcome.match(matchMessage);
    }

    ...
}
```

getMatchOutcome(ConditionContext,AnnotatedTypeMetadata)方法的实现相当复杂，除了处理以上两个 Bean 条件注解，@ConditionalOnSingleCandidate 也被纳入其中。这三个注解基本上处理逻辑类似，主要的匹配结果由getMatchingBeans(Condition Context,BeanSearchSpec)方法决定：

```
@Order(Ordered.LOWEST_PRECEDENCE)
class OnBeanCondition extends SpringBootCondition implements ConfigurationCondition {
...
private MatchResult getMatchingBeans(ConditionContext context, BeanSearchSpec beans) {
    ConfigurableListableBeanFactory beanFactory = context.getBeanFactory();
    if (beans.getStrategy() == SearchStrategy.ANCESTORS) {
        BeanFactory parent = beanFactory.getParentBeanFactory();
        Assert.isInstanceOf(ConfigurableListableBeanFactory.class, parent,
                "Unable to use SearchStrategy.PARENTS");
```

```
            beanFactory = (ConfigurableListableBeanFactory) parent;
        }
        MatchResult matchResult = new MatchResult();
        boolean considerHierarchy = beans.getStrategy() != SearchStrategy.CURRENT;
        List<String> beansIgnoredByType = getNamesOfBeansIgnoredByType(
                beans.getIgnoredTypes(), beanFactory, context, considerHierarchy);
        for (String type : beans.getTypes()) {
            Collection<String> typeMatches =
                    getBeanNamesForType(beanFactory, type,
                    context.getClassLoader(), considerHierarchy);
            typeMatches.removeAll(beansIgnoredByType);
            if (typeMatches.isEmpty()) {
                matchResult.recordUnmatchedType(type);
            }
            else {
                matchResult.recordMatchedType(type, typeMatches);
            }
        }
        ...
        return matchResult;
    }
    ...
}
```

方法参数 BeanSearchSpec 是 Bean 条件注解的包装对象，当 SearchStrategy 为 SearchStrategy.ANCESTORS时，beanFactory切换为其Parent ConfigurableListableBeanFactory。其中，getNamesOfBeansIgnoredByType(List,ListableBeanFactory,ConditionContext,boolean)方法计算排除的 Bean 名称，后续被 typeMatches 排除，而核心的匹配逻辑在 getBeanNamesForType(ListableBeanFactory,String,ClassLoader,boolean)方法中完成：

```
@Order(Ordered.LOWEST_PRECEDENCE)
class OnBeanCondition extends SpringBootCondition implements ConfigurationCondition {
 ...
 private Collection<String> getBeanNamesForType(ListableBeanFactory beanFactory,
         String type, ClassLoader classLoader, boolean considerHierarchy)
         throws LinkageError {
     try {
         Set<String> result = new LinkedHashSet<>();
```

```java
            collectBeanNamesForType(result, beanFactory,
                    ClassUtils.forName(type, classLoader), considerHierarchy);
            return result;
        }
        catch (ClassNotFoundException | NoClassDefFoundError ex) {
            return Collections.emptySet();
        }
    }

    private void collectBeanNamesForType(Set<String> result,
            ListableBeanFactory beanFactory, Class<?> type, boolean considerHierarchy) {
        result.addAll(BeanTypeRegistry.get(beanFactory).getNamesForType(type));
        if (considerHierarchy && beanFactory instanceof HierarchicalBeanFactory) {
            BeanFactory parent = ((HierarchicalBeanFactory) beanFactory)
                    .getParentBeanFactory();
            if (parent instanceof ListableBeanFactory) {
                collectBeanNamesForType(result, (ListableBeanFactory) parent, type,
                        considerHierarchy);
            }
        }
    }
    ...
}
```

而 `getBeanNamesForType(ListableBeanFactory,String,ClassLoader,boolean)`方法的计算结果由 `collectBeanNamesForType(Set,ListableBeanFactory, Class,boolean)`方法完成。当前 Bean 搜索策略不为当前上下文搜索时，即执行 `boolean considerHierarchy = beans.getStrategy() != SearchStrategy.CURRENT` 语句后，`considerHierarchy` 等于 `true`，`collectBeanNamesForType(Set,Listable BeanFactory,Class,boolean)`方法，将递归地调用收集 Bean 名称集合。其中当前上下文 Bean 名称集合由 `BeanTypeRegistry#getNamesForType(Claass)`方法获取：

```java
final class BeanTypeRegistry implements SmartInitializingSingleton {
...
Set<String> getNamesForType(Class<?> type) {
    updateTypesIfNecessary();
    return this.beanTypes.entrySet().stream()
            .filter((entry) -> entry.getValue() != null
```

```
                    && type.isAssignableFrom(entry.getValue()))
                .map(Map.Entry::getKey)
                .collect(Collectors.toCollection(LinkedHashSet::new));
    }
    ...
    private void updateTypesIfNecessary() {
        if (this.lastBeanDefinitionCount != this.beanFactory.getBeanDefinitionCount()) {
            Iterator<String> names = this.beanFactory.getBeanNamesIterator();
            while (names.hasNext()) {
                String name = names.next();
                if (!this.beanTypes.containsKey(name)) {
                    addBeanType(name);
                }
            }
            this.lastBeanDefinitionCount = this.beanFactory.getBeanDefinitionCount();
        }
    }
    ...
    private void addBeanType(String name) {
        if (this.beanFactory.containsSingleton(name)) {
            this.beanTypes.put(name, this.beanFactory.getType(name));
        }
        else if (!this.beanFactory.isAlias(name)) {
            addBeanTypeForNonAliasDefinition(name);
        }
    }
    ...
    }
```

根据 getNamesForType(Claass)方法的调用链路，首先执行 updateTypesIfNecessary()方法，其中 addBeanType(String)方法将更新 beanTypes 的内容，其方法参数 name 作为 Key，Value 为 BeanFactory#getType(String)方法的执行结果：

```
    public abstract class AbstractBeanFactory extends FactoryBeanRegistrySupport implements
ConfigurableBeanFactory {
    ...
    @Override
    @Nullable
```

```
    public Class<?> getType(String name) throws NoSuchBeanDefinitionException {
        String beanName = transformedBeanName(name);

        // Check manually registered singletons.
        Object beanInstance = getSingleton(beanName, false);
        if (beanInstance != null && beanInstance.getClass() != NullBean.class) {
            if (beanInstance instanceof FactoryBean
&& !BeanFactoryUtils.isFactoryDereference(name)) {
                return getTypeForFactoryBean((FactoryBean<?>) beanInstance);
            }
            else {
                return beanInstance.getClass();
            }
        }

        // No singleton instance found -> check bean definition.
        BeanFactory parentBeanFactory = getParentBeanFactory();
        if (parentBeanFactory != null && !containsBeanDefinition(beanName)) {
            // No bean definition found in this factory -> delegate to parent.
            return parentBeanFactory.getType(originalBeanName(name));
        }

        RootBeanDefinition mbd = getMergedLocalBeanDefinition(beanName);

        // Check decorated bean definition, if any: We assume it'll be easier
        // to determine the decorated bean's type than the proxy's type.
        BeanDefinitionHolder dbd = mbd.getDecoratedDefinition();
        if (dbd != null && !BeanFactoryUtils.isFactoryDereference(name)) {
            RootBeanDefinition tbd =
            getMergedBeanDefinition(dbd.getBeanName(), dbd.getBeanDefinition(), mbd);
            Class<?> targetClass = predictBeanType(dbd.getBeanName(), tbd);
            if (targetClass != null
            && !FactoryBean.class.isAssignableFrom(targetClass)) {
                return targetClass;
            }
        }

        Class<?> beanClass = predictBeanType(beanName, mbd);
```

```
        // Check bean class whether we're dealing with a FactoryBean.
        if (beanClass != null &&
            FactoryBean.class.isAssignableFrom(beanClass)) {
            if (!BeanFactoryUtils.isFactoryDereference(name)) {
                // If it's a FactoryBean, we want to look at what it creates, not at the factory
class.
                return getTypeForFactoryBean(beanName, mbd);
            }
            else {
                return beanClass;
            }
        }
        else {
            return (!BeanFactoryUtils.isFactoryDereference(name) ? beanClass : null);
        }
    }
    ...
    }
```

该方法首先执行 getSingleton(String,boolean)方法，Bean 条件装配 OnBeanCondition 实现了 ConfigurationCondition 接口：

```
 */
@Order(Ordered.LOWEST_PRECEDENCE)
class OnBeanCondition extends SpringBootCondition implements ConfigurationCondition {
...
@Override
public ConfigurationPhase getConfigurationPhase() {
    return ConfigurationPhase.REGISTER_BEAN;
}
...
}
```

> ConfigurationPhase 用于 ConditionEvaluator#shouldSkip(AnnotatedTypeMetadata, ConfigurationPhase)方法评估，请读者自行回顾“@Conditional 条件装配原理”一节。

该实现指示 ConditionEvaluator 在注册 Bean 阶段（ConfigurationPhase#REGISTER_BEAN）进行评估。因此，@ConditionalOnBean 和@ConditionalOnMissingBean 的 JavaDoc 强烈建议开

发人员仅在自动装配中使用该条件注解：

> it is strongly recommended to use this condition on auto-configuration classes only.

从而确保 `getSingleton(String,boolean)`方法返回 `null`，不参与 Bean 类型的计算。随后，通过 Bean 名称获取 `RootBeanDefinition`，再从 `RootBeanDefinition` 中计算 Bean 类型，于是与 `@ConditionalOnBean`、`@ConditionalOnMissingBean` 的 JavaDoc 声明遥相呼应：

> The condition can only match the bean definitions that have been processed by the application context

简言之，`@ConditionalOnBean` 和`@ConditionalOnMissingBean` 基于 `BeanDefinition` 进行名称或类型的匹配。

整合 Bean 条件注解——重构 formatter-spring-boot-starter

在"整合 Class 条件注解——`formatter-spring-boot-starter`"一节中，曾讨论 `ObjectMapper` Bean 会被 `JacksonAutoConfiguration` 自动装配，不过存在一定的前提：

```
@Configuration
@ConditionalOnClass(ObjectMapper.class)
public class JacksonAutoConfiguration {
...
@Configuration
@ConditionalOnClass(Jackson2ObjectMapperBuilder.class)
static class JacksonObjectMapperConfiguration {

    @Bean
    @Primary
    @ConditionalOnMissingBean
    public ObjectMapper jacksonObjectMapper(Jackson2ObjectMapperBuilder builder) {
        return builder.createXmlMapper(false).build();
    }

}
...
}
```

结合目前已掌握的知识，能够得出 `ObjectMapper` Bean 的初始化需要满足以下条件：

（1）ObjectMapper 必须存在于 Class Path 中。

（2）Jackson2ObjectMapperBuilder 必须存在于 Class Path 中。

（3）ObjectMapper Bean 必须在所有的 Spring 应用上下文中不存在。

前两个条件容易满足，比如 **auto-configuration-sample** 工程已增加 com.fasterxml.jackson.core:jackson-databind，因此 ObjectMapper 必然存在。而 Jackson2ObjectMapperBuilder 源于 org.springframework:spring-web 4.1.1，工程只需依赖 org.springframework.boot:spring-boot-starter-web 1.2.0 及以上版本即可，而当前工程的 Spring Boot 版本为 2.0.2，故直接添加该依赖就能工作。

条件 3 则相对复杂，因为它需要判断当前应用所有的 Spring 应用上下文，先不论所有层次性，即使是单一层次，假设其他自动装配 Class 声明了 ObjectMapper Bean，那么 JacksonAutoConfiguration 将不会自动装配 ObjectMapper Bean，因此 FormatterAutoConfiguration 不能声明该 Bean，完整的重构步骤如下。

- 增加 Jackson2ObjectMapperBuilder Maven 依赖

将 org.springframework.boot:spring-boot-starter-web 添加到 **auto-configuration-sample** 工程的 pom.xml 文件中：

```
<dependencies>
    ...
    <!-- Spring Boot Web 依赖 -->
    <dependency>
        <groupId>org.springframework.boot</groupId>
        <artifactId>spring-boot-starter-web</artifactId>
    </dependency>
    ...
</dependencies>
```

- 增加 JsonFormatter 构造器

在 JsonFormatter 中增加一个带 ObjectMapper 参数的构造器，以便外部注入 ObjectMapper Bean：

```
public class JsonFormatter implements Formatter {

    private final ObjectMapper objectMapper;

    public JsonFormatter() {
        this(new ObjectMapper());
```

```
    }

    public JsonFormatter(ObjectMapper objectMapper) {
        this.objectMapper = objectMapper;
    }
...
}
```

- 新增 JsonFormatter Bean

在 FormatterAutoConfiguration 中添加新的 JsonFormatter Bean 声明，该 Bean 依赖于可能存在的 ObjectMapper Bean：

```
@Configuration
public class FormatterAutoConfiguration {
...
    @Bean
    @ConditionalOnBean(ObjectMapper.class)
    public Formatter objectMapperFormatter(ObjectMapper objectMapper) {
        return new JsonFormatter(objectMapper);
    }
}
```

- 调整 jsonFormatter() Bean 的声明方式

由于新增 objectMapperFormatter Bean 的存在，故 jsonFormatter Bean 应调整为互斥声明方式：

```
@Configuration
public class FormatterAutoConfiguration {
...
    @Bean
    @ConditionalOnClass(name = "com.fasterxml.jackson.databind.ObjectMapper")
    @ConditionalOnMissingBean(type = "com.fasterxml.jackson.databind.ObjectMapper")
    public Formatter jsonFormatter() {
        return new JsonFormatter();
    }

    @Bean
    @ConditionalOnBean(ObjectMapper.class)
    public Formatter objectMapperFormatter(ObjectMapper objectMapper) {
```

```
        return new JsonFormatter(objectMapper);
    }
}
```

目前，`Formatter` Bean 装配条件的情况变得复杂起来，分为以下三种情况：

- 当 `ObjectMapper` Class 不存在时，Bean 为 `DefaultFormatter` 实例，其名称为“defaultFormatter”；
- 当 `ObjectMapper` Class 存在且其 Bean 不存在时，Bean 为 `JsonFormatter` 默认构造器创建 `ObjectMapper` 实例，其名称为“jsonFormatter”；
- 当 `ObjectMapper` Class 存在且其 Bean 也存在时，Bean 为 `JsonFormatter` 构造器注入 `ObjectMapper` Bean，其名称为“objectMapperFormatter”。

为了检验当前引导类 `FormatterBootstrap` 具体使用哪个 Bean，需要做出以下调整。

- 调整 `FormatterBootstrap` 输出 `Formatter` Bean 名称

```
@EnableAutoConfiguration
public class FormatterBootstrap {

    public static void main(String[] args) {
        ConfigurableApplicationContext context = new
SpringApplicationBuilder(FormatterBootstrap.class)
                .web(WebApplicationType.NONE)  // 非 Web 应用
                .run(args);                    // 运行
        // 待格式化对象
        Map<String, Object> data = new HashMap<>();
        data.put("name", "小马哥");
        // 获取 Formatter，来自 FormatterAutoConfiguration
        Map<String, Formatter> beans = context.getBeansOfType(Formatter.class);
        if (beans.isEmpty()) { // 如果 Bean 不存在，则抛出异常
            throw new NoSuchBeanDefinitionException(Formatter.class);
        }
        beans.forEach((beanName, formatter) -> {
            System.out.printf("[Bean name : %s] %s.format(data) : %s\n", beanName,
formatter.getClass().getSimpleName(),
                    formatter.format(data));
        });
        //  关闭当前上下文
        context.close();
```

```
    }
}
```

运行 FormatterBootstrap 的结果，观察日志的输出：

```
(...部分内容被省略...)
[Bean name : objectMapperFormatter] JsonFormatter.format(data) : {"name":"小马哥"}
(...部分内容被省略...)
```

该运行结果符合期望，然而 formatter-spring-boot-starter 仅整合 Class 和 Bean 两种条件注解就使自动装配逻辑变得略微复杂。随着条件化自动装配的深入整合，它需要开发人员具备更严谨的思维方式和全局的洞察力，毕竟那时的场景只会更加错综复杂。下面讨论的“属性条件注解”就符合这种情况！

9.4.3 属性条件注解

@ConditionalOnProperty 作为属性条件注解，其属性来源于 Spring Environment。在 Spring Framework 场景中，Java 系统属性和环境变量是典型的 Spring Environment 属性配置来源（PropertySource）。而在 Spring Boot 场景中，application.properties 也是其中来源之一。换言之，Spring Environment 允许多个属性配置来源（PropertySource）并存，详细的讨论请读者翻阅“外部化配置”章节。

@ConditionalOnProperty 由 Spring Boot 1.1 引入，其属性方法声明从 1.2 版本开始开始稳定，其中属性方法 relaxedNames()在 2.0 版本中被删除，其完整的注解声明如下：

> 在 Spring Boot 1.1 之前，属性条件的判断采用注解@ConditionalOnExpression，即“Spring 表达式条件注解”，在后面的同名章节中有详细的讨论。

```
@Retention(RetentionPolicy.RUNTIME)
@Target({ ElementType.TYPE, ElementType.METHOD })
@Documented
@Conditional(OnPropertyCondition.class)
public @interface ConditionalOnProperty {

...
String[] value() default {};

...
```

```
    String prefix() default "";

    ...
    String[] name() default {};

    ...
    String havingValue() default "";

    ...
    boolean matchIfMissing() default false;

}
```

> 以上实现源码源于 Spring Boot 2.0.2.RELEASE。

@ConditionalOnProperty 属性方法的组合情况较为复杂，将其逐一说明，如下表所示。

属性方法	使用说明	默认值	多值属性	起始版本
prefix()	配置属性名称前缀	""	否	1.1
value()	name()的别名，参考 name()说明	空数组	是	1.1
name()	如果 prefix()不为空，则完整配置属性名称为 prefix()+name()，否则为 name()的内容	空数组	是	1.2
havingValue()	表示期望的配置属性值，并且禁止使用 false	""	否	1.2
matchIfMissing()	用于判断当属性值不存在时是否匹配	false	否	1.2

整合属性条件注解——重构 formatter-spring-boot-starter

接下来将@ConditionalOnProperty 整合到 FormatterAutoConfiguration 中，加深对属性条件注解的理解：

```
@Configuration
@ConditionalOnProperty(prefix = "formatter", name = "enabled", havingValue = "true")
public class FormatterAutoConfiguration {
 ...
}
```

> 源码位置：以上示例代码可通过查找 spring-boot-2.0-samples/formatter-spring-boot-starter 工程获取。

以上 `@ConditionalOnProperty` 的语义是，当 Spring `Environment` 的属性 `formatter.enabled` 为“true”时，`FormatterAutoConfiguration` 才会自动装配。否则，`FormatterAutoConfiguration` 和“defaultFormatter” `Formatter` Bean 均不会存在于 Spring 应用上下文中。

重启 spring-boot-2.0-samples/auto-configuration-sample 工程中的引导类 `Formatter Bootstrap`，检验以上结论的正确性：

```
(...部分内容被省略...)
Exception in thread "main" org.springframework.beans.factory.NoSuchBeanDefinitionException:
No qualifying bean of type 'thinking.in.spring.boot.samples.autoconfigure.formatter.Formatter'
available
(...部分内容被省略...)
```

`NoSuchBeanDefinitionException` 异常信息的出现，证明对`@ConditionalOnProperty` 的语义理解的正确性。那么如何修复这个问题，使引导程序运行如初呢？下面提供两种方案供读者参考。

方法一：增加属性配置

既然缺少 `formatter.enabled` 的配置，那么可以直接填补这个缺失。此处可以利用 `SpringApplicationBuilder#properties(String...)` API 来配置默认属性：

```
@EnableAutoConfiguration
public class FormatterBootstrap {

    public static void main(String[] args) {
        ConfigurableApplicationContext context = new
SpringApplicationBuilder(FormatterBootstrap.class)
                .web(WebApplicationType.NONE)  // 非 Web 应用
                .properties("formatter.enabled=true") // 配置默认属性，“=”前后
                                                       // 不能有空格
                .run(args);                     // 运行
        ...
    }
}
```

其中，`.properties("formatter.enabled = true")`语句即为变更的代码部分，为当前

SpringApplication 提供默认属性配置，请注意“=”前后不能有空格，否则会配置失效。

> 关于 SpringApplication 更多的细节，将在“SpringApplication”章节中讨论。

再次运行 FormatterBootstrap，观察日志的变化：

```
(...部分内容被省略...)
[Bean name : objectMapperFormatter] JsonFormatter.format(data) : {"name":"小马哥"}
(...部分内容被省略...)
```

formatter.enabled 属性配置后的 FormatterBootstrap 运行如初。

以上实现通过 SpringApplicationBuilder#properties(String...)API 配置 formatter.enabled 属性，这种方式属于内部化配置，当然 Spring Boot 外部化配置同样能对同名属性进行配置。换言之，FormatterBootstrap 也可以通过 application.properties 文件配置 formatter.enabled 属性：

```
# FormatterAutoConfiguration 属性配置
# 与.properties("formatter.enabled=true")配置相反
formatter.enabled = false
```

重启 FormatterBootstrap，观察运行结果是否存在变化：

```
(...部分内容被省略...)
Exception in thread "main" org.springframework.beans.factory.NoSuchBeanDefinitionException:
No qualifying bean of type 'thinking.in.spring.boot.samples.autoconfigure.formatter.Formatter'
available
(...部分内容被省略...)
```

NoSuchBeanDefinitionException 的异常再次出现，说明 application.properties 文件中的 formatter.enabled 属性覆盖了引导类中的.properties("formatter.enabled=true")的设置。此处是有意为之，其目的无非是想提醒读者，在实践 @ConditionalOnProperty 的过程中，应对 Spring Environment 关联的属性配置源（PropertySource）的优先次序予以高度的关注，在后面的“外部化配置”章节中将有详细的讨论。

假设该引导类不增加属性配置的方式，如何让其运行无误呢？接下来讨论调整 @ConditionalOnProperty 属性的实现方案。

方法二：调整@ConditionalOnProperty#matchIfMissing()属性

从实现成本而言，调整@ConditionalOnProperty#matchIfMissing()属性的方案更低廉，因为接入 Spring Boot 应用不需要做出调整，如下所示。

```
@Configuration
@ConditionalOnProperty(prefix = "formatter", name = "enabled", havingValue = "true",
      matchIfMissing = true) // 当属性配置不存在时，同样视作匹配
public class FormatterAutoConfiguration {
...
}
```

以上代码变更的部分仅增加了属性配置 `matchIfMissing = true`，其语义为当属性 `formatter.enabled` 不存在时，同样视作匹配。因此，即使 **spring-boot-2.0-samples/auto-configuration-sample** 工程中的引导类 `FormatterBootstrap` 不增加配置，同样能正常获取 "defaultFormatter" `Formatter` Bean，并且执行成功。

第一步将`.properties("formatter.enabled=true")`和 `application.properties` 的配置注释，以还原配置前的状态，`FormatterBootstrap.java` 调整如下：

```
@EnableAutoConfiguration
public class FormatterBootstrap {

    public static void main(String[] args) {
        ConfigurableApplicationContext context = new
SpringApplicationBuilder(FormatterBootstrap.class)
                .web(WebApplicationType.NONE)  // 非 Web 应用
                // .properties("formatter.enabled=true") // 配置默认属性，"="
                                                          // 前后不能有空格
                .run(args);                     // 运行
        // 待格式化对象
        Map<String, Object> data = new HashMap<>();
        data.put("name", "小马哥");
        // 获取 Formatter 来自 FormatterAutoConfiguration
        Formatter formatter = context.getBean(Formatter.class);
        System.out.printf("formatter.format(data) : %s\n", formatter.format(data));
        //  关闭当前上下文
        context.close();
    }
}
```

`application.properties` 调整如下：

```
# formatter.enabled = false
```

再次运行 FormatterBootstrap，观察日志变化：

```
(...部分内容被省略...)
[Bean name : objectMapperFormatter] JsonFormatter.format(data) : {"name":"小马哥"}
(...部分内容被省略...)
```

运行结果符合预期。再将 application.properties 文件中的配置还原如下：

```
formatter.enabled = false
```

重启 FormatterBootstrap，其运行结果再次出现异常：

```
(...部分内容被省略...)
Exception in thread "main" org.springframework.beans.factory.NoSuchBeanDefinitionException: No qualifying bean of type 'thinking.in.spring.boot.samples.autoconfigure.formatter.Formatter' available
(...部分内容被省略...)
```

因此，此时 formatter.enabled 为 false，无法与期望值 true 匹配，故条件不成立，装配动作不再执行。

通常，@ConditionalOnProperty 注解作为 Spring Boot 自动装配组件的属性条件开关，当自动装配组件需要默认装配时，不妨将 matchIfMissing() 属性值调整为 true，这样能够减少 Spring Boot 应用接入的配置成本，尤其在 Spring Boot Starter 中效果明显。当应用需要关闭其组件装配时，可以通过属性配置进行调整。这种实现方式在 Spring Boot 内建的自动装配组件中尤为常见，比如 JMX 自动装配：

```
@Configuration
@ConditionalOnClass({ MBeanExporter.class })
@ConditionalOnProperty(prefix = "spring.jmx", name = "enabled", havingValue = "true", matchIfMissing = true)
public class JmxAutoConfiguration implements EnvironmentAware, BeanFactoryAware {
 ...
}
```

当开发人员配置属性 spring.jmx.enabled =false 时，JmxAutoConfiguration 自动装配失效。类似的内建自动装配 Class 有 AopAutoConfiguration、ElasticsearchAutoConfiguration 等。

再次将 application.properties 文件中的配置 formatter.enabled 注释，使其能够正常工作，接下来继续讨论并整合 Resource 条件注解。

9.4.4 Resource 条件注解

按照 Conditional 条件注解的实现规律，@ConditionalOnResource 的 Condition 实现为 OnResourceCondition：

```
@Target({ ElementType.TYPE, ElementType.METHOD })
@Retention(RetentionPolicy.RUNTIME)
@Documented
@Conditional(OnResourceCondition.class)
public @interface ConditionalOnResource {

 /**
  * The resources that must be present.
  * @return the resource paths that must be present.
  */
 String[] resources() default {};

}
```

其中，属性方法 resources()的 JavaDoc 指示只有资源必须存在时条件方可成立，因此，结合 OnResourceCondition 实现加以分析：

```
@Order(Ordered.HIGHEST_PRECEDENCE + 20)
class OnResourceCondition extends SpringBootCondition {

 private final ResourceLoader defaultResourceLoader = new DefaultResourceLoader();

 @Override
 public ConditionOutcome getMatchOutcome(ConditionContext context,
         AnnotatedTypeMetadata metadata) {
     MultiValueMap<String, Object> attributes = metadata
             .getAllAnnotationAttributes(ConditionalOnResource.class.getName(), true);
     ResourceLoader loader = (context.getResourceLoader() != null
             ? context.getResourceLoader() : this.defaultResourceLoader);
     List<String> locations = new ArrayList<>();
     ...
     List<String> missing = new ArrayList<>();
     for (String location : locations) {
```

```
            String resource = context.getEnvironment().resolvePlaceholders(location);
            if (!loader.getResource(resource).exists()) {
                missing.add(location);
            }
        }
        if (!missing.isEmpty()) {
            return ConditionOutcome.noMatch(ConditionMessage
                    .forCondition(ConditionalOnResource.class)
                    .didNotFind("resource", "resources").items(Style.QUOTE, missing));
        }
        return ConditionOutcome
                .match(ConditionMessage.forCondition(ConditionalOnResource.class)
                        .found("location", "locations").items(locations));
    }
    ...
}
```

以上实现逻辑看似并不复杂，大致分为如下步骤：

- 获取@ConditionalOnResource 注解元属性信息 attributes；
- 获取 ResourceLoader 对象 loader；
- 解析@ConditionalOnResource#resources()属性中可能存在的占位符；
- 通过 ResourceLoader 对象 loader 逐一判断解析后的资源位置是否存在；
    - 如果均已存在，则说明条件成立；
    - 否则，条件不成立。

不过，ResourceLoader loader = (context.getResourceLoader() != null ? context.getResourceLoader() : this.defaultResourceLoader)让问题变得复杂了。假设 context.getResourceLoader()不返回 null，那么返回对象具体是哪种 ResourceLoader？已知 Condition 实现均被 ConditionEvaluator#shoudSkip(AnnotatedTypeMetadata, ConfigurationPhase)方法调用，评估条件是否成立：

```
class ConditionEvaluator {

private final ConditionContextImpl context;


/**
```

```
     * Create a new {@link ConditionEvaluator} instance.
     */
    public ConditionEvaluator(BeanDefinitionRegistry registry, Environment environment,
ResourceLoader resourceLoader) {
        this.context = new ConditionContextImpl(registry, environment, resourceLoader);
    }

    public boolean shouldSkip(AnnotatedTypeMetadata metadata, ConfigurationPhase phase) {
        ...
        for (String[] conditionClasses : getConditionClasses(metadata)) {
            for (String conditionClass : conditionClasses) {
                ...
                if (requiredPhase == null || requiredPhase == phase) {
                    if (!condition.matches(context, metadata)) {
                        return true;
                    }
                }
            }
        }

        return false;
    }
    ...
    }
```

其中，Condition#matches(ConditionContext,AnnotatedTypeMetadata)方法的首参关联的 ResourceLoader 在内置类 ConditionEvaluator.ConditionContextImpl 的构造器中完成初始化：

```
    private static class ConditionContextImpl implements ConditionContext {
    ...
    public ConditionContextImpl(BeanDefinitionRegistry registry, Environment environment,
ResourceLoader resourceLoader) {
        ...
        this.resourceLoader = (resourceLoader != null ? resourceLoader :
deduceResourceLoader(registry));
    }
    private ResourceLoader deduceResourceLoader(BeanDefinitionRegistry source) {
            if (source instanceof ResourceLoader) {
                return (ResourceLoader) source;
```

```
            }
            return null;
        }
    ...
}
```

明显地，resourceLoader 属性的来源有两个，一个是来自外部参数传递，另一个是获取 BeanDefinitionRegistry 和 ResourceLoader 双接口实现对象（如果存在）。简言之，ConditionContext#getResourceLoader()的返回值来源于 ConditionEvaluator 构造参数 ResourceLoader 或 BeanDefinitionRegistry。然而构造 ConditionEvaluator 实例的实现分布在四处，以 Spring Framework 5.0.6.RELASE 实现为例：

- AnnotatedBeanDefinitionReader 构造器

```
public class AnnotatedBeanDefinitionReader {
    ...
    public AnnotatedBeanDefinitionReader(BeanDefinitionRegistry registry, Environment environment) {
        ...
        this.conditionEvaluator = new ConditionEvaluator(registry, environment, null);
        ...
    }
    ...
}
```

- AnnotatedBeanDefinitionReader#setEnvironment(Environment) 方法

```
public class AnnotatedBeanDefinitionReader {
    ...
    public void setEnvironment(Environment environment) {
        this.conditionEvaluator = new ConditionEvaluator(this.registry, environment, null);
    }
    ...
}
```

结合分析，AnnotatedBeanDefinitionReader 并没有在 ConditionEvaluator 构造中传递 ResourceLoader 实例。

- ClassPathScanningCandidateComponentProvider#isConditionMatch(MetadataRea der)方法

```
    public class ClassPathScanningCandidateComponentProvider implements EnvironmentCapable,
ResourceLoaderAware {
    ...
    private boolean isConditionMatch(MetadataReader metadataReader) {
        if (this.conditionEvaluator == null) {
            this.conditionEvaluator =
                    new ConditionEvaluator(getRegistry(), this.environment,
this.resourcePatternResolver);
        }
        return !this.conditionEvaluator.shouldSkip(metadataReader.getAnnotationMetadata());
    }
    ...
    @Override
    public void setResourceLoader(@Nullable ResourceLoader resourceLoader) {
        this.resourcePatternResolver =
ResourcePatternUtils.getResourcePatternResolver(resourceLoader);
        ...
    }
    }
```

由于 ClassPathScanningCandidateComponentProvider 实现了 ResourceLoaderAware 接口，当其作为 Spring Bean 时，resourcePatternResolver 字段将被 Spring 应用上下文初始化。如果是开发人员自定义实现，则该字段的赋值情况存在变数。总之，resourcePatternResolver 字段的状态无法确定。

- ConfigurationClassBeanDefinitionReader 构造器

```
    class ConfigurationClassBeanDefinitionReader {
    ...
    ConfigurationClassBeanDefinitionReader(BeanDefinitionRegistry registry, SourceExtractor
sourceExtractor,
            ResourceLoader resourceLoader, Environment environment, BeanNameGenerator
importBeanNameGenerator,
            ImportRegistry importRegistry) {

        ...
        this.resourceLoader = resourceLoader;
```

```
        ...
        this.conditionEvaluator = new ConditionEvaluator(registry, environment,
resourceLoader);
    }
    ...
}
```

在 ConditionEvaluator 的构造过程中，其所需的 ResourceLoader 对象来自 ConfigurationClassBeanDefinitionReader 构造参数，故需要再跟踪其构造地点，即 ConfigurationClassPostProcessor#processConfigBeanDefinitions(BeanDefinitionRegistry)：

```
public class ConfigurationClassPostProcessor implements
BeanDefinitionRegistryPostProcessor,
        PriorityOrdered, ResourceLoaderAware, BeanClassLoaderAware, EnvironmentAware {
    ...
    public void processConfigBeanDefinitions(BeanDefinitionRegistry registry) {
        ...
            // Read the model and create bean definitions based on its content
            if (this.reader == null) {
                this.reader = new ConfigurationClassBeanDefinitionReader(
                        registry, this.sourceExtractor, this.resourceLoader,
this.environment,
                        this.importBeanNameGenerator, parser.getImportRegistry());
            }
        ...
    }
    ...
    @Override
    public void setResourceLoader(ResourceLoader resourceLoader) {
        Assert.notNull(resourceLoader, "ResourceLoader must not be null");
        this.resourceLoader = resourceLoader;
        ...
    }
    ...
}
```

同样 ConfigurationClassPostProcessor 作为 ResourceLoaderAware 的实现类，其 resourceLoader 的初始化来源于覆盖方法 setResourceLoader(ResourceLoader)，又由于 ConfigurationClassPostProcessor 是 Spring Framework 默认的内建 BeanDefinitionRegistry PostProcessor Bean 组件：

```
public class AnnotationConfigUtils {
    ...
    public static Set<BeanDefinitionHolder> registerAnnotationConfigProcessors(
            BeanDefinitionRegistry registry, @Nullable Object source) {
        ...
        if (!registry.containsBeanDefinition(CONFIGURATION_ANNOTATION_PROCESSOR_BEAN_NAME)) {
            RootBeanDefinition def = new
RootBeanDefinition(ConfigurationClassPostProcessor.class);
            def.setSource(source);
            beanDefs.add(registerPostProcessor(registry, def,
CONFIGURATION_ANNOTATION_PROCESSOR_BEAN_NAME));
        }
        ...
        return beanDefs;
    }
    ...
}
```

因此，ConditionEvaluator 关联的 ResourceLoader 来自 Spring 应用上下文。

- ConfigurationClassParser 构造器

```
class ConfigurationClassParser {
    ...
    public ConfigurationClassParser(MetadataReaderFactory metadataReaderFactory,
            ProblemReporter problemReporter, Environment environment, ResourceLoader
resourceLoader,
            BeanNameGenerator componentScanBeanNameGenerator, BeanDefinitionRegistry
registry) {

        ...
        this.resourceLoader = resourceLoader;
        ...
        this.conditionEvaluator = new ConditionEvaluator(registry, environment,
```

```
resourceLoader);
    }
    ...
    }
```

同样 ConditionEvaluator 所需的 ResourceLoader 来自 ConfigurationClassParser 构造参数，并且该构造器同样被 ConfigurationClassPostProcessor#processConfigBeanDefinitions 方法调用：

```
    public class ConfigurationClassPostProcessor implements
BeanDefinitionRegistryPostProcessor,
        PriorityOrdered, ResourceLoaderAware, BeanClassLoaderAware, EnvironmentAware {
    ...
    public void processConfigBeanDefinitions(BeanDefinitionRegistry registry) {
        ...
        ConfigurationClassParser parser = new ConfigurationClassParser(
                this.metadataReaderFactory, this.problemReporter, this.environment,
                this.resourceLoader, this.componentScanBeanNameGenerator, registry);
        ...
    }
    ...
    }
```

因此，ConditionEvaluator 关联的 ResourceLoader 同样来自 Spring 应用上下文。

综上所述，默认情况下，ConditionContext#getResourceLoader()的返回值存在两种可能：

- 来源于 Spring 应用上下文 ResourceLoaderAware 回调；
- 为 null。

当 OnResourceCondition#getMatchOutcome(ConditionContext,AnnotatedTypeMetadata)方法在执行 ResourceLoaderloader = (context.getResourceLoader() != null ? context.getResourceLoader():this.defaultResourceLoader)语句时，loader 不是来源于 Spring 应用上下文 ResourceLoaderAware 回调，就是 DefaultResourceLoader 类型的 defaultResourceLoader 字段。那么，ResourceLoaderAware 回调的内容需要具体明确。

凡是任意实现 ResourceLoaderAware 接口的 Bean，在其生命周期工程中，会被 Spring 应用上下文设置 ResourceLoader 对象，即在 Bean 初始化之前，执行 ResourceLoaderAware 回调工作。默认情况下，Spring Framework 采用的是标准 BeanPostProcessor 接口的实现 ApplicationContextAwareProcessor：

```
class ApplicationContextAwareProcessor implements BeanPostProcessor {

    private final ConfigurableApplicationContext applicationContext;
    ...
    public ApplicationContextAwareProcessor(ConfigurableApplicationContext applicationContext)
{
        this.applicationContext = applicationContext;
        ...
    }


    @Override
    @Nullable
    public Object postProcessBeforeInitialization(final Object bean, String beanName) throws
BeansException {
        AccessControlContext acc = null;
        ...
        if (acc != null) {
            AccessController.doPrivileged((PrivilegedAction<Object>) () -> {
                invokeAwareInterfaces(bean);
                return null;
            }, acc);
        }
        else {
            invokeAwareInterfaces(bean);
        }

        return bean;
    }

    private void invokeAwareInterfaces(Object bean) {
        if (bean instanceof Aware) {
            ...
            if (bean instanceof ResourceLoaderAware) {
                ((ResourceLoaderAware) bean).setResourceLoader(this.applicationContext);
            }
            ...
        }
```

```
    }
    ...
}
```

其中，生命周期方法 postProcessBeforeInitialization(Object,String) 委派给 invokeAwareInterfaces(Object)方法执行 ResourceLoaderAware 接口回调。不过，该方法传递的 ResourceLoader 对象却是构造参数 ConfigurableApplicationContext 实例。换言之，ConfigurableApplicationContext 是 ResourceLoader 子接口。再跟踪 ApplicationContextAwareProcessor 构造位置，即 AbstractApplicationContext#prepareBeanFactory(ConfigurableListableBeanFactory)方法：

```
public abstract class AbstractApplicationContext extends DefaultResourceLoader
        implements ConfigurableApplicationContext {
    ...
    protected void prepareBeanFactory(ConfigurableListableBeanFactory beanFactory) {
        ...
        beanFactory.addBeanPostProcessor(new ApplicationContextAwareProcessor(this));
        ...
    }
    ...
}
```

ApplicationContextAwareProcessor 构造参数恰好是当前 Spring 应用上下文对象，同时 prepareBeanFactory(ConfigurableListableBeanFactory)方法也处于 Spring 应用上下文启动过程中（AbstractApplicationContext#refresh()）的准备 ConfigurableListableBeanFactory 阶段。因此，在通常情况下，Spring Framework 的 ConditionContext#getResourceLoader()方法所返回的 ResourceLoader 对象即当前 Spring 应用上下文实例。由于 Spring Framework 内建多种 Spring 应用上下文存的实现，目前还无法对 ResourceLoader 的具体类型做出定论，仍需分析 ResourceLoader 在 Spring Framework 内部的类的层级关系，如下图所示。

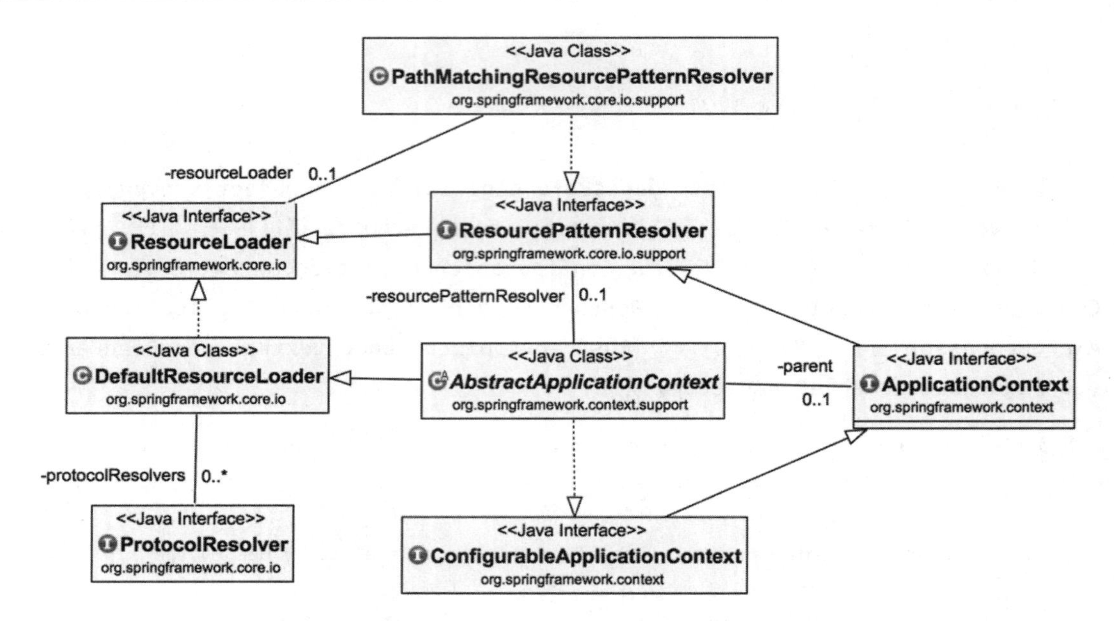

在 Spring Framework 中，`ResourcePatternResolver` 是 `ResourceLoader` 唯一的扩展接口，且其实现类仅为 `PathMatchingResourcePatternResolver`。其中，`PathMatchingResourcePatternResolver` 需关联零到一个 `ResourceLoader` 实例，同时它又被 `AbstractApplicationContext` 关联，作为 `AbstractApplicationContext#getResources(String)`方法的底层实现：

```
public abstract class AbstractApplicationContext extends DefaultResourceLoader
        implements ConfigurableApplicationContext {
    ...
    public AbstractApplicationContext() {
        this.resourcePatternResolver = getResourcePatternResolver();
    }
    ...
    protected ResourcePatternResolver getResourcePatternResolver() {
        return new PathMatchingResourcePatternResolver(this);
    }
    ...
    @Override
    public Resource[] getResources(String locationPattern) throws IOException {
        return this.resourcePatternResolver.getResources(locationPattern);
    }
    ...
}
```

而 resourcePatternResolver 所关联的 ResourceLoader 又是 AbstractApplicationContext 对象本身。值得注意的是，AbstractApplicationContext 扩展了 ResourceLoader 接口实现类 DefaultResourceLoader。也就是说，AbstractApplicationContext 对象也是 ResourceLoader 对象。同时 AbstractApplicationContext 实现 ConfigurableApplicationContext 接口，而该接口又扩展了 ResourcePatternResolver，故 AbstractApplicationContext 实例也是 ResourcePatternResolver 实例，不过它实现 getResources(String)方法时，委派给 resourcePatternResolver 对象，并且 AbstractApplicationContext 没有覆盖父类 DefaultResourceLoader#getResource(String)方法。综上所述，DefaultResourceLoader 实际上是 Spring Framework 中唯一的 ResourceLoader 实现。这也意味着，在语句执行 ResourceLoaderloader = (context.getResourceLoader() != null ? context.getResourceLoader():this.defaultResourceLoader) 时，默认情况下，loader 的行为与普通的 DefaultResourceLoader 实例并无差别。

1. Spring Framework 4.3 协议扩展机制——ProtocolResolver

请读者注意以上结论的用词——默认情况下。实际上，AbstractApplicationContext 对象与 this.defaultResourceLoader 会存在行为不一致的情况。请读者再次仔细观察以上 UML 图，DefaultResourceLoader 的下方存在一个名为 ProtocolResolver 的接口，两者的关联关系是 0 对 N。实际上，这个风险正是由于 DefaultResourceLoader API 变更所致的。从 Spring Framework 4.3 开始，DefaultResourceLoader 增添了 addProtocolResolver(ProtocolResolver)方法，允许开发人员自定义 ProtocolResolver 实现，以扩展自定义协议的 Resource 的获取逻辑。

在 Spring Framework 4.3 之前，DefaultResourceLoader 默认的协议资源获取逻辑一分为二，“classpath”协议的资源由 ClassPathResource 获取，其他协议则由 UrlResource 获取：

```
public class DefaultResourceLoader implements ResourceLoader {
...
@Override
public Resource getResource(String location) {
     Assert.notNull(location, "Location must not be null");
     if (location.startsWith("/")) {
          return getResourceByPath(location);
     }
     else if (location.startsWith(CLASSPATH_URL_PREFIX)) {
          return new ClassPathResource(location.substring(CLASSPATH_URL_PREFIX.length()),
getClassLoader());
     }
     else {
```

```
            try {
                // Try to parse the location as a URL...
                URL url = new URL(location);
                return new UrlResource(url);
            }
            catch (MalformedURLException ex) {
                // No URL -> resolve as resource path.
                return getResourceByPath(location);
            }
        }
    }
    ...
    protected Resource getResourceByPath(String path) {
        return new ClassPathContextResource(path, getClassLoader());
    }
    ...
}
```

> 以上实现源码的 Maven GAV 坐标为：org.springframework:spring-core:4.2.8.RELEASE。

其中，UrlResource 利用 Java API URL 获取资源内容。通常资源连接由 URL#openConnection() 方法桥接获取：

```
public final class URL implements java.io.Serializable {
...
transient URLStreamHandler handler;
...
    public URLConnection openConnection() throws java.io.IOException {
        return handler.openConnection(this);
    }
...
}
```

具体实现则委派给 URLStreamHandler#openConnection(URL) 发布方法，随之调用 URLConnection#getInputStream()方法获取资源的字节流（InputStream），代码如下所示。

```
public class URLBootstrap {
```

```
        public static void main(String[] args) throws Exception {
            // 构建URL对象
            URL url = new URL("https://github.com/mercyblitz");
            // 获取URLConnection对象
            URLConnection urlConnection = url.openConnection();
           // 自动关闭InputStream
            try (InputStream inputStream = urlConnection.getInputStream()) {
                // 复制资源流到标准输出流
                StreamUtils.copy(inputStream, System.out);
            }
        }
    }
```

> 源码位置：以上示例代码可通过查找 spring-boot-2.0-samples/auto-configuration-sample 工程获取。
>
> URLBootstrap 运行后的控制台输出：
>
> (...部分内容被省略...)
>
> <title>mercyblitz (小马哥) · GitHub</title>
>
> (...部分内容被省略...)

2. Java 标准协议扩展机制——URLStreamHandler

以上示例通过 URL API 轻松地获取 HTTPS 资源。通过 URL 协议(Protocol)与 URLStreamHandler 建立实现映射关系，JDK 内建了最常用的通信协议资源获取实现，如下表所示。

URL 协议（Protocol）	URLStreamHandler 实现类
file	sun.net.www.protocol.file.Handler
jar	sun.net.www.protocol.jar.Handler
http	sun.net.www.protocol.http.Handler
https	sun.net.www.protocol.https.Handler
ftp	sun.net.www.protocol.ftp.Handler

以上实现均遵循 JDK URLStreamHandler 扩展机制。默认情况下，URLStreamHandler 实现类约定存放在 sun.net.www.protocol package 下，全类名模式为 sun.net.www.protocol.${protocol}.Handler，其中${protocol}为 URL 中的协议（Protocol）。开发人员也可以根据该模式实现 URL 协议 URLStreamHandler 的扩展。除此之外，实现 URLStreamHandlerFactory 接口则是另一种选择，如 Apache Tomcat JDNI 实现

DirContextURLStreamHandlerFactory。

尽管 URLStreamHandler 或 URLStreamHandlerFactory 接口具备资源的特性，然而在 API 体验上，URLStreamHandler#openConnection(URL)方法的访问修饰符为 protected，程序无法直接调用，并且 URL 主要的建树在资源的信息描述上，并不能很好地表述资源特性。盖因如此，Spring Framework 抽象出统一资源接口 Resource，该对象通常由 ResourceLoader#getResource(String) 方法获取。因此，在 Spring 应用中经常出现 ResourceLoader.getResource("classpath:/...") 的实现代码。

由于"classpath"并非通用资源协议，JDK 自然不会为该协议提供 URLStreamHandler 实现。换言之，URL 无法获取该协议的资源。所以，Spring Framework 扩展其协议实现 ClassPathResource 和 ClassPathContextResource。所以，在 Spring Framework 4.3 之前，由于均使用 JDK URLStreamHandler 扩展机制，所以 AbstratApplication 实例和任意 DefaultResoureLoader 对象并没有差别。

当 ProtocolResolver 出现并能够在运行时增加到 AbstratApplication 实例中后，两者的差异就出现了。不过，ProtocolResolver 的出现也有积极的价值，它能够简化复杂的 URLStreamHandler 及 URLStreamHandlerFactory 的扩展机制。因此，读者在使用 Spring Framework 4.3+或 Spring Boot 1.4+的过程中，应予以关注。

3. 整合 Resource 条件注解——重构 formatter-spring-boot-starter

FormatterAutoConfiguration 的自动装配是由 Spring 工程加载机制辅助完成的，因此此处可增加条件注解@ConditionalOnResource，指向 META-INF/spring.factories 资源，代码如下所示。

```
@Configuration
@ConditionalOnProperty(prefix = "formatter", name = "enabled", havingValue = "true",
      matchIfMissing = true) // 当属性配置不存在时，同样视作匹配
@ConditionalOnResource(resources = "META-INF/spring.factories")
public class FormatterAutoConfiguration {
...
}
```

> 源码位置：以上示例代码可通过查找 spring-boot-2.0-samples/formatter-spring-boot-starter 工程获取。

运行引导类 FormatterBootstrap，观察日志输出：

```
(...部分内容被省略...)
```

```
[Bean name : objectMapperFormatter] JsonFormatter.format(data) : {"name":"小马哥"}
(...部分内容被省略...)
```

结果依旧如故，说明@ConditionalOnResource(resources = "META-INF/spring.factories")已经生效。不过这并非是一个严谨的实现，因为META-INF/spring.factories有可能在其他Starter jar中出现，此处仅为整合示例，请读者留意。

当前的引导类FormatterBootstrap被程序（.web(WebApplicationType.NONE)）限定在非Web应用场景下，接下来一同探讨如何将FormatterAutoConfiguration也限定在同样的场景下。

9.4.5 Web应用条件注解

Web应用条件注解同样是一对语义相反的注解：@ConditionalOnWebApplication和@ConditionalOnNotWebApplication。

@ConditionalOnWebApplication作为当前应用是否为Web类型的条件判断注解，从Spring Boot 1.0开始就被引入。由于Spring Framework 5.0兴起的Web Flux技术，导致Servlet技术和容器不再是Web应用开发和运行的唯一选择，进而使得Spring Boot 2.0嵌入式容器新增了Netty Web Server的选项。因此，@ConditionalOnWebApplication在Spring Boot 2.0中增添了ConditionalOnWebApplication.Type类型的type()属性方法，用于限定判断的范围：

```
@Target({ ElementType.TYPE, ElementType.METHOD })
@Retention(RetentionPolicy.RUNTIME)
@Documented
@Conditional(OnWebApplicationCondition.class)
public @interface ConditionalOnWebApplication {

 /**
  * The required type of the web application.
  * @return the required web application type
  */
 Type type() default Type.ANY;

 /**
  * Available application types.
  */
 enum Type {

     /**
```

```
         * Any web application will match.
         */
        ANY,

        /**
         * Only servlet-based web application will match.
         */
        SERVLET,

        /**
         * Only reactive-based web application will match.
         */
        REACTIVE

    }

}
```

与其相反语义的注解@ConditionalOnNotWebApplication则从Spring Boot 1.0～2.0未发生声明变化，且不需依赖任何属性方法：

```
@Target({ ElementType.TYPE, ElementType.METHOD })
@Retention(RetentionPolicy.RUNTIME)
@Documented
@Conditional(OnWebApplicationCondition.class)
public @interface ConditionalOnNotWebApplication {

}
```

根据注解的声明，OnWebApplicationCondition 即为@ConditionalOnWebApplication 的 Condition 实现类，也用作@ConditionalOnNotWebApplication 注解的判断：

```
@Order(Ordered.HIGHEST_PRECEDENCE + 20)
class OnWebApplicationCondition extends SpringBootCondition {
...
@Override
public ConditionOutcome getMatchOutcome(ConditionContext context,
        AnnotatedTypeMetadata metadata) {
    boolean required = metadata
```

```
            .isAnnotated(ConditionalOnWebApplication.class.getName());
    ConditionOutcome outcome = isWebApplication(context, metadata, required);
    if (required && !outcome.isMatch()) {
        return ConditionOutcome.noMatch(outcome.getConditionMessage());
    }
    if (!required && outcome.isMatch()) {
        return ConditionOutcome.noMatch(outcome.getConditionMessage());
    }
    return ConditionOutcome.match(outcome.getConditionMessage());
}
...
}
```

当执行 getMatchOutcome(ConditionContext,AnnotatedTypeMetadata)方法时，首先判断 required 的情况，当 required 为 true 时，说明当前 Configuration Class 添加了 @ConditionalOnWebApplication 的声明。由于 OnWebApplicationCondition 是非公开(public）类，所以 required 为 false 时，说明@ConditionalOnNotWebApplication 标注在 Configuration Class 上。接下来调用 isWebApplication(ConditionContext,AnnotatedTypeMetadata, boolean)方法，判断当前 Spring 应用是否为 Web 应用：

```
class OnWebApplicationCondition extends SpringBootCondition {
...
private ConditionOutcome isWebApplication(ConditionContext context,
        AnnotatedTypeMetadata metadata, boolean required) {
    ConditionMessage.Builder message = ConditionMessage.forCondition(
            ConditionalOnWebApplication.class, required ? "(required)" : "");
    Type type = deduceType(metadata);
    if (Type.SERVLET == type) {
        return isServletWebApplication(context);
    }
    else if (Type.REACTIVE == type) {
        return isReactiveWebApplication(context);
    }
    else {
        ConditionOutcome servletOutcome = isServletWebApplication(context);
        if (servletOutcome.isMatch() && required) {
            return new ConditionOutcome(servletOutcome.isMatch(),
                    message.because(servletOutcome.getMessage()));
```

```
            }
            ConditionOutcome reactiveOutcome = isReactiveWebApplication(context);
            if (reactiveOutcome.isMatch() && required) {
                return new ConditionOutcome(reactiveOutcome.isMatch(),
                        message.because(reactiveOutcome.getMessage()));
            }
            boolean finalOutcome = (required
                    ? servletOutcome.isMatch() && reactiveOutcome.isMatch()
                    : servletOutcome.isMatch() || reactiveOutcome.isMatch());
            return new ConditionOutcome(finalOutcome,
                    message.because(servletOutcome.getMessage()).append("and")
                            .append(reactiveOutcome.getMessage()));
        }
    }
    ...
}
```

首先，deduceType(AnnotatedTypeMetadata)方法推断应用类型 Type，如果 type()指定类型，则使用指定类型，否则采用 Type.ANY。根据类型来具体判断：

- Type.SERVLET→isServletWebApplication(ConditionContext)；
- Type.REACTIVE→isReactiveWebApplication(ConditionContext)；
- Type.ANY→isServletWebApplication(ConditionContext)优先，结合 isReactiveWebApplication(ConditionContext)方法综合判断。

isServletWebApplication(ConditionContext)与 isReactiveWebApplication(ConditionContext)的判断逻辑类似，以 isServletWebApplication(ConditionContext)为例：

```
class OnWebApplicationCondition extends SpringBootCondition {

    private static final String WEB_CONTEXT_CLASS = "org.springframework.web.context."
            + "support.GenericWebApplicationContext";
    ...
    private ConditionOutcome isServletWebApplication(ConditionContext context) {
        ConditionMessage.Builder message = ConditionMessage.forCondition("");
        if (!ClassUtils.isPresent(WEB_CONTEXT_CLASS, context.getClassLoader())) {
            return ConditionOutcome
                    .noMatch(message.didNotFind("web application classes").atAll());
        }
```

```java
        if (context.getBeanFactory() != null) {
            String[] scopes = context.getBeanFactory().getRegisteredScopeNames();
            if (ObjectUtils.containsElement(scopes, "session")) {
                return ConditionOutcome.match(message.foundExactly("'session' scope"));
            }
        }
        if (context.getEnvironment() instanceof ConfigurableWebEnvironment) {
            return ConditionOutcome
                    .match(message.foundExactly("ConfigurableWebEnvironment"));
        }
        if (context.getResourceLoader() instanceof WebApplicationContext) {
            return ConditionOutcome.match(message.foundExactly("WebApplicationContext"));
        }
        return ConditionOutcome.noMatch(message.because("not a servlet web application"));
    }
    ...
}
```

该方法判断维度较多，首先判断 `org.springframework.web.context.support.GenericWebApplicationContext`（用 `WEB_CONTEXT_CLASS` 字段表示）是否存在于当前 Class Path 中，如果不存在，则说明当前应用并非 Servlet Web 应用场景。随后判断上下文中是否存在 scope 为“session”的 Bean，如果存在，则说明当前应用属于 Servlet Web 应用。后续两则判断分别为判断 Spring 应用上下文所关联 `Environment` 是否为 `ConfigurableWebEnvironment`，以及 `ResourceLoader` 是否为 `WebApplicationContext`。前文曾讨论 `AbstractApplication` 实例是 `ResourceLoader` 对象，因此，当 Spring 应用上下文属于 `WebApplicationContext` 时，说明条件成立，即当前 Spring 应用属于 Servlet Web 类型。类似的逻辑判断同样出现在 `isReactiveWebApplication(ConditionContext)`方法中：

```java
@Order(Ordered.HIGHEST_PRECEDENCE + 20)
class OnWebApplicationCondition extends SpringBootCondition {
    ...
    private ConditionOutcome isReactiveWebApplication(ConditionContext context) {
        ConditionMessage.Builder message = ConditionMessage.forCondition("");
        if (context.getEnvironment() instanceof ConfigurableReactiveWebEnvironment) {
            return ConditionOutcome
                    .match(message.foundExactly("ConfigurableReactiveWebEnvironment"));
        }
        if (context.getResourceLoader() instanceof ReactiveWebApplicationContext) {
```

```
            return ConditionOutcome
                    .match(message.foundExactly("ReactiveWebApplicationContext"));
        }
        return ConditionOutcome
                .noMatch(message.because("not a reactive web application"));
    }
    ...
}
```

回顾示例 FormatterBootstrap 的实现：

```
@EnableAutoConfiguration
public class FormatterBootstrap {

    public static void main(String[] args) {
        ConfigurableApplicationContext context = new
SpringApplicationBuilder(FormatterBootstrap.class)
                .web(WebApplicationType.NONE)  // 非 Web 应用
                ...
    }
}
```

当 SpringApplicationBuilder.web(WebApplicationType.NONE)执行后，当前应用被显式地设置为非 Web 应用，此时的 OnWebApplicationCondition 是无法成立的。原因在于 Spring 应用上下文既非 WebApplicationContext 类型，也不是 ReactiveWebApplicationContext 实例，而是普通的 AnnotationConfigApplicationContext，其中细节的讨论将在下一篇“理解 SpringApplication”中展开。

整合 Web 应用条件注解——重构 formatter-spring-boot-starter

目前，需要将 FormatterAutoConfiguration 限定在非 Web 应用场景中，实际所需的调整仅一行代码，即添加注解@ConditionalOnNotWebApplication 即可：

```
@Configuration
@ConditionalOnProperty(prefix = "formatter", name = "enabled", havingValue = "true",
        matchIfMissing = true) // 当属性配置不存在时，同样视作匹配
@ConditionalOnResource(resources = "META-INF/spring.factories")
@ConditionalOnNotWebApplication
public class FormatterAutoConfiguration {
...
```

```
}
```

再次运行引导类 FormatterBootstrap，日志输出没有变化：

```
(...部分内容被省略...)
[Bean name : objectMapperFormatter] JsonFormatter.format(data) : {"name":"小马哥"}
(...部分内容被省略...)
```

9.4.6　Spring 表达式条件注解

前文讨论的所有 Spring Boot 条件注解在功能性上相对单一，无法提供复杂语义的条件判断，因此，Spring 表达式条件注解是其中一种可选方案。当然，自定义@Conditional 注解的方式更具有弹性，如“自定义@Conditional 条件装配”一节中的@ConditionalOnSystemProperty。

Spring 表达式英文全名为 Spring Expression Language，简写为 SpEL。就功能特性而言，它与其他 Java 表达式语言没有明显的差别，如 OGNL 和 JSPEL 等。按照 Spring Framework 官方文档的描述，Spring 表达式能够适用于 Spring 旗下的所有产品：

> the Spring Expression Language was created to provide the Spring community with a single well supported expression language that can be used across all the products in the Spring portfolio.

具体细节可参考 Spring Framework 官方文档的“4. Spring Expression Language (SpEL)”章节：https://docs.spring.io/spring-framework/docs/current/spring-framework-reference/core.html# expressions.

Spring 表达式条件注解声明非常简单，其中属性方法 value()用于评估表达式的真伪：

```
@Retention(RetentionPolicy.RUNTIME)
@Target({ ElementType.TYPE, ElementType.METHOD })
@Documented
@Conditional(OnExpressionCondition.class)
public @interface ConditionalOnExpression {

    /**
     * The SpEL expression to evaluate. Expression should return {@code true} if the
     * condition passes or {@code false} if it fails.
     * @return the SpEL expression
     */
    String value() default "true";
```

```
}
```

该注解同样贯穿于 Spring Boot 1.0～2.0 之中，其条件判断实现 OnExpressionCondition 的逻辑相对简单：

```
@Order(Ordered.LOWEST_PRECEDENCE - 20)
class OnExpressionCondition extends SpringBootCondition {

    @Override
    public ConditionOutcome getMatchOutcome(ConditionContext context,
            AnnotatedTypeMetadata metadata) {
        String expression = (String) metadata
                .getAnnotationAttributes(ConditionalOnExpression.class.getName())
                .get("value");
        expression = wrapIfNecessary(expression);
        String rawExpression = expression;
        expression = context.getEnvironment().resolvePlaceholders(expression);
        ConfigurableListableBeanFactory beanFactory = context.getBeanFactory();
        BeanExpressionResolver resolver = (beanFactory != null
                ? beanFactory.getBeanExpressionResolver() : null);
        BeanExpressionContext expressionContext = (beanFactory != null
                ? new BeanExpressionContext(beanFactory, null) : null);
        if (resolver == null) {
            resolver = new StandardBeanExpressionResolver();
        }
        Object result = resolver.evaluate(expression, expressionContext);
        boolean match = result != null && (boolean) result;
        return new ConditionOutcome(match, ConditionMessage
                .forCondition(ConditionalOnExpression.class, "(" + rawExpression + ")")
                .resultedIn(result));
    }

    /**
     * Allow user to provide bare expression with no '#{}' wrapper.
     * @param expression source expression
     * @return wrapped expression
     */
    private String wrapIfNecessary(String expression) {
        if (!expression.startsWith("#{")) {
```

```
            return "#{" + expression + "}";
        }
        return expression;
    }

}
```

首先，从注解@ConditionalOnExpression 元信息中，程序获取 value()属性作为表达式内容 expression。如果 expression 未包含“#{}”时，则将其补充。该内容随即被 Spring 应用上下文关联的 Environment 进行占位符处理，然后 BeanExpressionResolver 对象评估表达式的真伪，当评估结果为 true 时，条件成立。需要关注的是，当 Spring 应用上下文中的 ConfigurableListableBeanFactory 所关联 BeanExpressionResolver 行为调整后，可能会影响评估的结果。

@ConditionalOnExpression 在 Spring Boot 内部的使用情况也是较少的，由于 @ConditionalOnProperty 从 Spring Boot 1.1 才引入，所以@ConditionalOnExpression 在早期的版本中用于配置属性的条件判断，比如 AopAutoConfiguration 在 Spring Boot 1.0.2.RELEASE 中：

```
@Configuration
@ConditionalOnClass({ EnableAspectJAutoProxy.class, Aspect.class, Advice.class })
@ConditionalOnExpression("${spring.aop.auto:true}")
public class AopAutoConfiguration {
 ...
}
```

> 以上实现源码的 Maven GAV 坐标为：org.springframework.boot:spring-boot-autoconfigure:1.0.2.RELEASE。

该实现在 1.2.8.RELEASE 中调整为：

```
@Configuration
@ConditionalOnClass({ EnableAspectJAutoProxy.class, Aspect.class, Advice.class })
@ConditionalOnProperty(prefix = "spring.aop", name = "auto", havingValue = "true",
matchIfMissing = true)
public class AopAutoConfiguration {
 ...
}
```

> 以上实现源码的 Maven GAV 坐标为：org.springframework.boot:spring-boot-autoconfigure:1.2.8.RELEASE。

尽管@ConditionalOnProperty 在配置属性的语义上要优于@ConditionalOnExpression，不过它要表达多组配置属性则较为烦琐，然而@ConditionalOnExpression 就能轻松解决，比如 EndpointMBeanExportAutoConfiguration：

```
@Configuration
@ConditionalOnExpression("${endpoints.jmx.enabled:true} && ${spring.jmx.enabled:true}")
@AutoConfigureAfter({ EndpointAutoConfiguration.class, JmxAutoConfiguration.class })
@EnableConfigurationProperties(EndpointMBeanExportProperties.class)
public class EndpointMBeanExportAutoConfiguration {
...
}
```

> 以上实现源码的 Maven GAV 坐标为：org.springframework.boot:spring-boot-actuator: 1.2.8.RELEASE。

当然，从维护性、扩展性和灵活性上考虑，自定义 Condition 更适合，如 EndpointMBeanExportAutoConfiguration 从 Spring Boot 1.3 开始就调整如下：

```
@Configuration
@Conditional(JmxEnabledCondition.class)
@AutoConfigureAfter({ EndpointAutoConfiguration.class, JmxAutoConfiguration.class })
@EnableConfigurationProperties(EndpointMBeanExportProperties.class)
public class EndpointMBeanExportAutoConfiguration {
...
static class JmxEnabledCondition extends SpringBootCondition {

    @Override
    public ConditionOutcome getMatchOutcome(ConditionContext context,
            AnnotatedTypeMetadata metadata) {
        boolean jmxEnabled = isEnabled(context, "spring.jmx.");
        boolean jmxEndpointsEnabled = isEnabled(context, "endpoints.jmx.");
        return new ConditionOutcome(jmxEnabled && jmxEndpointsEnabled,
                "JMX Endpoints");
    }

    private boolean isEnabled(ConditionContext context, String prefix) {
        RelaxedPropertyResolver resolver = new RelaxedPropertyResolver(
                context.getEnvironment(), prefix);
```

```
        return resolver.getProperty("enabled", Boolean.class, true);
    }

  }
  ...
}
```

留心观察会发现 Spring Boot 的自定义 Condition 基本上均扩展 SpringBootCondition，而非直接实现 Condition 接口。读者不妨回顾 OnBeanCondition、OnClassCondition 和 OnResourceCondition 等实现。

整合 Spring 表达式条件注解——重构 formatter-spring-boot-starter

同样地，在检验 Spring Boot 1.1 之前的版本中使用@ConditionalOnExpression 进行属性真伪判断，调整 FormatterAutoConfiguration 如下：

```
@Configuration
@ConditionalOnProperty(prefix = "formatter", name = "enabled", havingValue = "true",
        matchIfMissing = true) // 当属性配置不存在时，同样视作匹配
@ConditionalOnResource(resources = "META-INF/spring.factories")
@ConditionalOnNotWebApplication
@ConditionalOnExpression("${formatter.enabled:true}")
public class FormatterAutoConfiguration {
...
}
```

重启引导类 FormatterBootstrap，结果如下：

```
(...部分内容被省略...)
[Bean name : objectMapperFormatter] JsonFormatter.format(data) : {"name":"小马哥"}
(...部分内容被省略...)
```

9.5　小马哥有话说

“走向自动装配”部分花费不少的篇幅讨论 Spring 自动装配的特性，这些特性能够让我们明显地感受到 Spring Framework 为此曾付出的努力。Spring Framework 从 1.x 版本到最新的 5.0 版本，核心特性在每个版本中渐进式地提升，尤其是注解驱动方面的能力。不难看出，Spring Framework 到了 5.x 时代注解特性趋于稳定，仅引入了为数不多的特性注解。可以说，在注解驱动和自动装配方面，Spring Framework 的 4.3 版本和 5.0 版本并没有明显的差异。尽管如此，这些特性却是 Spring Boot

自动装配的基石。换言之，虽然庞大的 Spring Boot API 体系形同“百万雄师”，然而当深入了解并掌握 Spring Framework 核心特性之后，做到得其要领则如同探囊取物般简单。同时，我们通过 Spring Framework 学习到了 API 设计的重要性，就兼容性而论，Spring Boot 只能望 Spring Framework 的“项背”。Spring Boot 在 1.x 版本之间出现间歇性兼容问题，以及升级到 2.0 版本后的部分 API “重新洗牌”。这些破坏性升级让应用的稳定性存在变数，无形中伤害了业界对 Spring 社区的信任度。当然这些问题并非空穴来风，它们将在后续的章节讨论中层出不穷。或许有读者会问：“既然 Spring Boot 存在这样和那样的问题，那么小马哥为什么还要写书呢？”。答案非常简单，不可否认的是，Spring Boot 是非常优秀的框架，然而作为一个独立的开发人员，“不以物喜，优劣并叙，功过并陈”，才是“格物致知”的精神，也是科学和理性地讨论问题的原则。这也是为什么小马哥选择深入讨论的原因之一。总之，“君子不以言举人，不以人废言”，不因 Spring Boot 存在某些不足，从而对其全盘否定。

9.6 下一站：理解 SpringApplication

当“走向自动装配”的讨论即将进入尾声之际，对以上内容稍作回顾。Spring Boot 的自动装配所依赖的注解驱动、`@Enable` 模块驱动、条件装配、Spring 工厂加载机制等特性均来自 Spring Framework。无论在 Spring Framework 中，还是 Spring Boot 应用场景中，这些功能特性均围绕 Spring 应用上下文及其管理的 Bean 生命周期展开介绍。不同的是，在 Spring Framework 时代，Spring 应用上下文通常由容器启动，如 `ContextLoaderListener` 或 `WebApplicationInitializer` 的实现类由 Servlet 容器装载并驱动。到了 Spring Boot 时代，Spring 应用上下文的启动则通过调用 `SpringApplication#run(Object,String...)`或 `SpringApplicationBuilder#run (String...)` 方法并配合`@SpringBootApplication` 或`@EnableAutoConfiguration` 注解的方式完成。那么，`SpringApplication` 和 `SpringApplicationBuilder` 底层到底施加了哪些“魔法”，使得启动方式发生了逆转？解开其中的疑惑正是“理解 `SpringApplication`”中讨论的重点。

第 3 部分 理解SpringApplication

> “纸上得来终觉浅，绝知此事要躬行。”——陆游《冬夜读书示子聿》

如果 Spring Boot 自动装配是源于 Spring Framework 而构建的，那么 `SpringApplication` 则是全新的 Spring 应用 API。在前几章的讨论中，几乎所有 Spring Boot 示例都使用 `SpringApplication` 或 `SpringApplicationBuilder` API，执行 `run` 方法引导 Spring Boot 应用。从 Spring Boot 功能特性而言，本部分才算真正深入探讨的开始。Spring Boot 2.0 官方文档将讨论的内容安排在“23. SpringApplication”章节，按照其章节的安排，划分为如下小节：

- 23.1 Startup Failure（启动失败）。
- 23.2 Customizing the Banner（自定义 Banner）。
- 23.3 Customizing SpringApplication（自定义 `SpringApplication`）。
- 23.4 Fluent Builder API（流式 Builder API）。
- 23.5 Application Events and Listeners（Spring 应用事件和监听器）。
- 23.6 Web Environment（Spring 应用 Web 环境）。
- 23.7 Accessing Application Arguments（存储 Spring 应用启动参数）。
- 23.8 Using the ApplicationRunner or CommandLineRunner（使用 `ApplicationRunner` 或 `CommandLineRunner` 接口）。
- 23.9 Application Exit（Spring 应用退出）。
- 23.10 Admin Features（Spring 应用管理特性）。

Spring Boot 官方文档基本上遵照以上功能特性来划分章节，然而这样的安排恐怕给开发人员的系统性思考带来不利的影响。以 `SpringApplication` 为例，官方并没有告知其运行过程中的细节，导致开发人员遇到错综复杂的情况时无从下手。因此，这部分的讨论范围将基本覆盖以上知识点，并且议题也有所取舍，按照由浅入深的方式逐一讨论。

从 Spring Boot 应用进程来看，整体的生命周期大体上总结如下：

- `SpringApplication` 初始化阶段；

- `SpringApplication` 运行阶段；
- `SpringApplication` 结束阶段；
- Spring Boot 应用退出。

按照以上生命周期中各个阶段的顺序，接下来展开对 `SpringApplication` 的讨论。

第10章 SpringApplication 初始化阶段

SpringApplication 初始化阶段属于运行前的准备阶段，大多数 Spring Boot 应用直接或间接地使用 SpringApplication API 驱动 Spring 应用，SpringApplication 允许指定应用的类型，大体上包括 Web 应用和非 Web 应用。从 Spring Boot 2.0 开始，Web 应用又可分为 Servlet Web 和 Reactive Web。当然 SpringApplication 也可以调整 Banner 的输出，配置默认属性的内容等。这些状态变更的操作只要在 run()方法之前指定即可。简单而言，SpringApplication 的准备阶段主要由两阶段完成：构造阶段和配置阶段。接下来分别深入讨论。

10.1 SpringApplication 构造阶段

顾名思义，SpringApplication 构造阶段自然由其构造器完成，不过在日常的 Spring Boot 应用开发过程中，开发人员很少直接与 SpringApplication 构造器打交道，而是调用其静态方法 run(Class,String...)来启动应用。实际上，如果稍微留心观察，则会发现其构造过程就在其中：

```
public class SpringApplication {
    ...
    public static ConfigurableApplicationContext run(Class<?> primarySource,
            String... args) {
        return run(new Class<?>[] { primarySource }, args);
```

```
    }

    ...
    public static ConfigurableApplicationContext run(Class<?>[] primarySources,
            String[] args) {
        return new SpringApplication(primarySources).run(args);
    }
    ...
}
```

以如下极简的引导类为例：

```
@EnableAutoConfiguration
public class SpringApplicationBootstrap {

    public static void main(String[] args) {
        // 启动 Spring Boot 应用
        SpringApplication.run(SpringApplicationBootstrap.class,args);
    }
}
```

> 源码位置：以上示例代码可通过查找 spring-boot-2.0-samples/spring-application-sample 工程获取。

按照 SpringApplication 的实现，该类的引导语句 SpringApplication.run(SpringApplicationBootstrap.class,args) 等价于 new SpringApplication(SpringApplicationBootstrap.class).run(args)。无论 SpringApplication#run(Class, String...) 方法，还是其构造器，均需要 Class 类型的 primarySources 参数，那么这个参数有什么实际意义呢？

10.1.1 理解 SpringApplication 主配置类

SpringApplication 主配置类的概念和实现是从 Spring Boot 2.0 开始引入的。通常情况下，引导类将作为 primarySources 参数的内容，如上例的 SpringApplicationBootstrap。这些引导类基本上具备一个特点，不是标注 @EnableAutoConfiguration，就是标注 @SpringBootApplication。根据前面的结论，由于 @SpringBootApplication 元标注 @EnableAutoConfiguration，因此，两者均被底层 Spring 应用上下文视作 @EnableAuto

Configuration 处理。在 SpringApplicationBootstrap 类声明中移除@EnableAutoConfiguration 的定义，再次运行程序，控制台输出似乎没有明显的差异，不过移除前的输出则多了以下内容：

```
(...部分内容被省略...)
o.s.j.e.a.AnnotationMBeanExporter: Registering beans for JMX exposure on startup
(...部分内容被省略...)
o.s.j.e.a.AnnotationMBeanExporter: Unregistering JMX-exposed beans on shutdown
```

已知@EnableAutoConfiguration 能够激活 Spring Boot 内建和自定义组件的自动装配特性。当它从 SpringApplicationBootstrap 中移除后，自然 AnnotationMBeanExporter 没有被 Spring 应用上下文注册。如果程序强行从 Spring 应用上下文中获取该 Bean，则 NoSuchBeanDefinitionException 会随之抛出。按此推论，调整示例加以验证：

```
public class SpringApplicationBootstrap {

    public static void main(String[] args) {
        // 启动 Spring Boot 应用
        ConfigurableApplicationContext context =
SpringApplication.run(SpringApplicationBootstrap.class, args);
        // 获取 AnnotationMBeanExporter Bean
        AnnotationMBeanExporter beanExporter =
context.getBean(AnnotationMBeanExporter.class);
        // 输出 AnnotationMBeanExporter 对象
        System.out.println("AnnotationMBeanExporter Bean : " + beanExporter);
        // 关闭上下文
        context.close();
    }
}
```

运行 SpringApplicationBootstrap，观察日志输出是否符合期望：

```
(...部分内容被省略...)
Exception in thread "main" org.springframework.beans.factory.NoSuchBeanDefinitionException:
No qualifying bean of type 'org.springframework.jmx.export.annotation.AnnotationMBeanExporter'
available
(...部分内容被省略...)
```

调整后的引导类运行结果验证了推论的正确性。如果该引导类再次恢复

`@EnableAutoConfiguration` 的标注，则工作正常：

```
(...部分内容被省略...)
AnnotationMBeanExporter Bean :
org.springframework.jmx.export.annotation.AnnotationMBeanExporter@6e15fe2
(...部分内容被省略...)
```

当引导类作为 `primarySources` 参数时，假设 Spring 应用上下文将其视为 Configuration Class 处理，那么该类型的 Bean 必然存在。再次调整实现加以验证：

```java
@EnableAutoConfiguration
public class SpringApplicationBootstrap {

    public static void main(String[] args) {
        // 启动 Spring Boot 应用
        ConfigurableApplicationContext context =
SpringApplication.run(SpringApplicationBootstrap.class, args);
        // 获取 AnnotationMBeanExporter Bean
        AnnotationMBeanExporter beanExporter =
context.getBean(AnnotationMBeanExporter.class);
        // 输出 AnnotationMBeanExporter 对象
        System.out.println("AnnotationMBeanExporter Bean : " + beanExporter);
        // 输出 SpringApplicationBootstrap 对象
        System.out.println("SpringApplicationBootstrap Bean : " +
context.getBean(SpringApplicationBootstrap.class));
        // 关闭上下文
        context.close();
    }
}
```

运行输出如下：

```
(...部分内容被省略...)
SpringApplicationBootstrap Bean :
thinking.in.spring.boot.samples.spring.application.bootstrap.SpringApplicationBootstrap@69868
52
(...部分内容被省略...)
```

运行结果再次证明之前的假设，总之 `primarySources` 参数实际为 Spring Boot 应用上下文的 Configuration Class。换言之，该配置类也不一定非得使用引导类，例如抽取独立的 Configuration Class ：

```
@Configuration
@EnableAutoConfiguration
public class SpringApplicationConfiguration {
}
```

随后，将 SpringApplicationBootstrap 替换为 SpringApplicationConfiguration，并移除 @EnableAutoConfiguration，同时获取 SpringApplicationConfiguration Bean：

```
public class SpringApplicationBootstrap {

    public static void main(String[] args) {
        // 启动 Spring Boot 应用
        ConfigurableApplicationContext context =
SpringApplication.run(SpringApplicationConfiguration.class, args);
        // 获取 AnnotationMBeanExporter Bean
        AnnotationMBeanExporter beanExporter =
context.getBean(AnnotationMBeanExporter.class);
        // 输出 AnnotationMBeanExporter 对象
        System.out.println("AnnotationMBeanExporter Bean : " + beanExporter);
        // 输出 SpringApplicationConfiguration 对象
        System.out.println("SpringApplicationConfiguration Bean : " +
context.getBean(SpringApplicationConfiguration.class));
        // 关闭上下文
        context.close();
    }
}
```

再次运行 SpringApplicationBootstrap，观察是否存在异常：

```
AnnotationMBeanExporter Bean :
org.springframework.jmx.export.annotation.AnnotationMBeanExporter@6f6745d6
SpringApplicationConfiguration Bean :
thinking.in.spring.boot.samples.spring.application.config.SpringApplicationConfiguration$$Enh
ancerBySpringCGLIB$$c853d76a@27508c5d
```

调整后的引导类工作正常，输入内容的差别仅在于 SpringApplicationConfiguration Bean 的替换。因此，主配置类 primarySources 与传统 Spring Configuration Class 并无差异。

主配置类属性 primarySources 除初始化构造器参数外，还能通过 SpringApplication#addPrimarySources(Collection)方法追加修改：

```
public class SpringApplication {
 ...
 private Set<Class<?>> primarySources;
 ...
 public void addPrimarySources(Collection<Class<?>> additionalPrimarySources) {
     this.primarySources.addAll(additionalPrimarySources);
 }
 ...
}
```

10.1.2 SpringApplication 的构造过程

已知 SpringApplication#run(Class,String...)方法的执行会伴随 SpringApplication 对象的构造，其调用的构造器为 SpringApplication(Class...)：

```
public class SpringApplication {
    ...
    public SpringApplication(Class<?>... primarySources) {
     this(null, primarySources);
 }

 public SpringApplication(ResourceLoader resourceLoader, Class<?>... primarySources) {
     this.resourceLoader = resourceLoader;
     Assert.notNull(primarySources, "PrimarySources must not be null");
     this.primarySources = new LinkedHashSet<>(Arrays.asList(primarySources));
     this.webApplicationType = deduceWebApplicationType();
     setInitializers((Collection) getSpringFactoriesInstances(
             ApplicationContextInitializer.class));
     setListeners((Collection) getSpringFactoriesInstances(ApplicationListener.class));
     this.mainApplicationClass = deduceMainApplicationClass();
 }
    ...
}
```

根据调用链路，实际执行的构造器为 SpringApplication(ResourceLoader, Class<?>...)，其中主配置类 primarySources 被 SpringApplication 对象“primarySources”属性存储，随后依次调用 deduceWebApplicationType()、setInitializers(Collection)、setListeners(Collection) 和 deduceMainApplicationClass()方法。按照方法的中文语义，它们分别为：推断 Web 应用类

型、加载 Spring 应用上下文初始化器、加载 Spring 应用事件监听器和推断应用引导类，下面将逐一讨论。

10.1.3 推断 Web 应用类型

推断 Web 应用类型属于当前 Spring Boot 应用 Web 类型的初始化过程，因为该类型可在 SpringApplication 构造后及 run 方法之前，再通过 setWebApplicationType(WebApplicationType)方法调整。在推断 Web 应用类型的过程中，由于当前 Spring 应用上下文尚未准备，所以实现采用的是检查当前 ClassLoader 下基准 Class 的存在性判断：

```
public class SpringApplication {
    ...
    private static final String[] WEB_ENVIRONMENT_CLASSES = { "javax.servlet.Servlet",
            "org.springframework.web.context.ConfigurableWebApplicationContext" };
    ...
    private static final String REACTIVE_WEB_ENVIRONMENT_CLASS = "org.springframework."
            + "web.reactive.DispatcherHandler";

     private static final String MVC_WEB_ENVIRONMENT_CLASS = "org.springframework."
            + "web.servlet.DispatcherServlet";
    ...
    private WebApplicationType deduceWebApplicationType() {
     if (ClassUtils.isPresent(REACTIVE_WEB_ENVIRONMENT_CLASS, null)
                && !ClassUtils.isPresent(MVC_WEB_ENVIRONMENT_CLASS, null)) {
            return WebApplicationType.REACTIVE;
     }
     for (String className : WEB_ENVIRONMENT_CLASSES) {
            if (!ClassUtils.isPresent(className, null)) {
                return WebApplicationType.NONE;
            }
     }
     return WebApplicationType.SERVLET;
 }
    ...
}
```

deduceWebApplicationType()利用 ClassUtils#isPresent(String,ClassLoader)方法依次判断 DispatcherHandler、ConfigurableWebApplicationContext、Servlet 和

DispatcherServlet 的存在性组合情况，从而推断 Web 应用类型：

（1）当 DispatcherHandler 存在，并且 DispatcherServlet 不存在时，换言之，Spring Boot 仅依赖 WebFlux 存在时，此时的 Web 应用类型为 Reactive Web，即 WebApplicationType.REACTIVE。

（2）当 Servlet 和 ConfigurableWebApplicationContext 均不存在时，当前应用为非 Web 应用，即 WebApplicationType.NONE，因为这些 API 均是 Spring Web MVC 必需的依赖。

（3）当 Spring WebFlux 和 Spring Web MVC 同时存在时，Web 应用类型同样是 Servlet Web，即 WebApplicationType.SERVLET。

当 deduceWebApplicationType()执行完毕后，SpringApplication 的构造进入“加载 Spring 应用上下文初始化器”的过程。

10.1.4 加载 Spring 应用上下文初始化器（ApplicationContextInitializer）

这个过程包含两个动作，依次为 getSpringFactoriesInstances(Class) 方法和 setInitializers(Collection) 方法。前者的执行委派给 getSpringFactoriesInstances (Class,Class[],Object...)方法：

```
public class SpringApplication {
    ...
    private <T> Collection<T> getSpringFactoriesInstances(Class<T> type) {
        return getSpringFactoriesInstances(type, new Class<?>[] {});
    }

    private <T> Collection<T> getSpringFactoriesInstances(Class<T> type,
            Class<?>[] parameterTypes, Object... args) {
        ClassLoader classLoader = Thread.currentThread().getContextClassLoader();
        // Use names and ensure unique to protect against duplicates
        Set<String> names = new LinkedHashSet<>(
                SpringFactoriesLoader.loadFactoryNames(type, classLoader));
        List<T> instances = createSpringFactoriesInstances(type, parameterTypes,
                classLoader, args, names);
        AnnotationAwareOrderComparator.sort(instances);
        return instances;
    }
    ...
```

```
}
```

此处同样运用了 Spring 工厂加载机制方法 SpringFactoriesLoader.loadFactoryNames(Class,ClassLoader)。结合本例的场景，该方法将返回所有 META-INF/spring.factories 资源中配置的 ApplicationContextInitializer 实现类名单，例如 org.springframework.boot:spring-boot:2.0.2.RELEASE 中的内容：

```
# Application Context Initializers
org.springframework.context.ApplicationContextInitializer=\
org.springframework.boot.context.ConfigurationWarningsApplicationContextInitializer,\
org.springframework.boot.context.ContextIdApplicationContextInitializer,\
org.springframework.boot.context.config.DelegatingApplicationContextInitializer,\
org.springframework.boot.web.context.ServerPortInfoApplicationContextInitializer
```

当 getSpringFactoriesInstances(Class,Class[],Object...)方法获取实现类名单后，调用 createSpringFactoriesInstances(Class,Class[],ClassLoader,Object[],Set)方法初始化这些实现类：

```
public class SpringApplication {
    ...
    private <T> List<T> createSpringFactoriesInstances(Class<T> type,
            Class<?>[] parameterTypes, ClassLoader classLoader, Object[] args,
            Set<String> names) {
        List<T> instances = new ArrayList<>(names.size());
        for (String name : names) {
            try {
                Class<?> instanceClass = ClassUtils.forName(name, classLoader);
                Assert.isAssignable(type, instanceClass);
                Constructor<?> constructor = instanceClass
                        .getDeclaredConstructor(parameterTypes);
                T instance = (T) BeanUtils.instantiateClass(constructor, args);
                instances.add(instance);
            }
            catch (Throwable ex) {
                throw new IllegalArgumentException(
                        "Cannot instantiate " + type + " : " + name, ex);
            }
        }
        return instances;
```

```
    }
        ...
}
```

按照 getSpringFactoriesInstances(Class) 方法的实现，即调用重载方法 getSpringFactoriesInstances(Class,Class[],Object...)，并传递空数组的 parameterTypes 和 args 方法参数。换言之，当创建 ApplicationContextInitializer 实例集合（名称集合用 names 表示）时，**ApplicationContextInitializer 实现类必须存在默认构造器**。挑选 ContextIdApplicationContextInitializer 进行检验，情况也是如此：

```java
public class ContextIdApplicationContextInitializer implements
        ApplicationContextInitializer<ConfigurableApplicationContext>, Ordered {

    private int order = Ordered.LOWEST_PRECEDENCE - 10;

    public void setOrder(int order) {
        this.order = order;
    }

    @Override
    public int getOrder() {
        return this.order;
    }

    @Override
    public void initialize(ConfigurableApplicationContext applicationContext) {
        ContextId contextId = getContextId(applicationContext);
        applicationContext.setId(contextId.getId());
        applicationContext.getBeanFactory().registerSingleton(ContextId.class.getName(),
                contextId);
    }

    private ContextId getContextId(ConfigurableApplicationContext applicationContext) {
        ApplicationContext parent = applicationContext.getParent();
        if (parent != null && parent.containsBean(ContextId.class.getName())) {
            return parent.getBean(ContextId.class).createChildId();
        }
        return new ContextId(getApplicationId(applicationContext.getEnvironment()));
```

```
    }

    private String getApplicationId(ConfigurableEnvironment environment) {
        String name = environment.getProperty("spring.application.name");
        return (StringUtils.hasText(name) ? name : "application");
    }
    ...
}
```

不过ContextIdApplicationContextInitializer实现Ordered接口的举动让我们下意识地认为会出现排序的操作，这个动作在 createSpringFactoriesInstances 方法执行后发生，当多个ApplicationContextInitializer实例来自不同的META-INF/spring.factories资源声明时，对它们进行排序是再自然不过的行为了。当然 ApplicationContextInitializer 实现类实现Ordered接口也并非是必须的。当所有的ApplicationContextInitializer实例集合执行排序后，下一个动作是将它们关联到 SpringApplication#initializers 属性，供后续操作使用：

```
public class SpringApplication {
    ...
    public void setInitializers(
            Collection<? extends ApplicationContextInitializer<?>> initializers) {
        this.initializers = new ArrayList<>();
        this.initializers.addAll(initializers);
    }
    ...
}
```

setInitializers(Collection)方法的实现非常简单，不过它属于覆盖性更新。换言之，在执行 SpringApplication#run 方法前，这些在构造过程中加载的 ApplicationContextInitializer 实例集合存在被 setInitializers(Collection)方法覆盖的可能。

当 SpringApplication 构建器执行该方法后，“加载 Spring 应用事件监听器”的动作立即执行。

10.1.5　加载 Spring 应用事件监听器（ApplicationListener）

本过程的实现手段与“加载 Spring 应用上下文初始化器”基本一致，先执行 getSpringFactoriesInstances 方法，再设置实例集合，只不过初始化的对象类型从 ApplicationContextInitializer 变成 ApplicationListener.setListeners(Collection)方法同样是覆盖更新：

```
public class SpringApplication {
    ...
   public void setListeners(Collection<? extends ApplicationListener<?>> listeners) {
        this.listeners = new ArrayList<>();
        this.listeners.addAll(listeners);
   }
    ...
}
```

10.1.6 推断应用引导类

推断应用引导类是 SpringApplication 构造过程的末尾动作，其执行方法为 deduceMainApplicationClass()：

```
private Class<?> deduceMainApplicationClass() {
    try {
        StackTraceElement[] stackTrace = new RuntimeException().getStackTrace();
        for (StackTraceElement stackTraceElement : stackTrace) {
            if ("main".equals(stackTraceElement.getMethodName())) {
                return Class.forName(stackTraceElement.getClassName());
            }
        }
    }
    catch (ClassNotFoundException ex) {
        // Swallow and continue
    }
    return null;
}
```

该方法根据当前线程执行栈来判断其栈中哪个类包含 main 方法。尽管这个方法的实现并不严谨，不过可覆盖绝多大数以 Java 的标准 main 方法引导的情况。

至此，在 SpringApplication 构造的过程中，SpringApplication 属性 primarySources、webApplicationType、initializers、listeners 和 mainApplicationClass 均得到了初始化，下面将深入探讨它们在不同阶段所扮演的角色。

10.2 SpringApplication 配置阶段

配置阶段位于构造阶段和运行阶段之间，该阶段是可选的，主要用于调整或补充构造阶段的状态、左右运行时行为，以 SpringApplication Setter 方法为代表，用于调整 SpringApplication 的行为，补充行为则以 add*方法为主，其部分简要的描述在官方文档的“23.3 Customizing SpringApplication”一节。下面的章节也会采用“自定义 SpringApplication”作为讨论的议题。或许通过 Setter 方法配置 SpringApplication 的行为过于烦琐，因此，Spring Boot 引入 SpringApplicationBuilder 以提升 API 的便利性，这方面的内容在官方文档的“23.4 Fluent Builder API”章节有着墨。大多数情况下，开发人员无须调整 SpringApplication 的默认状态，而本节讨论的目的是不仅让读者熟悉 SpringApplication 和 SpringApplicationBuilder API，而且为后续阅读和理解 Spring Cloud 或 Spring Cloud Data Flow 的实现源码提供技术储备，两者均基于 Spring Boot 开发，并且深度地使用 SpringApplication API 来构建应用。

10.2.1 自定义 SpringApplication

在官方文档的“23.3 Customizing SpringApplication”一节中实际告诉开发人员三个内容：

- 调整 SpringApplication 设置；
- 增加 SpringApplication 配置源；
- 调整 Spring Boot 外部化配置（Externalized Configuration）。

10.2.2 调整 SpringApplication 设置

官方文档的示例演示了如何关闭 Spring Boot 应用中的 Banner：

```
public static void main(String[] args) {
 SpringApplication app = new SpringApplication(MySpringConfiguration.class);
 app.setBannerMode(Banner.Mode.OFF);
 app.run(args);
}
```

关闭 Banner 的 API 方法在不同的 Spring Boot 版本中存在差异，如从 Spring Boot 1.0～1.3 时，SpringApplication 可通过 setShowBanner(boolean)方法实现 Banner 信息的屏蔽，而该方法从 1.4 版本开始被删除，使用 setBannerMode(Banner.Mode)代替，即示例中的方式。

几乎所有 SpringApplication 调整应用行为的方法都与 SpringApplicationBuilder 方法一一对应，结合以下表格中的内容来综合理解。

SpringApplication 方法	SpringApplication Builder 方法	场景说明	默认值	起始版本
setAddCommandLineProperties	addCommandLineProperties	是否添加命令行参数	true	1.0
setAdditionalProfiles	additionalProfiles	添加附加 SpringProfile	空 Set	1.0
setApplicationContextClass	contextClass	关联当前应用的 ApplicationContext 实现类	null	1.0
setBanner	banner	设置 Banner 实现	null	1.2
setBannerMode	bannerMode	设置 Bander.Mode，包括日志（LOG）、控制台（CONSOLE）和关闭	CONSOLE	1.3
setBeanNameGenerator	beanNameGenerator	设置@Configuration Bean 名称的生成器实现	null	1.0
setDefaultProperties	properties	多个重载方法，配置默认的配置项	null	1.0
setEnvironment	environment	关联当前应用的 Environment 实现类	null	1.0
setHeadless	headless	Java 系统属性 java.awt.headless 的配置值，当值为 true 时，键盘、鼠标等图形化界面交互方式失效	true	1.0
setInitializers	无	覆盖更新 ApplicationContextInitializer 集合	所有 META-INF/spring.factories 资源中声明的 ApplicationContextInitializer 集合	1.0
addInitializers	initializers	追加 ApplicationContextInitializer 集合	同上	1.0
setListeners	无	覆盖更新 ApplicationListener 集合	所有 META-INF/spring.factories 资源中声明的 ApplicationListener 集合	1.0

续表

SpringApplication 方法	SpringApplication Builder 方法	场 景 说 明	默 认 值	起始版本
addListeners	listeners	追加 ApplicationListener 集合	同上	1.0
setLogStartupInfo	logStartupInfo	是否日志输出启动时信息	true	1.0
setMainApplicationClass	main	设置 Main Class，主要用于调整日志输出	由 deduceMainApplicationClass()方法推断	1.0
setRegisterShutdownHook	registerShutdownHook	设置是否让 ApplicationContext 注册 ShutdownHook 线程	true	1.0
setResourceLoader	resourceLoader	设置当前 ApplicationContext 的 ResourceLoader	null	1.0
setSources	sources	增加 SpringApplication 配置源	空 Set	1.0
setWebApplicationType	web	设置 WebApplicationType	由 deduceWebApplicationType()方法推断	2.0
setWebEnvironment	web	2.0 版本不推荐使用，设置是否为 Web 类型	False（在 1.x 版本中）	1.0
addPrimarySources	无	添加主配置类	由 SpringApplication 构造参数决定	2.0

除 setInitializers 和 setListeners 方法外，基本上 SpringApplication 配置方法的命名规则都属于 Setter 方式，而 SpringApplicationBuilder 则是属性命名规则，例如以下 SpringApplication 示例：

```
    SpringApplication springApplication = new
SpringApplication(ThinkingInSpringBootApplication.class);
    springApplication.setBannerMode(Banner.Mode.CONSOLE);
    springApplication.setWebApplicationType(WebApplicationType.NONE);
```

```
springApplication.setAdditionalProfiles("prod");
springApplication.setHeadless(true);
```

从代码简洁性上观察，以上方式似乎略显烦琐，将其调整成 SpringApplicationBuilder 方式后，明显地感觉简明流畅：

```
new SpringApplicationBuilder(DiveInSpringBootApplication.class)
    .bannerMode(Banner.Mode.CONSOLE)
    .web(WebApplicationType.NONE)
    .profiles("prod")
    .headless(true)
    .run(args);
```

这大概是官方称 SpringApplicationBuilder 为“23.4 Fluent Builder API”的原因。同时，SpringApplicationBuilder 也是 Builder 模式的实现，所以其配置方法均返回 this 对象，如 SpringApplicationBuilder#sources(Set)方法：

```
public class SpringApplicationBuilder {
...
 public SpringApplicationBuilder sources(Class<?>... sources) {
     this.sources.addAll(new LinkedHashSet<>(Arrays.asList(sources)));
     return this;
 }
...
}
```

从底层实现上分析，该配置方法实际作用于 SpringApplication 对象。值得一提的是，SpringApplication#setSources(Set)或 SpringApplicationBuilder#sources(Set)方法用于增加 SpringApplication 配置源，也是下一节需要讨论的议题。

10.2.3 增加 SpringApplication 配置源

在“理解 SpringApplication 主配置类”一节中，了解了 SpringApplication 构造参数 primarySources 是从 Spring Boot 2.0 开始引入的，换言之，它在 Spring Boot 1.x 中并不存在。通常开发人员将引导类作为主配置源，所以 SpringApplication 在前后版本中的使用方式是相同的，并且运行时也无法察觉其中变化，其主要原因在于 Spring Boot 1.x 的主配置类以 SpringApplication 配置源成员的形式存在，用属性 sources 存储：

```
public class SpringApplication {
```

```
    ...
    private final Set<Object> sources = new LinkedHashSet<Object>();
    ...
    public SpringApplication(Object... sources) {
        initialize(sources);
    }
    ...
    private void initialize(Object[] sources) {
        if (sources != null && sources.length > 0) {
            this.sources.addAll(Arrays.asList(sources));
        }
        ...
    }
    ...
}
```

> 以上实现源码源于 Spring Boot `1.5.10.RELEASE`。
> Maven GAV 的坐标为：`org.springframework.boot:spring-boot:1.5.10.RELEASE`。

而属性 `sources` 并没有在 Spring Boot 2.0 中移除，仍旧充当配置源的角色，不过 Spring Boot 官方文档在迁移到 2.0 版本的过程中，工作人员忽视了 1.x 版本和 2.0 版本在实现上的差别，导致“23.3 Customizing SpringApplication”一节中的提示文字存在一些瑕疵：

> The constructor arguments passed to `SpringApplication` are configuration sources for Spring beans. In most cases, these are references to `@Configuration` classes, but they could also be references to XML configuration or to packages that should be scanned.

前半部分的文字描述是正确的，然而最后一句提到“XML 配置或 `package` 也能作为引用”则在 Spring Boot 2.0 中存在问题，因为从该版本开始，`SpringApplication` 构造器的签名发生了变化，即方法 `SpringApplication(Object...)` 和 `SpringApplication (ResourceLoader, Object...)` 的“`Object...`”参数被替换为“`Class...`”，所以 XML 配置文件或 packages 无法作为 `SpringApplication` 构造器参数传递，只能选择 `SpringApplication#setSources(Set)` 方法传递。

尽管文档出现这样的错误，不过它却透漏了一些信息。换言之，除 `@Configuration` Class 作为 Spring Bean 的配置源外，至少还有 XML 配置文件和 `package` 作为其他配置源。因此，可根据以上线索展开对 `SpringApplication` 配置源的讨论。

理解 SpringApplication 配置源

Spring Boot 2.0 的 SpringApplication 配置源分别来自主配置类、@Configuration Class、XML 配置文件和 package，而在 Spring Boot 1.x 中并不区分主配置类与@Configuration Class，无论哪个版本，它们统称为“配置源”。在 Spring Boot 2.0 中用 SpringApplication#getAllSources() 方法表示：

```
public class SpringApplication {
    ...
    private Set<Class<?>> primarySources;

    private Set<String> sources = new LinkedHashSet<>();
    ...
    public Set<Object> getAllSources() {
        Set<Object> allSources = new LinkedHashSet<>();
        if (!CollectionUtils.isEmpty(this.primarySources)) {
            allSources.addAll(this.primarySources);
        }
        if (!CollectionUtils.isEmpty(this.sources)) {
            allSources.addAll(this.sources);
        }
        return Collections.unmodifiableSet(allSources);
    }
    ...
}
```

而在 Spring Boot 1.x 中则用 SpringApplication#getSources()方法描述：

```
public class SpringApplication {
    ...
    private final Set<Object> sources = new LinkedHashSet<Object>();
    ...
    public Set<Object> getSources() {
        return this.sources;
    }
    ...
}
```

如果仅从获取“配置源”的语义分析，则两个方法并没有明显的区别。无论 primarySources 属性，还是 sources 属性，均为 LinkedHashSet 实例，所以不但包含配置源去重语义，而且说明配置源是有序的。不过 Spring Boot 2.0 和 1.x 的实现在设置配置源上出现了差异，后者允许参数为 Class、Class 名称、Package、Package 名称或 XML 配置资源：

```
public class SpringApplication {
...
public void setSources(Set<Object> sources) {
     Assert.notNull(sources, "Sources must not be null");
     this.sources.addAll(sources);
}
...
}
```

而前者的语义范围有所缩小，故方法签名也发生了变化：

```
public class SpringApplication {
...
public void setSources(Set<String> sources) {
     Assert.notNull(sources, "Sources must not be null");
     this.sources = new LinkedHashSet<>(sources);
}
...
}
```

该方法退化为仅允许 Class 名称、Package 名称和 XML 配置资源配置，再结合官方文档，进一步说明了 SpringApplication 配置源的来源。不过这些并非直接的证据，仍需结合代码加以分析，这部分的内容将在“准备 ApplicationContext 阶段”章节中进行讨论。

10.2.4　调整 Spring Boot 外部化配置

调整 Spring Boot 外部化配置作为自定义 SpringApplication 的能力之一，源于官方文档的描述：

> It is also possible to configure the SpringApplication by using an application.properties file. See *Chapter 24, Externalized Configuration* for details.

文档如此描述似乎过于草率，它完全可以描述得更直白，如文中的 application.properties 文件，实际上它可覆盖 SpringApplication#setDefaultProperties 方法的设置，从而影响 SpringApplication 的行为。同时，SpringApplication#setDefaultProperties 也能影响

application.properties 文件搜索的路径，通过属性 spring.config.location 或 spring.config.additional-location 实现，详情请参考“外部化配置”章中的“来源于 Spring Boot 应用属性文件（application.properties）”一节。

当 SpringApplication 构建并配置妥当后，其初始化阶段完成，下一步进入运行阶段。

第 11 章 SpringApplication 运行阶段

SpringApplication 运行阶段属于核心过程，完整地围绕 run(String...)方法展开。该过程结合初始化阶段完成的状态，进一步完善了运行时所需要准备的资源，随后启动 Spring 应用上下文，在此期间伴随 Spring Boot 和 Spring 事件的触发，形成完整的 SpringApplication 生命周期。因此，下面将围绕以下三个子议题进行讨论：

- SpringApplication 准备阶段；
- ApplicationContext 启动阶段；
- ApplicationContext 启动后阶段。

11.1 SpringApplication 准备阶段

本阶段属于 ApplicationContext 启动的前一阶段，以 Spring Boot 2.0.2.RELEASE 实现为例，涉及的范围从 run(String...) 方法调用开始，到 refreshContext(ConfigurableApplicationContext)调用前：

```
public class SpringApplication {
...
public ConfigurableApplicationContext run(String... args) {
```

```
        StopWatch stopWatch = new StopWatch();
        stopWatch.start();
        ConfigurableApplicationContext context = null;
        Collection<SpringBootExceptionReporter> exceptionReporters = new ArrayList<>();
        configureHeadlessProperty();
        SpringApplicationRunListeners listeners = getRunListeners(args);
        listeners.starting();
        try {
            ApplicationArguments applicationArguments = new DefaultApplicationArguments(
                    args);
            ConfigurableEnvironment environment = prepareEnvironment(listeners,
                    applicationArguments);
            configureIgnoreBeanInfo(environment);
            Banner printedBanner = printBanner(environment);
            context = createApplicationContext();
            exceptionReporters = getSpringFactoriesInstances(
                    SpringBootExceptionReporter.class,
                    new Class[] { ConfigurableApplicationContext.class }, context);
            prepareContext(context, environment, listeners, applicationArguments,
                    printedBanner);
            refreshContext(context);
            ...
        }
        ...
    }
    ...
}
```

除 StopWatch 等少数无足轻重的对象外，该过程依次准备的核心对象为：ApplicationArguments、SpringApplicationRunListeners、ConfigurableEnvironment、Banner、ConfigurableApplicationContext 和 SpringBootExceptionReporter 集合，或许仅对 ConfigurableApplicationContext 印象深刻，ConfigurableEnvironment 次之，而对其他对象无感。为了加深对它们的理解，接下来按照以上脉络逐一讨论。

11.1.1 理解 SpringApplicationRunListeners

按照代码实现逻辑，SpringApplicationRunListeners 是由 getRunListeners (String[])

方法创建的：

```
public class SpringApplication {
 ...
 private SpringApplicationRunListeners getRunListeners(String[] args) {
     Class<?>[] types = new Class<?>[] { SpringApplication.class, String[].class };
     return new SpringApplicationRunListeners(logger, getSpringFactoriesInstances(
             SpringApplicationRunListener.class, types, this, args));
 }
 ...
}
```

其中 SpringApplicationRunListeners 属于组合模式的实现，内部关联了 SpringApplicationRunListener 的集合：

```
class SpringApplicationRunListeners {

 private final Log log;

 private final List<SpringApplicationRunListener> listeners;

 SpringApplicationRunListeners(Log log,
         Collection<? extends SpringApplicationRunListener> listeners) {
     this.log = log;
     this.listeners = new ArrayList<>(listeners);
 }

 public void starting() {
     for (SpringApplicationRunListener listener : this.listeners) {
         listener.starting();
     }
 }
 ...
}
```

SpringApplicationRunListeners 的结构简单，因此理解并不困难。相反，SpringApplicationRunListener 仍旧是一个“黑盒”，需要将其“拆解”。

11.1.2 理解 SpringApplicationRunListener

按照字面意思，SpringApplicationRunListener 可理解为 Spring Boot 应用的运行时监听器，其监听方法被 SpringApplicationRunListeners 迭代地执行，如以上 starting()方法，其他方法还包括：

- environmentPrepared(ConfigurableEnvironment)；
- contextPrepared(ConfigurableApplicationContext)；
- contextLoaded(ConfigurableApplicationContext)；
- started(ConfigurableApplicationContext)；
- running(ConfigurableApplicationContext)；
- failed(ConfirableApplicationContext context, Throwableexception)。

以上监听方法与运行阶段的对应关系如下表所示。

监 听 方 法	运行阶段说明	Spring Boot 起始版本
starting()	Spring 应用刚启动	1.0
environmentPrepared(ConfigurableEnvironment)	ConfigurableEnvironment 准备妥当，允许将其调整	1.0
contextPrepared(ConfigurableApplicationContext)	ConfigurableApplicationContext 准备妥当，允许将其调整	1.0
contextLoaded(ConfigurableApplicationContext)	ConfigurableApplicationContext 已装载，但仍未启动	1.0
started(ConfigurableApplicationContext)	ConfigurableApplicationContext 已启动，此时 Spring Bean 已初始化完成	2.0
running(ConfigurableApplicationContext)	Spring 应用正在运行	2.0
failed(ConfigurableApplicationContext,Throwable)	Spring 应用运行失败	2.0

依照 getRunListeners(String[])方法的实现，SpringApplicationRunListener 集合则来自 getSpringFactoriesInstances(Class,Class[],Object...)方法，尽管它看起来似曾相识，如“加载 Spring 应用上下文初始化器（ApplicationContextInitializer）”一节中曾涉及的 getSpringFactoriesInstances(Class) 方法。明显地，两者属于 getSpringFactories Instances 的重载方法，也符合参数较少的方法调用较多者的原则：

```
public class SpringApplication {
...
```

```
    private <T> Collection<T> getSpringFactoriesInstances(Class<T> type) {
        return getSpringFactoriesInstances(type, new Class<?>[] {});
    }

    private <T> Collection<T> getSpringFactoriesInstances(Class<T> type,
            Class<?>[] parameterTypes, Object... args) {
        ClassLoader classLoader = Thread.currentThread().getContextClassLoader();
        // Use names and ensure unique to protect against duplicates
        Set<String> names = new LinkedHashSet<>(
                SpringFactoriesLoader.loadFactoryNames(type, classLoader));
        List<T> instances = createSpringFactoriesInstances(type, parameterTypes,
                classLoader, args, names);
        AnnotationAwareOrderComparator.sort(instances);
        return instances;
    }
    ...
}
```

当时的讨论曾提到:"当创建 ApplicationContextInitializer 实例集合(名称集合用 names 表示)时，**ApplicationContextInitializer 实现类必须存在默认构造器**"。换言之，SpringApplicationRunListener 的构造器参数必须依次为 SpringApplication 和 String[]类型，回顾 getRunListeners(String[])实现:

```
public class SpringApplication {
    ...
    private SpringApplicationRunListeners getRunListeners(String[] args) {
        Class<?>[] types = new Class<?>[] { SpringApplication.class, String[].class };
        return new SpringApplicationRunListeners(logger, getSpringFactoriesInstances(
                SpringApplicationRunListener.class, types, this, args));
    }
    ...
}
```

同时，结合 SpringFactoriesLoader 机制，Spring Boot SpringApplicationRunListener 内建实现非常轻松地被定位:

```
# Run Listeners
org.springframework.boot.SpringApplicationRunListener=\
org.springframework.boot.context.event.EventPublishingRunListener
```

> 以上 SpringApplicationRunListener 位于 org.springframework.boot:spring- boot:2.0.2.RELEASE 内的 META-INF/spring.factories 资源中。

EventPublishingRunListener 作为 Spring Boot 唯一的内建实现，完全符合上述构造器参数签名的约束：

```
public class EventPublishingRunListener implements SpringApplicationRunListener, Ordered {

    private final SpringApplication application;

    private final String[] args;

    private final SimpleApplicationEventMulticaster initialMulticaster;

    public EventPublishingRunListener(SpringApplication application, String[] args) {
        this.application = application;
        this.args = args;
        this.initialMulticaster = new SimpleApplicationEventMulticaster();
        for (ApplicationListener<?> listener : application.getListeners()) {
            this.initialMulticaster.addApplicationListener(listener);
        }
    }

    @Override
    public int getOrder() {
        return 0;
    }

    @Override
    public void starting() {
        this.initialMulticaster.multicastEvent(
                new ApplicationStartingEvent(this.application, this.args));
    }
    ...
}
```

该构造器将参数 application 和 args 均与属性关联，并且将根据 SpringApplication 已关联的 ApplicationListener 实例列表动态地添加到 SimpleApplicationEventMulticaster 对象

中。SimpleApplicationEventMulticaster 源于 Spring Framework，实现 ApplicationEventMulticaster 接口，用于发布 Spring 应用事件（ApplicationEvent）。因此，EventPublishingRunListener 实际上充当 Spring Boot 事件发布者的角色，如在 starting()方法中发布 ApplicationStartingEvent。

11.1.3 理解 Spring Boot 事件

尽管 SpringApplicationRunListener 和 Spring Boot 事件（SpringApplicationEvent）从 Spring Boot 1.0 开始引入，然而纵贯 Spring Boot 1.x~2.0 的发展，监听方法与 Spring Boot 事件的对应关系也发生了变化，如下表所示。

监 听 方 法	Spring Boot 事件	Spring Boot 起始版本
starting()	ApplicationStartingEvent	1.5
environmentPrepared(ConfigurableEnvironment)	ApplicationEnvironmentPreparedEvent	1.0
contextPrepared(ConfigurableApplicationContext)		1.0
contextLoaded(ConfigurableApplicationContext)	ApplicationPreparedEvent	1.0
started(ConfigurableApplicationContext)	ApplicationStartedEvent	1.0
running(ConfigurableApplicationContext)	ApplicationReadyEvent	1.3
failed(ConfigurableApplicationContext,Throwable)	ApplicationFailedEvent	1.0

值得注意的是，SpringApplicationRunListener 是 Spring Boot 应用运行时监听器，并非 Spring Boot 事件监听器。不过 Spring Boot 官方文档对 SpringApplicationRunListener 的部分只字未提，开发人员只要遵照 SpringApplicationRunListener 构造器参数约定，以及结合 SpringFactoriesLoader 机制，完全能够将该接口进行扩展。

换言之，以上 Spring Boot 事件所对应的 ApplicationListener 实现是由 SpringApplication 构造器参数关联并添加到属性 SimpleApplicationEventMulticaster#initialMulticaster 中的。比如 SpringApplicationRunListener#starting()方法运行后，ApplicationStartingEvent 随即触发，此时 initialMulticaster 同步地执行 ApplicationListener<ApplicationStartingEvent>集合的监听回调方法 onApplicationEvent(ApplicationStartingEvent)，这些行为保证均源于 Spring Framework 事件/监听器的机制。因此，对于 Spring Boot 应用而言，Spring Boot 事件和 Spring 事件是存在差异的，按照官方文档“23.5 Application Events and Listeners”的说法是：

> In addition to the usual Spring Framework events, such as ContextRefreshedEvent, a `SpringApplication` sends some additional application events.

官方进一步解释了这些事件的顺序和时机：

> 1. An `ApplicationStartingEvent` is sent at the start of a run but before any processing, except for the registration of listeners and initializers.
> 2. An `ApplicationEnvironmentPreparedEvent` is sent when the `Environment` to be used in the context is known but before the context is created.
> 3. An `ApplicationPreparedEvent` is sent just before the refresh is started but after bean definitions have been loaded.
> 4. An `ApplicationStartedEvent` is sent after the context has been refreshed but before any application and command-line runners have been called.
> 5. An `ApplicationReadyEvent` is sent after any application and command-line runners have been called. It indicates that the application is ready to service requests.
> 6. An `ApplicationFailedEvent` is sent if there is an exception on startup.

如果熟悉Spring事件机制，则不难理解Spring事件是由Spring应用上下文`ApplicationContext`对象触发的。然而 Spring Boot 事件的发布者则是 `SpringApplication.initialMulticaster` 属性（`SimpleApplicationEventMulticaster` 类型），并且 `SimpleApplicationEventMulticaster` 也来自 Spring Framework，那么，Spring Boot 事件与 Spring 事件必然存在某种联系，同时两者也存在差异。因此，带着这些疑惑，接下来深入讨论 Spring 和 Spring Boot 的事件/监听机制。

11.1.4 理解 Spring 事件/监听机制

从实现的角度分析，Spring 事件/监听机制属于事件/监听器模式，可视为观察者模式（Observer Pattern）的扩展。早在 Java 1.0 时，观察者模式就被 Java SE API `java.util.Observable` 和 `java.util.Observer` 实现，前者可认为是数据的发布者，后者为数据的接受者。`Observable` 和 `Observer` 的关联关系为一对多或多对多。观察者模式中传播的数据更为抽象，如 `Observable` 可发布任意 `Object`，而事件/监听器模式所发布的内容则有类型限制，在 Java 中，它必须是 `java.util.EventObject` 对象。虽然在 Java 编程语言层面并无此限制，不过这是行之有年的“业界”规则，所以，无论 Java Beans，还是 Java AWT/Swing，都遵照该规则，Spring 事件自然也不会例外，所以 Spring 事件抽象类 `ApplicationEvent` 必然扩展 `EventObject`：

```
public abstract class ApplicationEvent extends EventObject {
```

```
    ...
    /**
     * Create a new ApplicationEvent.
     * @param source the object on which the event initially occurred (never {@code null})
     */
    public ApplicationEvent(Object source) {
        super(source);
        this.timestamp = System.currentTimeMillis();
    }
    ...
}
```

然而 EventObject 并不提供默认构造器，它需要外部传递一个名为 source 的构造器参数，用于记录并跟踪事件的来源，如 Spring 事件 ContextRefreshedEvent，其事件源为当前 ApplicationContext 对象。

同时，Java 事件监听者必须是 EventListener 实例，不过 EventListener 仅为标签接口，内部并没有提供任何实现方法：

```
/**
 * A tagging interface that all event listener interfaces must extend.
 * @since JDK1.1
 */
public interface EventListener {
}
```

正如同其 JavaDoc 所述："所有的事件监听器接口必须扩展本标签接口"。当然 EventListener 如此设计并非毫无道理，因为它很难适用于所有事件监听器场景，如 Java AWT 鼠标事件监听器 MouseListener，它就监听了鼠标点击、鼠标按下及鼠标释放等事件：

```
public interface MouseListener extends EventListener {

    /**
     * Invoked when the mouse button has been clicked (pressed
     * and released) on a component.
     */
    public void mouseClicked(MouseEvent e);

    /**
     * Invoked when a mouse button has been pressed on a component.
```

```
     */
    public void mousePressed(MouseEvent e);

    /**
     * Invoked when a mouse button has been released on a component.
     */
    public void mouseReleased(MouseEvent e);

    /**
     * Invoked when the mouse enters a component.
     */
    public void mouseEntered(MouseEvent e);

    /**
     * Invoked when the mouse exits a component.
     */
    public void mouseExited(MouseEvent e);
}
```

MouseListener 这样的设计有利有弊，其利在于一个实现可以集中处理不同的事件，其弊则是未来接口需要增加监听方法时，客户端不得不为此变更而适配。大多数情况下，实现类没有必要实现所有方法，通常 API 层面又提供了监听接口的适配器实现，如 MouseAdapter，它提供监听空实现的适配方法。当然 Java 8 interface 的 default 方法特性能够解决以上问题，不过这是后话了。

明显地，Spring 事件监听器实现采取了相反的设计理念，通过限定监听方法数量，仅抽象单一方法 onApplicationEvent(ApplicationEvent)，将其用于监听 Spring 事件 ApplicationEvent。

然而由于 ApplicationEvent 并非具体类，自然会存在不同的派生实现，那么作为单一监听方法的 onApplicationEvent(ApplicationEvent)应该如何区分不同类型的 ApplicationEvent 实例呢？Spring Framework 3.0 之前的 ApplicationListener 基本是无解的，它必须监听所有 ApplicationContext，如果要过滤不同类型的事件，则需借助 instanceof 方式进行筛选：

```
public interface ApplicationListener extends EventListener {

 /**
  * Handle an application event.
  * @param event the event to respond to
  */
 void onApplicationEvent(ApplicationEvent event);
```

```
}
```

> 以上实现源码源于 Spring Framework 2.5.6.SEC03。
> Maven GAV 坐标为：org.springframework:spring-context:2.5.6.SEC03。

这种情况从 Spring Framework 3.0 开始得到改善，ApplicationListener 支持 ApplicationEvent 泛型监听。当 ContextRefreshedEvent 事件发布后，ApplicationListener<ContextRefreshedEvent>实现的 onApplicationEvent 方法仅监听具体 ApplicationEvent 实现，不再监听所有的 Spring 事件，无须借助 instanceof 方式进行筛选：

```
public interface ApplicationListener<E extends ApplicationEvent> extends EventListener {

    /**
     * Handle an application event.
     * @param event the event to respond to
     */
    void onApplicationEvent(E event);

}
```

由于泛型参数的限制，泛型化的 ApplicationListener 无法监听不同类型的 ApplicationEvent，如 ApplicationListener<ContextRefreshedEvent>无法同时监听 ContextStartedEvent。如果继续使用 ApplicationListener<ApplicationEvent>，那么又回到 3.0 版本前的时代。为此，Spring Framework 3.0 又引入了 SmartApplicationListener 接口：

```
 */
public interface SmartApplicationListener extends ApplicationListener<ApplicationEvent>,
Ordered {

    /**
     * Determine whether this listener actually supports the given event type.
     */
    boolean supportsEventType(Class<? extends ApplicationEvent> eventType);

    /**
     * Determine whether this listener actually supports the given source type.
     */
    boolean supportsSourceType(Class<?> sourceType);
```

```
}
```

该接口通过 supports*方法过滤需要监听的 ApplicationEvent 类型和事件源类型，从而达到监听不同类型的 ApplicationEvent 的目的，如 Spring Boot 外部化应用配置文件（application.properties）的事件监听器实现 ConfigFileApplicationListener，它也是 SmartApplicationListener 的实现类，并监听 ApplicationEnvironmentPreparedEvent 和 ApplicationPreparedEvent 这两个 Spring Boot 事件，具体实现细节请参考“外部化配置”章节。

当然目前所讨论的范围局限于 Spring 事件/监听器的编程模型接口，即 ApplicationEvent 和 ApplicationListener，下面讨论 Spring 事件发布。

1. Spring 事件发布

在讨论 EventPublishingRunListener 时，曾提到 Spring Boot 事件是由 SimpleApplicationEventMulticaster 发布的，它也来自 Spring Framework，并且是 ApplicationEventMulticaster 接口的实现类。该接口主要承担两种职责，一是关联 ApplicationListener，二是广播 ApplicationEvent：

```java
public interface ApplicationEventMulticaster {

    /**
     * Add a listener to be notified of all events.
     * @param listener the listener to add
     */
    void addApplicationListener(ApplicationListener<?> listener);

    /**
     * Add a listener bean to be notified of all events.
     * @param listenerBeanName the name of the listener bean to add
     */
    void addApplicationListenerBean(String listenerBeanName);

    /**
     * Remove a listener from the notification list.
     * @param listener the listener to remove
     */
    void removeApplicationListener(ApplicationListener<?> listener);
```

```
    /**
     * Remove a listener bean from the notification list.
     * @param listenerBeanName the name of the listener bean to add
     */
    void removeApplicationListenerBean(String listenerBeanName);

    /**
     * Remove all listeners registered with this multicaster.
     * <p>After a remove call, the multicaster will perform no action
     * on event notification until new listeners are being registered.
     */
    void removeAllListeners();

    /**
     * Multicast the given application event to appropriate listeners.
     * <p>Consider using {@link #multicastEvent(ApplicationEvent, ResolvableType)}
     * if possible as it provides a better support for generics-based events.
     * @param event the event to multicast
     */
    void multicastEvent(ApplicationEvent event);

    /**
     * Multicast the given application event to appropriate listeners.
     * <p>If the {@code eventType} is {@code null}, a default type is built
     * based on the {@code event} instance.
     * @param event the event to multicast
     * @param eventType the type of event (can be null)
     * @since 4.2
     */
    void multicastEvent(ApplicationEvent event, @Nullable ResolvableType eventType);

}
```

1）ApplicationEventMulticaster 注册 ApplicationListener

从ApplicationEventMulticaster接口定义分析，前半部分方法均与ApplicationListener相关，包括直接关联的方法：addApplicationListener(ApplicationListener)、removeApplicationListener(ApplicationListener)和 removeAllListeners()。其实现类SimpleApplicationEventMulticaster与ApplicationListener关联关系如以下UML图所示。

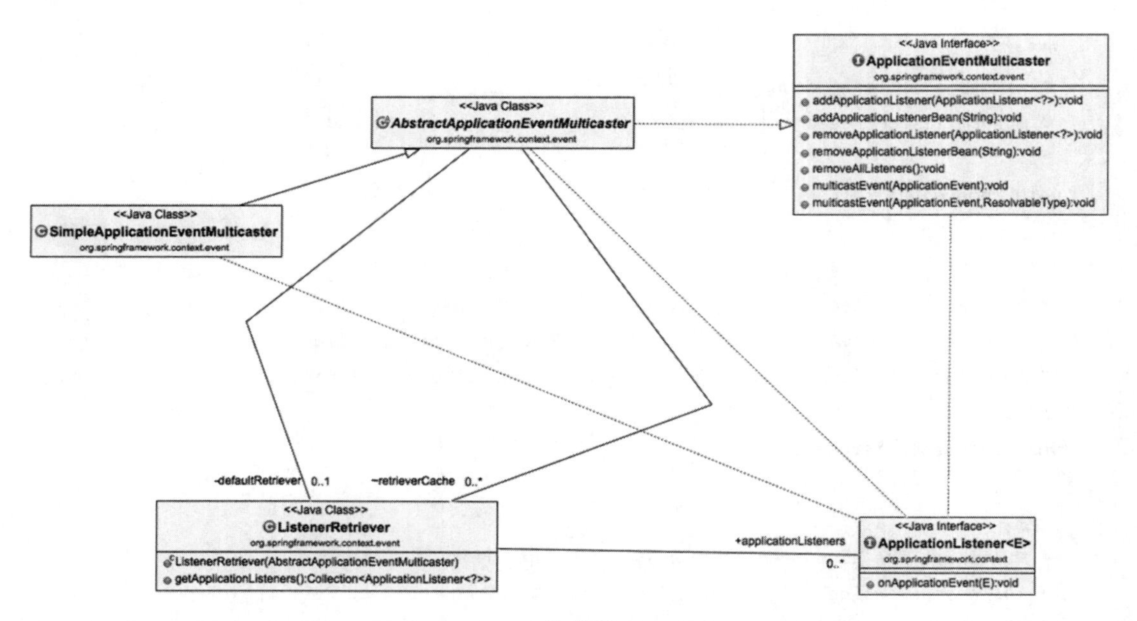

SimpleApplicationEventMulticaster 的基类 AbstractApplicationEventMulticaster 并未直接关联 ApplicationListener，而是通过属性 defaultRetriever 和 retrieverCache 关联的，映射数量分别是 0..1 和 0..N：

```
public abstract class AbstractApplicationEventMulticaster
        implements ApplicationEventMulticaster, BeanClassLoaderAware, BeanFactoryAware {

    private final ListenerRetriever defaultRetriever = new ListenerRetriever(false);

    final Map<ListenerCacheKey, ListenerRetriever> retrieverCache = new
ConcurrentHashMap<>(64);
    ...
}
```

> ListenerRetriever 和 ListenerCacheKey 均为 AbstractApplicationEventMulticaster 的内置类。

按照 ApplicationEventMulticaster 接口关联方法的定义，AbstractApplicationEventMulticaster 与 ApplicationListener 应该是一对多的关系。结合上图分析，ListenerRetriever 与 ApplicationListener 是一对多的关系，而 AbstractApplicationEventMulticaster 与 ListenerRetriever 也是一对多的关系。说明 AbstractApplicationEvent

Multicaster 将 ApplicationListener 做了分类，再结合 retrieverCache 的定义，它是一个 ListenerCacheKey 为 Key、ListenerRetriever 为 Value 的 Map 类型的缓存，同时 ListenerCacheKey 关联了事件类型和数据源类型：

```
public abstract class AbstractApplicationEventMulticaster
        implements ApplicationEventMulticaster, BeanClassLoaderAware, BeanFactoryAware {
 ...
    private static final class ListenerCacheKey implements Comparable<ListenerCacheKey> {

        private final ResolvableType eventType;

        @Nullable
        private final Class<?> sourceType;

        public ListenerCacheKey(ResolvableType eventType, @Nullable Class<?> sourceType) {
            Assert.notNull(eventType, "Event type must not be null");
            this.eventType = eventType;
            this.sourceType = sourceType;
        }

        @Override
        public boolean equals(Object other) {
            if (this == other) {
                return true;
            }
            ListenerCacheKey otherKey = (ListenerCacheKey) other;
            return (this.eventType.equals(otherKey.eventType) &&
                    ObjectUtils.nullSafeEquals(this.sourceType, otherKey.sourceType));
        }

        @Override
        public int hashCode() {
            return this.eventType.hashCode() * 29 +
ObjectUtils.nullSafeHashCode(this.sourceType);
        }
        ...
    }
    ...
```

```
}
```

构造参数 eventType 和 sourceType 作为 equals(Object)和 hashCode()方法的计算因子，帮助 AbstractApplicationEventMulticaster.retrieverCache 属性去重。换言之，通过事件类型和事件源类型的组合，AbstractApplicationEventMulticaster 是通过 ApplicationEvent 与 ListenerRetriever 进行关联的，实质上是将 ApplicationEvent 与 ApplicationListener 集合进行关联，因此 AbstractApplicationEventMulticaster#getApplicationListeners (ApplicationEvent,ResolvableType) 方法返回 ApplicationEvent 关联的 ApplicationListener 集合：

```
public abstract class AbstractApplicationEventMulticaster
        implements ApplicationEventMulticaster, BeanClassLoaderAware, BeanFactoryAware {
...
protected Collection<ApplicationListener<?>> getApplicationListeners(
            ApplicationEvent event, ResolvableType eventType) {

        Object source = event.getSource();
        Class<?> sourceType = (source != null ? source.getClass() : null);
        ListenerCacheKey cacheKey = new ListenerCacheKey(eventType, sourceType);

        // Quick check for existing entry on ConcurrentHashMap...
        ListenerRetriever retriever = this.retrieverCache.get(cacheKey);
        if (retriever != null) {
            return retriever.getApplicationListeners();
        }
        ...
}
...
}
```

该方法的实现逻辑远比当前展示的代码复杂，不过此处可以观其大意，假设 retrieverCache 命中，并且返回 ListenerRetriever 对象的 getApplicationListeners()方法内容：

```
public abstract class AbstractApplicationEventMulticaster
        implements ApplicationEventMulticaster, BeanClassLoaderAware, BeanFactoryAware {
...
private class ListenerRetriever {

        public final Set<ApplicationListener<?>> applicationListeners;
```

```
        public final Set<String> applicationListenerBeans;

        private final boolean preFiltered;

        public ListenerRetriever(boolean preFiltered) {
            this.applicationListeners = new LinkedHashSet<>();
            this.applicationListenerBeans = new LinkedHashSet<>();
            this.preFiltered = preFiltered;
        }

        public Collection<ApplicationListener<?>> getApplicationListeners() {
            LinkedList<ApplicationListener<?>> allListeners = new LinkedList<>();
            for (ApplicationListener<?> listener : this.applicationListeners) {
                allListeners.add(listener);
            }
            if (!this.applicationListenerBeans.isEmpty()) {
                BeanFactory beanFactory = getBeanFactory();
                for (String listenerBeanName : this.applicationListenerBeans) {
                    try {
                        ApplicationListener<?> listener =
beanFactory.getBean(listenerBeanName, ApplicationListener.class);
                        if (this.preFiltered || !allListeners.contains(listener)) {
                            allListeners.add(listener);
                        }
                    }
                    catch (NoSuchBeanDefinitionException ex) {
                        // Singleton listener instance (without backing bean definition)
disappeared -
                        // probably in the middle of the destruction phase
                    }
                }
            }
            AnnotationAwareOrderComparator.sort(allListeners);
            return allListeners;
        }
    }
    ...
```

```
    }
```

最终返回的 ApplicationListener 集合可能直接来源于 addApplicationListener (ApplicationListener)方法，以及 addApplicationListenerBean (String)与 BeanFactory 配合所获取的实例，这些合并的集合再经过 Ordered 或@Order 排序。换言之，Spring Boot 事件监听器均经过排序。

目前仅知晓 ApplicationEventMulticaster 与 ApplicationListener 的关联关系及相关接口方法，接下来介绍与广播事件相关的方法。

2）ApplicationEventMulticaster 广播事件

在分析 ApplicationEventMulticaster 接口定义时，提到了该接口有两个与广播事件相关的方法，分别是 multicastEvent(ApplicationEvent)和 multicastEvent(ApplicationEvent, ResolvableType)，两者均在 SimpleApplicationEventMulticaster 中实现，并且也是 Spring Framework 唯一实现。从这方面来看，Spring 事件也应由 SimpleApplicationEventMulticaster 广播，不过这仅是一种推测。

在实现上，multicastEvent(ApplicationEvent)复用 multicastEvent(ApplicationEvent, ResolvableType)的实现：

```
    public class SimpleApplicationEventMulticaster extends AbstractApplicationEventMulticaster
{
    ...
    @Override
    public void multicastEvent(ApplicationEvent event) {
        multicastEvent(event, resolveDefaultEventType(event));
    }

    @Override
    public void multicastEvent(final ApplicationEvent event, @Nullable ResolvableType eventType)
{
        ResolvableType type = (eventType != null ? eventType : resolveDefaultEventType(event));
        for (final ApplicationListener<?> listener : getApplicationListeners(event, type)) {
            Executor executor = getTaskExecutor();
            if (executor != null) {
                executor.execute(() -> invokeListener(listener, event));
            }
            else {
                invokeListener(listener, event);
```

```
            }
        }
    }

    private ResolvableType resolveDefaultEventType(ApplicationEvent event) {
        return ResolvableType.forInstance(event);
    }
    ...
}
```

> multicastEvent(ApplicationEvent,ResolvableType)方法从 Spring Framework 4.2 开始引入。实际上，前面讨论的getApplicationListeners(ApplicationEvent,ResolvableType)方法也是从 4.2 版本开始重构的。之前的版本没有 ResolvableType 参数——尽管 ResolvableType 是从 Spring Framework 4.0 开始出现的。此处提及 ResolvableType 的目的在于为读者阅读其他 Spring Framework 或 Spring Boot 源码提供一些指引。ResolvableType 是 Spring Framework 为简化 Java 反射 API 而提供的组件，能够轻松地获取泛型类型等。

multicastEvent(ApplicationEvent,ResolvableType) 再利用 getApplicationListeners(ApplicationEvent,ResolvableType)方法的返回值，迭代地执行 ApplicationListener#onApplicationEvent(ApplicationEvent)方法：

```
    public class SimpleApplicationEventMulticaster extends AbstractApplicationEventMulticaster
{
    ...
    protected void invokeListener(ApplicationListener<?> listener, ApplicationEvent event) {
        ErrorHandler errorHandler = getErrorHandler();
        if (errorHandler != null) {
            try {
                doInvokeListener(listener, event);
            }
            catch (Throwable err) {
                errorHandler.handleError(err);
            }
        }
        else {
            doInvokeListener(listener, event);
        }
    }
```

```
    @SuppressWarnings({"unchecked", "rawtypes"})
    private void doInvokeListener(ApplicationListener listener, ApplicationEvent event) {
        try {
            listener.onApplicationEvent(event);
        }
        ...
    }
    ...
}
```

此处有一个细节，当 SimpleApplicationEventMulticaster 允许事件广播时，ApplicationListener 异步处理监听事件。其中，属性 ExecutortaskExecutor 通过 setTaskExecutor(Executor)方法初始化：

```
public class SimpleApplicationEventMulticaster extends AbstractApplicationEventMulticaster
{

    @Nullable
    private Executor taskExecutor;
    ...
    public void setTaskExecutor(@Nullable Executor taskExecutor) {
        this.taskExecutor = taskExecutor;
    }
    ...
}
```

不过，无论 Spring Framework，还是 Spring Boot，均未使用该方法来提升为异步执行，并且由于 EventPublishingRunListener 的封装，使得 Spring Boot 事件监听器无法异步执行。

至此，已了解 ApplicationEventMulticaster 实现类 SimpleApplicationEventMulticaster 注册 ApplicationListener，以及广播 ApplicationEvent 时的处理手段。不过 Spring 事件机制依赖于 SimpleApplicationEventMulticaster，然而并不止于此。接下来继续讨论 Spring ApplicationContext 与 SimpleApplicationEventMulticaster 是如何关联的。

3）ApplicationEventMulticaster 与 ApplicationContext 之间的关系

在 Spring 场景中，类似于 Spring Boot 这样直接使用 SimpleApplicationEventMulticaster 的场景几乎没有。Spring Framework 文档也没有直接出现 SimpleApplicationEventMulticaster 的使用方法，仅提及 ApplicationEventMulticaster 一次：

> If another strategy for event publication becomes necessary, refer to the JavaDoc for Spring's `ApplicationEventMulticaster` interface.

同时，Spring Framework 文档在“1.15. Additional capabilities of the ApplicationContext”章节继续说明，开发人员可使用 `ApplicationEventPublisher` 发布 `ApplicationEvent`：

> To enhance `BeanFactory` functionality in a more framework-oriented style the context package also provides the following functionality:
>
> - ...
> - *Event publication* to namely beans implementing the `ApplicationListener` interface, through the use of the `ApplicationEventPublisher` interface.
> - ...

再比较接口 `ApplicationEventPublisher` 与 `ApplicationEventMulticaster` API 的定义，似乎感觉不到两者存在某种必然的联系：

```
@FunctionalInterface
public interface ApplicationEventPublisher {

    /**
     * Notify all <strong>matching</strong> listeners registered with this
     * application of an application event. Events may be framework events
     * (such as RequestHandledEvent) or application-specific events.
     * @param event the event to publish
     * @see org.springframework.web.context.support.RequestHandledEvent
     */
    default void publishEvent(ApplicationEvent event) {
        publishEvent((Object) event);
    }

    /**
     * Notify all <strong>matching</strong> listeners registered with this
     * application of an event.
     * <p>If the specified {@code event} is not an {@link ApplicationEvent},
     * it is wrapped in a {@link PayloadApplicationEvent}.
     * @param event the event to publish
     * @since 4.2
```

```
     * @see PayloadApplicationEvent
     */
    void publishEvent(Object event);

}
```

ApplicationEventPublisher 仅存在两个发布 ApplicationEvent 的重载方法 publishEvent，并无关联 ApplicationListener 的操作方法。不过 ApplicationEventPublisher 接口被 Spring 上下文接口 ApplicationContext 扩展，因此，无论哪种 Spring 应用上下文实例，均具备发布 ApplicationEvent 的能力。同时，ApplicationContext 的子接口 ConfigurableApplicationContext 提供了添加 ApplicationListener 实例的关联方法 addApplicationListener(ApplicationListener)，不过该方法相较于 ApplicationEventMulticaster 接口而言，其完备性明显不足。以上接口的关系如下图所示。

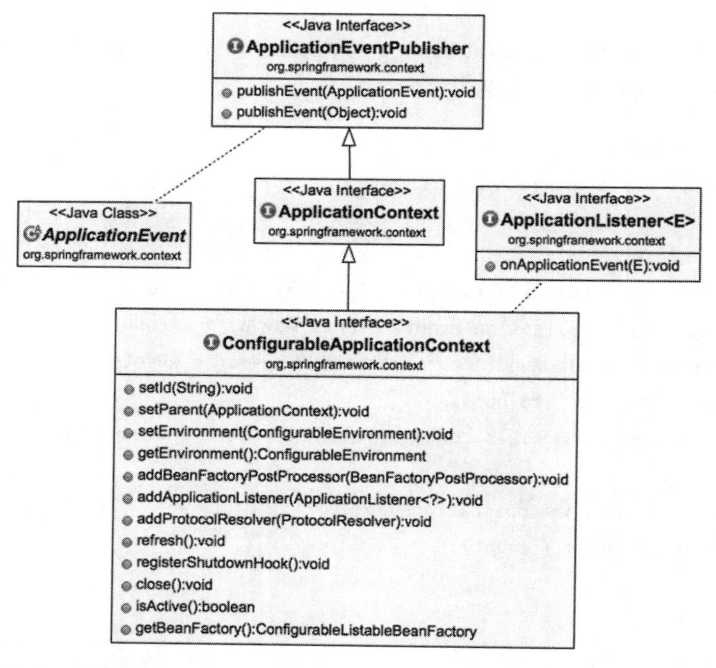

它们之间的关系不难理解，难在如何得到 ApplicationEventPublisher 实例。因此，官方文档又在“1.6.3. Other Aware interfaces”章节中告知开发人员 Spring Bean 实现 ApplicationEventPublisherAware 接口的语义，当 setApplicationEventPublisher(ApplicationEventPublisher) 方法生命周期回调时，获取 ApplicationEventPublisher 实例。下表为 Application EventPublisherAware 接口的编程方式。

Name	Injected Dependency	Explained in...
...		
ApplicationEventPublisherAware	Event publisher of the enclosing ApplicationContext	Additional capabilities of the ApplicationContext
...		

假设偏好注解驱动编程，则 ApplicationEventPublisher 实例可通过@Autowired 的方式注入 Spring Bean，这部分内容在文档的“1.9.2. @Autowired”章节做了补充：

> You can also use @Autowired for interfaces that are well-known resolvable dependencies: BeanFactory, ApplicationContext, Environment, ResourceLoader, ApplicationEventPublisher, and MessageSource.

不难发现，无论通过 Aware 接口回调，还是通过@Autowired 注入，ApplicationEventPublisher 的获取与 ApplicationContext 是紧密关联的。由于 ApplicationContext 扩展了 ApplicationEventPublisher，是否可认为 Spring 应用上下文就是 ApplicationEventPublisher 实例呢？实际上可将该问题再向外延伸，即 ApplicationEventPublisherAware 接口回调时的 ApplicationEventPublisher 实例是否为@Autowired 注入的对象？

或许仍对 ApplicationEventPublisher 没有太多感觉，不过文档中提到了 ResourceLoader 也能被@Autowired 注入，这不禁使人联想到在本书的“走向自动装配”章节中关于 ResourceLoaderAware 接口回调的讨论，当时接触到一个 BeanPostProcessor 实现类，名为 ApplicationContextAwareProcessor，在其 invokeAwareInterfaces(Object)方法执行时，如果当前 Bean 属于 ResourceLoaderAware 对象，则当前 ApplicationContext 对象将作为 setResourceLoader(ResourceLoader)方法的参数。ApplicationEventPublisherAware 接口回调的处理逻辑同样如此：

```
    class ApplicationContextAwareProcessor implements BeanPostProcessor {
    ...
    private void invokeAwareInterfaces(Object bean) {
        if (bean instanceof Aware) {
            ...
            if (bean instanceof ApplicationEventPublisherAware) {
                ((ApplicationEventPublisherAware)
bean).setApplicationEventPublisher(this.applicationContext);
            }
```

```
            ...
        }
    }
    ...
}
```

此处已经解释了在 ApplicationEventPublisherAware Bean 生命周期回调时，Bean 所获得的 ApplicationEventPublisher 实例就是当前 ApplicationContext。

> 在 Spring Framework 中，Aware Bean 生命周期回调均由 ApplicationContextAwareProcessor 处理。

“走向自动装配”章节也曾提到 ApplicationContextAwareProcessor 实例是在 AbstractApplicationContext#prepareBeanFactory(ConfigurableListableBeanFactory) 方法关联到上下文中的，同样地，ApplicationEventPublisher 对象的关联出现在该方法的下方：

```
public abstract class AbstractApplicationContext extends DefaultResourceLoader
        implements ConfigurableApplicationContext {
...
protected void prepareBeanFactory(ConfigurableListableBeanFactory beanFactory) {
        ...
        beanFactory.registerResolvableDependency(ResourceLoader.class, this);
        beanFactory.registerResolvableDependency(ApplicationEventPublisher.class, this);
        beanFactory.registerResolvableDependency(ApplicationContext.class, this);
        ...
        beanFactory.addBeanPostProcessor(new ApplicationListenerDetector(this));
        ...
}
...
}
```

由于以上方法在 Spring 应用上下文启动时调用，同时执行了 ResourceLoader 和 ApplicationEventPublisher Bean 的关联及 ApplicationListenerDetector 的添加操作。因此 ApplicationEventPublisher Bean 与 ApplicationEventPublisherAware 回调实例均为当前 ApplicationContext 对象，同样的结论也适用于 ResourceLoader 和 ResourceLoaderAware。尽管如此，仍旧没有解释 ApplicationEventMulticaster 与 ApplicationContext 的关系。不过距离答案已经越来越近。既然 ApplicationContext 对象是 ApplicationEventPublisher 实例，则不难发现 AbstractApplicationContext 完全实现了 ApplicationEventPublisher 接口：

```
public abstract class AbstractApplicationContext extends DefaultResourceLoader
        implements ConfigurableApplicationContext {
    ...
    @Override
    public void publishEvent(ApplicationEvent event) {
        publishEvent(event, null);
    }
    ...
    @Override
    public void publishEvent(Object event) {
        publishEvent(event, null);
    }
    ...
    protected void publishEvent(Object event, @Nullable ResolvableType eventType) {
        ...
        ApplicationEvent applicationEvent;
        if (event instanceof ApplicationEvent) {
            applicationEvent = (ApplicationEvent) event;
        }
        else {
            applicationEvent = new PayloadApplicationEvent<>(this, event);
            if (eventType == null) {
                eventType = ((PayloadApplicationEvent)
applicationEvent).getResolvableType();
            }
        }

        // Multicast right now if possible - or lazily once the multicaster is initialized
        if (this.earlyApplicationEvents != null) {
            this.earlyApplicationEvents.add(applicationEvent);
        }
        else {
            getApplicationEventMulticaster().multicastEvent(applicationEvent, eventType);
        }

        // Publish event via parent context as well...
        if (this.parent != null) {
            if (this.parent instanceof AbstractApplicationContext) {
```

```
                ((AbstractApplicationContext) this.parent).publishEvent(event, eventType);
            }
            else {
                this.parent.publishEvent(event);
            }
        }
    }
    ...
    ApplicationEventMulticaster getApplicationEventMulticaster() throws IllegalStateException
{
        ...
        return this.applicationEventMulticaster;
    }
    ...
}
```

明显地，在 publishEvent(Object,ResolvableType) 方法中显式地调用了 getApplicationEventMulticaster().multicastEvent(applicationEvent, eventType) 方法，而 applicationEventMulticaster 属性又由 initApplicationEventMulticaster()方法初始化，并且该方法又被“refresh()”：

```
public abstract class AbstractApplicationContext extends DefaultResourceLoader
        implements ConfigurableApplicationContext {
...
@Override
public void refresh() throws BeansException, IllegalStateException {
    synchronized (this.startupShutdownMonitor) {
        ...
            // Initialize event multicaster for this context.
            initApplicationEventMulticaster();
        ...
}
...
protected void initApplicationEventMulticaster() {
    ConfigurableListableBeanFactory beanFactory = getBeanFactory();
    if
    (beanFactory.containsLocalBean(APPLICATION_EVENT_MULTICASTER_BEAN_NAME)) {
        this.applicationEventMulticaster =
```

```
                    beanFactory.getBean(APPLICATION_EVENT_MULTICASTER_BEAN_NAME,
ApplicationEventMulticaster.class);
            ...
        }
        else {
            this.applicationEventMulticaster = new
SimpleApplicationEventMulticaster(beanFactory);
            beanFactory.registerSingleton(APPLICATION_EVENT_MULTICASTER_BEAN_NAME,
this.applicationEventMulticaster);
            ...
        }
    }
    ...
    }
```

在 applicationEventMulticaster 属性的初始化过程中，首先判断当前 Spring 应用上下文是否存在名为"applicationEventMulticaster"且类型为 ApplicationEventMulticaster 的 Spring Bean，如果不存在，则将其构造为 SimpleApplicationEventMulticaster 对象，并且注册为 Spring Bean。

> 之所以如此设计，个人臆断为让 Spring 应用上下文具备自定义 ApplicationEventMulticaster 的弹性。尽管 SimpleApplicationEventMulticaster 是框架内部唯一的 ApplicationEventMulticaster 实现，然而 SimpleApplicationEventMulticaster 能够关联 Executor 对象，从而支持异步事件广播。当然，开发人员也可完全自主开发 ApplicationEventMulticaster 接口的实现。

不过，从 Spring Framework 4.2 开始，相较于 ApplicationEventMulticaster 对象，ApplicationEventPublisher 实例允许发布 Object 对象为负载的事件 PayloadApplicationEvent，随后的流程与其他 ApplicationContext 一致，均通过 ApplicationEventMulticaster#multicastEvent(ApplicationEvent,ResolvableType)方法传播。总之，SimpleApplicationEventMulticaster 既是 Spring Boot 事件广播实现，也是 Spring 事件发布实现。

前面曾提到，ApplicationEventMulticaster 对象提供 ApplicationListener 实例增删查的能力，并且 ConfigurableApplicationContext 也具备注册 ApplicationListener 的方法。那么，是否意味着 ConfigurableApplicationContext 实例存在复用 ApplicationEventMulticaster 对象的可能性呢？答案是肯定的。

AbstractApplicationContext 作为 ConfigurableApplicationContext 的抽象实现类，在

内部已实现了 addApplicationListener(ApplicationListener)方法，并且其子类没有再次覆盖该实现：

```
public abstract class AbstractApplicationContext extends DefaultResourceLoader
        implements ConfigurableApplicationContext {
 ...
 @Override
 public void addApplicationListener(ApplicationListener<?> listener) {
        Assert.notNull(listener, "ApplicationListener must not be null");
        if (this.applicationEventMulticaster != null) {
              this.applicationEventMulticaster.addApplicationListener(listener);
        }
        else {
              this.applicationListeners.add(listener);
        }
 }
 ...
}
```

该实现存在两种可能，当 applicationEventMulticaster 存在时，将 ApplicationListener 参数追加其中，否则添加到 applicationListeners 属性中。因此，applicationListeners 属性是否最终被 applicationEventMulticaster 属性使用尚不清晰，需要分析该属性的使用情况。不难发现，在 registerListeners()方法中，applicationListeners 属性由 getApplicationListeners()方法返回，被添加到 applicationEventMulticaster 属性中：

```
public abstract class AbstractApplicationContext extends DefaultResourceLoader
        implements ConfigurableApplicationContext {
 ...
 public Collection<ApplicationListener<?>> getApplicationListeners() {
        return this.applicationListeners;
 }

 @Override
 public void refresh() throws BeansException, IllegalStateException {
        synchronized (this.startupShutdownMonitor) {
              ...
                    initApplicationEventMulticaster();
```

```
            // Initialize other special beans in specific context subclasses.
            onRefresh();

            // Check for listener beans and register them.
            registerListeners();

            // Instantiate all remaining (non-lazy-init) singletons.
            finishBeanFactoryInitialization(beanFactory);
            ...
    }
    ...
    protected void registerListeners() {
        // Register statically specified listeners first.
        for (ApplicationListener<?> listener : getApplicationListeners()) {
            getApplicationEventMulticaster().addApplicationListener(listener);
        }

        // Do not initialize FactoryBeans here: We need to leave all regular beans
        // uninitialized to let post-processors apply to them!
        String[] listenerBeanNames = getBeanNamesForType(ApplicationListener.class, true,
false);
        for (String listenerBeanName : listenerBeanNames) {
            getApplicationEventMulticaster().addApplicationListenerBean(listenerBeanName);
        }
        ...
    }
    ...
}
```

不仅如此，通过调用 `ApplicationEventMulticaster#addApplicationListenerBean(String)`方法，当前 Spring 应用上下文将 `ApplicationListener` Bean 名称集合也追加至 `applicationEventMulticaster` 属性中，因此也验证了 Spring Framework 官方文档在“1.15.2. Standard and custom events”中的描述：

> If a bean that implements the `ApplicationListener` interface is deployed into the context, every time an `ApplicationEvent` gets published to the `ApplicationContext`, that bean is notified. Essentially, this is the standard *Observer* design pattern.

同时，文档也提到 ApplicationListener 与 ApplicationContext 组合的编程模型为“观察者模式”（Observer）。之所以 applicationEventMulticaster 属性没有直接关联 Application Listener Bean 集合，是因为此时的上下文尚未初始化 Bean。从实现上分析，registerListeners() 在 refresh() 方法中的调用次序处于 initApplicationEventMulticaster() 和 finishBeanFactoryInitialization(ConfigurableListableBeanFactory) 方法之间，其中 initApplicationEventMulticaster()方法为初始化 applicationEventMulticaster 属性（前文已讨论），finishBeanFactoryInitialization(ConfigurableListableBeanFactory)方法则是初始化 Spring 容器中的 Bean。由此看来，ConfigurableApplicationContext#addApplication Listener(ApplicationListener)方法的底层实现还是由其关联 ApplicationEventMulticaster 对象提供的。

总览全局，ApplicationEventMulticaster、ApplicationListener、ApplicationEvent、ApplicationEventPublisher 与 ConfigurableApplicationContext 等之间的关系如下图所示。

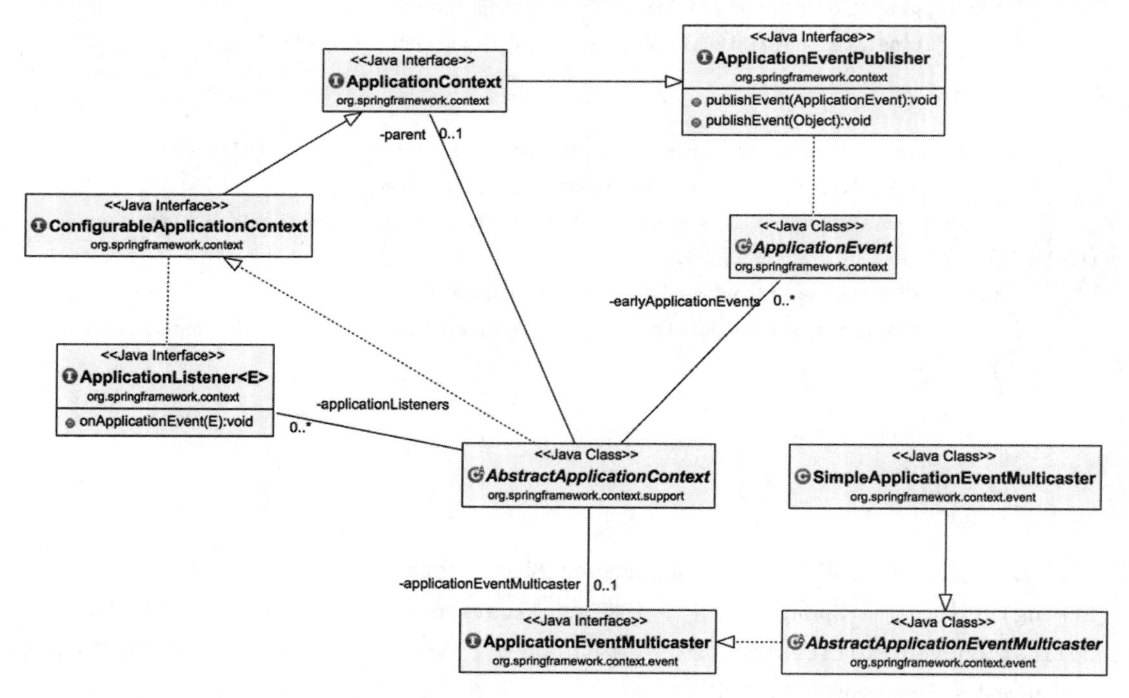

SimpleApplicationEventMulticaster 作为 Spring Framework 唯一的 ApplicationEvent Multicaster 实现，默认情况下，无论在 Spring Boot 使用场景中，还是在传统 Spring 应用中，均充当同步广播事件对象的角色。开发人员只需关注 ApplicationEvent 类型及对应的 ApplicationListener 的实现即可。接下来继续讨论 Spring Framework 内建和自定义事件。

2. Spring 内建事件

Spring 内建事件在 Spring Framework 官方文档的“1.15.2. Standard and custom events”章节中被提到，将 Spring 内建事件绘制成如下表格。

Event	Explanation
ContextRefreshedEvent	Published when the ApplicationContext is initialized or refreshed, for example, using the refresh() method on the ConfigurableApplicationContext interface. "Initialized" here means that all beans are loaded, post-processor beans are detected and activated, singletons are pre-instantiated, and the ApplicationContext object is ready for use. As long as the context has not been closed, a refresh can be triggered multiple times, provided that the chosen ApplicationContext actually supports such "hot" refreshes. For example, XmlWebApplicationContext supports hot refreshes, but GenericApplicationContextdoes not.
ContextStartedEvent	Published when the ApplicationContext is started, using the start() method on the ConfigurableApplicationContext interface. "Started" here means that all Lifecycle beans receive an explicit start signal. Typically this signal is used to restart beans after an explicit stop, but it may also be used to start components that have not been configured for autostart , for example, components that have not already started on initialization.
ContextStoppedEvent	Published when the ApplicationContext is stopped, using the stop() method on the ConfigurableApplicationContext interface. "Stopped" here means that all Lifecycle beans receive an explicit stop signal. A stopped context may be restarted through astart() call.
ContextClosedEvent	Published when the ApplicationContext is closed, using the close() method on the ConfigurableApplicationContext interface. "Closed" here means that all singleton beans are destroyed. A closed context reaches its end of life; it cannot be refreshed or restarted.
RequestHandledEvent	A web-specific event telling all beans that an HTTP request has been serviced. This event is published *after* the request is complete. This event is only applicable to web applications using Spring's DispatcherServlet.

除 RequestHandledEvent 外，其他 Spring 内建事件均与 ConfigurableApplicationContext 的方法调用有关，简要说明，它们分别如下。

- ContextRefreshedEvent：Spring 应用上下文就绪事件；
- ContextStartedEvent：Spring 应用上下文启动事件；
- ContextStoppedEvent：Spring 应用上下文停止事件；

- ContextClosedEvent：Spring 应用上下文关闭事件。

其中，ContextRefreshedEvent 最常见。接下来逐一讨论。

1）Spring 应用上下文就绪事件——ContextRefreshedEvent

当 ConfigurableApplicationContext#refresh()方法执行到 finishRefresh()方法时，Spring 应用上下文发布 ContextRefreshedEvent：

```
public abstract class AbstractApplicationContext extends DefaultResourceLoader
        implements ConfigurableApplicationContext {
    ...
    @Override
    public void refresh() throws BeansException, IllegalStateException {
            ...
                // Last step: publish corresponding event.
                finishRefresh();
            ...
    }
    ...
    protected void finishRefresh() {
        ...
        // Publish the final event.
        publishEvent(new ContextRefreshedEvent(this));
        ...
    }
    ...
}
```

此时，Spring 应用上下文中 Bean 均已完成初始化，并能投入使用。通常 ApplicationListener<ContextRefreshedEvent>实现类监听该事件，用于获取需要的 Bean，防止出现 Bean 提早初始化所带来的潜在风险。

> 通常 BeanPostProcessor 也能用于获取指定的 Bean 对象，由于 BeanFactory 错综复杂的 Bean 处理逻辑，选择 ApplicationListener<ContextRefreshedEvent>是更安全的实践。

2）Spring 应用上下文启停事件——ContextStartedEvent 和 ContextStoppedEvent

Spring 事件 ContextStartedEvent 和 ContextStoppedEvent 较为罕见，两者分别在 AbstractApplicationContext 的 start()和 stop()方法中发布：

```
public abstract class AbstractApplicationContext extends DefaultResourceLoader
      implements ConfigurableApplicationContext {
  ...
  @Override
  public void start() {
      getLifecycleProcessor().start();
      publishEvent(new ContextStartedEvent(this));
  }

  @Override
  public void stop() {
      getLifecycleProcessor().stop();
      publishEvent(new ContextStoppedEvent(this));
  }
  ...
  }
```

在绝大多数场景中，以上两个方法不会被调用，然而并不意味它们没有存在的价值。假设 `AbstractApplicationContext#start()`方法被调用后，其关联的 `LifecycleProcessor` 实例将触发所有 `Lifecycle` Bean 的 `start()`方法调用。同理，当 `AbstractApplicationContext#stop()`方法执行时，触发 `Lifecycle#stop()`方法的调用。这两种方法的行为可以理解成 Spring Framework 为开发人员提供的两种控制 `ApplicationContext` 启停的操作方法，不仅能回调 `Lifecycle` Bean 的生命周期方法，而且能发布 `ContextStartedEvent` 和 `ContextStoppedEvent` 事件，供自定义 `ApplicationListener` 监听。

> 关于 `Lifecycle` Bean 的具体详情，请读者自行参考 Spring Framework 官方文档“Startup and shutdown callbacks”章节。

尽管 `AbstractApplicationContext` 也是 `Lifecycle` 的实现类，然而其实例并非 `Lifecycle` Bean，因此在 Spring Framework 框架实现中，`AbstractApplicationContext` 的 `start()`和 `stop()`方法不会执行。通常它们也不会在 Spring Boot 使用场景中执行。不过，它们却出现在并不显眼的 Spring Cloud“resume”和“pause”Endpoint 中：

```
@Endpoint(id = "restart", enableByDefault = false)
public class RestartEndpoint implements ApplicationListener<ApplicationPreparedEvent> {
 ...
 @Endpoint(id = "pause")
 public class PauseEndpoint {
```

```
        @WriteOperation
        public Boolean pause() {
            if (isRunning()) {
                doPause();
                return true;
            }
            return false;
        }
    }

    @Endpoint(id = "resume")
    @ConfigurationProperties("management.endpoint.resume")
    public class ResumeEndpoint {

        @WriteOperation
        public Boolean resume() {
            if (!isRunning()) {
                doResume();
                return true;
            }
            return false;
        }
    }
    ...
    public synchronized void doPause() {
        if (this.context != null) {
            this.context.stop();
        }
    }

    // @ManagedOperation
    public synchronized void doResume() {
        if (this.context != null) {
            this.context.start();
        }
    }
    ...
```

```
}
```

> 以上实现源码源于 Spring Cloud 2.0.0.RELEASE。
> Maven GAV 坐标为：org.springframework.cloud:spring-cloud-context:2.0.0.RELEASE。

同时，Spring Cloud 的实现也间接地说明 ContextStartedEvent 和 ContextStoppedEvent 的传播必然在 ContextRefreshedEvent 之后。

> 对于不同 ApplicationEvent 的执行次序，请读者予以高度的重视，这不但关乎对 Spring 事件/监听机制的深刻理解，也有助于合理地设计 ApplicationListener 的实现。在后续 Spring Boot 事件/监听机制的讨论中，会明显地感受到这一点。

3）Spring 应用上下文关闭事件——ContextClosedEvent

Spring 应用上下文关闭事件 ContextClosedEvent 与 ContextRefreshedEvent 正好相反，它由 ConfigurableApplicationContext#close() 方法调用时触发，同样由子类 AbstractApplicationContext 实现，且其实现类均继承该实现：

```
public abstract class AbstractApplicationContext extends DefaultResourceLoader
        implements ConfigurableApplicationContext {
    ...
    @Override
    public void close() {
        synchronized (this.startupShutdownMonitor) {
            doClose();
            // If we registered a JVM shutdown hook, we don't need it anymore now:
            // We've already explicitly closed the context.
            if (this.shutdownHook != null) {
                try {
                    Runtime.getRuntime().removeShutdownHook(this.shutdownHook);
                }
                catch (IllegalStateException ex) {
                    // ignore - VM is already shutting down
                }
            }
        }
    }
```

```
    ...
    protected void doClose() {
        if (this.active.get() && this.closed.compareAndSet(false, true)) {
            if (logger.isInfoEnabled()) {
                logger.info("Closing " + this);
            }

            LiveBeansView.unregisterApplicationContext(this);

            try {
                // Publish shutdown event.
                publishEvent(new ContextClosedEvent(this));
            }
            ...

            // Stop all Lifecycle beans, to avoid delays during individual destruction.
            if (this.lifecycleProcessor != null) {
                try {
                    this.lifecycleProcessor.onClose();
                }
                ...
            }
            ...
        }
    }
    ...
}
```

当 ConfigurableApplicationContext#close()方法调用后，Spring 应用上下文除了发布 ContextClosedEvent，Lifecycle Bean 也进入 onClose()生命周期回调，同时在 Runtime 中移除 ShutdownHook 线程，这也意味着当前 Spring 应用上下文完全被销毁，无法像调用 stop()方法那样还能通过 start()方法恢复，所以不难理解为什么它们在 Spring Cloud 中分别用于 PauseEndpoint（暂停）和 ResumeEndpoint（恢复）。

至此，了解了 Spring 四种内建事件的触发时机及关联的 ConfigurableApplicationContext 方法。值得一提的是，以上四种 Spring 内建事件均继承于抽象类 ApplicationContextEvent，并将 ApplicationContext 作为事件源，因此又可称它们为 Spring 上下文事件。它们之间的层次关系将在后面讨论。

4）Spring 应用上下文事件——`ApplicationContextEvent`

Spring 应用上下文事件 `ApplicationContextEvent` 与 Spring 事件 `ApplicationEvent`，以及与其四个内建子事件类型的层析关系如以下 UML 图所示。

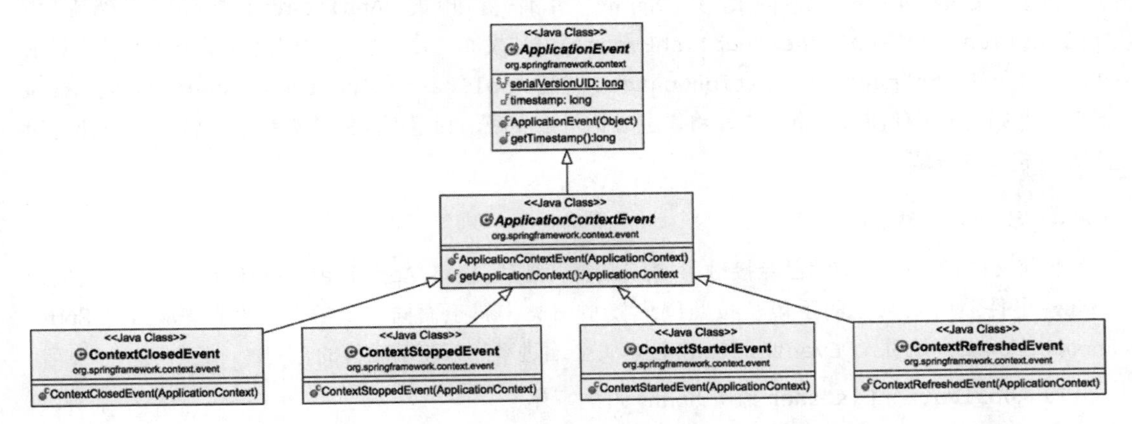

当然，Spring Framework 事件/监听机制并不限于“自娱自乐”，同样具备自定义实现事件和监听的弹性。

3. 自定义 Spring 事件

自定义 Spring 事件的方法同样被 Spring Framework 官方文档安排在“1.15.2. Standard and custom events”章节中：

> You can also create and publish your own custom events. This example demonstrates a simple class that extends Spring’s `ApplicationEvent` base class:

```
public class BlackListEvent extends ApplicationEvent {

    private final String address;
    private final String test;

    public BlackListEvent(Object source, String address, String test) {
        super(source);
        this.address = address;
        this.test = test;
    }

    // accessor and other methods...
}
```

> To publish a custom ApplicationEvent, call the publishEvent() method on an ApplicationEventPublisher.

按照文档的提示，自定义 Spring 事件需扩展 ApplicationEvent，然后由 ApplicationEventPublisher#publishEvent()方法发布。尽管官方文档没有说明，不过经过前文讨论，已知 AbstractApplicationContext 就是 ApplicationEventPublisher 的实现。在接下来的“Spring 事件监听”章节中，将以上讨论付诸实践，利用不同的监听手段，加深对 Spring 事件/监听机制的理解。

4. Spring 事件监听

在前文的讨论中，实际已经接触了 Spring 事件监听接口 ApplicationListener，它是实现 Spring 事件监听最常见的手段，按照编程模型归类，属于面向接口编程。在此基础上，Spring Framework 4.2 开始引入@EventListner 注解，全面进军注解驱动编程的行列。

1）ApplicationListener 监听 Spring 内建事件

已知 AbstractApplictionContext 提供发布 ApplicationEvent 和关联 ApplicationListener 实现的机制，并且任意 Spring Framework 内建 ConfigurableApplicationContext 实现类均继承 AbstractApplictionContext 的事件/监听行为，故选择 GenericApplicationContext 检验之：

```
public class ApplicationListenerOnSpringEventsBootstrap {

    public static void main(String[] args) {
        // 创建 ConfigurableApplicationContext 实例
        ConfigurableApplicationContext context = new GenericApplicationContext();
        System.out.println("创建 Spring 应用上下文 : " + context.getDisplayName());
        // 添加 ApplicationListener 非泛型实现
        context.addApplicationListener(event ->
                System.out.println("触发事件 : " + event.getClass().getSimpleName())
        );

        // refresh(): 初始化应用上下文
        System.out.println("应用上下文准备初始化...");
        context.refresh(); // 发布 ContextRefreshedEvent
        System.out.println("应用上下文已初始化...");

        // stop(): 停止应用上下文
        System.out.println("应用上下文准备停止启动...");
```

```
        context.stop();    // 发布 ContextStoppedEvent
        System.out.println("应用上下文已停止启动...");

        // start(): 启动应用上下文
        System.out.println("应用上下文准备启动启动...");
        context.start();  // 发布 ContextStartedEvent
        System.out.println("应用上下文已启动启动...");

        // close(): 关闭应用上下文
        System.out.println("应用上下文准备关闭...");
        context.close();  // 发布 ContextClosedEvent
        System.out.println("应用上下文已关闭...");
    }

}
```

> 源码位置：以上示例代码可通过查找 spring-framework-samples/spring-framework-5.0.x-sample 工程获取。

该引导类运行后的结果符合前文源码分析的结论，ContextRefreshedEvent、ContextStoppedEvent、ContextStartedEvent 和 ContextClosedEvent 依次输出：

```
    (...部分内容被省略...)
    创建 Spring 应用上下文 :
org.springframework.context.support.GenericApplicationContext@6438a396
    应用上下文准备初始化...
    触发事件: ContextRefreshedEvent
    应用上下文已初始化...
    应用上下文准备停止启动...
    触发事件: ContextStoppedEvent
    应用上下文已停止启动...
    应用上下文准备启动启动...
    触发事件: ContextStartedEvent
    应用上下文已启动启动...
    应用上下文准备关闭...
    触发事件: ContextClosedEvent
    应用上下文已关闭...
    (...部分内容被省略...)
```

2）ApplicationListener 监听自定义 Spring 泛型事件

由于前面通过 addApplicationListener()方法关联了 ApplicationListener 对象，下面的示例将替换为以泛型 ApplicationListener Bean 注册的方式检验自定义 Spring 事件。同时，在前文对 AbstractApplicationContext#close()的分析中，未见 Spring 应用上下文销毁关联 ApplicationEventMulticaster 属性，所以本实例仍需证明，即使在 close()方法执行后，仍可以发布 Spring 事件：

```
    public class ApplicationListenerBeanOnCustomEventBootstrap {

        public static void main(String[] args) {
            // 创建 Spring 应用上下文 GenericApplicationContext
            GenericApplicationContext context = new GenericApplicationContext();
            // 注册 ApplicationListener<MyApplicationEvent>实现
           // MyApplicationListener
            context.registerBean(MyApplicationListener.class); // registerBean
           // 方法从 Spring 5 引入
            // 初始化上下文
            context.refresh();
            // 发布自定义事件 MyApplicationEvent
            context.publishEvent(new MyApplicationEvent("Hello World"));
            // 关闭上下文
            context.close();
            // 再次发布事件
            context.publishEvent(new MyApplicationEvent("Hello World Again"));
        }

        public static class MyApplicationEvent extends ApplicationEvent {

            public MyApplicationEvent(String source) {
                super(source);
            }
        }

        public static class MyApplicationListener implements
ApplicationListener<MyApplicationEvent> {

            @Override
```

```
        public void onApplicationEvent(MyApplicationEvent event) {
            System.out.println(event.getClass().getSimpleName());
        }
    }
}
```

> 源码位置：以上示例代码可通过查找 spring-framework-samples/spring-framework-5.0.x-sample 工程获取。

MyApplicationEventListener 仅监听事件 MyApplicationEvent，故只输出该事件内容：

```
(...部分内容被省略...)
MyApplicationEvent
(...部分内容被省略...)
```

不过日志并未输出发布上下文关闭后的事件，并且在 AbstractApplicationContext 层面确实没有移除 ApplicationEventMulticaster 属性，那么 ApplicationListener 监听实现原理究竟是怎样的呢？

3）ApplicationListener 监听实现原理

Spring 应用上下文的关闭通常由 AbstractApplicationContext.close()方法调用触发，所关联的动作基本由 doClose()方法完成：

```
public abstract class AbstractApplicationContext extends DefaultResourceLoader
        implements ConfigurableApplicationContext {
    ...
    @Override
    public void close() {
        synchronized (this.startupShutdownMonitor) {
            doClose();
            ...
        }
    }
    ...
    protected void doClose() {
        if (this.active.get() && this.closed.compareAndSet(false, true)) {
            ...
            // Destroy all cached singletons in the context's BeanFactory.
            destroyBeans();
```

```
            // Close the state of this context itself.
            closeBeanFactory();

            // Let subclasses do some final clean-up if they wish...
            onClose();

            this.active.set(false);
        }
    }
    ...
    protected void destroyBeans() {
        getBeanFactory().destroySingletons();
    }
    ...
}
```

在其方法调用链路中，destroyBeans()方法是核心，它将 ApplicationListener Bean 从 ApplicationEventMulticaster 关联缓存中移除：

```
class ApplicationListenerDetector implements DestructionAwareBeanPostProcessor,
MergedBeanDefinitionPostProcessor {
    ...
    @Override
    public void postProcessBeforeDestruction(Object bean, String beanName) {
        if (bean instanceof ApplicationListener) {
            try {
                ApplicationEventMulticaster multicaster =
this.applicationContext.getApplicationEventMulticaster();
                multicaster.removeApplicationListener((ApplicationListener<?>) bean);
                multicaster.removeApplicationListenerBean(beanName);
            }
            catch (IllegalStateException ex) {
                // ApplicationEventMulticaster not initialized yet - no need to remove a
listener
            }
        }
    }
    ...
```

```
}
```

其中，缓存删除逻辑位于 AbstractApplicationEventMulticaster：

```
public abstract class AbstractApplicationEventMulticaster
        implements ApplicationEventMulticaster, BeanClassLoaderAware, BeanFactoryAware {
    ...
    @Override
    public void removeApplicationListener(ApplicationListener<?> listener) {
        synchronized (this.retrievalMutex) {
            this.defaultRetriever.applicationListeners.remove(listener);
            this.retrieverCache.clear();
        }
    }

    @Override
    public void removeApplicationListenerBean(String listenerBeanName) {
        synchronized (this.retrievalMutex) {
            this.defaultRetriever.applicationListenerBeans.remove(listenerBeanName);
            this.retrieverCache.clear();
        }
    }
    ...
}
```

因此，在 Spring 应用上下文关闭后，AbstactApplicationContext 即使能够广播任意事件，然而所关联的 ApplicationListener 却无法找到。最终导致 MyApplicationListener 无法再监听 MyApplicationEvent 事件。

不过，我们更迫切地想知道在 Spring 应用上下文关闭之前，事件监听是如何实现的呢？经过前文的铺陈，已知 SimpleApplicationEventMulticaster 既是 Spring Boot 事件广播的实现，也是 Spring 事件广播的实现。因此，参考其 multicastEvent 方法实现：

```
    public class SimpleApplicationEventMulticaster extends AbstractApplicationEventMulticaster
{
    ...
    @Override
    public void multicastEvent(final ApplicationEvent event, @Nullable ResolvableType eventType)
{
        ResolvableType type = (eventType != null ? eventType : resolveDefaultEventType(event));
```

```
        for (final ApplicationListener<?> listener : getApplicationListeners(event, type)) {
            Executor executor = getTaskExecutor();
            if (executor != null) {
                executor.execute(() -> invokeListener(listener, event));
            }
            else {
                invokeListener(listener, event);
            }
        }
    }
    ...
    protected void invokeListener(ApplicationListener<?> listener, ApplicationEvent event) {
        ErrorHandler errorHandler = getErrorHandler();
        if (errorHandler != null) {
            try {
                doInvokeListener(listener, event);
            }
            catch (Throwable err) {
                errorHandler.handleError(err);
            }
        }
        else {
            doInvokeListener(listener, event);
        }
    }

    @SuppressWarnings({"unchecked", "rawtypes"})
    private void doInvokeListener(ApplicationListener listener, ApplicationEvent event) {
        try {
            listener.onApplicationEvent(event);
        }
        ...
    }
    ...
}
```

在“ApplicationEventMulticaster 广播事件”一节中，曾提到 SimpleApplicationEventMulticaster 可关联 Executor 对象，实现异步事件广播。然而，SimpleApplicationEvent

Multicaster 默认采用同步广播事件的方式。简言之，Spring 事件通过调用 SimpleApplicationEventMulticaster#multicastEvent 方法广播，根据 ApplicationEvent 具体类型查找匹配的 ApplicationListener 列表，然后逐一同步或异步地调用 ApplicationListener#onApplicationEvent(Application Event)方法，实现 ApplicationListener 事件监听。

目前讨论的范围仅限于面向接口编程的 Spring 事件/监听模型，不过从 Spring Framework 4.2 开始引入了新的编程模型，即注解驱动的@EventListener。

4）注解驱动 Spring 事件监听——@EventListener

Spring 注解驱动事件监听的内容出现在 Spring Framework 官方文档的"Annotation-based event listeners"一节：

> As of Spring 4.2, an event listener can be registered on any public method of a managed bean via the EventListener annotation. The BlackListNotifier can be rewritten as follows:
>
> ```
> public class BlackListNotifier {
>
>     private String notificationAddress;
>
>     public void setNotificationAddress(String notificationAddress) {
>         this.notificationAddress = notificationAddress;
>     }
>
>     @EventListener
>     public void processBlackListEvent(BlackListEvent event) {
>         // notify appropriate parties via notificationAddress...
>     }
> }
> ```

总结而言，@EventListener 必须标记在 **Spring 托管 Bean 的 public 方法中**，并且支持单一类型事件监听，如 processBlackListEvent(BlackListEvent)方法，不过，在严谨性方面，相较于 JSR（Java Specification Requests）而言，Spring Framework 官方文档存在差距：

> As you can see above, the method signature once again declares the event type it listens to, but this time with a flexible name and without implementing a specific listener interface. The event type can also be narrowed through generics as long as the actual event type resolves your generic parameter in its implementation hierarchy.

描述中并没有告知标注@EventListener 的方法能不能有返回值，并且如果该方法定义在抽象类上，当子类作为 Spring Bean 时，这些方法是否还能成功地监听？为此，将新建示例，解释以上困惑。前例选择 GenericApplicationContext 实例作为 ApplicationEventPublisher 对象，以下示例将替换为其子类 AnnotationConfigApplicationContext 来注册@EventListener Bean：

```
public class AnnotatedEventListenerBootstrap {

    public static void main(String[] args) {
        // 创建注解驱动 Spring 应用上下文
        AnnotationConfigApplicationContext context = new
AnnotationConfigApplicationContext();
        // 注册@EventListener 类 MyEventListener
        context.register(MyEventListener.class);
        // 初始化上下文
        context.refresh();
        // 关闭上下文
        context.close();
    }

    /**
     * {@link EventListener} 抽象类
     */
    public static abstract class AbstractEventListener {
        @EventListener(ContextRefreshedEvent.class)
        public void onContextRefreshedEvent(ContextRefreshedEvent event) {
            System.out.println("AbstractEventListener : " +
event.getClass().getSimpleName());
        }
    }

    /**
     * 具体 {@link EventListener}类，作为 Spring Bean，继承 {@link AbstractEventListener}
     */
    public static class MyEventListener extends AbstractEventListener {

        @EventListener(ContextClosedEvent.class)
        public boolean onContextClosedEvent(ContextClosedEvent event) {
            System.out.println("MyEventListener : " + event.getClass().getSimpleName());
```

```
            return true;
        }
    }
}
```

> 源码位置：以上示例代码可通过查找 spring-framework-samples/spring-framework-5.0.x-sample 工程获取。

假设 `onContextRefreshedEvent(ContextRefreshedEvent)`方法执行，则说明声明在抽象类 `AbstractEventListener` 中的`@EventListener` public 方法被 `MyEventListener` 继承。同时，假设 `onContextClosedEvent(ContextClosedEvent)`执行，则说明非 `void` `@EventListener` public 方法也是合法的。因此，执行 `AnnotatedEventListenerBootstrap`，检验这些假设是否成立：

```
(...部分内容被省略...)
AbstractEventListener : ContextRefreshedEvent
MyEventListener : ContextClosedEvent
(...部分内容被省略...)
```

从执行结果来看，以上两种假设均已成立，即`@EventListener` public 方法既可声明在抽象类中，也可以是非 `void` 方法。

> Spring Framework 支持`@EventListener` public 方法允许为非 `void`，`@EventListener` 违背了事件监听器注解的设计原则。无论接口驱动，还是注解驱动，事件监听器的监听方法是公开且没有返回值的，同时不会“`throws`”异常。读者可自行参考 JSR 中的相关注解设计，如 JPA 中`@EntityListeners` 和`@PrePersist` 等，以及 WebSocket 的`@OnMessage` 等。

除此之外，单个`@EventListener` 方法也支持多种事件类型的监听。

5）`@EventListener` 方法监听多 `ApplicationEvent`

如文档所述：

> If your method should listen to several events or if you want to define it with no parameter at all, the event type(s) can also be specified on the annotation itself:
>
> ```
> @EventListener({ContextStartedEvent.class, ContextRefreshedEvent.class})
> public void handleContextStart() {
>     ...
> }
> ```

以上描述存在一定的模糊性，当@EventListener方法监控多种ApplicationEvent类型时，方法能否支持获取单个 ApplicationEvent 参数，或者分别为 ContextStartedEvent 和 ContextRefreshedEvent 的参数形式。如果情况如后半段所言，仅支持无参监听方法，则无法获取对应的事件或事件源，这样的设计并非良好。对比 JPA @EntityListeners 监听实体操作时，尽管@PrePersist 或 @PreUpdate 等注解不是监听事件对象，然而它们可监控事件源，所以它们能如此操作：

```
public class UserCreatedDateListener {

    @PrePersist
    @PreUpdate
    public void setCreatedDate(User user) {
        user.setCreatedDate(new Date());
    }
}
```

换言之，假设 @EventListener 监控的是 Spring 应用上下文事件，如官方示例那样（@EventListener({ContextStartedEvent.class, ContextRefreshedEvent. class})），那么 Spring Framework 能否支持 ApplicationContextEvent 作为参数，或者两个参数分别为 ContextStartedEvent 和 ContextRefreshedEvent 实例呢？这个问题有待检验，因此，继续新建示例加以观察：

```
public class AnnotatedEventListenerOnMultiEventsBootstrap {

    public static void main(String[] args) {
        // 创建注解驱动 Spring 应用上下文
        AnnotationConfigApplicationContext context = new
AnnotationConfigApplicationContext();
        // 注册@EventListener 类 MyMultiEventsListener
        context.register(MyMultiEventsListener.class);
        // 初始化上下文
        context.refresh();
        // 关闭上下文
        context.close();
    }

    /**
     * 具体 {@link EventListener}类，提供不同监听多 Spring 事件方法
```

```
     */
    public static class MyMultiEventsListener {

        /**
         * 无参数监听 {@link ContextRefreshedEvent} 和 {@link ContextClosedEvent} 事件
         */
        @EventListener({ContextRefreshedEvent.class, ContextClosedEvent.class})
        public void onEvent() {
            System.out.println("onEvent");
        }

        /**
         * 单一 {@link ApplicationContextEvent} 参数监听 {@link ContextRefreshedEvent} 和
{@link ContextClosedEvent} 事件
         *
         * @param event {@link ApplicationContextEvent}
         */
        @EventListener({ContextRefreshedEvent.class, ContextClosedEvent.class})
        public void onApplicationContextEvent(ApplicationContextEvent event) {
            System.out.println("onApplicationContextEvent : " +
event.getClass().getSimpleName());
        }

        /**
         * {@link ContextRefreshedEvent} 和 {@link ContextClosedEvent} 参数监听
         *
         * @param refreshedEvent     {@link ContextRefreshedEvent}
         * @param contextClosedEvent {@link ContextClosedEvent}
         */
        @EventListener({ContextRefreshedEvent.class, ContextClosedEvent.class})
        public void onEvents(ContextRefreshedEvent refreshedEvent, ContextClosedEvent
contextClosedEvent) {
            System.out.println("onEvents : " + refreshedEvent.getClass().getSimpleName()
                    + " , " + contextClosedEvent.getClass().getSimpleName());
        }
    }
}
```

> 源码位置：以上示例代码可通过查找 spring-framework-samples/spring-framework-5.0.x-sample 工程获取。

以上方法均监听 ContextRefreshedEvent 和 ContextClosedEvent 事件，其中 onEvent 代表无参数监听方法，onApplicationContextEvent 是单一 ApplicationContextEvent 参数监听方法，而 onEvents 为双参数监听方法。运行该引导类，观察运行结果：

```
    (...部分内容被省略...)
    Caused by: java.lang.IllegalStateException: Maximum one parameter is allowed for event
listener method: public void
thinking.in.spring.boot.samples.spring5.context.event.AnnotatedEventListenerOnMultiEventsBoot
strap$MyMultiEventsListener.onEvents(org.springframework.context.event.ContextRefreshedEvent,
org.springframework.context.event.ContextClosedEvent)
    (...部分内容被省略...)
```

运行日志提醒@EventListener 方法最大支持一个方法参数，说明 onEvents 的声明非法，将其注释并重新运行：

```
    (...部分内容被省略...)
    onEvent
    onApplicationContextEvent : ContextRefreshedEvent
    onEvent
    onApplicationContextEvent : ContextClosedEvent
    (...部分内容被省略...)
```

观察运行结果并结合前面的结论，综合如下：**“@EventListener 方法必须是 Spring Bean 中的 public 方法，并支持返回类型为非 void 的情况。当它监听一个或多个 ApplicationEvent 时，其参数可为零到一个 ApplicationEvent”。**

经此讨论使得 Spring Framework 官方文档中的“模糊地带”变得清晰。按照官方文档“发展”的脉络，下一个环节将讨论@EventListener 方法异步化的支持。

6）@EventListener 异步方法

援引 Spring Framework 官方文档“Asynchronous Listeners”章节的描述：

> If you want a particular listener to process events asynchronously, simply reuse the regular @Async support:
>
> ```
> @EventListener
> @Async
> ```

```
public void processBlackListEvent(BlackListEvent event) {
    // BlackListEvent is processed in a separate thread
}
```

按照文档指引，@EventListener 方法的异步化非常简单，在原有方法基础上增加注解@Async 即可。假设监听方法由同步切换为异步，当它被执行时，必然发生线程的切换，因此可在执行方法内部增加当前线程的名称的辅助信息，故新建示例如下：

```
public class AnnotatedAsyncEventListenerBootstrap {

    public static void main(String[] args) {
        // 创建注解驱动 Spring 应用上下文
        AnnotationConfigApplicationContext context = new
AnnotationConfigApplicationContext();
        // 注册异步@EventListener 类 MyAsyncEventListener
        context.register(MyAsyncEventListener.class);
        println(" Spring 应用上下文正在初始化...");
        // 初始化上下文
        context.refresh();
        // 关闭上下文
        context.close();
    }

    @EnableAsync // 需要激活异步，否则@Async 无效
    public static class MyAsyncEventListener {

        @EventListener(ContextRefreshedEvent.class)
        @Async
        public boolean onContextRefreshedEvent(ContextRefreshedEvent event) {
            println(" MyAsyncEventListener : " + event.getClass().getSimpleName());
            return true;
        }
    }

    /**
     * 输出内容并附加当前线程信息
     *
     * @param content 输出内容
```

```
     */
    private static void println(String content) {
        // 当前线程名称
        String threadName = Thread.currentThread().getName();
        System.out.println("[ 线程 " + threadName + " ] : " + content);
    }
}
```

> 源码位置：以上示例代码可通过查找 spring-framework-samples/spring-framework-5.0.x-sample 工程获取。

执行并观察日志输出：

```
[ 线程 main ] :  Spring 应用上下文正在初始化...
(...部分内容被省略...)
Exception in thread "main" org.springframework.aop.AopInvocationException: Null return value
from advice does not match primitive return type for: public boolean
thinking.in.spring.boot.samples.spring5.context.event.AnnotatedAsyncEventListenerBootstrap$My
AsyncEventListener.onContextRefreshedEvent(org.springframework.context.event.ContextRefreshed
Event)
(...部分内容被省略...)
[ 线程 SimpleAsyncTaskExecutor-1 ] :  MyAsyncEventListener : ContextRefreshedEvent
```

尽管@Async 执行输出正常，并且线程也不再是 main，然而在日志中出现异常，大意是指 onContextRefreshedEvent 方法的返回类型不应该为原生（**primitive**）类型，故其从 boolean 调整为 Boolean：

```
        @EventListener(ContextRefreshedEvent.class)
        @Async
        public Boolean onContextRefreshedEvent(ContextRefreshedEvent event) {
            println(" MyAsyncEventListener : " + event.getClass().getSimpleName());
            return true;
        }
```

再次运行，结果正常：

```
[ 线程 main ] :  Spring 应用上下文正在初始化...
(...部分内容被省略...)
[ 线程 SimpleAsyncTaskExecutor-1 ] :  MyAsyncEventListener : ContextRefreshedEvent
```

因此，为了确保@EventListener 方法在同步和异步执行时的兼容性，它的返回类型应该为 void。尽管它支持非 void，然而其价值几乎为零。

不过，从运行结果观察，@Async 确实使@EventListener 方法异步化。然而，尚不清楚是否与事件广播器实现 SimpleApplicationEventMulticaster 关联 Executor 有关，该部分的讨论将在“@EventListener 方法实现原理”中展开。

综上所述，将@EventListener 同步和异步方法进行对比，如下表所示。

方法类型	访问修饰符	返回类型	参数数量	参数类型
同步	非 private	任意	0 或 1	监听事件类型或子类
异步	非 private	非原生类型	同上	同上

不仅如此，官方文档还向开发人员透漏@EventListener 方法可控制其执行顺序。

7）@EventListener 方法执行顺序

继续援引 Spring Framework 官方文档“Ordering listeners”章节的描述：

> If you need the listener to be invoked before another one, just add the @Order annotation to the method declaration:
>
> ```
> @EventListener
> @Order(42)
> public void processBlackListEvent(BlackListEvent event) {
>     // notify appropriate parties via notificationAddress...
> }
> ```

类似地，ApplicationListener 接口在监听 ApplicationEvent 时，也能通过 Ordered 接口和标注@Order 注解来实现监听次序：

```
public abstract class AbstractApplicationEventMulticaster
        implements ApplicationEventMulticaster, BeanClassLoaderAware, BeanFactoryAware {
...
private class ListenerRetriever {
    ...
    public Collection<ApplicationListener<?>> getApplicationListeners() {
        LinkedList<ApplicationListener<?>> allListeners = new LinkedList<>();
        ...
        AnnotationAwareOrderComparator.sort(allListeners);
        return allListeners;
```

```
        }
    }
    ...
}
```

因此，在@EventListener 方法上标注@Order 注解来控制监听方法的执行次序就不难理解了。至于这是如何实现的？以及如果它们不标注@Order，则其排序逻辑又会是怎样的呢？还是将这些悬念留在“@EventListener 方法实现原理”一节中解开。

8）*@EventListener 方法监听泛型 ApplicationEvent*

在@EventListener 对泛型 ApplicationEvent 的支持方面，官方文档在“Generic events”章节的着墨有所微增，涉及前文讨论的 API ResolvableType。按其代码指引，泛型 ApplicationEvent 需要实现 ResolvableTypeProvider 接口：

> In such a case, you can implement ResolvableTypeProvider to *guide* the framework beyond what the runtime environment provides:
>
> ```
> public class EntityCreatedEvent<T>
>         extends ApplicationEvent implements ResolvableTypeProvider {
>
>     public EntityCreatedEvent(T entity) {
>         super(entity);
>     }
>
>     @Override
>     public ResolvableType getResolvableType() {
>         return ResolvableType.forClassWithGenerics(getClass(),
>                 ResolvableType.forInstance(getSource()));
>     }
> }
> ```

该接口出现在 Spring Framework4.2 中，PayloadApplicationEvent 是其唯一的实现类：

```
    public class PayloadApplicationEvent<T> extends ApplicationEvent implements
ResolvableTypeProvider {
    ...
    @Override
    public ResolvableType getResolvableType() {
        return ResolvableType.forClassWithGenerics(getClass(),
```

```
ResolvableType.forInstance(getPayload()));
    }
    ...
    }
```

前面曾讨论 PayloadApplicationEvent 用于实现 ApplicationEventPublisher#publishEvent(Object) 方法（同样从 Spring Framework 4.2 开始引入），意在广播非 ApplicationEvent 类型的事件源，最终由 AbstractApplicationContext#publishEvent(Object event)负责实施：

```
    public abstract class AbstractApplicationContext extends DefaultResourceLoader
        implements ConfigurableApplicationContext {
    ...
    @Override
    public void publishEvent(Object event) {
        publishEvent(event, null);
    }
    ...
    protected void publishEvent(Object event, @Nullable ResolvableType eventType) {
        ...
        if (event instanceof ApplicationEvent) {
            ...
        }
        else {
            applicationEvent = new PayloadApplicationEvent<>(this, event);
            if (eventType == null) {
                eventType = ((PayloadApplicationEvent)
applicationEvent).getResolvableType();
            }
        }
        ...
    }
    ...
    }
```

换言之，无论注解驱动的@EventListener 方法，还是面向接口编程的 ApplicationListener，均能监控泛型 ApplicationEvent。为检验以上结论的正确性，构建类似的事件/监听场景：

- 构建用户实体类型

```
public static class User {

    private final String name;

    public User(String name) {
        this.name = name;
    }

    @Override
    public String toString() {
        return "User{name='" + name + "\'}";
    }
}
```

- 构建泛型事件

```
public static class UserEventListener implements ApplicationListener<GenericEvent<User>> {

    @EventListener
    public void onUserEvent(GenericEvent<User> event) {
        System.out.println("onUserEvent : " + event.getSource());
    }

    @Override
    public void onApplicationEvent(GenericEvent<User> event) {
        System.out.println("onApplicationEvent : " + event.getSource());
    }
}
```

- 构建用户事件监听器

实现 ApplicationListener，也包含@EventListener 方法：

```
public static class UserEventListener implements ApplicationListener<GenericEvent<User>> {

    @EventListener
    public void onUserEvent(GenericEvent<User> event) {
        System.out.println("onUserEvent : " + event.getSource());
    }
```

```
    @Override
    public void onApplicationEvent(GenericEvent<User> event) {
        System.out.println("onApplicationEvent : " + event.getSource());
    }
}
```

- 构建引导类

```
public class GenericEventListenerBootstrap {

    public static void main(String[] args) {
        // 创建注解驱动 Spring 应用上下文
        AnnotationConfigApplicationContext context = new
AnnotationConfigApplicationContext();
        // 注册 UserEventListener，即实现 ApplicationListener，也包含
        // @EventListener 方法
        context.register(UserEventListener.class);
        // 初始化上下文
        context.refresh();
        // 构造泛型事件
        GenericEvent<User> event = new GenericEvent(new User("小马哥"));
        // 发送泛型事件
        context.publishEvent(event);
        // 发送 User 对象作为事件源
        context.publishEvent(new User("mercyblitz"));
        // 关闭上下文
        context.close();
    }
    ...
}
```

> 源码位置：以上示例代码可通过查找 spring-framework-samples/spring-framework-5.0.x-sample 工程获取。

运行并观察日志输出：

```
(...部分内容被省略...)
onUserEvent : User{name='小马哥'}
onApplicationEvent : User{name='小马哥'}
```

```
(...部分内容被省略...)
```

事实证明，@EventListener 和 ApplicationListener 均能监听泛型 ApplicationEvent。

在“Generic events”一节的结尾部分增加了一行提示性文字：

> This works not only for ApplicationEvent but any arbitrary object that you’d send as an event.

如果不铺陈以上内容，那么这句话很难被理解。结合以上示例，@EventListener 方法和 ApplicationListener 均监听的是 GenericEvent<User>，那么这句话是否能理解为 @EventListener 方法可选择 User 对象作为方法参数？为检验假设的正确性，再次调整 UserEventListener 实现，新增一个@EventListener 方法 onUser(User)，如下所示。

```
public static class UserEventListener implements ApplicationListener<GenericEvent<User>> {

    @EventListener
    public void onUser(User user) {
        System.out.println("onUser : " + user);
    }
...
}
```

重启引导类 GenericEventListenerBootstrap，观察日志变化：

```
(...部分内容被省略...)
onUserEvent : User{name='小马哥'}
onApplicationEvent : User{name='小马哥'}
onUser : User{name='mercyblitz'}
(...部分内容被省略...)
```

运行结果证明了假设的正确性，@EventListener 方法的确也可直接监听事件源 User 对象，类似于 JPA 中的@PrePersist 方法。不过该功能依赖 Spring Framework 4.2 及更高版本，并且由于 ApplicationListener 接口泛型范围的限制，其方法参数必须是 ApplicationEvent 的对象，因此无法支持该特性。因此，在“@EventListener 方法对比 ApplicationListener 接口”一节中，将对比@EventListener 方法与 ApplicationListener 接口的异同。

> 根据 Spring Boot 与 Spring Framework 的依赖关系，该特性能在 Spring Boot 1.3.x 中使用。

9）@EventListener 方法对比 ApplicationListener 接口

对比情况如下表所示。

监 听 类 型	访问性	顺 序 控 制	返回类型	参数数量	参数类型	泛型事件
@EventListener 同步方法	public	@Order	任意	0 或 1	事件类型或泛型参数类型	支持
@EventListener 异步方法	public	Order	非原生类型	0 或 1	事件类型或泛型参数类型	支持
ApplicationListener	public	Order 或 Ordered	void	1	事件类型	不支持

至此，有关@EventListener 方法特性的细节尚未清晰，后面针对其原理做出分析。

10）@EventListener 方法实现原理

首先，@EventListener 方法必须在 Spring Bean 中，并且访问修饰符为 public，前者是方法来源，后者是筛选的条件。Spring 事件/监听机制围绕 ApplicationEvent、ApplicationListener 和 ApplicationEventMulticaster 三者展开，这很容易让人联想到@EventListener 方法的执行应该与 ApplicationListener 接口有关。

从 Spring Framework 4.2 开始，框架层面在 AnnotationConfigUtils#registerAnnotationConfigProcessors(BeanDefinitionRegistry,Object)方法中增加了@EventListener 方法处理的相关 Bean 的注册：

```
public class AnnotationConfigUtils {
    ...
    public static Set<BeanDefinitionHolder> registerAnnotationConfigProcessors(
            BeanDefinitionRegistry registry, @Nullable Object source) {
        ...
        if (!registry.containsBeanDefinition(EVENT_LISTENER_PROCESSOR_BEAN_NAME)) {
            RootBeanDefinition def = new
RootBeanDefinition(EventListenerMethodProcessor.class);
            def.setSource(source);
            beanDefs.add(registerPostProcessor(registry, def,
EVENT_LISTENER_PROCESSOR_BEAN_NAME));
        }
        if (!registry.containsBeanDefinition(EVENT_LISTENER_FACTORY_BEAN_NAME)) {
            RootBeanDefinition def = new
RootBeanDefinition(DefaultEventListenerFactory.class);
            def.setSource(source);
```

```
            beanDefs.add(registerPostProcessor(registry, def,
EVENT_LISTENER_FACTORY_BEAN_NAME));
        }

        return beanDefs;
    }
    ...
}
```

其中，EventListenerMethodProcessor 是@EventListener 方法的生命周期处理器，而 DefaultEventListenerFactory 则是@EventListener 方法与 ApplicationListener 适配器的工程类。参考 Spring Framework 5.0.6.RELEASE 的实现，EventListenerMethodProcessor 作为 SmartInitializingSingleton 接口实现。

11）SmartInitializingSingleton 生命周期回调

SmartInitializingSingleton 是从 Spring Framework 4.1 开始引入的回调接口，其 JavaDoc 描述如下：

> Callback interface triggered at the end of the singleton pre-instantiation phase during BeanFactory bootstrap. This interface can be implemented by singleton beans in order to perform some initialization after the regular singleton instantiation algorithm, avoiding side effects with accidental early initialization (e.g. from ListableBeanFactory.getBeansOfType calls). In that sense, it is an alternative to InitializingBean which gets triggered right at the end of a bean's local construction phase.
> This callback variant is somewhat similar to org.springframework.context.event.ContextRefreshedEvent but doesn't require an implementation of org.springframework.context.ApplicationListener, with no need to filter context references across a context hierarchy etc.

大致意思是说，SmartInitializingSingleton 类似于 ApplicationListener<ContextRefreshedEvent>的回调接口，在所有的“**单体 Bean 预实例化阶段**（**singleton pre-instantiation phase**）”之后触发，避免类似于调用 ListableBeanFactory#getBeansOfType(Class)方法所造成的过早的 Bean 初始化。此处需要解释的是“单体 Bean 预实例化阶段”的概念。简单地说，SmartInitializingSingleton#afterSingletonsInstantiated()方法将在 ConfigurableListableBeanFactory#preInstantiateSingletons()方法执行时调用，而后者则由 DefaultListableBeanFactory#preInstantiateSingletons()方法实现：

```
public class DefaultListableBeanFactory extends AbstractAutowireCapableBeanFactory
```

```
        implements ConfigurableListableBeanFactory, BeanDefinitionRegistry, Serializable {
    ...
    @Override
    public void preInstantiateSingletons() throws BeansException {
        ...
        List<String> beanNames = new ArrayList<>(this.beanDefinitionNames);

        // Trigger initialization of all non-lazy singleton beans...
        for (String beanName : beanNames) {
            RootBeanDefinition bd = getMergedLocalBeanDefinition(beanName);
            if (!bd.isAbstract() && bd.isSingleton() && !bd.isLazyInit()) {
                if (isFactoryBean(beanName)) {
                    ...
                        if (isEagerInit) {
                            getBean(beanName);
                        }
                    ...
                }
                else {
                    getBean(beanName);
                }
            }
        }

        // Trigger post-initialization callback for all applicable beans...
        for (String beanName : beanNames) {
            Object singletonInstance = getSingleton(beanName);
            if (singletonInstance instanceof SmartInitializingSingleton) {
                final SmartInitializingSingleton smartSingleton =
(SmartInitializingSingleton) singletonInstance;
                if (System.getSecurityManager() != null) {
                    AccessController.doPrivileged((PrivilegedAction<Object>) () -> {
                        smartSingleton.afterSingletonsInstantiated();
                        return null;
                    }, getAccessControlContext());
                }
                else {
                    smartSingleton.afterSingletonsInstantiated();
```

```
                }
            }
        }
    }
    ...
}
```

明显地，所有 SmartInitializingSingleton Bean 的 afterSingletonsInstantiated()方法在当前 Spring 容器中的 Bean 初始化之后执行，并在 DefaultListableBeanFactory#preInstantiateSingletons()方法执行结束前调用。同时，从完整的 Spring 应用上下文生命周期分析，以上 preInstantiateSingletons() 方法处于 AbstractApplicationContext#finishBeanFactoryInitialization(ConfigurableListableBeanFactory)调用周期的最后一个操作，并且也接近于 refresh()方法的完成阶段：

```
public abstract class AbstractApplicationContext extends DefaultResourceLoader
        implements ConfigurableApplicationContext {
    ...
    @Override
    public void refresh() throws BeansException, IllegalStateException {
        synchronized (this.startupShutdownMonitor) {
            ...
            try {
                ...
                // Instantiate all remaining (non-lazy-init) singletons.
                finishBeanFactoryInitialization(beanFactory);

                // Last step: publish corresponding event.
                finishRefresh();
            }
            ...
        }
    }
    protected void finishBeanFactoryInitialization(ConfigurableListableBeanFactory beanFactory)
{
        ...
        beanFactory.freezeConfiguration();

        // Instantiate all remaining (non-lazy-init) singletons.
```

```
        beanFactory.preInstantiateSingletons();
    }
    ...
}
```

以上方法的调用关系如下图所示。

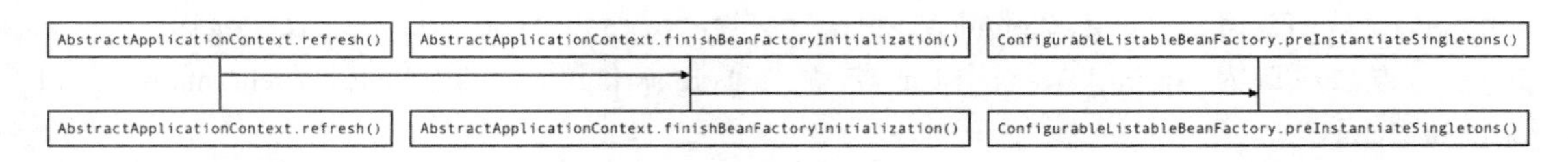

12）EventListenerMethodProcessor 的实现原理

回到 EventListenerMethodProcessor 的实现：

```
public class EventListenerMethodProcessor implements SmartInitializingSingleton,
ApplicationContextAware {
    ...
    @Override
    public void afterSingletonsInstantiated() {
        List<EventListenerFactory> factories = getEventListenerFactories();
        ConfigurableApplicationContext context = getApplicationContext();
        String[] beanNames = context.getBeanNamesForType(Object.class);
        for (String beanName : beanNames) {
            ...
                    try {
                        processBean(factories, beanName, type);
                    }
            ...
        }
    }
    ...
    protected List<EventListenerFactory> getEventListenerFactories() {
        Map<String, EventListenerFactory> beans =
getApplicationContext().getBeansOfType(EventListenerFactory.class);
        List<EventListenerFactory> factories = new ArrayList<>(beans.values());
        AnnotationAwareOrderComparator.sort(factories);
        return factories;
    }
    ...
```

```
    }
```

按照 afterSingletonsInstantiated()方法的生命周期，getEventListenerFactories()方法根据类型在 Spring 应用上下文中查找所有 EventListenerFactory Bean 的操作不会引起过早初始化的问题。随后，该方法将排序并返回所有的 EventListenerFactory Bean。在 Spring Framework 中，DefaultEventListenerFactory 作为 EventListenerFactory 接口实现类，它被 AnnotationConfigUtils#registerAnnotationConfigProcessors(BeanDefinitionRegistry, Object)方法注册为 Spring Bean，Bean 名称为“org.springframework.context.event.internalEventListenerFactory”：

```
    public class AnnotationConfigUtils {
    ...
     public static final String EVENT_LISTENER_FACTORY_BEAN_NAME =
            "org.springframework.context.event.internalEventListenerFactory";
    ...
     public static Set<BeanDefinitionHolder> registerAnnotationConfigProcessors(
            BeanDefinitionRegistry registry, @Nullable Object source) {
        ...
        if (!registry.containsBeanDefinition(EVENT_LISTENER_FACTORY_BEAN_NAME)) {
            RootBeanDefinition def = new
RootBeanDefinition(DefaultEventListenerFactory.class);
            def.setSource(source);
            beanDefs.add(registerPostProcessor(registry, def,
EVENT_LISTENER_FACTORY_BEAN_NAME));
        }

        return beanDefs;
    }
    ...
    }
```

因此，getEventListenerFactories()方法默认返回仅包含 DefaultEventListenerFactory 实例的集合，随后作为 processBean(List,String,Class)方法的首参：

```
    public class EventListenerMethodProcessor implements SmartInitializingSingleton,
ApplicationContextAware {
    ...
    protected void processBean(
            final List<EventListenerFactory> factories, final String beanName, final Class<?>
```

```
targetType) {

        if (!this.nonAnnotatedClasses.contains(targetType)) {
            Map<Method, EventListener> annotatedMethods = null;
            try {
                annotatedMethods = MethodIntrospector.selectMethods(targetType,
                        (MethodIntrospector.MetadataLookup<EventListener>) method ->
                                AnnotatedElementUtils.findMergedAnnotation(method,
EventListener.class));
            }
            ...
            else {
                // Non-empty set of methods
                ConfigurableApplicationContext context = getApplicationContext();
                for (Method method : annotatedMethods.keySet()) {
                    for (EventListenerFactory factory : factories) {
                        if (factory.supportsMethod(method)) {
                            Method methodToUse = AopUtils.selectInvocableMethod(method,
context.getType(beanName));
                            ApplicationListener<?> applicationListener =
                                    factory.createApplicationListener(beanName,
targetType, methodToUse);
                            if (applicationListener instanceof
ApplicationListenerMethodAdapter) {
                                ((ApplicationListenerMethodAdapter)
applicationListener).init(context, this.evaluator);
                            }
                            context.addApplicationListener(applicationListener);
                            break;
                        }
                    }
                }
                ...
            }
        }
    }

    }
```

该方法实现较为简单，首先从指定 Bean 类型中查找所有标注@EventListener 的方法，由于AnnotatedElementUtils.findMergedAnnotation(method, EventListener. class)语句的作用，方法筛选的规则是仅判断 Bean 中所有 public 方法是否标注@EventListener，所以即使方法存在返回值也会纳入候选集合，从而解释了前文示例中返回类型为 boolean 的方法能够执行的原因。

> 至于@Async 和@EventListener 方法返回类型为什么不能为原生类型的原因在于，@Async 标注的方法被 Spring AOP 提升，最终导致 JdkDynamicAopProxy 在判断方法返回类型时异常，不过它并不会影响方法执行，详情请读者自行参考 JdkDynamicAopProxy#invoke 方法，@Async 方法拦截处理请参考 AnnotationAsyncExecutionInterceptor。

随后，EventListenerMethodProcessor 将候选的@EventListener 方法集合逐一经过EventListenerFactory 实例列表（首参）的匹配。而默认情况下，EventListenerFactory 实例列表仅为 DefaultEventListenerFactory 对象，故将@EventListener 方法适配为ApplicationListener 对象：

```
public class DefaultEventListenerFactory implements EventListenerFactory, Ordered {
    ...
    public boolean supportsMethod(Method method) {
        return true;
    }

    @Override
    public ApplicationListener<?> createApplicationListener(String beanName, Class<?> type,
Method method) {
        return new ApplicationListenerMethodAdapter(beanName, type, method);
    }

}
```

因此，@EventListener 方法实现的确与 ApplicationListener 有关，从而证明了前文的假设。

上述内容较为完整地讨论了@EventListener 方法的使用方法和实现原理。同时完善了“Spring 事件/监听机制”系统性的讨论。不过它与 Spring Boot 又有怎样的关系呢？接下来将继续讨论 Spring Boot 事件/监听机制。

5. 总结 Spring 事件/监听机制

在正式进入讨论之前，需要对以上 Spring 事件监听的手段进行总结，包括 Spring 事件、Spring

事件监听器、Spring 事件广播器，以及它们与 Spring 应用上下文之间的关系。

1）总结 Spring 事件

Spring 事件的 API 表述为 ApplicationEvent，继承于 Java 规约的抽象类 java.util.EventObject，并需要显式地调用父类构造器传递事件源参数。Spring Framework 内建事件有五种，包括 ContextRefreshedEvent、ContextStartedEvent、ContextStoppedEvent、ContextClosedEvent 和 RequestHandledEvent。其中，前四种为 Spring 应用上下文事件，均继承于抽象类 ApplicationContextEvent，以 ApplicationContext 作为事件源。

除了以上 Spring 内建事件，Spring Framework 允许应用自定义 ApplicationEvent，并且能够实现自定义泛型 Spring 事件，不过这类事件需要实现 ResolvableTypeProvider 接口，通过该接口暴露其泛型元信息，搭建与 Spring 事件/监听机制“沟通的桥梁”。从广义上来看，无论 Spring Boot 事件，还是 Spring Cloud 事件，均属于 ApplicationEvent 的扩展，是特定领域的自定义 Spring 事件，如 Spring Boot 事件基类 SpringApplicationEvent 和 Spring Cloud Bus 基类 RemoteApplicationEvent。

2）总结 Spring 事件监听手段

关于 Spring 事件监听的手段，Spring Framework 层面提供了两种途径，其一是面向接口编程的 ApplicationListener，其二是注解驱动编程的@EventListener 方法。两种途径均能监听一种或多种事件，并且支持泛型事件。从实现层面分析，@EventListener 方法属于 Bean 方法与 ApplicationListener 接口的适配，即 ApplicationListenerMethodAdapter。两者的差异在于，@EventListener 方法必须依赖于 Spring 应用上下文，而 ApplicationListener 对象尽管默认也与 ApplicationContext 关联，但实际上它是 ApplicationEventMulticaster 实例的组成元素。因此，这也是 Spring 事件/监听机制与 Spring Boot 的差异，当然实际情况没有如此单一，将在后面的“理解 Spring Boot 事件/监听机制”一节详加讨论。

3）总结 Spring 事件广播器

Spring 事件广播在 Spring Framework 中有两类 API 表达方式，一是 ApplicationEventPublisher#publishEvent 方法，二是 ApplicationEventMulticaster#multicastEvent。默认情况下，Spring Framework 仅提供一种实现，前者由 AbstractApplicationContext 实现，后者由 SimpleApplicationEventMulticaster 实现。其中，SimpleApplicationEventMulticaster 不但是 AbstractApplicationContext 实现 ApplicationEventPublisher 接口语义的基础，而且也为 AbstractApplicationContext 提供关联 ApplicationListener 的实现存储（可参考其 addApplicationListener (ApplicationListener)方法实现）。

综上所述，SimpleApplicationEventMulticaster 是 ApplicationEvent、ApplicationListener 和 ConfigurableApplicationContext 之间连接的纽带。再回顾前文关于

EventPublishingRunListener 的讨论，在 SpringApplicationRunListener 生命周期方法中，SimpleApplicationEventMulticaster 为 EventPublishingRunListener 发布 Spring Boot 事件，包括 ApplicationStartingEvent 和 ApplicationEnvironmentPreparedEvent 等。换言之，SimpleApplicationEventMulticaster 不仅是 Spring 事件与应用上下文的纽带，而且它也在 Spring Boot 事件与 SpringApplication 之间充当相同的角色。那么，值得思考的是，SpringApplication 对象关联的 SimpleApplicationEventMulticaster 与其关联 Spring 应用上下文存在何种关系呢？这个问题的答案将在下一节中揭晓。

11.1.5 理解 Spring Boot 事件/监听机制

理解 Spring 事件/监听机制是一项非常重要的技术储备，因为 Spring Boot 事件/监听机制同样基于 ApplicationEventMulticaster、ApplicationEvent 和 ApplicationListener 实现，所以理解它并非难事。假设 Spring Boot 事件/监听机制只是单纯地利用 Spring 事件/监听 API，那么情况则变得简单明了，它和 Spring 应用上下文事件发布和监听器管理属于“各自为政，互不干涉”的关系。如果两者相互关联，那么情况将变得异常复杂。然而不巧的是，这两种情况同时存在。或许读者不禁要问：“这两种情形明显相互矛盾，怎么可能并存呢？”。这样的质疑在逻辑上是合理的，不过请读者**务必心存版本的意识**。在 Spring Boot 1.4 之前，SpringApplication 所管理的 Spring 应用上下文复用了 EventPublishingRunListener 关联的 SimpleApplicationEventMulticaster。随后的版本则不再延续这样的设计，其中必然存在某些原因，接下来的内容将以上情况单独讨论，从中尝试寻找问题的答案。下面将以 Spring Boot 1.4 为分水岭，将 Spring Boot 事件/监听机制分为“早期 Spring Boot 事件/监听机制”和“当前 Spring Boot 事件/监听机制”。

1. 早期 Spring Boot 事件/监听机制

前文“ApplicationEventMulticaster 与 ApplicationConext 之间的关系”章节曾讨论：

```
public abstract class AbstractApplicationContext extends DefaultResourceLoader
        implements ConfigurableApplicationContext {
    ...
    @Override
    public void refresh() throws BeansException, IllegalStateException {
        synchronized (this.startupShutdownMonitor) {
            ...
                // Initialize event multicaster for this context.
                initApplicationEventMulticaster();
            ...
    }
```

```
    ...
    protected void initApplicationEventMulticaster() {
        ConfigurableListableBeanFactory beanFactory = getBeanFactory();
        if (beanFactory.containsLocalBean(APPLICATION_EVENT_MULTICASTER_BEAN_NAME)) {
            this.applicationEventMulticaster =
                    beanFactory.getBean(APPLICATION_EVENT_MULTICASTER_BEAN_NAME,
ApplicationEventMulticaster.class);
            ...
        }
        else {
            this.applicationEventMulticaster = new
SimpleApplicationEventMulticaster(beanFactory);
            beanFactory.registerSingleton(APPLICATION_EVENT_MULTICASTER_BEAN_NAME,
this.applicationEventMulticaster);
            ...
        }
    }
    ...
}
```

在 applicationEventMulticaster 属性的初始化过程中，首先判断当前 Spring 应用上下文是否存在名为“applicationEventMulticaster”且类型为 ApplicationEventMulticaster 的 Spring Bean，如果不存在，则将其构造为 SimpleApplicationEventMulticaster 对象，并且注册为 Spring Bean。

在 Spring Boot 1.0～1.3 的实现中，EventPublishingRunListener 在 contextPrepared (ConfigurableApplicationContext)方法执行阶段中将 ApplicationEventMulticaster 属性注册为 ApplicationEventMulticaster Bean：

```
public class EventPublishingRunListener implements SpringApplicationRunListener, Ordered {
    ...
    @Override
    public void contextPrepared(ConfigurableApplicationContext context) {
        registerApplicationEventMulticaster(context);
    }

    private void registerApplicationEventMulticaster(
            ConfigurableApplicationContext context) {
        context.getBeanFactory().registerSingleton(
```

```
                AbstractApplicationContext.APPLICATION_EVENT_MULTICASTER_BEAN_NAME,
                this.multicaster);
        if (this.multicaster instanceof BeanFactoryAware) {
            ((BeanFactoryAware) this.multicaster)
                    .setBeanFactory(context.getBeanFactory());
        }
    }
    ...
}
```

> 以上实现源码源于 Spring Boot 1.3.8.RELEASE。
> Maven GAV 坐标为：org.springframework.boot:spring-boot:1.3.8.RELEASE。

依照 SpringApplicationRunListener 生命周期的执行顺序，contextPrepared(ConfigurableApplicationContext)方法在 AbstractApplicationContext#refresh()方法之前执行，此 SpringApplication 与其关联的 Spring AppliationContext 使用相同的 ApplicationEventMulticaster 对象，即 SimpleApplicationEventMulticaster 实例，并将其注册为 Spring Bean。随后的 contextLoaded(ConfigurableApplicationContext)方法同样在 AbstractApplicationContext#refresh()方法之前执行，将 SpringApplication 所关联的 ApplicationListener 实例列表添加到当前 Spring 应用上下文 ConfigurableApplicationContext 对象中：

```
public class EventPublishingRunListener implements SpringApplicationRunListener, Ordered {
    ...
    @Override
    public void contextLoaded(ConfigurableApplicationContext context) {
        for (ApplicationListener<?> listener : this.application.getListeners()) {
            if (listener instanceof ApplicationContextAware) {
                ((ApplicationContextAware) listener).setApplicationContext(context);
            }
            context.addApplicationListener(listener);
        }
        publishEvent(new ApplicationPreparedEvent(this.application, this.args, context));
    }
    ...
}
```

> 以上实现源码源于 Spring Boot 1.3.8.RELEASE。

> Maven GAV 坐标为：org.springframework.boot:spring-boot:1.3.8.RELEASE。

又已知在 ConfigurableApplicationContext#refresh()调用 refresh()的过程中，ConfigurableApplicationContext 将其关联的 ApplicationListener 实例与其管理的 ApplicationListener 的 Bean 实例整合为新的集合：

```
public abstract class AbstractApplicationContext extends DefaultResourceLoader
        implements ConfigurableApplicationContext {
    ...
    protected void registerListeners() {
        // Register statically specified listeners first.
        for (ApplicationListener<?> listener : getApplicationListeners()) {
            getApplicationEventMulticaster().addApplicationListener(listener);
        }

        // Do not initialize FactoryBeans here: We need to leave all regular beans
        // uninitialized to let post-processors apply to them!
        String[] listenerBeanNames = getBeanNamesForType(ApplicationListener.class, true,
false);
        for (String listenerBeanName : listenerBeanNames) {
            getApplicationEventMulticaster().addApplicationListenerBean(listenerBeanName);
        }

        // Publish early application events now that we finally have a multicaster...
        Set<ApplicationEvent> earlyEventsToProcess = this.earlyApplicationEvents;
        this.earlyApplicationEvents = null;
        if (earlyEventsToProcess != null) {
            for (ApplicationEvent earlyEvent : earlyEventsToProcess) {
                getApplicationEventMulticaster().multicastEvent(earlyEvent);
            }
        }
    }
    ...
}
```

> 以上实现源码源于 Spring Framework 4.2.8.RELEASE。
> Maven GAV 坐标为：org.springframework:spring-context:4.2.8.RELEASE。

此时的 AbstractApplicationContext#getApplicationEventMulticaster()方法所返回的对象恰好是 EventPublishingRunListener 所关联的 ApplicationEventMulticaster 对象，即SimpleApplicationEventMulticaster实例。因此，那些配置在META-INF/spring.factories 资源中的 ApplicationListener 实例列表将被重复插入以上 SimpleApplicationEventMulticaster 对象：

```
public abstract class AbstractApplicationEventMulticaster
        implements ApplicationEventMulticaster, BeanClassLoaderAware, BeanFactoryAware {
    ...
    @Override
    public void addApplicationListener(ApplicationListener<?> listener) {
        synchronized (this.retrievalMutex) {
            this.defaultRetriever.applicationListeners.add(listener);
            this.retrieverCache.clear();
        }
    }
    ...
    private class ListenerRetriever {

        public final Set<ApplicationListener<?>> applicationListeners;
        ...
        public ListenerRetriever(boolean preFiltered) {
            this.applicationListeners = new LinkedHashSet<ApplicationListener<?>>();
            ...
        }
        ...
    }
    ...
}
```

不过，由于 AbstractApplicationEventMulticaster 在底层实现采用了 LinkedHashSet 的集合存储结构，并且通常 ApplicationListener 实现类不会覆盖 Object 的 hashCode 和 equals 方法，所以相同对象的重复插入不会引起事件的重复监听。反之，如果将 ApplicationListener 实现类的 hashCode 实现不稳定数据返回，则同一 ApplicationListener 两次监听的现象将会重现，如下例所示。

ContextRefreshedEventListener 是 ApplicationListener<ContextRefreshedEvent>

的实现类，其 id 属性使用的是当前的纳秒数值，如果对象不同，则 id 值随之变化，而其 hashCode() 方法同样依赖于当前纳秒时间，不过方法每次调用时，其 hashcoole 数值都不同，实现如下所示。

```
    public class ContextRefreshedEventListener implements
ApplicationListener<ContextRefreshedEvent> {

        private final long id;

        public ContextRefreshedEventListener() {
            // id 使用构造时的纳秒时间
            this.id = System.nanoTime();
        }

        @Override
        public void onApplicationEvent(ContextRefreshedEvent event) {
            System.out.printf("ContextRefreshedEventListener[id :%d] 接收到事件:%s\n",
                    id, event.getClass().getSimpleName());
        }

        public int hashCode() {
            // 返回执行时的纳秒时间
            return (int) System.nanoTime();
        }
    }
```

将 ContextRefreshedEventListener 配置到 META-INF/spring.factories 文件中：

```
    # ApplicationListener 实现配置
    org.springframework.context.ApplicationListener=\
    thinking.in.spring.boot.samples.spring.application.event.listener.ContextRefreshedEventLi
stener
```

实现引导类 DuplicatedEventListenerBootstrap：

```
    public class DuplicatedEventListenerBootstrap {

        public static void main(String[] args) {
            new
            SpringApplicationBuilder(DuplicatedEventListenerBootstrap.class)
                    .web(false)            // 非 Web 应用
```

```
                .run(args)              // 运行 SpringApplication
                .close();               // 关闭 ConfigurableApplicationContext
    }
}
```

> 源码位置：以上示例代码可通过查找 spring-boot-1.x-samples/spring-boot-1.3.x-project 工程获取。

运行并观察日志输出：

```
(...部分内容被省略...)
ContextRefreshedEventListener[id :4724162738326] 接收到事件 : ContextRefreshedEvent
ContextRefreshedEventListener[id :4724162738326] 接收到事件 : ContextRefreshedEvent
(...部分内容被省略...)
```

DuplicatedEventListenerBootstrap 的运行结果证实了在 Spring Boot 1.3.8.RELEASE 下 ApplicationListener 实例出现重复添加的情况。当然，这并不意味着是一种设计缺陷，毕竟绝大多数 ApplicationListener 实现类不会覆盖 equals 和 hashCode 方法，所以 AbstractApplicationEventMulticaster 采用 LinkedHashSet 的存储结构是合理的，不但能实现去重，还可保持其插入的顺序性（当 ApplicationListener 实现类没有实现 Ordered 接口或标注 @Order 时，插入顺序即为后续的监听执行顺序）。

核心的问题并不在于重复添加 ApplicationListener 所带来的隐患，而是其中设计的考量。简言之，SpringApplication 从 META-INF/spring.factories 资源中加载的 ApplicationListener 实例列表是否有必要关联到 Spring 应用上下文 ConfigurableApplicationContext 对象。如同 Spring Boot 1.0～1.3 版本中的实现，不但 SpringApplication 中的 ApplicationListener 能够监听 ConfigurableApplicationContext 所发送的事件，同样地，Spring 应用上下文所关联的 ApplicationListener 也能监听部分 Spring Boot 事件。之所以不是所有的 Spring Boot 事件，是因为部分 Spring Boot 事件在 Spring 应用上下文初始化之前就发布了，因此它们无法被监听。换言之，在 Spring 应用上下文初始化之后，它无法监听后续的 Spring Boot 事件。EventPublishingRunListener 与 Spring Boot 事件的关系，以及与 AbstractApplicationContext 调用的前后关系如下表所示。

监听方法	refresh() 方法执行顺序	Spring Boot 事件	Spring Boot 起始版本
started()	调用前	ApplicationStartedEvent	1.0

environmentPrepared()	调用前	ApplicationEnvironmentPreparedEvent	1.0
contextLoaded()	调用前	ApplicationPreparedEvent	1.0
running()	调用后	ApplicationReadyEvent	1.3
failed()	调用后	ApplicationFailedEvent	1.0

> 由于 ApplicationStartingEvent 出现在 Spring Boot 1.5 中，高于当前讨论的版本，故予以排除。

如果以上结论是正确的，且在 Spring Boot 1.0～1.3 中，由于 SpringApplication 与其管理的 Spring 应用上下文共用同一个 SimpleApplicationEventMulticaster 对象，理论上，ConfigurableApplicationContext 关联的 ApplicationListener 可监听 ContextRefreshedEvent 之后的事件：ApplicationReadyEvent 和 ApplicationFailedEvent。因此，可通过一个 ApplicationListener 实现类监听 ApplicationReadyEvent 和 ApplicationFailedEvent 事件，并将其声明为 Spring Bean，当 Spring Boot 正常启动时，ApplicationReadyEvent 事件将被传播并监听，否则，ApplicationFailedEvent 代替之。

1）SmartApplicationListener 监听多 Spring Boot 事件

实际上，多 Spring Boot 事件监听器完全可使用 ApplicationListener 实现，然而由于多处 Spring Boot 事件监听器内建实现选择的是 SmartApplicationListener 接口，如加载 application.proeprties 文件的 ConfigFileApplicationListener 及日志配置的 LoggingApplicationListener 等，因此当前实现也基于 SmartApplicationListener 接口。

> 由于 SmartApplicationListener 从 Spring Framework 3.0 开始被引入，所以它能在任何 Spring Boot 版本中使用。

- 多 Spring Boot 事件 SmartApplicationListener 实现：MultipleSpringBootEventsListener

```java
public class MultipleSpringBootEventsListener implements SmartApplicationListener {

    @Override
    public boolean supportsEventType(Class<? extends ApplicationEvent> eventType) {
        // 支持事件的类型
        return ApplicationReadyEvent.class.equals(eventType) ||
                ApplicationFailedEvent.class.equals(eventType);
    }
```

```
    @Override
    public boolean supportsSourceType(Class<?> sourceType) {
        // SpringApplicationEvent 均以 SpringApplication 作为配置源
        return SpringApplication.class.equals(sourceType);
    }

    @Override
    public void onApplicationEvent(ApplicationEvent event) {
        if (event instanceof ApplicationReadyEvent) {
            // 当事件为 ApplicationReadyEvent 时，随机地抛出异常
            if (new Random().nextBoolean()) {
                throw new RuntimeException("ApplicationReadyEvent 事件监听异常!");
            }
        }
        System.out.println("MultipleSpringBootEventsListener 监听到事件 : " +
event.getClass().getSimpleName());
    }

    @Override
    public int getOrder() {
        return Ordered.HIGHEST_PRECEDENCE;
    }
}
```

当 onApplicationEvent(ApplicationEvent)方法监听到 ApplicationReadyEvent 事件时，随机地抛出异常，用于检验该实例对 ApplicationFailedEvent 的监听。

- MultipleSpringBootEventsListener 引导类

```
public class MultipleSpringBootEventsListenerBootstrap {

    public static void main(String[] args) {
        new
        SpringApplicationBuilder(MultipleSpringBootEventsListener.class) // 注册为 Spring
Bean
                .web(false)             // 非 Web 应用
                .run(args)              // 运行 SpringApplication
                .close();               // 关闭 ConfigurableApplicationContext
    }
```

```
}
```

> 源码位置：以上示例代码可通过查找 spring-boot-1.x-samples/spring-boot-1.3.x-project 工程获取。

运行并观察日志输出：

- 正常运行输出：

```
(...部分内容被省略...)
MultipleSpringBootEventsListener 收到事件: ApplicationReadyEvent
(...部分内容被省略...)
```

- 执行异常内容：

```
(...部分内容被省略...)
java.lang.RuntimeException: ApplicationReadyEvent 事件监听异常!
(...部分内容被省略...)
MultipleSpringBootEventsListener 监听到事件 : ApplicationFailedEvent
(...部分内容被省略...)
```

同样地，@EventListener 方法也支持监听以上两种 Spring Boot 事件。

2）@EventListener 方法监听多 Spring Boot 事件

在事件监听器实现 MultipleSpringBootEventsListener 中增加@EventListener 方法，如下所示。

```
public class MultipleSpringBootEventsListener implements SmartApplicationListener {
  ...
    @EventListener({ApplicationReadyEvent.class, ApplicationFailedEvent.class})
    public void onSpringBootEvent(SpringApplicationEvent event) {
        System.out.println("@EventListener 监听到事件 : " +
event.getClass().getSimpleName());
    }
}
```

> 源码位置：以上示例代码可通过查找 spring-boot-1.x-samples/spring-boot-1.3.x-project 工程获取。

重启 MultipleSpringBootEventsListenerBootstrap 引导类，观察日志输出。

- 正常运行输出：

```
(...部分内容被省略...)
MultipleSpringBootEventsListener 监听到事件: ApplicationReadyEvent
@EventListener 监听到事件: ApplicationReadyEvent
(...部分内容被省略...)
```

- 执行异常内容：

```
(...部分内容被省略...)
java.lang.RuntimeException: ApplicationReadyEvent 事件监听异常!
(...部分内容被省略...)
MultipleSpringBootEventsListener 监听到事件 : ApplicationFailedEvent
(...部分内容被省略...)
@EventListener 监听到事件 : ApplicationFailedEvent
(...部分内容被省略...)
```

以上两种运行结果证明了之前的推论，即由于 SpringApplication 与 ConfigurableApplicationContext 共用 SimpleApplicationEventMulticaster 对象的原因，Spring 应用上下文中 ApplicationListener Bean 和@EventListener 方法均能监听 Spring Boot ApplicationReadyEvent 或 ApplicationFailedEvent 事件。

> 由于注解@EventListener 从 Spring Framework 4.2 才开始出现，因此，Spring Boot 版本需要 1.3 或更高。结合本节所涉及的 Spring Boot 的版本范围，类似以上示例的现象仅在 Spring Boot 1.3 中出现，请读者尤其关注版本间差异对应用架构的影响。

不过，在现实的使用场景中，即使开发人员熟悉 Spring 事件/监听机制，也鲜有机会使用 ApplicationListener Bean 去监听 Spring Boot 事件。然而，这样的设计的确让人苦恼，尤其在纯 ApplicationListener 接口非泛型的编程场景下。或许 Spring Boot 官方也考虑到以上因素，在 1.4 版本中，对 Spring Boot 事件/监听的实现策略做出了相应的调整，不再将 SpringApplication 关联的 SimpleApplicationEventMulticaster 对象复用到其管理的 ConfigurableApplicationContext 中。那么，这样的调整又会有哪些方面的影响呢？接下来，继续讨论 Spring Boot 1.4 时代的事件/监听机制。

2. 当前 Spring Boot 事件/监听机制

Spring Boot 1.4 就如同 Spring Framework 3.1 那样，属于 Spring Boot 分水岭式的版本。从此版本开始，Spring Boot 的核心 API 基本趋于稳定。同样，它在 Spring Boot 事件/监听机制方面也做出了微调，让其一直延续到最新的 2.0 版本。Spring Boot 1.4～2.0 版本应该是当前生产环境中分布最广泛的版本范围。前文曾提及，从 Spring Boot 1.4 开始，框架层面采取的是 Spring Boot 事件与 Spring 事件“各自为政，互不干涉”的设计原则。在实现上，前后设计的差异主要体现在 EventPublishingRunListener#contextPrepared(ConfigurableApplicationContext)方法上。从 Spring Boot 1.4 开始，该方法不再将 EventPublishingRunListener 关联的 SimpleApplicationEventMulticaster 属性对象注册为 Spring Bean：

```
public class EventPublishingRunListener implements SpringApplicationRunListener, Ordered {
    ...
    @Override
    public void contextPrepared(ConfigurableApplicationContext context) {

    }
    ...
}
```

> 以上实现源码源于 Spring Boot 1.4.7.RELEASE。
> Maven GAV 坐标为：org.springframework.boot:spring-boot:1.4.7.RELEASE。

如此实现就杜绝了 Spring 应用上下文所关联的 ApplicationListener 接受 Spring Boot 事件的广播，从而各自监听所在环境的事件。值得注意的是，EventPublishingRunListener 仍在调用其 contextLoaded(ConfigurableApplicationContext)方法时，将 SpringApplication 所关联的 ApplicationListener 添加至 ConfigurableApplicationContext 中：

```
public class EventPublishingRunListener implements SpringApplicationRunListener, Ordered {
    ...
    @Override
    public void contextLoaded(ConfigurableApplicationContext context) {
        for (ApplicationListener<?> listener : this.application.getListeners()) {
            if (listener instanceof ApplicationContextAware) {
                ((ApplicationContextAware) listener).setApplicationContext(context);
            }
            context.addApplicationListener(listener);
        }
        this.initialMulticaster.multicastEvent(
                new ApplicationPreparedEvent(this.application, this.args, context));
    }
    ...
}
```

> 以上实现源码源于 Spring Boot 1.4.7.RELEASE。
> Maven GAV 坐标为：org.springframework.boot:spring-boot:1.4.7.RELEASE。

根据 SpringApplicationRunListener 生命周期回调的特性，此时 Spring 应用上下文尚未初始化，因此，以上添加操作最终会追加到 AbstractApplicationContext 所关联的 SimpleApplicationEventMulticaster 属性中，当 Spring 应用上下文发布 Spring 事件后，这些被 contextLoaded 方法添加的 ApplicationListener 集合能够将它们监听。接下来继续通过示例检验以上结论。

Spring Boot ApplicationListener 监听 Spring 事件

直接构建 Spring Boot 引导类，并且结合 SpringApplication#listeners(ApplicationListener...)方法添加 ApplicationListener 实例：

```
public class SpringEventListenerBootstrap {

    public static void main(String[] args) {
        new SpringApplicationBuilder(Object.class) // 非@Configuration 充当
                                                   // 配置源
                .listeners(event ->               // 添加 ApplicationListener
                        System.out.println("SpringApplication 事件监听器 : "
                                + event.getClass().getSimpleName())
                )
                .web(false)                       // 非 Web 应用
                .run(args)                        // 运行 SpringApplication
                .close();                         // 关闭 ConfigurableApplicationContext
    }
}
```

运行该引导类并观察日志输出：

```
    监听到事件: ApplicationStartedEvent , 事件源 :
org.springframework.boot.SpringApplication@2ac273d3
    监听到事件: ApplicationEnvironmentPreparedEvent,事件源:
org.springframework.boot.SpringApplication@2ac273d3
    (...部分内容被省略...)
    监听到事件: ApplicationPreparedEvent , 事件源:
org.springframework.boot.SpringApplication@2ac273d3
    (...部分内容被省略...)
```

```
    监听到事件: ContextRefreshedEvent,事件源:
org.springframework.context.annotation.AnnotationConfigApplicationContext@49ec71f8
    监听到事件: ApplicationReadyEvent ,事件源:
org.springframework.boot.SpringApplication@2ac273d3
    (...部分内容被省略...)
    监听到事件: ContextClosedEvent , 事件源 :
org.springframework.context.annotation.AnnotationConfigApplicationContext@49ec71f8
    (...部分内容被省略...)
```

以上运行结果可谓一举多得，不但监听到不同 Spring Boot 事件和 Spring 事件，并且不同事件的发布顺序一览无余。同时，两类事件的源不同，Spring Boot 事件源为 `SpringApplication`，而此处 Spring 事件源则是 `AnnotationConfigApplicationContext` 对象，那么这个 Spring 应用上下文的实现类是如何配置的呢？后续的“创建 Spring 应用上下文”一节将会讨论。

明显地，Lambda 语句所创建的 `ApplicationListener` 对象的确监听到 Spring 事件，确切地说是 Spring 上下文事件。总之，从 Spring Boot 1.4 开始，`SpringApplication` 采用了独立的 `ApplicationEventMulticaster` 对象。默认情况下，`SpringApplication` 与 `AppliationContext` 仍使用 `SimpleApplicationEventMulticaster` 实例，不过两者不再是相同的对象。无论 Spring Framework，还是 Spring Boot，`SimpleApplicationEventMulticaster` 都是默认的 Spring 事件或 Spring Boot 事件广播器实现。

尽管 Spring Boot `ApplicationListener` 能够监听 Spring 事件，然而它绝大多数的事件场景在监听 Spring Boot 事件方面。接下来继续讨论内建 Spring Boot 事件监听器实现。

3. Spring Boot 内建事件监听器

在 Spring Boot 场景中，无论 Spring 事件监听器，还是 Spring Boot 事件监听器，均配置在 `META-INF/spring.factories` 资源中，并以 `org.springframework.context.ApplicationListener` 作为属性名称，属性值为 `ApplicationListener` 的实现类。以 Spring Boot `2.0.2.RELEASE` 实现为例，其 `ApplicationListener` 配置声明分布在 `spring-boot` 和 `spring-boot-autoconfigure` jar 文件中，如以下 GAV 模块所示。

- `org.springframework.boot:spring-boot:2.0.2.RELEASE`

```
# Application Listeners
org.springframework.context.ApplicationListener=\
org.springframework.boot.ClearCachesApplicationListener,\
org.springframework.boot.builder.ParentContextCloserApplicationListener,\
org.springframework.boot.context.FileEncodingApplicationListener,\
org.springframework.boot.context.config.AnsiOutputApplicationListener,\
```

```
org.springframework.boot.context.config.ConfigFileApplicationListener,\
org.springframework.boot.context.config.DelegatingApplicationListener,\
org.springframework.boot.context.logging.ClassPathLoggingApplicationListener,\
org.springframework.boot.context.logging.LoggingApplicationListener,\
org.springframework.boot.liquibase.LiquibaseServiceLocatorApplicationListener
```

- org.springframework.boot:spring-boot-autoconfigure:2.0.2.RELEASE

```
# Application Listeners
org.springframework.context.ApplicationListener=\
org.springframework.boot.autoconfigure.BackgroundPreinitializer
```

将以上 ApplicationListener 汇总至表格加以说明，如下表所示。

ApplicationListener 实现	监听事件	场景说明	引入版本
ClearCachesApplicationListener	ContextRefreshedEvent	清除 ReflectionUtils 和 ClassLoader 缓存	1.4
ParentContextCloserApplicationListener	ParentContextAvailableEvent	当 SpringApplication 关联上下文初始化并设置双亲上下文（parent）时	1.0
FileEncodingApplicationListener	ApplicationEnvironmentPreparedEvent	检测 spring.mandatoryFileEncoding 属性是否 file.encoding 匹配	1.0
AnsiOutputApplicationListener	ApplicationEnvironmentPreparedEvent	生成 ANSI	1.2
ConfigFileApplicationListener	ApplicationEnvironmentPreparedEvent 和 ApplicationPreparedEvent	加载应用配置文件，默认为 application.properties 或 application.yml	1.0
DelegatingApplicationListener	ApplicationEnvironmentPreparedEvent	将 Spring 事件委派给配置 context.listener.classes 所指定的多个 ApplicationListener 实现类	1.0

续表

ApplicationListener	监听事件	场景说明	引入

实现			版　本
ClassPathLoggingApplicationListener	ApplicationEnvironmentPreparedEvent 或 ApplicationFailedEvent	DEBUG 级别日志记录当前应用的 Class Path	1.0（2.0 包名被重构）
LoggingApplicationListener	ApplicationStartingEvent 或 ApplicationEnvironmentPreparedEvent 或 ApplicationPreparedEvent 或 ContextClosedEvent 或 ApplicationFailedEvent	识别日志框架，并加载日志配置文件	1.0（2.0 包名被重构）
LiquibaseServiceLocatorApplicationListener	ApplicationStartingEvent 或 ApplicationStartedEvent	取代 liquibase ServiceLocator	1.0
BackgroundPreinitializer	ApplicationEnvironmentPreparedEvent	异步线程触发早起初始化操作	1.3

以上最重要的 Spring Boot 内建事件监听器莫过于 ConfigFileApplicationListener 和 LoggingApplicationListener，前者负责 Spring Boot 应用配置属性文件的加载，后者用于 Spring Boot 日志系统的初始化（日志框架识别、日志配置文件加载等）。

> ConfigFileApplicationListener 部分的详情将在后续章节"外部化配置"中详细讨论，而 LoggingApplicationListener 的讨论将在下一册中展开。

关于 Spring Boot 事件/监听机制的讨论即将进入尾声，为此总结一二。

4. 总结 Spring Boot 事件/监听机制

Spring Boot 事件/监听机制继承于 Spring 事件/监听机制，同样涵盖 Spring Boot 事件、Spring Boot 事件监听手段、Spring Boot 事件广播器等。

1）总结 Spring Boot 事件

Spring Boot 事件类型继承 Spring 事件类型 ApplicationEvent，并且也是 SpringApplicationEvent 的子类。大多数 Spring 内建事件为 Spring 应用上下文事件，即 ApplicationContextEvent，其事件源为 ApplicationContext。而 Spring Boot 事件源则是 SpringApplication，其内建事件根据 EventPublishingRunListener 的生命周期回调方法依次发布。以 Spring Boot 2.0.2.RELEASE 实现为例，它们为 ApplicationStartingEvent、ApplicationEnvironmentPreparedEvent、ApplicationPreparedEvent、ApplicationStartedEvent、ApplicationReadyEvent 和 ApplicationFailedEvent。其中，ApplicationReadyEvent

和 ApplicationFailedEvent 在 Spring 应用上下文初始化后发布，即在 ContextRefreshedEvent 之后发布。

再结合 Spring Boot 内建事件 API 出现的版本分析，ApplicationReadyEvent 是从 Spring Boot 1.3 开始引入的，而 ApplicationStartingEvent 则是从 Spring Boot 1.5 开始出现的，替代 1.5 版本之前的 ApplicationStartedEvent。

2）Spring Boot 事件监听手段

尽管 Spring Boot 事件监听器仍继承于 Spring Framework，不过由于 Spring Boot 在 1.4 版本前后采取了不同的设计，因此其监听手段需要区别对待。

在早期 Spring Boot 事件/监听机制（Spring Boot 1.0～1.3 版本）中，由于 SpringApplication 与 Spring 应用上下文 ConfigurableApplicationContext 采用相同的 SimpleApplicationEventMulticaster 实例，因此 ConfigurableApplicationContext 关联的 ApplicationListener 和 @EventListener 方法能够监听 Spring Boot 事件 ApplicationReadyEvent 和 ApplicationFailedEvent。随着 Spring Boot 1.4 开始采用 SpringApplication 与 ConfigurableApplicationContext 隔离 SimpleApplicationEventMulticaster 实例的设计，这种监听手段不再奏效。换言之，从此时开始，Spring Boot 事件监听手段仅为 SpringApplication 关联 ApplicationListener 对象集合。其关联途径有二，一为 SpringApplication 构造阶段在 Class Path 下所加载所有 META-INF/spring.factories 资源中的 ApplicationListener 对象集合，二是通过方法 SpringApplication#addListeners(ApplicationListener...) 或 SpringApplicationBuilder#listeners(ApplicationListener...)显式地装配。

3）Spring Boot 事件广播器

Spring Boot 事件广播器同样来源于 Spring Framework 的实现类 SimpleApplicationEventMulticaster，其广播行为与 Spring 事件广播毫无差别，只不过 Spring Boot 中发布的事件类型是特定的。因此可以完整地做出以下结论，Spring Boot 事件/监听机制继承于 Spring 事件/监听机制，其事件类型继承于 Spring ApplicationEvent，事件监听器仍通过 ApplicationListener 实现，而广播器实现 SimpleApplicationEventMulticaster 将它们关联起来。

至此，关于 Spring 和 Spring Boot 事件/监听机制的讨论告一段落，由于 EventPublishingRunListener 作为 Spring Boot 事件发布的执行者，也是 Spring Boot 框架内部的唯一的 SpringApplicationRunListener 实现类，换言之就结束了 SpringApplicationRunListeners 的讨论，而重回“SpringApplication 准备阶段”的讨论。

11.1.6 装配 ApplicationArguments

当执行 SpringApplicationRunListeners#starting()方法后，SpringApplication 运行进入装

配 ApplicationArguments 逻辑：

```
public class SpringApplication {
    ...
    public ConfigurableApplicationContext run(String... args) {
        ...
        SpringApplicationRunListeners listeners = getRunListeners(args);
        listeners.starting();
        try {
            ApplicationArguments applicationArguments = new DefaultApplicationArguments(
                    args);
        }
        ...
    }
    ...
}
```

ApplicationArguments 实例的创建源于 Spring Boot 1.3，其实现类为 DefaultApplicationArguments，一个用于简化 Spring Boot 应用启动参数的封装接口，它的底层实现基于 Spring Framework 中的命令行配置源 SimpleCommandLinePropertySource：

```
public class DefaultApplicationArguments implements ApplicationArguments {

    private final Source source;

    private final String[] args;

    public DefaultApplicationArguments(String[] args) {
        Assert.notNull(args, "Args must not be null");
        this.source = new Source(args);
        this.args = args;
    }
    ...
    private static class Source extends SimpleCommandLinePropertySource {
        ...
    }

}
```

按照 SimpleCommandLinePropertySource JavaDoc 的描述：

> command line arguments are broken into two distinct groups: option arguments and non-option arguments

SimpleCommandLinePropertySource 将命令行参数分为两组，一为“选项参数”，二为“非选项参数”，两者均依赖于 SimpleCommandLineArgsParser 解析。其中，合法的选项参数如其 JavaDoc 所述：

> Working with option arguments Option arguments must adhere to the exact syntax: --optName[=optValue] That is, options must be prefixed with "--", and may or may not specify a value.

选项参数必须以“--”为前缀，其值可选。反之，非选项参数则未包含“--”前缀的命令行参数：

> Working with non-option arguments Any and all arguments specified at the command line without the "--" option prefix will be considered as "non-option arguments" and made available through the CommandLineArgs.getNonOptionArgs() method.

根据以上规则，命令行参数“--name=小马哥”将被 SimpleCommandLinePropertySource 解析为“name: 小马哥”的键值属性，因此 SimpleCommandLinePropertySource 作为 DefaultApplicationArguments 的底层实现 ApplicationArguments 接口的语义如下：

```java
public class DefaultApplicationArguments implements ApplicationArguments {
...
@Override
public String[] getSourceArgs() {
    return this.args;
}

@Override
public Set<String> getOptionNames() {
    String[] names = this.source.getPropertyNames();
    return Collections.unmodifiableSet(new HashSet<>(Arrays.asList(names)));
}

@Override
```

```
    public boolean containsOption(String name) {
        return this.source.containsProperty(name);
    }

    @Override
    public List<String> getOptionValues(String name) {
        List<String> values = this.source.getOptionValues(name);
        return (values != null ? Collections.unmodifiableList(values) : null);
    }

    @Override
    public List<String> getNonOptionArgs() {
        return this.source.getNonOptionArgs();
    }
    ...
}
```

其中，source 的类型为 Source，扩展于 SimpleCommandLinePropertySource，并将其 protected 方法的访问性提升为 public。

当 ApplicationArguments 实例在 SpringApplication 的准备阶段构造完毕后，它将投入 ApplicationRunner 回调方法参数的运用：

```
@FunctionalInterface
public interface ApplicationRunner {

    /**
     * Callback used to run the bean.
     * @param args incoming application arguments
     * @throws Exception on error
     */
    void run(ApplicationArguments args) throws Exception;

}
```

因此 ApplicationArguments 简化了 ApplicationRunner 参数的解析，进而使得 ApplicationRunner 作为 CommandLineRunner 的补充。由于 ApplicationRunner 和 CommandLineRunner 是 SpringApplication 运行阶段的讨论议题，届时将一并实践。

当 ApplicationArguments 实例准备完毕后，SpringApplication 的执行操作进入准备

ConfigurableEnvironment 的阶段。

11.1.7 准备 ConfigurableEnvironment

为了更好地理解 ConfigurableEnvironment 准备的过程，需要完整的 Environment 上下文背景和生命周期，因此该部分的讨论将在“外部化配置操作”的“理解 Spring Boot Environment 生命周期”一节展开，在此不再赘述。当 ConfigurableEnvironment 准备妥当后，进入创建 Spring 应用上下文阶段。

11.1.8 创建 Spring 应用上下文（ConfigurableApplicationContext）

SpringApplication 通过 createApplicationContext()方法创建 Spring 应用上下文，实际上 Spring 应用上下文才是驱动整体 Spring Boot 应用组件的核心引擎：

```
public class SpringApplication {
    ...
    public static final String DEFAULT_CONTEXT_CLASS = "org.springframework.context."
            + "annotation.AnnotationConfigApplicationContext";
    ...
    public static final String DEFAULT_WEB_CONTEXT_CLASS = "org.springframework.boot."
            + "web.servlet.context.AnnotationConfigServletWebServerApplicationContext";
    ...
    public static final String DEFAULT_REACTIVE_WEB_CONTEXT_CLASS = "org.springframework."
            +
"boot.web.reactive.context.AnnotationConfigReactiveWebServerApplicationContext";
    ...
    protected ConfigurableApplicationContext createApplicationContext() {
        Class<?> contextClass = this.applicationContextClass;
        if (contextClass == null) {
            try {
                switch (this.webApplicationType) {
                case SERVLET:
                    contextClass = Class.forName(DEFAULT_WEB_CONTEXT_CLASS);
                    break;
                case REACTIVE:
                    contextClass = Class.forName(DEFAULT_REACTIVE_WEB_CONTEXT_CLASS);
                    break;
```

```
                default:
                    contextClass = Class.forName(DEFAULT_CONTEXT_CLASS);
                }
            }
            catch (ClassNotFoundException ex) {
                throw new IllegalStateException(
                        "Unable create a default ApplicationContext, "
                                + "please specify an ApplicationContextClass",
                        ex);
            }
        }
        return (ConfigurableApplicationContext) BeanUtils.instantiateClass(contextClass);
    }
    ...
}
```

1. 根据 WebApplicationType 创建 Spring 应用上下文

默认情况下，createApplicationContext()方法根据 SpringApplication 构造阶段所推断的 Web 应用类型进行 ConfigurableApplicationContext 的创建。回顾 Web 应用类型推断的过程，SpringApplication#deduceWebApplicationType()方法通过 ClassLoader 判断核心类是否存在来推断 Web 应用类型，当 Spring WebFlux 类的 DispatcherHandler 存在，并且 Spring Web MVC 类 DispatcherServlet 不存在时，当前应用类型为 WebApplicationType.REACTIVE。当以上两个类均不存在时，则视作 WebApplicationType.NONE，否则为 WebApplicationType.SERVLET。换言之，如果 Spring Web MVC 和 Spring WebFlux 类库同时存在于 Class Path 中时，则应用类型为 WebApplicationType.SERVLET，因此当前 Spring 应用上下文对象为 AnnotationConfigServletWebServerApplicationContext。尽管应该避免出现这样的情况，然而由于不合理的传递依赖，这种情况时常发生。

当然，SpringApplication 也允许显式地调整 Web 应用类型，从而忽视 Web 应用类型推断的行为。本书中的大量示例就是这种情况，通过调用 SpringApplicationBuilder#web(WebApplicationType) 方法或 SpringApplication#setWebApplicationType(WebApplicationType)方法实现，比如在“走向自动装配”章节中的示例 EnableAutoConfigurationBootstrap：

```
@EnableAutoConfiguration(exclude = SpringApplicationAdminJmxAutoConfiguration.class)
public class EnableAutoConfigurationBootstrap {

    public static void main(String[] args) {
```

```java
        new SpringApplicationBuilder(EnableAutoConfigurationBootstrap.class)
                .web(WebApplicationType.NONE) // 非 Web 应用
                .run(args)                    // 运行
                .close();                     //  关闭当前上下文
    }
}
```

> 源码位置：以上示例代码可通过查找 spring-boot-2.0-samples/auto-configuration-sample 工程获取，该工程依赖 org.springframework.boot:spring-boot-starter。

结合 createApplicationContext()方法的实现逻辑，该 SpringApplication 所创建的 ConfigurableApplicationContext 对象类型为 AnnotationConfigApplicationContext，它与传统的 Spring Framework 注解驱动上下文无异。然而该引导类在运行后，其自动装配特性执行无误，从而说明了 Spring Boot 自动装配的原生能力同样来自 Spring Framework。

值得注意的是，以上讨论的范围集中在 Spring Boot 2.0 的实现上，由于 Spring Boot 1.x 尚未支持 Spring WebFlux，因此对于 Web 应用类型的判断相对简单，只有是与否两种结果。这一点在讨论 Spring Boot 条件注解@ConditionalOnWebApplication 时也曾提到，所以当时的 SpringApplication 中不存在 setWebApplicationType(WebApplicationType)方法，仅提供 setWebEnvironment (boolean)方法及便利的 SpringApplicationBuilder#web(boolean)方法，设置当前应用是否为 Web 类型。以 Spring Boot 1.5.10.RELEASE 实现为例，其 createApplicationContext()方法的逻辑更简单：

```java
public class SpringApplication {
...
public static final String DEFAULT_CONTEXT_CLASS = "org.springframework.context."
        + "annotation.AnnotationConfigApplicationContext";
...
public static final String DEFAULT_WEB_CONTEXT_CLASS = "org.springframework."
        + "boot.context.embedded.AnnotationConfigEmbeddedWebApplicationContext";
...
protected ConfigurableApplicationContext createApplicationContext() {
    Class<?> contextClass = this.applicationContextClass;
    if (contextClass == null) {
        try {
            contextClass = Class.forName(this.webEnvironment
                    ? DEFAULT_WEB_CONTEXT_CLASS : DEFAULT_CONTEXT_CLASS);
        }
```

```
            catch (ClassNotFoundException ex) {
                throw new IllegalStateException(
                        "Unable create a default ApplicationContext, "
                                + "please specify an ApplicationContextClass",
                        ex);
            }
        }
        return (ConfigurableApplicationContext) BeanUtils.instantiate(contextClass);
    }
    ...
}
```

> 以上实现源码源于 Spring Boot 1.5.10.RELEASE。
> Maven GAV 坐标为：org.springframework.boot:spring-boot:1.5.10.RELEASE。

在非 Web 和 Spring Web MVC 类型 ConfigurableApplicationContext 的创建逻辑中，该方法与Spring Boot 2.0的实现是一致的，因此打算迁移到Spring Boot 2.0的读者不必忧虑其中的兼容性。

无论在 Spring Boot 1.x，还是在 2.0 版本的实现中，以上讨论均属于默认场景的创建流程。换言之，当属性 applicationContextClass 不为 null 时，以上依赖 WebApplicationType 的创建逻辑不会执行。

2. 根据指定 ConfigurableApplicationContext 类型创建 Spring 应用上下文

从 Spring Boot 1.0 开始，通过 setApplicationContextClass(Class)方法，SpringApplication 使其保留了指定 ConfigurableApplicationContext 类型创建 Spring 应用上下文的弹性（当然使用 SpringApplicationBuilder#contextClass(Class)方法更便利）：

```
public class SpringApplication {
    ...
    public void setApplicationContextClass(
            Class<? extends ConfigurableApplicationContext> applicationContextClass) {
        this.applicationContextClass = applicationContextClass;
        if (!isWebApplicationContext(applicationContextClass)) {
            this.webApplicationType = WebApplicationType.NONE;
        }
    }
    ...
}
```

该方法值得关注的是，不但 applicationContextClass 属性被赋值，而且也有条件地设置了 webApplicationType 属性。或许读者会为此而感到困惑，为什么 applicationContextClass 属性已被设定，还需要设置 webApplicationType 属性呢？实际上，webApplicationType 属性还可以作为创建 ConfigurableEnvironment 对象具体类型的条件。

> 虽然 SpringApplication#setApplicationContextClass(Class)方法能够影响 Spring ConfigurableApplicationContext 的类型，但是由于实现 ConfigurableApplicationContext 的语义异常复杂——即使通过继承已有的模板抽象类 AbstractApplicationContext 及以上三种 Spring 应用上下文的具体实现。因此，该方法尽可能不要去触碰，以免节外生枝。

当 Spring Boot 应用的 Spring 应用上下文（ConfigurableApplicationContext）创建完毕后，SpringApplication 将在下一个阶段对其做运行前的准备。

11.1.9 Spring 应用上下文运行前准备

Spring 应用上下文运行前的准备工作由 SpringApplication#prepareContext 方法完成，根据 SpringApplicationRunListener 的生命周期回调又分为“Spring 应用上下文准备阶段”和“Spring 应用上下文装载阶段”。

1. Spring 应用上下文准备阶段

本阶段的执行从 prepareContext 方法开始，到 SpringApplicationRunListeners#contextPrepared 截止：

```
public class SpringApplication {
 ...
 private void prepareContext(ConfigurableApplicationContext context,
         ConfigurableEnvironment environment, SpringApplicationRunListeners listeners,
         ApplicationArguments applicationArguments, Banner printedBanner) {
     context.setEnvironment(environment);
     postProcessApplicationContext(context);
     applyInitializers(context);
     listeners.contextPrepared(context);
     ...
 }
 ...
}
```

按照以上实现逻辑，该过程又可由“设置 Spring 应用上下文 ConfigurableEnvironment”“Spring 应用上下文后置处理”“运用 Spring 应用上下文初始化器”和“Spring 应用上下文已准备生命周期回调”组成。

1）设置 Spring 应用上下文 ConfigurableEnvironment

> 在深入讨论之前，还是建议读者提前阅读“外部化配置”章节中 Environment 相关的内容，因为接下来的讨论有一定的技术壁垒。

本过程仅执行一行语句，即 context.setEnvironment(environment)，这或许看起来没有什么特别。从传统意义而言，Spring 应用上下文 ConfigurableApplicationContext 不仅通过其关联的 BeanFactory 对象管理 Bean 及它们的生命周期，而且属性也关联 ConfigurableEnvironment 实例。在 Spring Framework 中，该接口有两种实现：StandardEnvironment 或 StandardServletEnvironment，然而在 Spring Boot 2.0 中，新增了 StandardReactiveWebEnvironment 的实现。默认情况下，ConfigurableEnvironment 属性由模板方法 AbstractApplicationContext#createEnvironment()创建，因此当 Spring 应用上下文类型不同时，该方法返回的对象类型也是不同的。例如以上类型的实例分别由 AbstractApplicationContext、AbstractRefreshableWebApplicationContext 和 AnnotationConfigReactiveWebApplicationContext 创建。无论哪种具体类型的 ConfigurableEnvironment 对象，其功能特性与 getOrCreateEnvironment()方法返回值并无明显差异：

```
public class SpringApplication {
...
private ConfigurableEnvironment getOrCreateEnvironment() {
     if (this.environment != null) {
          return this.environment;
     }
     if (this.webApplicationType == WebApplicationType.SERVLET) {
          return new StandardServletEnvironment();
     }
     return new StandardEnvironment();
}
...
}
```

当 Spring 应用上下文类型为 AnnotationConfigReactiveWebApplicationContext 时，其 ConfigurableEnvironment 属性类型为 StandardReactiveWebEnvironment：

```
public class AnnotationConfigReactiveWebApplicationContext
        extends AbstractRefreshableConfigApplicationContext
        implements ConfigurableReactiveWebApplicationContext, AnnotationConfigRegistry {
    ...
    @Override
    protected ConfigurableEnvironment createEnvironment() {
        return new StandardReactiveWebEnvironment();
    }
    ...
}
```

StandardReactiveWebEnvironment 只是简单地继承 StandardEnvironment，并无任何修改：

```
public class StandardReactiveWebEnvironment extends StandardEnvironment
        implements ConfigurableReactiveWebEnvironment {

}
```

换言之，StandardReactiveWebEnvironment 等于 StandardEnvironment，所以 SpringApplication#getOrCreateEnvironment()方法返回值适用于以上三种不同的应用类型。既然 AbstractApplicationContext#createEnvironment()与 SpringApplication#getOrCreateEnvironment()的返回结果基本相同，为何 SpringApplication 要提前设置呢？其根本原因在于 AbstractApplicationContext#createEnvironment()方法的执行时机。该方法的执行调用链路为 refresh()→prepareRefresh()→getEnvironment()→createEnvironment()：

```
public abstract class AbstractApplicationContext extends DefaultResourceLoader
        implements ConfigurableApplicationContext {
    ...
    @Override
    public void refresh() throws BeansException, IllegalStateException {
        synchronized (this.startupShutdownMonitor) {
            // Prepare this context for refreshing.
            prepareRefresh();
            ...
        }
        ...
    }
    ...
```

```
    protected void prepareRefresh() {
        ...
        getEnvironment().validateRequiredProperties();
        ...
    }
    ...
    public ConfigurableEnvironment getEnvironment() {
        if (this.environment == null) {
            this.environment = createEnvironment();
        }
        return this.environment;
    }
    ...
}
```

createEnvironment()方法的执行不仅需要 environment 属性从未初始化，并且依赖于 Spring 上下文启动（refresh()）的生命周期。如果 environment 属性在此时创建，则其配置属性源（PropertySource）的装载极有可能通过 BeanFactoryPostProcessor 实现完成，然而在众多 BeanFactoryPostProcessor 集合中，程序很难保证该负责装载的 BeanFactoryPostProcessor 实现以最高优先级执行。因此，ConfigurableEnvironment 对象的装配工作需在 refresh()方法调用前完成。这方面的详尽讨论，请读者参考“外部化配置”的“理解 Spring Boot Environment 生命周期”一节。

> 看似简单的 ConfigurableEnvironment 对象赋值语句，其背后有着复杂的设计意图，间接地说明 Spring Framework API 良好的设计是一把双刃剑，一方面能够充分地保证扩展的弹性，另一方面又引入一定的复杂度。理解 Spring Framework 设计思想是打开 Spring Boot 这座“神秘花园”的钥匙。

按照 prepareContext 方法的执行顺序，下一步执行“Spring 应用上下文后置处理”。

2）Spring 应用上下文后置处理

Spring 应用上下文后置处理是根据 SpringApplication#postProcessApplicationContext(ConfigurableApplicationContext)方法的命名而来的，首先观察其声明：

```
public class SpringApplication {
...
/**
 * Apply any relevant post processing the {@link ApplicationContext}. Subclasses can
```

```
     * apply additional processing as required.
     * @param context the application context
     */
    protected void postProcessApplicationContext(ConfigurableApplicationContext context) {
        if (this.beanNameGenerator != null) {
            context.getBeanFactory().registerSingleton(
                    AnnotationConfigUtils.CONFIGURATION_BEAN_NAME_GENERATOR,
                    this.beanNameGenerator);
        }
        if (this.resourceLoader != null) {
            if (context instanceof GenericApplicationContext) {
                ((GenericApplicationContext) context)
                        .setResourceLoader(this.resourceLoader);
            }
            if (context instanceof DefaultResourceLoader) {
                ((DefaultResourceLoader) context)
                        .setClassLoader(this.resourceLoader.getClassLoader());
            }
        }
    }
    ...
}
```

按照该方法的 JavaDoc 的描述，其用于 ConfigurableApplicationContext 的后置处理，并且允许子类覆盖该实现，可能增加额外需要的附加功能。从当前实现可知，当 SpringApplication 存在自定义属性 beanNameGenerator 时，本方法将该对象注册成名为 CONFIGURATION_BEAN_NAME_GENERATOR 常量的 BeanNameGenerator Bean，最终将影响 ConfigurableApplicationContext 注解驱动 Bean 名的生成：

```
    public class ConfigurationClassPostProcessor implements
BeanDefinitionRegistryPostProcessor,
        PriorityOrdered, ResourceLoaderAware, BeanClassLoaderAware, EnvironmentAware {
    ...
    /* Using short class names as default bean names */
    private BeanNameGenerator componentScanBeanNameGenerator = new
AnnotationBeanNameGenerator();
    ...
    public void processConfigBeanDefinitions(BeanDefinitionRegistry registry) {
```

```
        ...
        SingletonBeanRegistry sbr = null;
        if (registry instanceof SingletonBeanRegistry) {
            sbr = (SingletonBeanRegistry) registry;
            if (!this.localBeanNameGeneratorSet) {
                BeanNameGenerator generator = (BeanNameGenerator)
sbr.getSingleton(CONFIGURATION_BEAN_NAME_GENERATOR);
                if (generator != null) {
                    this.componentScanBeanNameGenerator = generator;
                    this.importBeanNameGenerator = generator;
                }
            }
        }
        ...
        ConfigurationClassParser parser = new ConfigurationClassParser(
                this.metadataReaderFactory, this.problemReporter, this.environment,
                this.resourceLoader, this.componentScanBeanNameGenerator, registry);
        ...
    }
    ...
    }
```

同时，postProcessApplicationContext(ConfigurableApplicationContext)也覆盖当前 Spring 应用上下文默认所关联的 ResourceLoader 和 ClassLoader。不过，在设置 ResourceLoader 对象时，其前提条件是参数 context 是否为 GenericApplicationContext，前面曾讨论的两种不同 Spring Boot 应用上下文类型 AnnotationConfigApplicationContext 和 AnnotationConfigServletWebServerApplicationContext 均为 GenericApplicationContext 的子类，完整的类型层次关系如下图所示。

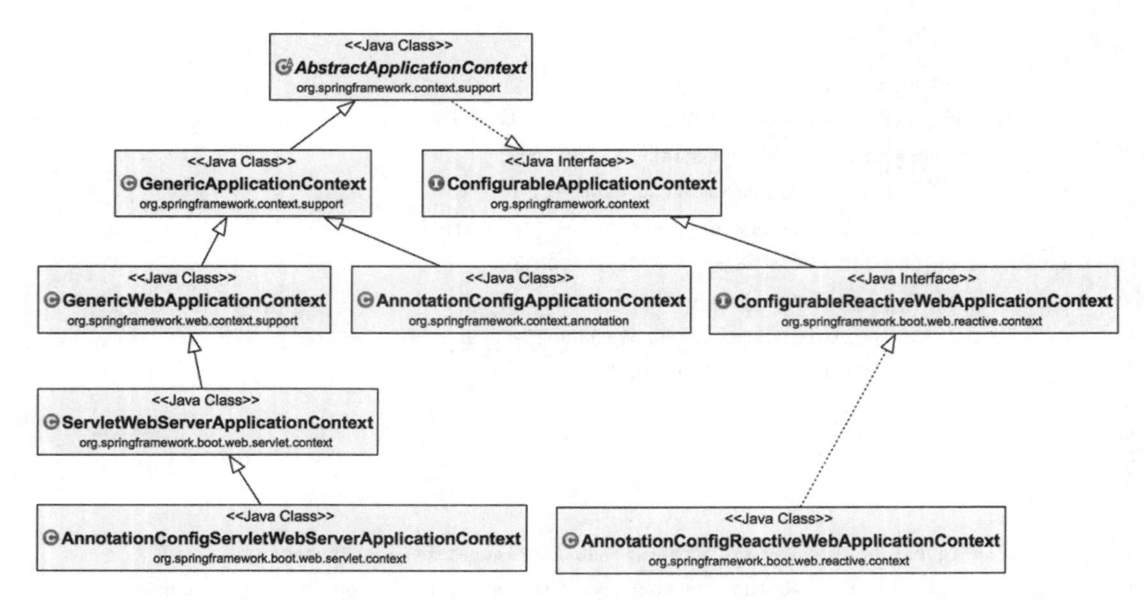

除 postProcessApplicationContext(ConfigurableApplicationContext) 方法外，applyInitializers(ConfigurableApplicationContext) 方法也能扩展 ConfigurableApplicationContext 实例，在“运用 Spring 应用上下文初始化器”中我们继续介绍。

3）运用 Spring 应用上下文初始化器（ApplicationContextInitializer）

前面曾讨论，SpringApplication 构造阶段所加载的 Spring 应用上下文初始化器存放在 SpringApplication 实例的 initializers 字段，该字段是 ApplicationContextInitializer 列表。在 Spring 应用上下文准备阶段时，它用于初始化 ConfigurableApplicationContext 实例：

```
public class SpringApplication {
...
protected void applyInitializers(ConfigurableApplicationContext context) {
    for (ApplicationContextInitializer initializer : getInitializers()) {
        Class<?> requiredType = GenericTypeResolver.resolveTypeArgument(
                initializer.getClass(), ApplicationContextInitializer.class);
        Assert.isInstanceOf(requiredType, context, "Unable to call initializer.");
        initializer.initialize(context);
    }
}
...
public Set<ApplicationContextInitializer<?>> getInitializers() {
    return asUnmodifiableOrderedSet(this.initializers);
```

```
    }
    ...
    private static <E> Set<E> asUnmodifiableOrderedSet(Collection<E> elements) {
        List<E> list = new ArrayList<>();
        list.addAll(elements);
        list.sort(AnnotationAwareOrderComparator.INSTANCE);
        return new LinkedHashSet<>(list);
    }
    ...
}
```

当 applyInitializers(ConfigurableApplicationContext)执行时，首先通过 getInitializers()方法获取 ApplicationContextInitializerSet，该 Set 将原有属性 initializers List 排序并去重。如果属性 initializers 中的实例完全来自 SpringApplication 构造器，那么该 List 已经过排序：

```
public class SpringApplication {
    ...
    public SpringApplication(ResourceLoader resourceLoader, Class<?>... primarySources) {
        ...
        setInitializers((Collection) getSpringFactoriesInstances(
                ApplicationContextInitializer.class));
        ...
    }
    ...
    private <T> Collection<T> getSpringFactoriesInstances(Class<T> type) {
        return getSpringFactoriesInstances(type, new Class<?>[] {});
    }

    private <T> Collection<T> getSpringFactoriesInstances(Class<T> type,
            Class<?>[] parameterTypes, Object... args) {
        ...
        List<T> instances = createSpringFactoriesInstances(type, parameterTypes,
                classLoader, args, names);
        AnnotationAwareOrderComparator.sort(instances);
        return instances;
    }
    ...
```

```
}
```

如此看来，asUnmodifiableOrderedSet(Collection)方法中的排序似乎没有存在的必要。其实不然，该方法还需要考虑 initializers 属性中来自 addInitializers (ApplicationContextInitializer...)方法的成员，随后将 initializers 属性添加至 LinkedHashSet 对象后返回，其目的在于不希望同一 ApplicationContextInitializer 初始化多次，不过无法保证 ApplicationContextInitializer 的实现类覆盖 hashCode 和 equals 方法，实现如下：

```
public class HelloWorldApplicationContextInitializer implements
        ApplicationContextInitializer<ConfigurableApplicationContext> {

    @Override
    public void initialize(ConfigurableApplicationContext applicationContext) {
        System.out.println("HelloWorldApplicationContextInitializer : Hello,World!");
    }
}
```

HelloWorldApplicationContextInitializer 未覆盖 hashCode 和 equals 方法，将其重复添加到 SpringApplication 中：

```
public class ApplicationContextInitializerBootstrap {

    public static void main(String[] args) {
        new SpringApplicationBuilder(Object.class)
                // 重复注册 HelloWorldApplicationContextInitializer
                .initializers(new HelloWorldApplicationContextInitializer(),
                        new HelloWorldApplicationContextInitializer())
                .run(args) // 运行 SpringApplication
                .close();  // 关闭 Spring 应用上下文
    }
}
```

> 源码位置：以上示例代码可通过查找 spring-boot-2.0-samples/spring-application-sample 工程获取。

执行并观察输出内容：

```
(...部分内容被省略...)
```

```
HelloWorldApplicationContextInitializer : Hello,World!
HelloWorldApplicationContextInitializer : Hello,World!
(...部分内容被省略...)
```

运行结果符合之前的推论，当 ApplicationContextInitializer 实现类未覆盖 hashCode 和 equals 方法时，多个同类实例重复添加到 SpringApplication 中，其 asUnmodifiableOrderedSet(Collection)方法无法去重。同理，当同一 ApplicationContextInitializer 实现类在不同的 META-INF/spring.factories 资源中声明时，重复执行的情况也会出现，因为 SpringFactoriesLoader#loadFactoryNames(Class,ClassLoader)方法也未去重。大多数场景下，ApplicationContextInitializer 对象为单例，通常可根据其类型判断唯一性，故调整 HelloWorldApplicationContextInitializer 如下：

```
public class HelloWorldApplicationContextInitializer implements
        ApplicationContextInitializer<ConfigurableApplicationContext> {

    @Override
    public void initialize(ConfigurableApplicationContext applicationContext) {
        System.out.println("HelloWorldApplicationContextInitializer : Hello,World!");
    }

    @Override
    public int hashCode() {
        return getClass().hashCode();
    }

    @Override
    public boolean equals(Object obj) {
        return getClass().equals(obj.getClass());
    }
}
```

> 凡是使用 Spring 工厂加载机制的场景，建议被加载实现类覆盖 hashCode 和 equals 方法，以免重复执行所带来的隐患。

再次运行 ApplicationContextInitializerBootstrap，结果不再出现重复输出：

```
(...部分内容被省略...)
HelloWorldApplicationContextInitializer : Hello,World!
(...部分内容被省略...)
```

由于 applyInitializers(ConfigurableApplicationContext) 方法迭代地执行 ApplicationContextInitializer 集合，所以对它们的顺序性和重复性应予以高度的关注。既然 ApplicationContextInitializer 提供初始化 ConfigurableApplicationContext 的能力，在职责上，自然与 postProcessApplicationContext(ConfigurableApplicationContext)方法在某种程度上重叠。尽管该方法确保优先于 ApplicationContextInitializer 执行，然而这并不能保证被该方法调整后的 ConfigurableApplicationContext 实例不被后续 ApplicationContextInitializer 对象覆盖性修改。即使在 Spring Cloud 场景中，也没有出现扩展 SpringApplication 类的情况，因此 postProcessApplicationContext(ConfigurableApplicationContext)方法被覆盖的概率基本为零。所以，建议开发人员以扩展 ApplicationContextInitializer 接口的方式实现 ConfigurableApplicationContext 的初始化。

> ApplicationContextInitializer 接口在 Spring Framework 时代并未得到广泛关注，原因在于 ApplicationContextInitializer 仅在 Spring Web MVC 场景（如 ContextLoader 和 FrameworkServlet）中初始化其关联的 ConfigurableApplicationContext 实例。尽管 ApplicationContextInitializer 和 SpringFactoriesLoader 分别从 Spring Framework 3.1 和 3.2 版本开始引入，并且其类文件各自分布在 spring-context 和 spring-core 模块，然而 ApplicationContextInitializer 却没有用非 Servlet Web 场景下的 Spring 上下文初始化，在设计上多少有些浪费。

接下来执行 Spring 应用上下文准备阶段中的最后一个方法，即 SpringApplicationRunListeners#contextPrepared(ConfigurableApplicationContext)。

> applyInitializers(ConfigurableApplicationContext)方法访问性的设计不是非常理想，它不应该为 protected。因为，当 SpringApplication 的子类覆盖该方法时，假设子类实现不调用 super 的实现，那么基于 Spring 工厂加载机制的 ApplicationContextInitializer 集合将不复存在。反之，即使子类复用 super 实现，无论 super.applyInitializers(ConfigurableApplicationContext)语句在扩展实现的前或后执行，基于 Spring 工厂加载机制的 ApplicationContextInitializer 集合都会独立执行（父类 SpringApplication 实现），假设子类方法自行扩展 ApplicationContextInitializer 来源，那么父类与子类的 ApplicationContextInitializer 集合执行不同调，这样无论给 SpringApplication 扩展实现的开发人员，还是使用该扩展的开发人员，均会面临风险。因此，建议不要覆盖 applyInitializers(ConfigurableApplicationContext) 方法的实现，仅使用 ApplicationContextInitializer 基于 Spring 工厂加载机制的扩展方式。

4）执行 SpringApplicationRunListener#contextPrepared 方法回调

按照SpringApplicationRunListeners#contextPrepared(ConfigurableApplicationContext) JavaDoc 的描述，当 Spring 应用上下文创建并准备完毕时，该方法被回调：

```
public interface SpringApplicationRunListener {
 ...
 /**
  * Called once the {@link ApplicationContext} has been created and prepared, but
  * before sources have been loaded.
  * @param context the application context
  */
 void contextPrepared(ConfigurableApplicationContext context);
 ...
}
```

文档进一步补充道："不过该方法在 ApplicationContext 加载源之前执行"。这里的"源"是指配置源，也是后续讨论的重点。

在讨论 SpringApplicationRunListener 的内容时，已知默认触发的方法来自实现类 EventPublishingRunListener，而该类在 Spring Boot 1.4～2.0 版本中提供了 contextPrepared (ConfigurableApplicationContext)方法的空实现：

```
public class EventPublishingRunListener implements SpringApplicationRunListener, Ordered {
 ...
 @Override
 public void contextPrepared(ConfigurableApplicationContext context) {

 }
 ...
}
```

> 在"早期 Spring Boot 事件/监听机制"一节中曾讨论 Spring Boot 1.0~1.3 版本的实现，contextPrepared(ConfigurableApplicationContext) 方法并非空实现，而是将 EventPublishingRunListener 的 ApplicationEventMulticaster 实例与 Spring 应用上下文关联。

默认情况下，尽管 Spring Boot 框架内建仅有一处 SpringApplicationRunListener 实现，不过开发人员可以自定义 SpringApplicationRunListener 实现。在"外部化配置"章节中，就存在基于 SpringApplicationRunListener#contextPrepared 方法而扩展外部化配置属性源的讨论。

至此，“Spring 应用上下文准备阶段”讨论结束，接下来讨论 Spring 应用上下文进入运行前准备的第二阶段，即“Spring 应用上下文装载阶段”。

2. Spring 应用上下文装载阶段

按照 SpringApplication#prepareContext 方法的实现，本阶段又可划分为四个过程，分别为：“注册 Spring Boot Bean”“合并 Spring 应用上下文配置源”“加载 Spring 应用上下文配置源”和“回调 SpringApplicationRunListener#contextLoaded 方法”。

1）注册 Spring Boot Bean

SpringApplication#prepareContext 方法将之前创建的 ApplicationArguments 对象和可能存在的 Banner 实例注册为 Spring 单体 Bean：

```
public class SpringApplication {
...
private void prepareContext(ConfigurableApplicationContext context,
        ConfigurableEnvironment environment, SpringApplicationRunListeners listeners,
        ApplicationArguments applicationArguments, Banner printedBanner) {
    ...
    // Add boot specific singleton beans
    context.getBeanFactory().registerSingleton("springApplicationArguments",
            applicationArguments);
    if (printedBanner != null) {
        context.getBeanFactory().registerSingleton("springBootBanner", printedBanner);
    }
    ...
}
...
}
```

其中，springBootBanner Bean 在 Spring Boot 官方参考文档的“23.2 Customizing the Banner”章节中有明文记载：

> The printed banner is registered as a singleton bean under the following name: springBootBanner.

相反，springApplicationArguments Bean 的内容却只字未提。因此，不得不纵观该 Spring Bean 在各个 Spring Boot 版本的实现情况。由于 ApplicationArguments 类从 Spring Boot 1.3 开始被引入，故检查的范围为 1.3 版本到 2.0 版本。经确认，springApplicationArguments Bean 从 Spring

Boot 1.3~2.0 版本均存在单体注册的实现，所以，开发人员可以放心大胆地在 Spring 应用上下文中依赖注入该 Bean。同样地，Spring Boot 1.3～1.5 的文档也没有提到该细节。

> 只要发现代码中的实现未出现在文档中，就需要特别谨慎地评估各个 Spring Boot 的实现情况，毕竟 Spring Boot 的兼容性存在不确定性。如本节的 `springBootBanner` Bean，Spring Boot 1.2 开始引入接口 `Banner` 和实现类 `SpringBootBanner`，然而 `springBootBanner` Bean 却从 Spring Boot 1.4 才开始注册到 Spring 应用上下文。

接下来继续讨论下一个过程——“合并 Spring 应用上下文配置源”。

2）合并 Spring 应用上下文配置源

合并 Spring 应用上下文配置源的操作由 `getAllSources()`方法实现，该方法是从 Spring Boot 2.0 开始引入的，且较为复杂：

```
public class SpringApplication {
...
private void prepareContext(ConfigurableApplicationContext context,
        ConfigurableEnvironment environment, SpringApplicationRunListeners listeners,
        ApplicationArguments applicationArguments, Banner printedBanner) {
    ...
    // Load the sources
    Set<Object> sources = getAllSources();
    ...
}
...
/**
 * Set additional sources that will be used to create an ApplicationContext. A source
 * can be: a class name, package name, or an XML resource location.
 * <p>
 * Sources set here will be used in addition to any primary sources set in the
 * constructor.
 * @param sources the application sources to set
 * @see #SpringApplication(Class...)
 * @see #getAllSources()
 */
public void setSources(Set<String> sources) {
    Assert.notNull(sources, "Sources must not be null");
    this.sources = new LinkedHashSet<>(sources);
```

```
    }

    /**
     * Return an immutable set of all the sources that will be added to an
     * ApplicationContext when {@link #run(String...)} is called. This method combines any
     * primary sources specified in the constructor with any additional ones that have
     * been {@link #setSources(Set) explicitly set}.
     * @return an immutable set of all sources
     */
    public Set<Object> getAllSources() {
        Set<Object> allSources = new LinkedHashSet<>();
        if (!CollectionUtils.isEmpty(this.primarySources)) {
            allSources.addAll(this.primarySources);
        }
        if (!CollectionUtils.isEmpty(this.sources)) {
            allSources.addAll(this.sources);
        }
        return Collections.unmodifiableSet(allSources);
    }
    ...
}
```

不难看出，getAllSources()方法返回值为只读 Set，它由两个子集合组成：属性 primarySources 和 sources，前者来自 SpringApplication 构造器参数（前文已讨论），后者来源于 setSources(Set<String>)方法，而该方法的参数为 Set 类型，其成员类型则是 String 对象。按照 JavaDoc 所述，这些成员可能是类名、包名或 XML 配置资源路径。不过，以上构造器和方法签名已被 Spring Boot 2.0 重新调整。以 Spring Boot 1.5.10.RELEASE 实现为例，SpringApplication 构造器配置源参数为 Object 数组：

```
public class SpringApplication {
    ...
    public SpringApplication(Object... sources) {
        initialize(sources);
    }
    ...
    private void initialize(Object[] sources) {
        if (sources != null && sources.length > 0) {
```

```
            this.sources.addAll(Arrays.asList(sources));
        }
        ...
    }
    ...
    private void prepareContext(ConfigurableApplicationContext context,
            ConfigurableEnvironment environment, SpringApplicationRunListeners listeners,
            ApplicationArguments applicationArguments, Banner printedBanner) {
        ...
        // Load the sources
        Set<Object> sources = getSources();
        ...
    }
    ...
    public Set<Object> getSources() {
        return this.sources;
    }
    ...
    public void setSources(Set<Object> sources) {
        Assert.notNull(sources, "Sources must not be null");
        this.sources.addAll(sources);
    }
    ...
}
```

> 以上实现源码源于 Spring Boot `1.5.10.RELEASE`。
> Maven GAV 坐标为：`org.springframework.boot:spring-boot:1.5.10.RELEASE`。

随即构造器将其参数传入 `initialize(Object[])`方法，随后参数被添加到 `sources` 字段中。除此之外，`sources` 字段还能被 `setSources(Set<Object>)`方法追加成员。换言之，在 Spring Boot 1.5 中，`SpringApplication` 并没有区分字段 `primarySources` 和 `sources`，所以不存在 `getAllSources()`方法，仅有 `getSources()`方法。实际上，所有 Spring Boot 1.x 实现亦是如此。不过，Spring Boot 2.0 版本中的 `getAllSources()`方法与 Spring Boot 1.x 中的 `getAllSources()`方法皆用于存储 Configuration Class、类名、包名及 Spring XML 配置资源路径。表面上，Spring Boot 1.x 和 2.0 版本中的配置源实现没有本质区别，然而在下一个阶段“加载 Spring 应用上下文配置源”中，将体会到两者的差异。

3）加载 Spring 应用上下文配置源

沿着方法调用的链路，load(ApplicationContext,Object[])方法将承担加载 Spring 应用上下文配置源的职责：

```
public class SpringApplication {
    ...
    protected void load(ApplicationContext context, Object[] sources) {
        if (logger.isDebugEnabled()) {
            logger.debug(
                    "Loading source " + StringUtils.arrayToCommaDelimitedString(sources));
        }
        BeanDefinitionLoader loader = createBeanDefinitionLoader(
                getBeanDefinitionRegistry(context), sources);
        if (this.beanNameGenerator != null) {
            loader.setBeanNameGenerator(this.beanNameGenerator);
        }
        if (this.resourceLoader != null) {
            loader.setResourceLoader(this.resourceLoader);
        }
        if (this.environment != null) {
            loader.setEnvironment(this.environment);
        }
        loader.load();
    }
    ...
}
```

该方法将 Spring 应用上下文 Bean 装载的任务交给了 BeanDefinitionLoader，且该实现类从 Spring Boot 1.0 就开始引入：

```
class BeanDefinitionLoader {

    private final Object[] sources;

    private final AnnotatedBeanDefinitionReader annotatedReader;

    private final XmlBeanDefinitionReader xmlReader;
```

```
    private BeanDefinitionReader groovyReader;

    private final ClassPathBeanDefinitionScanner scanner;

    private ResourceLoader resourceLoader;
    ...
    }
```

BeanDefinitionLoader组合了多个属性，第一个属性为SpringApplication#getAllSources()方法返回值，而属性 annotatedReader、xmlReader 和 groovy Reader 分别为注解驱动实现 AnnotatedBeanDefinitionReader、XML 配置实现 XmlBeanDefinitionReader 和 Groovy 实现 GroovyBeanDefinitionReader。其中，AnnotatedBeanDefinitionReader 从 Spring Framework 3.0 开始引入，并与 ClassPathBeanDefinitionScanner 配合，形成 AnnotationConfig ApplicationContext 扫描和注册配置类的基础，随后这些配置类将被解析为 Bean 定义 BeanDefinition：

```
    public class AnnotationConfigApplicationContext extends GenericApplicationContext
implements AnnotationConfigRegistry {

    private final AnnotatedBeanDefinitionReader reader;

    private final ClassPathBeanDefinitionScanner scanner;
    ...
    public void register(Class<?>... annotatedClasses) {
        Assert.notEmpty(annotatedClasses, "At least one annotated class must be specified");
        this.reader.register(annotatedClasses);
    }
    ...
    public void scan(String... basePackages) {
        Assert.notEmpty(basePackages, "At least one base package must be specified");
        this.scanner.scan(basePackages);
    }
    ...
    }
```

而 XmlBeanDefinitionReader 和 GroovyBeanDefinitionReader 则是注解@ImportResource 读取 BeanDefinition 的底层实现，按照其 JavaDoc 描述：

> By default, arguments to the value attribute will be processed using a GroovyBeanDefinitionReader if ending in ".groovy"; otherwise, an XmlBeanDefinitionReader will be used to parse Spring XML files.

当@ImportResource#locations 属性值以“.groovy”结尾时，采用 GroovyBeanDefinitionReader 读取 BeanDefinition，否则是 XmlBeanDefinitionReader。不难看出，Spring Boot 中的 BeanDefinitionLoader 是以上 BeanDefinition 读取的综合实现。当其 load()方法调用时，这些 BeanDefinitionReader 类型的属性各司其职，为 Spring 应用上下文从不同的配置源装载 Spring Bean 定义（BeanDefinition）。不过，执行 load()方法后，Spring 应用上下文并没有启动：

```java
class BeanDefinitionLoader {
    ...
    public int load() {
        int count = 0;
        for (Object source : this.sources) {
            count += load(source);
        }
        return count;
    }

    private int load(Object source) {
        Assert.notNull(source, "Source must not be null");
        if (source instanceof Class<?>) {
            return load((Class<?>) source);
        }
        if (source instanceof Resource) {
            return load((Resource) source);
        }
        if (source instanceof Package) {
            return load((Package) source);
        }
        if (source instanceof CharSequence) {
            return load((CharSequence) source);
        }
        throw new IllegalArgumentException("Invalid source type " + source.getClass());
    }
    ...
}
```

回顾 Spring Boot 2.0 中的 SpringApplication 构造器及其 setSources(Set<String>)方法，当 getAllSources()方法返回时，尽管返回值是 Set<Object>对象，但集合内部仅包含 Class 和 String 类型的元素，结合以上 BeanDefinitionLoader#load(Object)方法的实现，实际上 Resource 和 Package 类型判断的分支语句永远不会进入前面的代码逻辑分支。相反，在 Spring Boot 1.x 的实现中，由于 SpringApplication 构造器与其 setSources(Set<Object>)方法参数成员类型均为 Object，因此它的配置源范围并不限于 Class、类名、包名及 XML 配置资源路径，setSources(Set<Object>)方法参数还能够传递 Resource 或 Package 对象。然而 BeanDefinitionLoader 仅在 SpringApplication#createBeanDefinitionLoader(BeanDefinitionRegistry,Object[])方法中使用，或许作者忘记将 load(Object)方法实现随之调整。

> createBeanDefinitionLoader(BeanDefinitionRegistry,Object[])方法也是一个糟糕的设计：
>
> ```
> public class SpringApplication {
>     ...
>     protected BeanDefinitionLoader createBeanDefinitionLoader(
>
>             BeanDefinitionRegistry registry, Object[] sources) {
>
>         return new BeanDefinitionLoader(registry, sources);
>
>     }
>     ...
> }
> ```
>
> 该方法的访问性为 protected，这意味着它能被子类覆盖，然而 BeanDefinitionLoader 是一个非 public 类。换言之，如果需要覆盖，则 SpringApplication 子类必须与 BeanDefinitionLoader 在同一包下，即 org.springframework.boot。如果该方法的设计意图不在于被子类覆盖实现，那么其价值仅在于它能被同包的类访问。假设如此，最好使其方法的访问性为默认。然而，纵观 Spring Boot 1.0～2.0 的实现，这样的假设也是不成立的，因为它仅被 load(ApplicationContext, Object[])方法调用，而后者也是 protected 方法。总之，假设需要调整 ApplicationContext 装配行为，则可直接覆盖 load(ApplicationContext, Object[])方法，不必考虑覆盖 createBeanDefinitionLoader(BeanDefinitionRegistry,Object[])方法。

当 Spring 应用上下文配置源加载完毕后，紧接着执行 SpringApplicationRunListener#contextLoaded 方法回调。

4）执行 SpringApplicationRunListener#contextLoaded 方法回调

回顾前面的讨论，已知 SpringApplicationRunListener 唯一的实现类 EventPublishingRunListener 在此阶段将 SpringApplication 关联的 ApplicationListener 追加到 Spring 应用上下文，随后发布 ApplicationPreparedEvent 事件：

```
public class EventPublishingRunListener implements SpringApplicationRunListener, Ordered {
    ...
    @Override
    public void contextLoaded(ConfigurableApplicationContext context) {
        for (ApplicationListener<?> listener : this.application.getListeners()) {
            if (listener instanceof ApplicationContextAware) {
                ((ApplicationContextAware) listener).setApplicationContext(context);
            }
            context.addApplicationListener(listener);
        }
        this.initialMulticaster.multicastEvent(
                new ApplicationPreparedEvent(this.application, this.args, context));
    }
    ...
}
```

EventPublishingRunListener#contextLoaded(ConfigurableApplicationContext) 的实现在 Spring Boot 1.0～2.0 之间是一致的，由于此时的 ApplicationContext 尚未启动，故应用可通过监听事件 ApplicationPreparedEvent 来调整 SpringApplication 和 ApplicationContext 对象，如下所示。

```
public class ApplicationPreparedEventListener implements
ApplicationListener<ApplicationPreparedEvent> {

    @Override
    public void onApplicationEvent(ApplicationPreparedEvent event) {
        // 获取 Spring 应用上下文
        ConfigurableApplicationContext context = event.getApplicationContext();
        // 调整 Spring 应用上下文的 ID
        context.setId("context-mercyblitz");
        System.out.println("当前 Spring 应用上下文 ID 调整为：" + context.getId());
    }
}
```

并将以上实现配置在 Spring 工厂文件（META-INF/spring.factories）中：

```
# 配置 ApplicationListener
org.springframework.context.ApplicationListener = \
thinking.in.spring.boot.samples.spring.application.event.ApplicationPreparedEventListener
```

接着重用并运行 SpringApplicationBootstrap：

```
(...部分内容被省略...)
当前 Spring 应用上下文 ID 调整为：context-mercyblitz
(...部分内容被省略...)
s.c.a.AnnotationConfigApplicationContext : Refreshing
org.springframework.context.annotation.AnnotationConfigApplicationContext
(...部分内容被省略...)
```

> 源码位置：以上示例代码可通过查找 spring-boot-2.0-samples/spring-application-sample 工程获取。

调整 ID 日志在“Refreshing”日志之前输出，说明此时 ApplicationContext 尚未启动。类似地，开发人员也可选择实现 SpringApplicationRunListener#contextLoaded(ConfigurableApplicationContext)方法来扩展外部化配置，详情请参考“外部化配置”章节中“基于 SpringApplicationRunListener#contextLoaded 扩展外部化配置属性源”部分。

当 SpringApplicationRunListener#contextLoaded 方法执行后，Spring 应用上下文（ConfigurableApplicationContext）运行前准备的各个操作都执行完毕。接下来 Spring 应用上下文进入了实质性的启动阶段。

11.2　Spring 应用上下文启动阶段

本阶段的执行由 refreshContext(ConfigurableApplicationContext)方法实现：

```
public class SpringApplication {
...
protected void refresh(ApplicationContext applicationContext) {
    Assert.isInstanceOf(AbstractApplicationContext.class, applicationContext);
    ((AbstractApplicationContext) applicationContext).refresh();
}
...
private void refreshContext(ConfigurableApplicationContext context) {
```

```
        refresh(context);
        if (this.registerShutdownHook) {
            try {
                context.registerShutdownHook();
            }
            catch (AccessControlException ex) {
                // Not allowed in some environments.
            }
        }
    }
    ...
}
```

refreshContext(ConfigurableApplicationContext)首先调用 refresh(ApplicationContext)，执行 ApplicationContext 的启动（refresh()方法）。随后，默认情况下，Spring 应用上下文将注册 shutdownHook 线程，实现优雅的 Spring Bean 销毁生命周期回调（registerShutdownHook 的默认值为 true）。以上两种操作贯穿于 Spring Boot 1.0～2.0，不过从 Spring 1.4 开始，将它们重构至 refreshContext(ConfigurableApplicationContext)方法。

> 如果读者正在使用 Spring Boot 1.2 或更低的版本，则可能会发现 context.registerShutdownHook()并未在 refresh(ApplicationContext)方法后执行，而是在更早的阶段调用。实际上，前后版本的实现并无差异，因为 registerShutdownHook()方法属于 JVM shutdown hook 机制，具体执行并不由 Spring 应用上下文控制：

```
public abstract class AbstractApplicationContext extends DefaultResourceLoader
        implements ConfigurableApplicationContext {
    ...
    @Override
    public void registerShutdownHook() {
        if (this.shutdownHook == null) {
            // No shutdown hook registered yet.
            this.shutdownHook = new Thread() {
                @Override
                public void run() {
                    synchronized (startupShutdownMonitor) {
                        doClose();
                    }
```

```
                }
            };
            Runtime.getRuntime().addShutdownHook(this.shutdownHook);
        }
    }
    ...
}
```

> 同时，refresh(ApplicationContext)又是一个糟糕的设计，问题集中在方法参数类型上。从 Spring Boot 1.0 开始，SpringApplication#setApplicationContextClass(Class)方法的参数类型必须是 ConfigurableApplicationContext 的子类，同时 SpringApplication#createApplicationContext()方法的返回类型也是 ConfigurableApplicationContext。换言之，refresh(ApplicationContext) 的参数类型没有必要选择更抽象的 ApplicationContext（ConfigurableApplicationContext 的父接口），并且该方法仅被一处调用。同时，该方法的访问性同样为 protected，再次暗示开发人员子类可以覆盖该方法的实现，然而 ApplicationContext 接口并没有提供 refresh()方法，相反该方法出现在其子接口 ConfigurableApplicationContext 中。所以如此先判断 AbstractApplicationContext 类型，后调用其 refresh()方法的实现，反而将简单的问题复杂化。

随着 refreshContext(ConfigurableApplicationContext)方法的执行，Spring 应用上下文正式进入 Spring 生命周期，Spring Boot 核心特性也随之启动，如组件自动装配、嵌入式容器启动 Production-Ready 特性。在前面 Spring 事件机制的讨论中，已知当 Spring 应用上下文准备完毕后，事件 ContextRefreshedEvent 随之传播，此时仍处于生命周期中。因此，SpringApplication 并没有在当前方法中安插其他阶段的操作，接着它将执行 afterRefresh(ConfigurableApplicationContext, ApplicationArguments)方法，进入 ApplicationContext 启动后阶段。

11.3 Spring 应用上下文启动后阶段

实际上，SpringApplication#afterRefresh(ConfigurableApplicationContext, ApplicationArguments)方法并未给 Spring 应用上下文启动后阶段提供实现，而是将其交给开发人员自行扩展：

```
public class SpringApplication {
...
protected void afterRefresh(ConfigurableApplicationContext context,
        ApplicationArguments args) {
```

```
    }
    ...
}
```

如果仅从方法实现上来看，则这里没有任何值得讨论的地方。然而对比 Spring Boot 1.x 的实现，其设计的不确定性却值得引起我们高度重视，可从方法签名和语义两方面进行讨论。

11.3.1 afterRefresh 方法签名的变化

在前面“装备 ApplicationArguments”一节中曾提到 ApplicationArguments 是从 Spring Boot 1.3 开始引入的，因此，afterRefresh 方法在此版本前后出现了不兼容更新。从 Spring Boot 1.0～1.2，afterRefresh 方法签名如下：

```
public class SpringApplication {
    ...
    protected void afterRefresh(ConfigurableApplicationContext context, String[] args) {
        runCommandLineRunners(context, args);
    }
    ...
}
```

> 以上实现源码源于 Spring Boot 1.2.8.RELEASE。
> Maven GAV 坐标为：org.springframework.boot:spring-boot:1.2.8.RELEASE。

Spring Boot 1.3 在保留并标注@Deprecated 以上 afterRefresh 方法签名的同时，增加了其重载方法：

```
public class SpringApplication {
    ...
    protected void afterRefresh(ConfigurableApplicationContext context,
            ApplicationArguments args) {
        afterRefresh(context, args.getSourceArgs());
        callRunners(context, args);
    }
    ...
    @Deprecated
    protected void afterRefresh(ConfigurableApplicationContext context, String[] args) {
    }
```

```
    ...
}
```

> 以上实现源码源于 Spring Boot 1.3.8.RELEASE。
> Maven GAV 坐标为：org.springframework.boot:spring-boot:1.3.8.RELEASE。

因此，从 Spring Boot 1.4 开始，afterRefresh(ConfigurableApplicationContext,String[])的方法签名不复存在，后续的版本均由 afterRefresh(ConfigurableApplicationContext, ApplicationArguments)方法代替。不过这仅是变化的开始，该方法又从 Spring Boot 2.0 开始做出了语义调整的变化。

11.3.2 afterRefresh 方法语义的变化

以 Spring Boot 1.4 实现为例，afterRefresh(ConfigurableApplicationContext, ApplicationArguments)方法直接调用了 callRunners(ApplicationContext,Application Arguments)方法：

```
public class SpringApplication {
 ...
 protected void afterRefresh(ConfigurableApplicationContext context,
         ApplicationArguments args) {
     callRunners(context, args);
 }

 private void callRunners(ApplicationContext context, ApplicationArguments args) {
     List<Object> runners = new ArrayList<Object>();
     runners.addAll(context.getBeansOfType(ApplicationRunner.class).values());
     runners.addAll(context.getBeansOfType(CommandLineRunner.class).values());
     AnnotationAwareOrderComparator.sort(runners);
     for (Object runner : new LinkedHashSet<Object>(runners)) {
         if (runner instanceof ApplicationRunner) {
             callRunner((ApplicationRunner) runner, args);
         }
         if (runner instanceof CommandLineRunner) {
             callRunner((CommandLineRunner) runner, args);
         }
     }
```

```
    }
    ...
}
```

> 以上实现源码源于 Spring Boot `1.4.7.RELEASE`。
> Maven GAV 坐标为：`org.springframework.boot:spring-boot:1.4.7.RELEASE`。

以上方法的实现贯穿 Spring Boot 1.3～1.5，然而从 2.0 版本开始发生了变化：

```
public class SpringApplication {
    ...
    protected void afterRefresh(ConfigurableApplicationContext context,
            ApplicationArguments args) {
    }
    ...
}
```

> 以上实现源码源于 Spring Boot `2.0.2.RELEASE`。
> Maven GAV 坐标为：`org.springframework.boot:spring-boot:2.0.2.RELEASE`。

不难发现，Spring Boot 2.0 不再调用 `callRunners(ApplicationContext,ApplicationArguments)` 方法，因此，`afterRefresh(ConfigurableApplicationContext, ApplicationArguments)`方法不再具备执行 `ApplicationRunner` 或 `CommandLineRunner` Bean 的语义。假设 Spring Boot 1.5 应用曾扩展 `SpringApplication#callRunners(ApplicationContext, ApplicationArguments)`方法，倘若它需要执行 `ApplicationRunner` 或 `CommandLineRunner` Bean，则必然会在其方法内部执行 `super.callRunners` 语句。当该应用升级到 Spring Boot 2.0 后，显然这样的操作并无效果。幸运的是，`ApplicationRunner` 或 `CommandLineRunner` Bean 的执行并不会受到影响，不过它们的执行时机却有所延后：

```
public class SpringApplication {
    ...
    public ConfigurableApplicationContext run(String... args) {
        ...
        try {
            ...
            afterRefresh(context, applicationArguments);
            ...
            listeners.started(context);
```

```
            callRunners(context, applicationArguments);
        }
        ...
        return context;
    }
    ...
}
```

或许这并不意味着在实现上 Spring Boot 2.0 与之前的版本存在明显的差异，不过这两个方法之间还存在 `listeners.started(context)`语句，即 Spring Boot 2.0 为 `SpringApplicationRunListener` 新增的生命周期方法 `started(ConfigurableApplicationContext)`：

```
public interface SpringApplicationRunListener {
    ...
    /**
     * The context has been refreshed and the application has started but
     * {@link CommandLineRunner CommandLineRunners} and {@link ApplicationRunner
     * ApplicationRunners} have not been called.
     * @param context the application context.
     * @since 2.0.0
     */
    void started(ConfigurableApplicationContext context);
    ...
}
```

已知 `EventPublishingRunListener` 作为 Spring Boot 内建的唯一 `SpringApplicationRunListener` 实现，在其 `started(ConfigurableApplicationContext)`方法执行时，将发送 Spring Boot 事件 `ApplicationStartedEvent`。不过在 Spring Boot 版本升级的过程中，该事件的语义也随之发生变化。

11.3.3　Spring Boot 事件 ApplicationStartedEvent 语义的变化

在“理解 Spring Boot 事件”一节的讨论中，曾使用表格的方式列出引入 `ApplicationStartedEvent` 的 Spring Boot 版本为 1.0，不过还需做补充。在 Spring Boot 1.x 中，`ApplicationStartedEvent` 在 `EventPublishingRunListener#starting()`方法中广播：

```
public class EventPublishingRunListener implements SpringApplicationRunListener, Ordered {
    ...
```

```
    @Override
    @SuppressWarnings("deprecation")
    public void starting() {
        this.initialMulticaster
                .multicastEvent(new ApplicationStartedEvent(this.application, this.args));
    }
    ...
}
```

> 以上实现源码源于 Spring Boot `1.5.10.RELEASE`。
> Maven GAV 坐标为：`org.springframework.boot:spring-boot:1.5.10.RELEASE`。

以上方法为 Spring Boot 1.5 的实现，与更早版本的差异主要表现两个方面，一是较为明显的注解`@SuppressWarnings("deprecation")`，说明该方法使用了不推荐（`@Deprecated`）的 API，即 `ApplicationStartedEvent`：

```
@Deprecated
@SuppressWarnings("serial")
public class ApplicationStartedEvent extends ApplicationStartingEvent {

    /**
     * Create a new {@link ApplicationStartedEvent} instance.
     * @param application the current application
     * @param args the arguments the application is running with
     */
    public ApplicationStartedEvent(SpringApplication application, String[] args) {
        super(application, args);
    }

}
```

> 以上实现源码源于 Spring Boot `1.5.10.RELEASE`。
> Maven GAV 坐标为：`org.springframework.boot:spring-boot:1.5.10.RELEASE`。

第二处不同在于，Spring Boot 1.5 中的 `ApplicationStartedEvent` 不再直接继承 `SpringApplicationEvent`，而是扩展 1.5 版本新引入的事件 `ApplicationStartingEvent`，这意味着在 Spring Boot 1.5 中，当 `ApplicationStartedEvent` 被广播后，它可能被 `ApplicationStartedEvent` 和 `ApplicationStartingEvent` 的事件监听器（`ApplicationListener`）实现同

时监听到，运行以下示例：

```
public class ApplicationStartedEventListenerBootstrap {

    public static void main(String[] args) {
        new SpringApplicationBuilder(Object.class)
                .listeners((ApplicationListener<ApplicationStartedEvent>) event -> {
                    System.out.println("监听 Spring Boot 事件 ApplicationStartedEvent");
                }, (ApplicationListener<ApplicationStartingEvent>) event -> {
                    System.out.println("监听 Spring Boot 事件 ApplicationStartingEvent ");
                })
                .web(false) // 非 Web 应用
                .run(args)  // 运行 SpringApplication
                .close();   // 关闭 Spring 应用上下文
    }
}
```

> 源码位置：以上示例代码可通过查找 spring-boot-1.x-samples/spring-boot-1.5.x-project 工程获取。

其运行结果可检验推论：

```
监听 Spring Boot 事件 ApplicationStartedEvent
监听 Spring Boot 事件 ApplicationStartingEvent
(...部分内容被省略...)
```

Spring Boot 1.5 的变化是值得肯定的，至少它能引导部分开发人员将 Spring Boot 事件监听由 `ApplicationStartingEvent` 替换为 `ApplicationStartedEvent`，并在 `ApplicationStartedEvent` 的 JavaDoc 中给出了明确的指示：

> Deprecated as of 1.5 in favor of ApplicationStartingEvent.

同时，在英文语义和语法时态上，`ApplicationStartingEvent` 更符合 `EventPublishingRunListener#starting()`方法的语义。

> 在设计 API 命名时，建议读者关注动词的时态。

不过，Spring Boot 2.0 并没有将 `ApplicationStartedEvent` 移除，而是对其 JavaDoc 和代码实现重新修订：

```java
/**
 * Event published once the application context has been refreshed but before any
 * {@link ApplicationRunner application} and {@link CommandLineRunner command line}
 * runners have been called.
 *
 * @author Andy Wilkinson
 * @since 2.0.0
 */
@SuppressWarnings("serial")
public class ApplicationStartedEvent extends SpringApplicationEvent {

	private final ConfigurableApplicationContext context;

	/**
	 * Create a new {@link ApplicationStartedEvent} instance.
	 * @param application the current application
	 * @param args the arguments the application is running with
	 * @param context the context that was being created
	 */
	public ApplicationStartedEvent(SpringApplication application, String[] args,
			ConfigurableApplicationContext context) {
		super(application, args);
		this.context = context;
	}

	/**
	 * Return the application context.
	 * @return the context
	 */
	public ConfigurableApplicationContext getApplicationContext() {
		return this.context;
	}

}
```

> 根据以上 JavaDoc 的描述，小马哥一度认为`ApplicationStartedEvent`是从 Spring Boot 2.0 开始引入的，因其明确地描述了`@since 2.0.0`。实际上，这个类的包名在 Spring Boot 1.0～2.0 中均未发生变化，故“理解 Spring Boot 事件”一节中的表格写明其源于 Spring Boot 1.0。这

样的误会主要是作者对 JavaDoc 标签的误用，建议应该保留@since 1.0.0，并增加@revised 2.0.0 的描述。或许如此会让开发人员一目了然，因其标签存在于 Java Thread API 之中：

```
public
class Thread implements Runnable {
    ...
    /**
     * ...
     * @revised 6.0
     * @spec JSR-51
     */
    public void interrupt() {
        ...
    }
    ...
}
```

相较于 Spring Boot 1.x 的实现，ApplicationStartedEvent 增加了 ConfigurableApplicationContext 对象的关联，同时与 EventPublishingRunListener#started(ConfigurableApplicationContext)方法关联：

```
public class EventPublishingRunListener implements SpringApplicationRunListener, Ordered {
    ...
    @Override
    public void started(ConfigurableApplicationContext context) {
        context.publishEvent(
                new ApplicationStartedEvent(this.application, this.args, context));
    }
    ...
}
```

而 EventPublishingRunListener#starting()方法唯一地关联了 ApplicationStartingEvent。换言之，ApplicationStartedEvent 事件被延后触发，而 ApplicationStartingEvent 充当了前者过去的角色。

当 Spring Boot 事件 ApplicationStartedEvent 广播结束后，CommandLineRunner 和 ApplicationRunner Bean 随之被执行。

11.3.4 执行 CommandLineRunner 和 ApplicationRunner

Spring Boot 官方文档的“23.8 Using the ApplicationRunner or CommandLineRunner”章节提供了 CommandLineRunner 和 ApplicationRunner 的使用说明：

> If you need to run some specific code once the SpringApplication has started, you can implement the ApplicationRunner or CommandLineRunnerinterfaces. Both interfaces work in the same way and offer a single run method, which is called just before SpringApplication.run(…) completes.

按照官方文档的描述，ApplicationRunner 或 CommandLineRunner 均有 run 方法，在 SpringApplication.run(…)方法完成之前执行。其中，CommandLineRunner 接口提供简单的字符型数组作为参数，而 ApplicationRunner 则使用 ApplicationArguments：

> The CommandLineRunner interfaces provides access to application arguments as a simple string array, whereas the ApplicationRunner uses the ApplicationArguments interface discussed earlier.

文档接着说明，当 Spring 应用上下文出现多个 ApplicationRunner 或 CommandLineRunner Bean 时，通过实现 Ordered 接口或标注@Order 注解的方式来控制它们的执行顺序：

> If several CommandLineRunner or ApplicationRunner beans are defined that must be called in a specific order, you can additionally implement the org.springframework.core.Ordered interface or use the org.springframework.core.annotati on.Order annotation.

CommandLineRunner 和 ApplicationRunner 的使用场景

关于CommandLineRunner和ApplicationRunner的使用场景，官方文档并没有做过多的解释。尽管在 Spring Boot 2.0 中，ApplicationStartedEvent 事件监听回调略早于 CommandLineRunner 或 ApplicationRunner 的 run 方法的执行，然而它们均在 SpringApplication 和 Spring ConfigurableApplicationContext 准备妥当之后，并在 SpringApplication#run(String...) 执行完成之前，因此，两者的生命周期回调时机并没有本质的区别，那么为什么要引入 CommandLineRunner 和 ApplicationRunner 呢？难道仅为了读取启动参数？明显不是，毕竟所有 SpringApplicationEvent 实例均能获取启动参数，作为子类的 ApplicationStartedEvent 也不会例外。在前面“理解 Spring Boot 事件/监听机制”章节中曾讨论，Spring Boot 事件监听器均由 Spring

工厂加载机制加载并初始化，它们并非 Spring Bean，因此无法享受注解驱动和 Bean 生命周期回调接口的“福利”。然而这也并不意味着 ApplicationStartedEvent 事件 ApplicationListener 实现无法获取依赖的 Spring Bean，因为，ApplicationStartedEvent 同样关联 ConfigurableApplicationContext 对象。相反，CommandLineRunner 和 ApplicationRunner 能够获得这样的编程便利性，不过两者却无法获取 SpringApplication 对象。所以各有利弊，从编程结果上看，两者仍旧没有差异。

不过，分析 CommandLineRunner 和 ApplicationRunner 的使用场景，需要结合当时的“时空环境”。在 Spring Boot 1.x 中，ApplicationStartedEvent 事件过早地在 EventPublishingRunListener#starting()方法中触发，当时的 Spring ConfigurableApplicationContext 尚未创建，更谈不上获取依赖的 Bean，而 Spring Boot 1.0 引入的 CommandLineRunner 和 1.3 版本引入的 ApplicationRunner 在语义没有发生变化，即在 Spring 上下文完全准备完毕后执行。因为，它们属于的 Spring Boot 1.x 时代的产物，不过 Spring Boot 2.0 在 Spring Boot 事件上“加码”，使得监听 ApplicationStartedEvent 事件作为另外一种可选的技术手段。当然，ApplicationStartedEvent 并非唯一的 Spring 应用上下文启动后的 Spring Boot 事件，还包括在 SpringApplication 完成阶段触发的 Spring Boot 事件 ApplicationReadyEvent 和 ApplicationFailedEvent。接下来进入 SpringApplication 结束阶段的讨论。

第 12 章 SpringApplication 结束阶段

Spring Boot 1.x 和 2.0 版本对于 `SpringApplication` 结束阶段的实现逻辑是相对稳定的，存在 `SpringApplication` 正常结束和 `SpringApplication` 异常结束两种情况。不过在实现手段上，前后版本确实存在差异，下面进行讨论。

12.1 SpringApplication 正常结束

Spring Boot 2.0 为 `SpringApplication` 正常结束新引入了 `SpringApplicationRunListener` 的生命周期，即 `running(ConfigurableApplicationContext)`方法：

```
public interface SpringApplicationRunListener {
...
/**
 * Called immediately before the run method finishes, when the application context has
 * been refreshed and all {@link CommandLineRunner CommandLineRunners} and
 * {@link ApplicationRunner ApplicationRunners} have been called.
 * @param context the application context.
 * @since 2.0.0
 */
```

```
void running(ConfigurableApplicationContext context);
...
}
```

按照 JavaDoc 的描述，该方法在 Spring 应用上下文中已准备，并且 CommandLineRunner 和 ApplicationRunner Bean 均已执行完毕。EventPublishingRunListener 作为 Spring ApplicationRunListener 唯一内建实现，本方法中仅简单地广播 ApplicationReadyEvent 事件：

```
public class EventPublishingRunListener implements SpringApplicationRunListener, Ordered {
...
@Override
public void running(ConfigurableApplicationContext context) {
    context.publishEvent(
            new ApplicationReadyEvent(this.application, this.args, context));
}
...
}
```

不难看出，当 ApplicationReadyEvent 事件触发后，SpringApplication 的生命周期进入尾声，除非 SpringApplicationRunListeners#running(ConfigurableApplicationContext)方法执行异常：

```
public class SpringApplication {
...
public ConfigurableApplicationContext run(String... args) {
    ...
    try {
        listeners.running(context);
    }
    catch (Throwable ex) {
        handleRunFailure(context, ex, exceptionReporters, null);
        throw new IllegalStateException(ex);
    }
    return context;
}
...
}
```

换言之，开发人员有两种技术手段实现完成阶段的监听，一种为实现 SpringApplication

RunListener#running(ConfigurableApplicationContext)方法，另一种为实现 ApplicationReadyEvent 事件的 ApplicationListener。

在“理解 Spring Boot 事件”一节中曾讨论，ApplicationReadyEvent 从 Spring Boot 1.3 开始引入，这也意味着该事件也伴随该版本的 SpringApplication 实现而触发：

```
public class EventPublishingRunListener implements SpringApplicationRunListener, Ordered {
...
@Override
public void finished(ConfigurableApplicationContext context, Throwable exception) {
    publishEvent(getFinishedEvent(context, exception));
}

private SpringApplicationEvent getFinishedEvent(
        ConfigurableApplicationContext context, Throwable exception) {
    if (exception != null) {
        return new ApplicationFailedEvent(this.application, this.args, context,
                exception);
    }
    return new ApplicationReadyEvent(this.application, this.args, context);
}
...
}
```

> 以上实现源码源于 Spring Boot 1.3.8.RELEASE。
> Maven GAV 坐标为：org.springframework.boot:spring-boot:1.3.8.RELEASE。

ApplicationReadyEvent 事件由 EventPublishingRunListener#finished(ConfigurableApplicationContext,Throwable)方法触发，继承于 SpringApplicationRunListener 接口：

```
public interface SpringApplicationRunListener {
...
/**
 * Called immediately before the run method finishes.
 * @param context the application context or null if a failure occurred before the
 * context was created
 * @param exception any run exception or null if run completed successfully.
 */
void finished(ConfigurableApplicationContext context, Throwable exception);
```

```
    ...
}
```

> 以上实现源码源于 Spring Boot 1.3.8.RELEASE。
> Maven GAV 坐标为：org.springframework.boot:spring-boot:1.3.8.RELEASE。

该方法从 Spring Boot 1.0 开始引入，并在 2.0 版本中删除。不过 EventPublishingRunListener#finished(ConfigurableApplicationContext,Throwable)方法的实现在 Spring Boot 1.x 中存在细微的变化。

EventPublishingRunListener#finished(ConfigurableApplicationContext,Throwable)方法实现的变化

由于 ApplicationReadyEvent 是从 Spring Boot 1.3 开始引入的，因此 EventPublishingRunListener#finished(ConfigurableApplicationContext,Throwable)方法在 Spring Boot 1.0～1.2 中的实现是相同的：

```
public class EventPublishingRunListener implements SpringApplicationRunListener {
    ...
    @Override
    public void finished(ConfigurableApplicationContext context, Throwable exception) {
        if (exception != null) {
            ApplicationFailedEvent event = new ApplicationFailedEvent(this.application,
                    this.args, context, exception);
            publishEvent(event);
        }
    }
    ...
}
```

> 以上实现源码源于 Spring Boot 1.2.8.RELEASE。
> Maven GAV 坐标为：org.springframework.boot:spring-boot:1.2.8.RELEASE。

finished 方法仅处理异常情况，由于 Spring Boot 1.3 的实现前面已介绍，此处不再赘述。随后 Spring Boot 1.4 再次发生变化：

```
public class EventPublishingRunListener implements SpringApplicationRunListener, Ordered {
    ...
    @Override
```

```
    public void finished(ConfigurableApplicationContext context, Throwable exception) {
        SpringApplicationEvent event = getFinishedEvent(context, exception);
        if (context != null) {
            // Listeners have been registered to the application context so we should
            // use it at this point if we can
            context.publishEvent(event);
        }
        else {
            if (event instanceof ApplicationFailedEvent) {
                this.initialMulticaster.setErrorHandler(new LoggingErrorHandler());
            }
            this.initialMulticaster.multicastEvent(event);
        }
    }
    ...
}
```

> 以上实现源码源于 Spring Boot 1.4.7.RELEASE。
> Maven GAV 坐标为：org.springframework.boot:spring-boot:1.4.7.RELEASE。

Spring Boot 1.4 中的主要变化为，当正常执行时，事件的广播者不是 initialMulticaster，而是当前 Spring 应用上下文。而 Spring Boot 1.5 的做法更严谨：

```
public class EventPublishingRunListener implements SpringApplicationRunListener, Ordered {
    ...
    @Override
    public void finished(ConfigurableApplicationContext context, Throwable exception) {
        SpringApplicationEvent event = getFinishedEvent(context, exception);
        if (context != null && context.isActive()) {
            // Listeners have been registered to the application context so we should
            // use it at this point if we can
            context.publishEvent(event);
        }
        else {
            // An inactive context may not have a multicaster so we use our multicaster to
            // call all of the context's listeners instead
            if (context instanceof AbstractApplicationContext) {
                for (ApplicationListener<?> listener : ((AbstractApplicationContext) context)
```

```
                    .getApplicationListeners()) {
                this.initialMulticaster.addApplicationListener(listener);
            }
        }
        if (event instanceof ApplicationFailedEvent) {
            this.initialMulticaster.setErrorHandler(new LoggingErrorHandler());
        }
        this.initialMulticaster.multicastEvent(event);
    }
}
...
}
```

> 以上实现源码源于 Spring Boot `1.5.10.RELEASE`。
> Maven GAV 坐标为：`org.springframework.boot:spring-boot:1.5.10.RELEASE`。

正常执行的事件发送仍继承了 1.4 版本中的 Spring 应用上下文的实现，而异常处理将为 Spring 应用上下文所关联的 `ApplicationListener` 添加 `initialMulticaster` 属性，因其考虑到在 `SpringApplication#run(String...)`方法执行过程中，可能 `ConfigurableApplicationContext` 实例创建或初始化异常，甚至可能被显式地调用 `close()`方法，使其 `isActive()`方法返回 `false`。

不难看出，在 Spring Boot 1.x 中，`SpringApplicationRunListener#finished(ConfigurableApplicationContext,Throwable)`方法涵盖了正常结束和异常结束的语义。"异常结束"的部分则是接下来讨论的议题。

12.2 SpringApplication 异常结束

`SpringApplication` 异常结束阶段自然地让人联想到如何处理运行时相关的异常，与业务异常不同的是，这些异常一旦发生，基本上就宣告 Spring Boot 应用运行失败。与正常流程类似，异常流程同样作为 `SpringApplication` 生命周期的一个环节，将在 `SpringApplicationRunListener#finished(ConfigurableApplicationContext,Throwable)`方法中执行。不过前文也提到，这只是 Spring Boot 1.x 中的实现，而从 Spring Boot 2.0 开始，替换为 `SpringApplicationRunListener#failed(ConfigurableApplicationContext,Throwable)`方法。接下来逐一分析前后两个版本中的实现差异。

12.2.1 Spring Boot 异常处理

对于如何处理异常流程，Spring Boot 1.x 的实现也是反复调整，其基本策略是“catch”异常的方式。比如 Spring Boot 1.0 的实现为 catch Exception：

```
public class SpringApplication {
...
public ConfigurableApplicationContext run(String... args) {
    ...
    catch (Exception ex) {
        for (SpringApplicationRunListener runListener : runListeners) {
            finishWithException(runListener, context, ex);
        }
        if (context != null) {
            context.close();
        }
        ReflectionUtils.rethrowRuntimeException(ex);
        return context;
    }
    finally {
    }
}
...
private void finishWithException(SpringApplicationRunListener runListener,
        ConfigurableApplicationContext context, Exception exception) {
    try {
        runListener.finished(context, exception);
    }
    catch (Exception ex) {
        if (this.log.isDebugEnabled()) {
            this.log.error("Error handling failed", ex);
        }
        else {
            String message = ex.getMessage();
            message = (message == null ? "no error message" : message);
            this.log.warn("Error handling failed (" + message + ")");
        }
    }
```

```
    }
    ...
}
```

> 以上实现源码源于 Spring Boot 1.0.2.RELEASE。
> Maven GAV 坐标为：org.springframework.boot:spring-boot:1.0.2.RELEASE。

以上实现方法大致可拆为三个步骤：执行 finishWithException 方法、关闭 Spring 应用上下文和重抛异常。这样的实现方法持续到了 Spring Boot 1.2：

```
public class SpringApplication {
    ...
    public ConfigurableApplicationContext run(String... args) {
        ...
        catch (Throwable ex) {
            try {
                for (SpringApplicationRunListener runListener : runListeners) {
                    finishWithException(runListener, context, ex);
                }
                this.log.error("Application startup failed", ex);
            }
            finally {
                if (context != null) {
                    context.close();
                }
            }
            ReflectionUtils.rethrowRuntimeException(ex);
            return context;
        }
    }
    ...
}
```

> 以上实现源码源于 Spring Boot 1.2.8.RELEASE。
> MavenGAV 坐标为：org.springframework.boot:spring-boot:1.2.8.RELEASE。

两者的差异主要体现在 catch 的异常类型不同，官方应该意识到 Spring Boot 1.0 版本“catch”的 Exception 可能不全面，所以从 Spring Boot 1.1 开始，将 Exception 替换为 Throwable，因此

SpringApplicationRunListener#finished(ConfigurableApplicationContext,Throwable) 方法无法处理 Error 类型的情况。庆幸的是，该方法参数类型选择（Throwable）是合理的，因此其方法签名贯穿整个 Spring Boot 1.x。

从 Spring Boot 1.3 开始，异常处理逻辑则收拢到 handleRunFailure 方法中：

```
public class SpringApplication {
 ...
 public ConfigurableApplicationContext run(String... args) {
      ...
      catch (Throwable ex) {
           handleRunFailure(context, listeners, analyzers, ex);
           throw new IllegalStateException(ex);
      }
 }
 ...
private void handleRunFailure(ConfigurableApplicationContext context,
           SpringApplicationRunListeners listeners, Throwable exception) {
      if (logger.isErrorEnabled()) {
           logger.error("Application startup failed", exception);
           registerLoggedException(exception);
      }
      try {
           try {
                handleExitCode(context, exception);
                listeners.finished(context, exception);
           }
           finally {
                if (context != null) {
                     context.close();
                }
           }
      }
      catch (Exception ex) {
           logger.warn("Unable to close ApplicationContext", ex);
      }
      ReflectionUtils.rethrowRuntimeException(exception);
 }
 ...
```

```
}
```

> 以上实现源码源于 Spring Boot 1.3.8.RELEASE。
> Maven GAV 坐标为：org.springframework.boot:spring-boot:1.3.8.RELEASE。

故障分析器——FailureAnalyzers

handleRunFailure 方法在 Spring Boot 1.4 中增加了一个参数 FailureAnalyzers，并在上下文参数关闭之前作为 reportFailure(FailureAnalyzers,failure)方法执行错误分析并输出报告：

```
public class SpringApplication {
  ...
  private void handleRunFailure(ConfigurableApplicationContext context,
          SpringApplicationRunListeners listeners, FailureAnalyzers analyzers,
          Throwable exception) {
      try {
          ...
          finally {
              reportFailure(analyzers, exception);
              if (context != null) {
                  context.close();
              }
          }
      }
      ...
  }
  ...
  private void reportFailure(FailureAnalyzers analyzers, Throwable failure) {
      try {
          if (analyzers != null && analyzers.analyzeAndReport(failure)) {
              registerLoggedException(failure);
              return;
          }
      }
      ...
  }
  ...
```

```
}
```

> 以上实现源码源于 Spring Boot 1.4.7.RELEASE。
> Maven GAV 坐标为：org.springframework.boot:spring-boot:1.4.7.RELEASE。

可简单地认为 FailureAnalyzers 是 FailureAnalyzer 的组合类，在其构造阶段通过 Spring 工厂加载机制初始化并排序 FailureAnalyzer 列表：

```
public final class FailureAnalyzers {
    ...
    private final List<FailureAnalyzer> analyzers;
    ...
    FailureAnalyzers(ConfigurableApplicationContext context, ClassLoader classLoader) {
        ...
        this.analyzers = loadFailureAnalyzers(this.classLoader);
        prepareFailureAnalyzers(this.analyzers, context);
    }

    private List<FailureAnalyzer> loadFailureAnalyzers(ClassLoader classLoader) {
        List<String> analyzerNames = SpringFactoriesLoader
                .loadFactoryNames(FailureAnalyzer.class, classLoader);
        List<FailureAnalyzer> analyzers = new ArrayList<FailureAnalyzer>();
        for (String analyzerName : analyzerNames) {
            try {
                Constructor<?> constructor = ClassUtils.forName(analyzerName, classLoader)
                        .getDeclaredConstructor();
                ReflectionUtils.makeAccessible(constructor);
                analyzers.add((FailureAnalyzer) constructor.newInstance());
            }
            catch (Throwable ex) {
                log.trace("Failed to load " + analyzerName, ex);
            }
        }
        AnnotationAwareOrderComparator.sort(analyzers);
        return analyzers;
    }

    ...
```

```
}
```

> 以上实现源码源于 Spring Boot 1.4.7.RELEASE。
> Maven GAV 坐标为：org.springframework.boot:spring-boot:1.4.7.RELEASE。

加载后的 FailureAnalyzer 列表作为 FailureAnalyzers#analyze(FailureAnalysis, ClassLoader)方法的参数，随着 SpringApplication#reportFailure(FailureAnalyzers, Throwable)方法调用执行：

```
public final class FailureAnalyzers {
    ...
    public boolean analyzeAndReport(Throwable failure) {
        FailureAnalysis analysis = analyze(failure, this.analyzers);
        return report(analysis, this.classLoader);
    }

    private FailureAnalysis analyze(Throwable failure, List<FailureAnalyzer> analyzers) {
        for (FailureAnalyzer analyzer : analyzers) {
            try {
                FailureAnalysis analysis = analyzer.analyze(failure);
                if (analysis != null) {
                    return analysis;
                }
            }
            catch (Throwable ex) {
                log.debug("FailureAnalyzer " + analyzer + " failed", ex);
            }
        }
        return null;
    }

    private boolean report(FailureAnalysis analysis, ClassLoader classLoader) {
        List<FailureAnalysisReporter> reporters = SpringFactoriesLoader
                .loadFactories(FailureAnalysisReporter.class, classLoader);
        if (analysis == null || reporters.isEmpty()) {
            return false;
        }
        for (FailureAnalysisReporter reporter : reporters) {
```

```
            reporter.report(analysis);
        }
        return true;
    }
    ...
}
```

> 以上实现源码源于 Spring Boot 1.4.7.RELEASE。
> Maven GAV 坐标为：org.springframework.boot:spring-boot:1.4.7.RELEASE。

不难看出，FailureAnalyzer 仅分析故障，而故障报告则由 FailureAnalysisReporter 对象负责。

12.2.2 错误分析报告器——FailureAnalysisReporter

同样地，FailureAnalysisReporter 也由 Spring 工厂加载机制初始化并排序。

> 排序逻辑在 SpringFactoriesLoader#loadFactories(Class,ClassLoader)方法内部：

```
    public static <T> List<T> loadFactories(Class<T> factoryClass, ClassLoader classLoader)
{
        Assert.notNull(factoryClass, "'factoryClass' must not be null");
        ClassLoader classLoaderToUse = classLoader;
        if (classLoaderToUse == null) {
            classLoaderToUse = SpringFactoriesLoader.class.getClassLoader();
        }
        List<String> factoryNames = loadFactoryNames(factoryClass, classLoaderToUse);
        if (logger.isTraceEnabled()) {
            logger.trace("Loaded [" + factoryClass.getName() + "] names: " + factoryNames);
        }
        List<T> result = new ArrayList<T>(factoryNames.size());
        for (String factoryName : factoryNames) {
            result.add(instantiateFactory(factoryName, factoryClass,
classLoaderToUse));
        }
        AnnotationAwareOrderComparator.sort(result);
        return result;
    }
```

在 Spring Boot 框架中，仅存在一个内建 FailureAnalysisReporter 的实现 LoggingFailureAnalysisReporter。或许读者在使用 Spring Boot 过程中遇到过以下日志信息：

> APPLICATION FAILED TO START
> Description: ...

以上错误报告信息就是由 LoggingFailureAnalysisReporter 输出的：

```
public final class LoggingFailureAnalysisReporter implements FailureAnalysisReporter {
...
@Override
public void report(FailureAnalysis failureAnalysis) {
    ...
    if (logger.isErrorEnabled()) {
        logger.error(buildMessage(failureAnalysis));
    }
}

private String buildMessage(FailureAnalysis failureAnalysis) {
    StringBuilder builder = new StringBuilder();
    builder.append(String.format("%n%n"));
    builder.append(String.format("***************************%n"));
    builder.append(String.format("APPLICATION FAILED TO START%n"));
    builder.append(String.format("***************************%n%n"));
    builder.append(String.format("Description:%n%n"));
    builder.append(String.format("%s%n", failureAnalysis.getDescription()));
    if (StringUtils.hasText(failureAnalysis.getAction())) {
        builder.append(String.format("%nAction:%n%n"));
        builder.append(String.format("%s%n", failureAnalysis.getAction()));
    }
    return builder.toString();
}

}
```

> 以上实现源码源于 Spring Boot 1.4.7.RELEASE。
> Maven GAV 坐标为：org.springframework.boot:spring-boot:1.4.7.RELEASE。

与 FailureAnalysisReporter 不同的是，FailureAnalyzer 的内建实现相当丰富，下面以 org.springframework.boot:spring-boot-autoconfigure:1.4.7.RELEASE 中的 MEATA-INF/spring.factories 资源配置为例：

```
# Failure analyzers
org.springframework.boot.diagnostics.FailureAnalyzer=\
org.springframework.boot.autoconfigure.diagnostics.analyzer.NoSuchBeanDefinitionFailureAnalyzer,\
org.springframework.boot.autoconfigure.jdbc.DataSourceBeanCreationFailureAnalyzer,\
org.springframework.boot.autoconfigure.jdbc.HikariDriverConfigurationFailureAnalyzer
```

其中，NoSuchBeanDefinitionFailureAnalyzer 和 DataSourceBeanCreationFailureAnalyzer 在 Spring Boot 中经常出现。

当然，开发人员完全可以自定义实现 FailureAnalyzer 和 FailureAnalysisReporter。

12.2.3 自定义实现 FailureAnalyzer 和 FailureAnalysisReporter

首先，自定义 FailureAnalyzer 实现，仅处理 UnknownError 类型，分析并生成 FailureAnalysis：

```java
public class UnknownErrorFailureAnalyzer implements FailureAnalyzer {

    @Override
    public FailureAnalysis analyze(Throwable failure) {
        if (failure instanceof UnknownError) { // 判断上游异常类型判断
            return new FailureAnalysis("未知错误", "请重启尝试", failure);
        }
        return null;
    }
}
```

将 UnknownErrorFailureAnalyzer 配置到 MEATA-INF/spring.factories 资源中：

```
# FailureAnalyzer 配置
org.springframework.boot.diagnostics.FailureAnalyzer=\
thinking.in.spring.boot.samples.diagnostics.UnknownErrorFailureAnalyzer
```

然后，自定义 FailureAnalysisReporter 实现，控制台输出 UnknownErrorFailureAnalyzer 生成的 FailureAnalysis：

```
public class ConsoleFailureAnalysisReporter implements FailureAnalysisReporter {

    @Override
    public void report(FailureAnalysis analysis) {
        System.out.printf("故障描述：%s \n 执行动作：%s \n 异常堆栈：%s \n",
                analysis.getDescription(),
                analysis.getAction(),
                analysis.getCause());
    }
}
```

最后，实现简单 Spring Boot 引导类，在自定义 ApplicationContextListener 中抛出异常：

```
public class UnknownErrorSpringBootBootstrap {

    public static void main(String[] args) {
        new SpringApplicationBuilder(Object.class)
                .initializers(context -> {
                    throw new UnknownError("故意抛出异常");
                })
                .web(false) // 非 Web 应用
                .run(args)  // 运行 SpringApplication
                .close();   // 关闭 Spring 应用上下文
    }
}
```

> 源码位置：以上示例代码可通过查找 spring-boot-1.x-samples/spring-boot-1.4.x-project 工程获取。

运行后，观察控制台输出：

```
(...部分内容被省略...)
故障描述：未知错误
执行动作：请重启尝试
异常堆栈：java.lang.UnknownError：故意抛出异常
(...部分内容被省略...)
```

在实践过程中，由于 FailureAnalyzer#analyze(Throwable)方法参数类型为 Throwable，因此开发人员不得不进行异常类型判断，如同本例中的 UnknownErrorFailureAnalyzer。通常自定义实现 FailureAnalyzer 和 FailureAnalysisReporter 作为 Spring Boot Starter 核心特性之一，在大多数情况下，Starter 开发人员非常清楚异常的触发条件，从而引导 Starter 使用人员如何进行故障排查。

FailureAnalyzer 和 FailureAnalysisReporter 的特性从 Spring Boot 1.4 开始引入，到 Spring Boot 1.5 时，无论方法签名，还是实现，均未变化，因此无须再论。

不过，这两个方法在 Spring Boot 2.0 中被重新调整，幸运的是，由于它们均属于 private 方法，即使调整也不会影响 SpringApplication 扩展的兼容性，同时 FailureAnalyzer 和 FailureAnalysisReporter 的特性也予以支持，因此此番调整可视为一次重构，然而 Spring Boot 2.0 为什么要这么做呢？下一节将揭晓答案。

12.2.4 Spring Boot 2.0 重构 handleRunFailure 和 reportFailure 方法

handleRunFailure 和 reportFailure 方法首先的变化是方法签名，这两个方法不再直接传递 FailureAnalyzers，而是 SpringBootExceptionReporter 集合：

```
public class SpringApplication {
...
private void handleRunFailure(ConfigurableApplicationContext context,
        Throwable exception,
        Collection<SpringBootExceptionReporter> exceptionReporters,
        SpringApplicationRunListeners listeners) {
    try {
        try {
            handleExitCode(context, exception);
            if (listeners != null) {
                listeners.failed(context, exception);
            }
        }
        finally {
            reportFailure(exceptionReporters, exception);
            if (context != null) {
                context.close();
            }
        }
```

```
        }
        ...
    }

    private void reportFailure(Collection<SpringBootExceptionReporter> exceptionReporters,
            Throwable failure) {
        try {
            for (SpringBootExceptionReporter reporter : exceptionReporters) {
                if (reporter.reportException(failure)) {
                    registerLoggedException(failure);
                    return;
                }
            }
        }
        ...
    }
    ...
}
```

> 以上实现源码源于 Spring Boot `2.0.2.RELEASE`。
> Maven GAV 坐标为：`org.springframework.boot:spring-boot:2.0.2.RELEASE`。

SpringBootExceptionReporter 是 Spring Boot 2.0 新引入的函数式接口。

12.2.5 Spring Boot 2.0 的 SpringBootExceptionReporter 接口

`SpringBootExceptionReporter` 接口的内容介绍仅存于 JavaDoc，换言之，这又是一个未公开的特性：

```
@FunctionalInterface
public interface SpringBootExceptionReporter {

    /**
     * Report a startup failure to the user.
     * @param failure the source failure
     * @return {@code true} if the failure was reported or {@code false} if default
     * reporting should occur.
     */
```

```
boolean reportException(Throwable failure);

}
```

其中，handleRunFailure 和 reportFailure 方法所需的实例集合是经过 Spring 工厂加载机制初始化并排序而来的：

```
public class SpringApplication {
 ...
 public ConfigurableApplicationContext run(String... args) {
     ...
     Collection<SpringBootExceptionReporter> exceptionReporters = new ArrayList<>();
     ...
     try {
         ...
         exceptionReporters = getSpringFactoriesInstances(
                 SpringBootExceptionReporter.class,
                 new Class[] { ConfigurableApplicationContext.class }, context);
         ...
     }
     catch (Throwable ex) {
         handleRunFailure(context, ex, exceptionReporters, listeners);
             throw new IllegalStateException(ex);
     }

     try {
         listeners.running(context);
     }
         catch (Throwable ex) {
         handleRunFailure(context, ex, exceptionReporters, null);
         throw new IllegalStateException(ex);
     }
     return context;
 }
 ...
}
```

由于 SpringBootExceptionReporter 集合在初始化过程中，明确地执行了 getSpringFactoriesInstances(SpringBootExceptionReporter.class,newClass[]{ConfigurableApplication

Context.class},context)语句，所以当开发人员自定义 SpringBootExceptionReporter 时，类似于自定义 SpringApplicationRunListener，在其实现类中至少保留一个单 ConfigurableApplicationContext 类型的参数的构造器，比如 Spring Boot 2.0 内建唯一实现 FailureAnalyzers：

```
final class FailureAnalyzers implements SpringBootExceptionReporter {

    private static final Log logger = LogFactory.getLog(FailureAnalyzers.class);

    private final ClassLoader classLoader;

    private final List<FailureAnalyzer> analyzers;

    FailureAnalyzers(ConfigurableApplicationContext context) {
        this(context, null);
    }

    FailureAnalyzers(ConfigurableApplicationContext context, ClassLoader classLoader) {
        Assert.notNull(context, "Context must not be null");
        this.classLoader = (classLoader != null ? classLoader : context.getClassLoader());
        this.analyzers = loadFailureAnalyzers(this.classLoader);
        prepareFailureAnalyzers(this.analyzers, context);
    }
    ...
}
```

回顾前文所述，FailureAnalyzers 并非 Spring Boot 全新引入，它从 Spring Boot 1.4 就开始出现，不过在 Spring Boot 1.x 时代，它尚未实现 SpringBootExceptionReporter 接口。尽管 Spring Boot 升级到 2.0 版本后，FailureAnalyzers 的内部主要实现逻辑基本没有变化，因此，FailureAnalyzer 无论加载方式，还是执行，均能兼容。换言之，Spring Boot 2.0 为开发人员提供了一种全新的 Spring Boot 异常报告的机制。不过有一处细节：

```
public class SpringApplication {
    ...
    private void reportFailure(Collection<SpringBootExceptionReporter> exceptionReporters,
            Throwable failure) {
        try {
            for (SpringBootExceptionReporter reporter : exceptionReporters) {
                if (reporter.reportException(failure)) {
```

```
                    registerLoggedException(failure);
                    return;
                }
            }
        }
        ...
    }
    ...
}
```

> 以上实现源码源于 Spring Boot 2.0.2.RELEASE。
> Maven GAV 坐标为：org.springframework.boot:spring-boot:2.0.2.RELEASE。

当 SpringApplication 执行完成之后，Spring Boot 应用进程进入退出阶段。尽管相关操作仍在 SpringApplication#run(String...)方法内部完成，不过其依赖的核心特性属于 JVM 范畴，包括 Shutdown Hook、JVM 进程退出、System#exit(int)方法语义等，接下来将正式进入对“Spring Boot 应用退出”的讨论。

> 如果读者是 Spring Cloud Data Flow 的用户，则建议重点关注。

第 13 章 Spring Boot应用退出

关于 Spring Boot 应用退出的描述，Spring Boot 1.0～1.4 的官方文档基本相同，以 Spring Boot 1.4 为例，其中“23.9 Application Exit”章节的描述如下：

> Each `SpringApplication` registers a shutdown hook with the JVM to ensure that the `ApplicationContext` closes gracefully on exit. All the standard Spring lifecycle callbacks (such as the `DisposableBean` interface or the `@PreDestroy` annotation) can be used.

此处主要是强调 `SpringApplication` 注册 `shutdownhook` 线程，当 JVM 退出时，确保后续 Spring 应用上下文所管理的 Bean 能够在标准的 Spring 生命周期中回调，从而合理地销毁 Bean 所依赖的资源，如会话状态、JDBC 连接、网络连接等。不过以上文字描述曾在前面“Spring 应用上下文启动阶段”一节中有过讨论：

> 默认情况下，Spring 应用上下文将注册 `shutdownHook` 线程，实现优雅的 Spring Bean 销毁生命周期回调（`registerShutdownHook` 的默认值为 `true`）。

简言之，该特性是 `SpringApplication` 借助 `ConfigurableApplicationContext#registerShutdownHook` API 实现的。

文档继续补充：

> In addition, beans may implement the `org.springframework.boot.ExitCodeGenerator` interface if they wish to return a specific exit code when the application ends.

按照其字面意思，当 Spring Boot 程序执行结束时，ExitCodeGenerator Bean 将返回 getExitCode()方法实现的退出码，不过这也暗示着一个前提条件，即 Spring 应用上下文必须是活动的（ConfigurableApplicationContext#isActive()方法返回 true），说明此时 SpringApplication 属于正常结束。相反，当 SpringApplication 运行异常时，退出码又是如何影响 Spring Boot 应用的行为呢？

13.1 Spring Boot 应用正常退出

关于 ExitCodeGenerator Bean 的描述，Spring Boot 1.0～1.4 的官方文档与实际情况存在出入。经过仔细对比，发现即使 Spring Boot 应用运行结束，默认情况下，ExitCodeGenerator#getExitCode() 也不会执行。为什么会出现这样的纰漏，官方后续是否会加以修补呢？退出码到底存在哪些现实的价值呢？下面逐一解答这些疑问。

13.1.1 ExitCodeGenerator Bean 生成退出码

按照文档的描述，检验代码如下所示。

```
@EnableAutoConfiguration
public class ExitCodeGeneratorBootstrap {

    @Bean
    public ExitCodeGenerator exitCodeGenerator() {
        System.out.println("ExitCodeGenerator Bean 创建...");
        return () -> {
            System.out.println("执行退出码(88)生成...");
            return 88;
        };
    }

    public static void main(String[] args) {
        new SpringApplicationBuilder(ExitCodeGeneratorBootstrap.class)
                .web(false) // 非 Web 应用
                .run(args)  // 运行 Spring Boot 应用
                .close();   // 关闭应用上下文
    }
}
```

> 源码位置：以上示例代码可通过查找 spring-boot-1.x-samples/spring-boot-1.0.x-project 工程获取。

如果 `ExitCodeGenerator` Bean 创建成功，则控制台将输出“ExitCodeGenerator Bean 创建...”的内容。假设该 Bean `getExitCode()`方法执行，则控制台追加“执行退出码(88)生成...”的日志。将其运行，观察日志内容输出：

```
(...部分内容被省略...)
ExitCodeGenerator Bean 创建...
(...部分内容被省略...)
```

实际上，`ExitCodeGenerator#getExitCode()`方法并没有得到执行。或许 Spring Boot 官方也意识到以上问题，在随后的 Spring Boot 1.5 文档中加以调整并补充：

> In addition, beans may implement the `org.springframework.boot.ExitCodeGenerator` interface if they wish to return a specific exit code when `SpringApplication.exit()` is called. This exit code can then be passed to `System.exit()` to return it as a status code.

```
@SpringBootApplication
public class ExitCodeApplication {

   @Bean
   public ExitCodeGenerator exitCodeGenerator() {
        return new ExitCodeGenerator() {
             @Override
             public int getExitCode() {
                  return 42;
             }
        };
   }

   public static void main(String[] args) {
        System.exit(SpringApplication
                  .exit(SpringApplication.run(ExitCodeApplication.class, args)));
   }

}
```

此时需要程序显式地调用 SpringApplication#exit(ApplicationContext,ExitCodeGenerator...)方法。然而以上方法从 Spring Boot 1.0 开始就已存在：

```
public class SpringApplication {
 ...
 public static int exit(ApplicationContext context,
         ExitCodeGenerator... exitCodeGenerators) {
     int exitCode = 0;
     try {
         try {
             List<ExitCodeGenerator> generators = new ArrayList<ExitCodeGenerator>();
             generators.addAll(Arrays.asList(exitCodeGenerators));
             generators.addAll(context.getBeansOfType(ExitCodeGenerator.class)
                     .values());
             exitCode = getExitCode(generators);
         }
         finally {
             close(context);
         }


     }
     catch (Exception ex) {
         ex.printStackTrace();
         exitCode = (exitCode == 0 ? 1 : exitCode);
     }
     return exitCode;
 }
 ...
}
```

> 以上实现源码源于 Spring Boot 1.0.2.RELEASE。
> Maven GAV 坐标为：org.springframework.boot:spring-boot:1.0.2.RELEASE。

再结合 SpringApplication#exit(ApplicationContext,ExitCodeGenerator...)方法，将 ExitCodeGeneratorBootstrap 加以补充实现：

```
@EnableAutoConfiguration
public class ExitCodeGeneratorBootstrap {
```

```
    @Bean
    public ExitCodeGenerator exitCodeGenerator() {
        System.out.println("ExitCodeGenerator Bean 创建...");
        return () -> {
            System.out.println("执行退出码(88)生成...");
            return 88;
        };
    }

    public static void main(String[] args) {
//        重构前的实现
//        new SpringApplicationBuilder(ExitCodeGeneratorBootstrap.class)
//                .web(false) // 非 Web 应用
//                .run(args)  // 运行 Spring Boot 应用
//                .close();   // 关闭应用上下文

//        重构后的实现
        SpringApplication.exit(new
SpringApplicationBuilder(ExitCodeGeneratorBootstrap.class)
                .web(false) //非 Web 应用
                .run(args)  // 运行 Spring Boot 应用
        );
    }
}
```

> 源码位置：以上示例代码可通过查找 spring-boot-1.x-samples/spring-boot-1.0.x-project 工程获取。

重启 ExitCodeGeneratorBootstrap，观察日志输出：

```
(...部分内容被省略...)
ExitCodeGenerator Bean 创建...
(...部分内容被省略...)
执行退出码(88)生成...
(...部分内容被省略...)
```

调整后的 ExitCodeGeneratorBootstrap 尽管仍使用 Spring Boot 1.0 API 实现，但 ExitCodeGenerator Bean 的 getExitCode()方法却明显地得以执行。不过这个退出码用在何处呢？

正如官方文档所言，Spring Boot 应用需要显式地将该退出码传递到 `System#exit(int)`方法中，换言之，Spring Boot 框架并没有为开发人员隐式地实现：

> This exit code can then be passed to `System.exit()` to return it as a status code.

结合 `System#exit(int)`方法的 JavaDoc 描述加以理解：

> Terminates the currently running Java Virtual Machine. The argument serves as a status code; by convention, a nonzero status code indicates abnormal termination.

原来，退出码是一种约定，为非 0 值时，表示异常退出。同时，`ExitCodeGenerator` Bean 的正常工作依赖于 Spring 应用上下文必须活动的前提（`ConfigurableApplicationContext#isActive()`方法返回 `true`），属于 Spring Boot 正常结束流程。既然如此，通常情况下，其 JVM 进程退出码就是 0。如果 `ExitCodeGenerator#getExitCode()`方法也返回 0，那么这样的实现毫无价值。然而返回值为非 0 时，它又用在哪里呢？

13.1.2 ExitCodeGenerator Bean 退出码使用场景

按照官方示例代码的指示，为引导类 `ExitCodeGeneratorBootstrap` 增加 `System#exit(int)`方法的实现：

```
@EnableAutoConfiguration
public class ExitCodeGeneratorBootstrap {

    @Bean
    public ExitCodeGenerator exitCodeGenerator() {
        System.out.println("ExitCodeGenerator Bean 创建...");
        return () -> {
            System.out.println("执行退出码(88)生成...");
            return 88;
        };
    }

    public static void main(String[] args) {
//          重构前的实现
//          new SpringApplicationBuilder(ExitCodeGeneratorBootstrap.class)
//                  .web(false) // 非 Web 应用
```

```
//                .run(args)  // 运行 Spring Boot 应用
//                .close();   // 关闭应用上下文

//        重构后的实现
        int exitCode = SpringApplication.exit(new
SpringApplicationBuilder(ExitCodeGeneratorBootstrap.class)
                .web(false) // 非 Web 应用
                .run(args)  // 运行 Spring Boot 应用
        );
        // 传递退出码到 System#exit(int)方法
        System.exit(exitCode);
    }
}
```

> 源码位置：以上示例代码可通过查找 spring-boot-1.x-samples/spring-boot-1.0.x-project 工程获取。

实际上，单从程序日志输出分析的话，重构后的 `ExitCodeGeneratorBootstrap` 的运行结果并没有发生变化。如果将它在 IDEA 中运行，则 IDEA 会在控制台中增加一行退出码的提示：

```
Process finished with exit code 88
```

以上内容似乎在间接地提示用户，IDEA 通过运行子进程的方式启动引导类 `ExitCodeGeneratorBootstrap`。实际情况的确如此，IDEA 运行任一引导类时，控制台首句输出的就是引导类启动命令行：

```
/Library/Java/JavaVirtualMachines/jdk1.8.0_172.jdk/Contents/Home/bin/java
"-javaagent:/Applications/IntelliJ IDEA
CE.app/Contents/lib/idea_rt.jar=50000:/Applications/IntelliJ IDEA CE.app/Contents/bin"
-Dfile.encoding=UTF-8 -classpath "(...Class Path 内容被省略...)"
thinking.in.spring.boot.samples.spring.application.ExitCodeGeneratorBootstrap
```

换言之，IDEA 和引导程序同为 JVM 进程，前者为父进程，后者为子进程，被 IDEA 管理。所以，IDEA 能够获取子进程的退出码。因此，Spring Boot 框架退出码的设计也是基于此考虑的。当然，生产环境下的 Spring Boot 进程绝非通过 IDEA 这样的工具启动，通常经过命令行或启动脚本运行。不过，在真实的 Spring Boot 应用场景中，`SpringApplication#exit(ApplicationContext, ExitCodeGenerator...)`方法几乎没有被调用的理由，因为该方法最终会显式地关闭当前 Spring 应用上下文：

```java
public class SpringApplication {
 ...
 public static int exit(ApplicationContext context,
         ExitCodeGenerator... exitCodeGenerators) {
     int exitCode = 0;
     try {
         try {
             ...
             exitCode = getExitCode(generators);
         }
         finally {
             close(context);
         }

     }
     ...
     return exitCode;
 }
 ...
}
```

> 以上实现源码源于 Spring Boot `1.0.2.RELEASE`。
> Maven GAV 坐标为：`org.springframework.boot:spring-boot:1.0.2.RELEASE`。

一旦 `ConfigurableApplicationContext#close()`方法被调用，即使是 Web 类型的 Spring Boot 应用程序也不会阻塞主线程，导致应用直接关闭。如此一来，Spring Boot 应用却成了一闪而过的执行程序，同时退出码的捕获对 Java 程序而言并不友好。不过从 Spring Boot 1.3.2 开始，退出码事件 `ExitCodeEvent` 的引入使得情况有所变化。

> 尽管开发人员很少关注 JVM 进程退出码的返回，不过未来可能随着 Spring Cloud Data Flow 的流行，Spring Boot 退出码实现会被重视，因为在 Spring Cloud Data Flow 架构体系中，Spring Boot 应用将以 `jar` 文件的方式导入 Spring Cloud Data Flow 集群，随后作为 Stream，此时的 Spring Boot 应用将作为 JVM 子进程，被容器管理其生命周期。

`ExitCodeGenerator` Bean 退出码用于 `ExitCodeEvent` 事件监听

从编程模型上，Spring Boot 框架允许应用在 Spring `ConfigurableApplicationContext` 中增加 `ExitCodeEvent` 的监听器（`ApplicationListener`），前提是 `ExitCodeGenerator` Bean 返回非 0

的退出码：

```
public class SpringApplication {
...
public static int exit(ApplicationContext context,
        ExitCodeGenerator... exitCodeGenerators) {
    Assert.notNull(context, "Context must not be null");
    int exitCode = 0;
    try {
        try {
            ExitCodeGenerators generators = new ExitCodeGenerators();
            Collection<ExitCodeGenerator> beans = context
                    .getBeansOfType(ExitCodeGenerator.class).values();
            generators.addAll(exitCodeGenerators);
            generators.addAll(beans);
            exitCode = generators.getExitCode();
            if (exitCode != 0) {
                context.publishEvent(new ExitCodeEvent(context, exitCode));
            }
        }
        ...

    }
    ...
}
...
}
```

> 以上实现源码源于 Spring Boot `1.3.8.RELEASE`
> Maven GAV 坐标为：`org.springframework.boot:spring-boot:1.3.8.RELEASE`。

为此，新建引导类（需依赖 Spring Boot 1.3.2 及更高版本），检验 `ExitCodeEvent` 事件的监听：

```
public class ExitCodeEventOnExitBootstrap {

    @Bean
    public ExitCodeGenerator exitCodeGenerator() {
        return () -> {
            System.out.println("执行退出码(9)生成...");
```

```
            return 9;
        };
    }

    public static void main(String[] args) {
        System.exit(SpringApplication.exit(
                new
                SpringApplicationBuilder(ExitCodeEventOnExitBootstrap.class)
                        .listeners((ApplicationListener<ExitCodeEvent>) event ->
                                System.out.println("监听到退出码: " + event.getExitCode())
                        )
                        .web(false) // 非 Web 应用
                        .run(args)  // 运行 Spring Boot 应用
        ));
    }
}
```

> 源码位置：以上示例代码可通过查找 spring-boot-1.x-samples/spring-boot-1.3.x-project 工程获取。

通过 IDEA 运行，并观察控制台输出：

```
(...部分内容被省略...)
执行退出码(9)生成...
监听到退出码: 9
(...部分内容被省略...)
Process finished with exit code 9
```

ExitCodeGenerator Bean 生成的退出码 9 被 ApplicationListener<ExitCodeEvent>实例监听，同时影响 IDEA 对退出码的管理。当然以上关于退出码的讨论均基于 Spring Boot 应用正常退出，Spring Boot 框架显然不会仅考虑单一情况，下面进入“Spring Boot 应用异常退出”的讨论。

13.2 Spring Boot 应用异常退出

关于异常与退出码之间的关系，Spring Boot 1.5 和 2.0 版本官方文档均在“23.9 Application Exit”章节中做出了片面的说明：

> Also, the ExitCodeGenerator interface may be implemented by exceptions. When such an exception is encountered, Spring Boot returns the exit code provided by the implemented getExitCode() method.

如果将以上文字直译，或许有些不知所云，必须结合 SpringApplication 源码分析，该部分的内容在异常处理方法 handleRunFailure 的调用链路中出现：

```
public class SpringApplication {
  ...
  private void handleRunFailure(ConfigurableApplicationContext context,
        SpringApplicationRunListeners listeners, Throwable exception) {
      ...
            handleExitCode(context, exception);
      ...
  }
  ...
  private void handleExitCode(ConfigurableApplicationContext context,
        Throwable exception) {
      int exitCode = getExitCodeFromException(context, exception);
      ...
  }

  private int getExitCodeFromException(ConfigurableApplicationContext context,
        Throwable exception) {
      int exitCode = getExitCodeFromMappedException(context, exception);
      if (exitCode == 0) {
          exitCode = getExitCodeFromExitCodeGeneratorException(exception);
      }
      return exitCode;
  }
  ...
  private int getExitCodeFromExitCodeGeneratorException(Throwable exception) {
      if (exception == null) {
          return 0;
      }
      if (exception instanceof ExitCodeGenerator) {
          return ((ExitCodeGenerator) exception).getExitCode();
      }
```

```
        return
        getExitCodeFromExitCodeGeneratorException(exception.getCause());
    }
  }
```

> 以上实现源码源于 Spring Boot 1.3.8.RELEASE。
> Maven GAV 坐标为：org.springframework.boot:spring-boot:1.3.8.RELEASE。

当异常实现 ExitCodeGenerator 接口时，退出码直接采用 getExitCode()方法返回。尽管以上内容在 Spring Boot 1.5 的文档中才出现，实际上该特性从 Spring Boot 1.3.2 就予以支持。即便如此，官方文档的“只言片语”也无法完整地论述 ExitCodeGenerator 异常的使用场景。

13.2.1 ExitCodeGenerator 异常使用场景

根据前面的讨论，在异常处理方法的调用链路中，ExitCodeGenerator 异常获取退出码的逻辑在 handleRunFailure 方法中触发，而 handleRunFailure 方法仅在 SpringApplication#run(String...)方法下的异常 catch 流程中：

```
public class SpringApplication {
 ...
 public ConfigurableApplicationContext run(String... args) {
     ...
     SpringApplicationRunListeners listeners = getRunListeners(args);
     listeners.started();
     try {
          ...
     }
     catch (Throwable ex) {
          handleRunFailure(context, listeners, ex);
          throw new IllegalStateException(ex);
     }
 }
 ...
}
```

> 以上实现源码源于 Spring Boot 1.3.8.RELEASE。
> Maven GAV 坐标为：org.springframework.boot:spring-boot:1.3.8.RELEASE。

不难发现，SpringApplicationRunListeners#started()方法并不在 try catch 执行块之内。当 SpringApplication 管理的任一 SpringApplicationRunListener 实例在执行 started()方法异常时，ExitCodeGenerator 异常并不会被捕获，示例如下：

```
public class ExitCodeGeneratorExceptionBootstrap {

    public static void main(String[] args) {
        new SpringApplicationBuilder(Object.class)
                .listeners(event -> {
                    throw new ExitCodeGeneratorThrowable(event.getClass().getSimpleName());
                })
                .web(false)           // 非 Web 应用
                .run(args)            // 运行 SpringApplication
                .close();             // 关闭 ConfigurableApplicationContext
    }

    static class ExitCodeGeneratorThrowable extends RuntimeException implements
ExitCodeGenerator {

        public ExitCodeGeneratorThrowable(String message) {
            super(message);
        }

        @Override
        public int getExitCode() {
            return 95;
        }
    }
}
```

> 源码位置：以上示例代码可通过查找 spring-boot-1.x-samples/spring-boot-1.3.x-project 工程获取。

通过 IDEA 运行以上示例，控制台异常如期输出，然而退出码并未修改为 95：

```
(...部分内容被省略...)
Exception in thread "main"
thinking.in.spring.boot.samples.spring.application.ExitCodeGeneratorExceptionBootstrap$ExitCo
deGeneratorThrowable: ApplicationStartedEvent
```

```
(...部分内容被省略...)
Process finished with exit code 1
```

因为Spring Boot内建`SpringApplicationRunListener`实现`EventPublishingRunListener`在执行`started()`方法时，会广播Spring Boot事件`ApplicationStartedEvent`，该阶段并不在try catch逻辑中，因此即使异常实现`ExitCodeGenerator`也毫无效果。同时，从Spring Boot 1.5开始，`SpringApplicationRunListener`中的`started()`方法被移除，取而代之的是`starting()`方法，尽管此时事件仍旧是`ApplicationStartedEvent`（它也继承了1.5版本的新事件类型`ApplicationStartingEvent`），总之，以上问题也会在Spring Boot 1.5的`SpringApplicationRunListeners#starting()`方法中出现：

```
public class SpringApplication {
 ...
 public ConfigurableApplicationContext run(String... args) {
      ...
      SpringApplicationRunListeners listeners = getRunListeners(args);
      listeners.starting();
      try {
           ...
      }
      catch (Throwable ex) {
           handleRunFailure(context, listeners, analyzers, ex);
           throw new IllegalStateException(ex);
      }
 }
 ...
}
```

> 以上实现源码源于Spring Boot `1.5.10.RELEASE`。
> Maven GAV坐标为：`org.springframework.boot:spring-boot:1.5.10.RELEASE`。

当然Spring Boot 2.0也不例外：

```
public class SpringApplication {
 ...
 public ConfigurableApplicationContext run(String... args) {
      ...
      SpringApplicationRunListeners listeners = getRunListeners(args);
      listeners.starting();
```

```
        try {
            ...
        }
        catch (Throwable ex) {
            handleRunFailure(context, ex, exceptionReporters, listeners);
            throw new IllegalStateException(ex);
        }

        try {
            listeners.running(context);
        }
        catch (Throwable ex) {
            handleRunFailure(context, ex, exceptionReporters, null);
            throw new IllegalStateException(ex);
        }
        return context;
    }
    ...
}
```

> 以上实现源码源于 Spring Boot 2.0.2.RELEASE。
> Maven GAV 坐标为：org.springframework.boot:spring-boot:2.0.2.RELEASE。

因此，假设 Spring Boot 应用需要利用 ExitCodeGenerator 异常获取退出码，应该避免监听 ApplicationStartedEvent 和 ApplicationStartingEvent 事件。故而调整 ExitCodeGenerator ExceptionBootstrap 实现：

```
public class ExitCodeGeneratorExceptionBootstrap {

    public static void main(String[] args) {
        new SpringApplicationBuilder(Object.class)
                .listeners(
//                      event -> {   // 取消所有 Spring Boot 事件监听
                        (ApplicationListener<ApplicationReadyEvent>) event -> {
                            throw new
ExitCodeGeneratorThrowable(event.getClass().getSimpleName());
                        })
                .web(false)             // 非 Web 应用
```

```
                .run(args)           // 运行 SpringApplication
                .close();            // 关闭 ConfigurableApplicationContext
    }

    static class ExitCodeGeneratorThrowable extends RuntimeException implements
ExitCodeGenerator {

        public ExitCodeGeneratorThrowable(String message) {
            super(message);
        }

        @Override
        public int getExitCode() {
            return 95;
        }
    }
}
```

> 源码位置：以上示例代码可通过查找 spring-boot-1.x-samples/spring-boot-1.3.x-project 工程获取。

使用 IDEA 运行该引导类型，成功地使用退出码 95，异常信息中的事件类型名称也被替换为“ApplicationReadyEvent”：

```
(...部分内容被省略...)
thinking.in.spring.boot.samples.spring.application.ExitCodeGeneratorExceptionBootstrap$Ex
itCodeGeneratorThrowable: ApplicationReadyEvent
(...部分内容被省略...)
Process finished with exit code 95
```

> 以上引导类 `ExitCodeGeneratorExceptionBootstrap` 同样能在 Spring Boot 1.5 和 2.0 版本中运行，运行后的退出码与本例一致，请读者分别在 **spring-boot-1.x-samples/spring-boot-1.5.x-project** 工程和 **spring-boot-2.0-samples/spring-application-sample** 工程中查找副本。

本例尽管实现简单，然而复杂的 `SpringApplication` 生命周期很容易将问题变得难以理解。当然，除此之外，`SpringApplication` 获取退出码可能在更早的阶段实现，即 `getExitCodeFromMappedException(ConfigurableApplicationContext,Throwable)` 方法返

回：

```
public class SpringApplication {
...
private int getExitCodeFromMappedException(ConfigurableApplicationContext context,
        Throwable exception) {
    if (context == null) {
        return 0;
    }
    ExitCodeGenerators generators = new ExitCodeGenerators();
    Collection<ExitCodeExceptionMapper> beans = context
            .getBeansOfType(ExitCodeExceptionMapper.class).values();
    generators.addAll(exception, beans);
    return generators.getExitCode();
}
...
}
```

> 以上实现源码源于 Spring Boot `1.3.8.RELEASE`。
> Maven GAV 坐标为：`org.springframework.boot:spring-boot:1.3.8.RELEASE`。

此时，当 Spring 应用上下文活跃，并且包含 `ExitCodeExceptionMapper` Bean 的定义时，该方法将集合 `ExitCodeExceptionMapper` Bean 的 `getExitCode()`方法返回值，同样地，当退出码为非 0 时，Spring Boot 框架才会采纳该退出码。虽然 Spring Boot 官方文档并未提供任何关于 `ExitCodeExceptionMapper` 接口的说明，但并不妨碍下一节的讨论。

13.2.2 ExitCodeExceptionMapper Bean 映射异常与退出码

`ExitCodeExceptionMapper` 接口定义了维护异常与退出码的映射关系：

```
public interface ExitCodeExceptionMapper {

 /**
  * Returns the exit code that should be returned from the application.
  * @param exception the exception causing the application to exit
  * @return the exit code or {@code 0}.
  */
 int getExitCode(Throwable exception);
```

```
}
```

> 以上实现源码源于 Spring Boot `1.3.8.RELEASE`。
> Maven GAV 坐标为：`org.springframework.boot:spring-boot:1.3.8.RELEASE`。

该接口从 Spring Boot 1.3.2 引入，并在 Spring Boot 2.0 中得到持续支持。尽管文档对其只字未提，但可认为它将长期支持，至少目前如此。然而，`ExitCodeExceptionMapper` 的特性同样需要依赖 `ConfigurableApplicationContext` 依然活跃的前提，当 `ConfigurableApplicationContext#refresh()`过程执行失败时，`ExitCodeExceptionMapper` Bean 也不复存在。

不过要验证 `ExitCodeExceptionMapper` Bean 的效果并不困难，不过特别提醒的是由于 `ExitCodeExceptionMapper` Bean 和 `ExitCodeGenerator` 异常同属于 `SpringApplication#handleRunFailure` 方法生命周期，故方法 `SpringApplicationRunListeners#started()`或 `SpringApplicationRunListeners#starting()`的执行异常均无法捕获，并且还需要保证 Spring `ConfigurableApplicationContext` 正常运作。因此，异常发生的生命周期阶段非常有限，如 `SpringApplicationRunListener` 在 Spring Boot 1.0～1.5 中的 `finished(ConfigurableApplicationContext,Throwable)`方法，以及 Spring Boot 2.0 中的 `started(ConfigurableApplicationContext)`和 `running(ConfigurableApplicationContext)`，也包括 `CommandLineRunner` 或 `ApplicationRunner` Bean，如下例所示。

```
public class ExitCodeExceptionMapperBootstrap {

    @Bean
    public ExitCodeExceptionMapper exitCodeExceptionMapper() {
        return (throwable) -> 128;
    }

    public static void main(String[] args) {
        new
        SpringApplicationBuilder(ExitCodeExceptionMapperBootstrap.class)
                .listeners((ApplicationListener<ApplicationReadyEvent>) event -> {
                    throw new RuntimeException();
                })
                .web(false)             // 非 Web 应用
                .run(args)              // 运行 SpringApplication
                .close();               // 关闭 ConfigurableApplicationContext
    }
```

```
}
```

> 源码位置：以上示例代码可通过查找 spring-boot-1.x-samples/spring-boot-1.3.x-project 工程获取。

由于 `ExitCodeExceptionMapper` 从 Spring Boot 1.3.2 才予以支持，即使以上示例依赖于 Spring Boot 1.3 所引入的 `ApplicationReadyEvent` 事件也是没有问题的，故本实例可在 Spring Boot 1.3～2.0 环境下运行：

```
(...部分内容被省略...)
java.lang.RuntimeException: ApplicationReadyEvent
(...部分内容被省略...)
Process finished with exit code 128
```

退出码 128 被 IDEA 成功获取。

无论实现 `ExitCodeGenerator` 接口的 `Throwable` 实例，还是 `ExitCodeExceptionMapper` Bean，相同的疑问仍旧浮现。那么，异常下的退出码用在何处呢？换了一个角度思考，既然文档也没有相关说明，假设这块不太重要，为什么 Spring Boot 要不断地加码实现呢？接下来讨论退出码在 `SpringApplication` 异常结束时的使用场景。

> 对于 Spring Boot 官方文档的言之不详和部分瑕疵，或许读者已经有所体会，官方文档只能作为基础使用的参考资料。至于文档中未公开的特性，需要谨慎评估其支持连贯性，确认无误后，方可使用。

13.2.3 退出码用于 SpringApplication 异常结束

在 `SpringApplication` 异常结束时，Spring Boot 提供两种退出码与异常类型关联的方式，一是让 `Throwable` 对象实现 `ExitCodeGenerator` 接口，二是 `ExitCodeExceptionMapper` 实现退出码与 `Throwable` 的映射。前者不依赖于当前 Spring `ConfigurableApplicationContext` 是否活跃，后者则依赖。两者分别在 `getExitCodeFromExitCodeGeneratorException(Throwable)`方法和 `getExitCodeFromMappedException(ConfigurableApplicationContext,exception)` 方法中执行，都在 `handleExitCode(ConfigurableApplicationContext,exception)`方法内部执行：

> 当前源码分析仍采用 Spring Boot `1.3.8.RELEASE` 实现，这些特性从 Spring Boot 1.3.2 开始引入，后续版本的实现几乎没有变更。

```
public class SpringApplication {
...
private void handleExitCode(ConfigurableApplicationContext context,
        Throwable exception) {
    int exitCode = getExitCodeFromException(context, exception);
    if (exitCode != 0) {
        if (context != null) {
            context.publishEvent(new ExitCodeEvent(context, exitCode));
        }
        SpringBootExceptionHandler handler = getSpringBootExceptionHandler();
        if (handler != null) {
            handler.registerExitCode(exitCode);
        }
    }
}

private int getExitCodeFromException(ConfigurableApplicationContext context,
        Throwable exception) {
    int exitCode = getExitCodeFromMappedException(context, exception);
    if (exitCode == 0) {
        exitCode = getExitCodeFromExitCodeGeneratorException(exception);
    }
    return exitCode;
}
...
}
```

> 以上实现源码源于 Spring Boot `1.3.8.RELEASE`。
> Maven GAV 坐标为：`org.springframework.boot:spring-boot:1.3.8.RELEASE`。

结合前面的讨论，当 `getExitCodeFromException(ConfigurableApplicationContext, exception)`方法返回非 0 时，同样依赖 `ConfigurableApplicationContext` 发送 `ExitCodeEvent` 事件，与 `SpringApplication#exit(ApplicationContext,ExitCodeGenerator...)`方法不同的是，退出码将存储到 `SpringBootExceptionHandler` 对象中，而该对象来源于 `getSpringBootExceptionHandler()`方法：

```
public class SpringApplication {
...
```

```
SpringBootExceptionHandler getSpringBootExceptionHandler() {
    if (isMainThread(Thread.currentThread())) {
        return SpringBootExceptionHandler.forCurrentThread();
    }
    return null;
}

private boolean isMainThread(Thread currentThread) {
    return ("main".equals(currentThread.getName())
            || "restartedMain".equals(currentThread.getName()))
            && "main".equals(currentThread.getThreadGroup().getName());
}
...
}
```

> 以上实现源码源于 Spring Boot 1.3.8.RELEASE。
> Maven GAV 坐标为：org.springframework.boot:spring-boot:1.3.8.RELEASE。

当 isMainThread(Thread)方法认为当前线程为主线程时，调用 SpringBootExceptionHandler#forCurrentThread()方法获取 SpringBootExceptionHandler 实例。值得注意的是，其中存在判断当前线程名称是否为“restartedMain”的逻辑分支，这是因为应用依赖 org.springframework.boot:spring-boot-devtools 后，当 spring-boot-devtools 认为应用需要重启时，将启动 org.springframework.boot.devtools.restart.RestartLauncher 线程，该线程的名称为“restartedMain”：

```
class RestartLauncher extends Thread {

    private final String mainClassName;

    private final String[] args;

    private Throwable error;

    RestartLauncher(ClassLoader classLoader, String mainClassName, String[] args,
            UncaughtExceptionHandler exceptionHandler) {
        this.mainClassName = mainClassName;
        this.args = args;
        setName("restartedMain");
```

```
        setUncaughtExceptionHandler(exceptionHandler);
        setDaemon(false);
        setContextClassLoader(classLoader);
    }
}
```

> 以上实现源码源于 Spring Boot Devtools 1.3.8.RELEASE。
> Maven GAV 坐标为：org.springframework.boot:spring-boot-devtools:1.3.8.RELEASE。

综上所述，通常情况下 isMainThread(Thread)将返回 true，因此 getSpringBootExceptionHandler()方法返回 SpringBootExceptionHandler#forCurrentThread()执行的结果：

```
class SpringBootExceptionHandler implements UncaughtExceptionHandler {
    ...
    private static LoggedExceptionHandlerThreadLocal handler = new
LoggedExceptionHandlerThreadLocal();
    ...
    static SpringBootExceptionHandler forCurrentThread() {
        return handler.get();
    }

    /**
     * Thread local used to attach and track handlers.
     */
    private static class LoggedExceptionHandlerThreadLocal
            extends ThreadLocal<SpringBootExceptionHandler> {

        @Override
        protected SpringBootExceptionHandler initialValue() {
            SpringBootExceptionHandler handler = new SpringBootExceptionHandler(
                    Thread.currentThread().getUncaughtExceptionHandler());
            Thread.currentThread().setUncaughtExceptionHandler(handler);
            return handler;
        };

    }
```

```
}
```

> 以上实现源码源于 Spring Boot 1.3.8.RELEASE。
> Maven GAV 坐标为：org.springframework.boot:spring-boot:1.3.8.RELEASE。

按照 ThreadLocal 初始化的原理，当应用第一次执行 SpringBootExceptionHandler#forCurrentThread()方法时，LoggedExceptionHandlerThreadLocal#initialValue()方法将被调用，返回 SpringBootExceptionHandler 对象，而 SpringBootExceptionHandler 又是 Thread.UncaughtExceptionHandler 的扩展类，当执行线程遇到未捕获的异常时，Thread.UncaughtExceptionHandler#uncaughtException(Thread,Throwable)方法将处理该异常，示例如下：

```
public class UncaughtExceptionHandlerBootstrap {

    public static void main(String[] args) {
        // 获取当前线程 main
        Thread mainThread = Thread.currentThread();
        System.out.printf("当前执行线程 %s!\n", mainThread.getName());
        //为 main 线程设置 UncaughtExceptionHandler 实现
        mainThread.setUncaughtExceptionHandler((thread, throwable) - {
            System.out.printf("处理线程[%s]的非捕获异常，详情：%s\n"
                    , thread.getName(), throwable.getMessage());
        });
        // 抛出异常
        throw new RuntimeException("故意抛出异常，测试
                                        UncaughtExceptionHandler 是否处理！");
    }
}
```

该引导类运行后，RuntimeException 异常被 Thread.UncaughtExceptionHandler 实现捕获：

```
当前执行线程 main!
处理线程[main]的非捕获异常，详情：故意抛出异常，测试 UncaughtExceptionHandler 是否处理！
```

> 源码位置：以上示例代码可通过查找 spring-boot-1.x-samples/spring-boot-1.3.x-project 工程获取。

因此，当主线程执行异常时，将被 SpringBootExceptionHandler#uncaughtException(Thread,Throwable)方法处理：

```
class SpringBootExceptionHandler implements UncaughtExceptionHandler {
...
@Override
public void uncaughtException(Thread thread, Throwable ex) {
      try {
            if (isPassedToParent(ex) && this.parent != null) {
                  this.parent.uncaughtException(thread, ex);
            }
      }
      finally {
            this.loggedExceptions.clear();
            if (this.exitCode != 0) {
                  System.exit(this.exitCode);
            }
      }
}
...
}
```

> 以上实现源码源于 Spring Boot Devtools 1.3.8.RELEASE。
> Maven GAV 坐标为：org.springframework.boot:spring-boot-devtools:1.3.8.RELEASE。

至此，关于“理解 SpringApplication”的讨论将告一段落。

13.3 小马哥有话说

尽管 Spring Boot 官方文档并未过多介绍 SpringApplication，然而上述的讨论却显得异常复杂。总而言之，“理解 SpringApplication”是围绕 SpringApplication 生命周期来展开论述的，分为“初始化”“运行”和“结束”三个阶段，主要的核心特性包括 SpringApplicationRunListener、Spring Boot 事件和 Spring 应用上下文的生命周期管理等。然而有一个疑惑悬而未决，那就是 Spring Boot 引入 SpringApplication 的目的何在？在讨论“走向自动装配”部分时曾提到：

> 在 Spring Framework 时代，Spring 应用上下文通常由容器启动，如 ContextLoaderListener 或 WebApplicationInitializer 的实现类由 Servlet 容器装载并驱动。到了 Spring Boot 时代，Spring 应用上下文的启动则通过调用 SpringApplication#run(Object,String...)

> 或 SpringApplicationBuilder#run(String...)方法并配合@SpringBootApplication 或@EnableAutoConfiguration 注解的方式完成。

请注意上述行文，字里行间中并没有将 SpringApplication 限定在嵌入式 Web 应用场景，而是强调 Spring 应用上下文的启动，不仅因为 SpringApplication 可以引导非 Web 应用和嵌入式 Web 应用，而且它还能出现在 Spring Boot 应用部署在传统 Servlet 3.0+容器中的场景，不过，前文并没有提及，此处的讨论点到为止，详细的部分将在 Web 篇中展开。Spring Boot 官方文档在"87.1 Create a Deployable War File"章节中介绍了实现步骤：

> Typically, you should update your application's main class to extend SpringBootServlet Initializer

通常，只需将 Spring Boot 引导类扩展抽象类 SpringBootServletInitializer 即可。同时，SpringBootServletInitializer 是 WebApplicationInitializer 的实现类，因此，Spring Boot 应用在 Servlet 3.0+容器中部署时，其 onStartup(ServletContext)方法将在容器启动时回调：

```
public abstract class SpringBootServletInitializer implements WebApplicationInitializer {
    ...
    @Override
    public void onStartup(ServletContext servletContext) throws ServletException {
        ...
        WebApplicationContext rootAppContext = createRootApplicationContext(
                servletContext);
        ...
    }

    protected WebApplicationContext createRootApplicationContext(
            ServletContext servletContext) {
        SpringApplicationBuilder builder = createSpringApplicationBuilder();
        ...
        SpringApplication application = builder.build();
        ...
        return run(application);
    }
    ...
}
```

> 以上实现源码源于 Spring Boot 2.0.2.RELEASE。

> Maven GAV 的坐标为：`org.springframework.boot:spring-boot:2.0.2.RELEASE`。

源码证实 SpringApplication 同样运用在 Spring Boot 应用部署到传统 Servlet 3.0+容器的场景中，换言之，SpringApplication 并非为 Spring Boot 嵌入式 Web 应用“量身定制”。既然如此，无论在哪种 Spring Boot 应用场景下，功能组件均为 Spring Bean，那么为何不直接使用 ConfigurableApplicationContext 实现类来引导 Spring 应用呢？如注解驱动的 AnnotationConfigApplicationContext，尤其是 SpringApplication 在非 Web 应用场景下所创建的 ConfigurableApplicationContext 实例就是 AnnotationConfigApplicationContext。个人认为，Spring Boot 引入 SpringApplication 是对 Spring Framework 的应用上下文生命周期的补充。

传统的 Spring 应用上下文生命的起点源于 ConfigurableApplicationContext 对象的创建，运行则由其 refresh()方法引导，而终止于 close()方法的调用。Spring Framework 内建的 ConfigurableApplicationContext 实现类均继承于抽象类 AbstractApplicationContext。在 AbstractApplicationContext#refresh()方法执行过程中，伴随着组件 BeanFactory、Environment、ApplicationEventMulticaster 和 ApplicationListener 的创建，它们的职责分别涉及 Bean 容器、Spring 属性配置、Spring 事件广播和监听。实际上，SpringApplication 并未从本质上改变这些，因为 AbstractApplicationContext 提供了扩展接口，如 setEnvironment(ConfigurableEnvironment)方法允许替换默认的 Environment 对象，以及 initApplicationEventMulticaster() 和 registerListeners() 方法分别提供了获取 ApplicationEventMulticaster 和 ApplicationListener Bean 的机制。不过，这些扩展接口被 SpringApplication 在 Spring 应用上下文调用 refresh()方法之前予以运用，在 SpringApplicationRunListener 实现类 EventPublishingRunListener 的帮助下，全新地引入 Spring Boot 事件，并且间接地过渡到外部化配置，而后者则是运维篇“超越外部化配置”部分讨论的重点。

13.4 下一站：运维篇

至此，读者或许能对 SpringApplication 的理解更为系统化，同时核心篇也将进入尾声。通过总览“Spring Boot”“走向自动装配”和“理解 SpringApplication”三部分的讨论，试图将 Spring Boot 核心特性与 Spring Framework 做深层次的连接，反复地强调特性之间的关联性、版本的实现差异和设计思维的严谨性。人非圣贤，孰能无过，Spring 社区也不例外。Spring Boot 在发展的过程中既有内聚性，也存在反复性，即在 Spring Framework 特性上保持高内聚，几乎见不到重复的实现，然而在 Spring Boot 自身 API 设计上却出现了不少的反复，不免让用户在升级时“胆战心惊”。不过，这种破坏性的变更并不会因核心篇的完结而终止，在下一册运维篇中更是“罄竹难书”。或许这并非坏事，至少能够唤醒广大开发人员的敬畏之心，尤其对于基础设施的研发人员。无论设计还是实现，都应谋定而后动，切不可任意地删减接口，也不可画蛇添足。不过庆幸的是，Spring Boot 2.0 的实现趋于成熟，相信后续出现以上状况的概率较低，因此，建议如果是新的 Spring Boot 项目则直接使用 2.0 以上的版本，减少升级成本。如果还在为 Spring Boot 1.x 应用升级到 2.0 版本而感到困惑，不妨参考下一册运维篇的讨论，包括“超越外部化配置”和“简化 Spring 应用运维体系”两部分的内容，或许它能诠释 Spring Boot 强大的 Ops 能力。